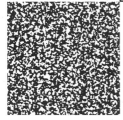

ENGINEERING BY
D E S I G N

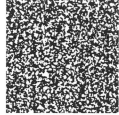

ENGINEERING BY
D E S I G N

G E R A R D   V O L A N D

N o r t h e a s t e r n   U n i v e r s i t y

ADDISON-WESLEY

Addison-Wesley is an imprint of Addison Wesley Longman, Inc.

Reading, Massachusetts ▪ Harlow, England ▪ Menlo Park, California
Berkeley, California ▪ Don Mills, Ontario ▪ Sydney ▪ Bonn ▪ Amsterdam
Tokoyo ▪ Mexico City

Senior Acquisitions Editor: Denise Olson
Project Manager: Phoebe Ling
Production Editor: Amy Willcutt
Production Assistant: Brooke Albright
Packager: Joan M. Flaherty
Compositor: Scott Silva
Cover Designer: Alwyn Velásquez
Text Designer: Melinda Grosser
Technical Illustrator: George Nichols
Copyeditor: Sarah Corey
Proofreader: Judith W. Strakalaiti

Cover credits
"City officials at Brooklyn Bridge" © UPI/Corbis-Bettmann
"Workmen cutting and tying tension cables" © Corbis-Bettman
"Brooklyn Bridge construction" © Museum of the City of New York
"Excavating under caisson" © Museum of the City of New York
"NYC Brooklyn Bridge from Brooklyn side" © 1992 Randy Duchaine

Access the latest information about Addison-Wesley books from our World Wide Web site:
http://www.awl.com/cseng/cad

Many of the designations used by manufacturers and sellers to distinguish their products are claimed as trademarks. Where those designations appear in this book, and Addison-Wesley was aware of a trademark claim, the designations have been printed in initial caps or all caps.

The programs and applications presented in this book have been included for their instructional value. They have been tested with care, but are not guaranteed for any particular purpose. The publisher does not offer any warranties or representations, nor does it accept any liabilities with respect to the programs or applications.

Library of Congress Cataloging-in-Publication Data

Voland, Gerard G. S.
  Engineering by design / Gerard Voland with Margaret Voland.
    p. cm.
  Includes bibliographical references and index.
  ISBN 0-201-49851-0
  1. Engineering design. I. Voland, Margaret. II. Title.
TA174.V65 1999
620'.0042--dc21   98-3135
                                                                 CIP

This book was typeset in QuarkXPress 3.32 on a Power Macintosh. The fonts used were Helvetica and Melior. It was printed on Rolland, a recycled paper.

1 2 3 4 5 6 7 8 9 10—MA—0201009998

I dedicate this work to my mother and father,
Eleanor and Norman Voland,
and to all those who have made the world a better place.

 **PREFACE**

Rationale for the Text

Engineering by Design is intended to serve a multiplicity of functions while conveying the excitement and sheer fun of solving technical problems in creative yet practical ways. It provides an introduction to the engineering profession through numerous case histories that illustrate various aspects of the design process. This is important because students can come to appreciate the interdisciplinary aspects of engineering problem solving as they work with case problems or review case histories in design which demonstrate (sometimes in a very startling but effective manner) valuable lessons in engineering practice.

Many engineering colleges are revising their curricula in response to the new ABET 2000 accreditation requirements. Various design and manufacturing topics are being integrated more into undergraduate engineering programs, and new mechanisms for delivering this course material are being developed with a general focus upon experiential learning by undergraduates. *Engineering by Design* has been developed as an aid for these important efforts. It is written in a way that I hope will be both engaging and accessible to students while maintaining the accuracy and rigor that one would expect of an engineering textbook.

Freshmen and sophomores often have little substantive knowledge of professional engineering practice; indeed, these students may have many misconceptions about the actual work that engineers perform and the types of problems with which they wrestle. The initial chapters and case histories of this book will give students a limited but informed understanding of the engineering profession. Later material (including case problems through which students can apply their newly acquired knowledge of design) then broadens and deepens this understanding.

Through this approach, students discover the need to

- formulate problems correctly
- work successfully in interdisciplinary teams
- develop their creativity, imagination, and analytical skills
- make informed ethical decisions
- hone their written and oral communication skills

Most important, they learn that engineering is a service profession, dedicated to satisfying humanity's needs through responsible, methodical, and creative problem solving.

▇ Structure

▨ Case Histories, Case Problems, and a Design Template

Engineering By Design introduces students to such critical design topics as needs assessment, problem formulation, modeling, patents, abstraction and synthesis, economic analysis, product liability, ergonomics, engineering ethics, hazards analysis, design for X, materials selection, and manufacturing processes. The engineering design process provides the skeletal structure for the text, around which are wrapped numerous case histories that illustrate both successes and failures in engineering design.

According to Larry Richards, Director of the Center for Computer Aided Engineering at the University of Virginia, engineering cases generally fall into one of three categories. They are case studies, case histories, and case problems.[1] A *case study* presents an ideal or benchmark solution that may serve as a model for future work. In contrast, a *case history* describes how a problem was solved, and points out the consequences of the decisions that were made. This text contains numerous case histories; the more extensive ones have been collected at the end of each chapter so that they will not interrupt the flow of material in the chapters. Each case history has been selected to illustrate a particular principle, procedure, or lesson in the text. Students and faculty can use these case histories as important resources for study, reflection, and discussion.

Finally, a *case problem* sets forth an open-ended (perhaps unsolved) situation that leaves the solution up to the reader. It can be a learning module designed to put students to work in teams to define the problem and solve it through research, discussion, and/or lab work. Four case problems have been prepared for use as active learning modules; they follow Chapter 11. Each of these case problems contains a substantial amount of background information, as well as proposed or existing solutions, so that students will be able to "hit the ground running" if any of these problems is chosen for a design project.

Moreover, these case problem modules serve as examples of the depth, breadth, and type of information that one should acquire about a technical problem before embarking upon the development of a design solution. The

---

1. Fitzgerald (1995).

A Sample Section of the Design Template

□□

DESIGN PHASE 3 Abstraction and Synthesis

You have, by now, formulated a problem statement, identified general and specific design goals, and acquired background technical knowledge. At this point you need to generate alternative design solutions. Initial concepts of a solution may fall into only a few categories or types. It is a good idea to expand the number of solution types, through the method of **abstraction**, in order to ensure that a better approach or view has not been overlooked. Next, **models** will help you to organize data that is important to your problem, structure your thoughts, describe relationships, and analyze the proposed design solutions.

Assignment	Section
3.1 To initiate **abstraction,** break up your problem into as many meaningful subproblems as is reasonable. Classify these subproblems under more general categories, for which the distinctive characteristic of the subproblem is a special case. In other words, broaden the definition of the subproblem to give yourself more latitude in thinking (even if it violates some constraints of the original problem statement). Consider the principles or approaches that could be used to achieve the objective of each subproblem. Join elements of the different partial solution categories in some advantageous manner, staying within the range in each category that would still satisfy the given problem.	6.1

background material in each problem provides sufficient information to understand the general parameters and factors that must be considered during a design effort; however, students should seek additional information beyond that contained in the case problems.

Immediately following the four case problems is a list of 50 *case problem topics,* which describe some situations that call for engineering design solutions. Since these descriptions are very brief, students will need to research the background and current status of each situation. Again, the four case problem modules described above can serve as examples of the type of background data that should be collected via such research.

The text begins with a unique *Design Project Assignment Template* (immediately following this preface) that the instructor can use to select and assign tasks to students who are working on a design project. With this template, instructors can directly correlate their students' efforts on a project with appropriate material from each chapter of this text. Of course, not all assignments need to be performed; the template provides a "menu" from which instructors can select tasks and topics that they wish to emphasize.

This template also provides an abbreviated summary of the text with key concepts highlighted in **bold** font. Students can use this template to review and correlate the critical elements of the design process as they develop an overall perspective of the material presented throughout the text. If a term or concept in the template is not understood completely, the student should recognize that lack of understanding or comfort with the concept as a warning flag. The student should then return to the (referenced) section in the text in which the concept is discussed and study further. Through such efforts, the template can serve as an important learning aid.

Flexibility

The text is designed to be used in a number of different course structures and environments. We recommend either of two general coverage options—Full Path or Fast Track—depending upon the time that is available in a particular course (see Figure P.1).

Among the courses for which this text has been prepared are the following:

Introduction to Engineering Design The entire text can be covered in a full-semester (15-week) introductory course on design. Such a schedule allows the instructor to incorporate many of the exercises and activities contained in the Topic Keys (described below) into the course.

Design courses offered on a quarter-system (10-week) schedule may need to abbreviate coverage of certain topics. In such cases, a variation of the Fast Track may be best, with inclusion of some material from Chapters 5, 8, 9, and 11 in accordance with the instructor's preferences.

Introduction to Engineering The text can be used in a general introductory course to the engineering profession by focusing upon an appropriate subset of the numerous case histories integrated throughout the book. (Of course, each instructor will decide which cases are appropriate for his or her class.) These cases provide a flexible framework for in-class discussions of engineering practice, and the students will become familiar with the major phases of the design process as they work their way through the text.

An alternative approach is to follow the Fast Track with Chapter 9 (Failure Analysis and Hazards Analysis) included in the mix. The material in Chapter 9 may be particularly instructive to students in a broad sense.

Engineering Graphics and Design This text can be integrated easily into the traditional engineering graphics and design courses, particularly in two-course sequences in which there is sufficient time to cover the multitude of topics (e.g., design, graphics, and CAD) often assigned to such classes. For one-semester or one-quarter courses, we recommend the Fast Track.

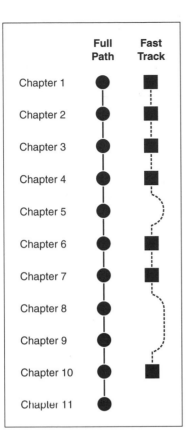

FIGURE P.1 Two of the possible routes through the material in this text.

Advanced Design Courses Many students are not formally introduced to design (in the form of a designated course) until their junior or senior years. In such cases, the entire text should be covered if time allows. Some instructors may want to devote more time to the advanced topics in later chapters, particularly Chapters 9 (Hazards Analysis and Failure Analysis) and 11 (implementation, especially the various design for X topics, fabrication processes, and materials selection).

Table P.1 summarizes these recommendations.

Annotated Overview

Chapter 1 This introductory chapter is critical because it provides an overview of all subsequent material. The design process in its entirety is reviewed, together with certain modern engineering practices. Students

TABLE P.1 Use of this text according to the course offered.

Course	Schedule	Recommended Coverage
Introduction to Engineering Design	Semester	Full Path (Chapters 1–11)
	Quarter	Fast Track + abbreviated coverage of Chapters 5, 8, 9, 11
Introduction to Engineering	Semester or Quarter	Case histories and readings selected by the instructor or Fast Track with Chapter 9 included
Engineering Graphics and Design	Two-course sequence	Full Path
	One-course (semester or quarter)	Fast Track
Advanced Design	Semester	Full Path with more time devoted to advanced topics (e.g., DFX, failure analysis, materials selection, fabrication processes)
	Quarter	Full Path

should be encouraged to focus upon Section 1.5 "Writing Technical Reports" as they prepare technical papers, design proposals, progress reports, and other documents.

Chapter 2 Again, this chapter should be incorporated into any version of the course because it is devoted to phase 1 of the design process, needs assessment.

Chapter 3 Students must become aware of how important it is, and how difficult it can be, to formulate the real problem that must be solved. Chapter 3 describes several techniques for problem formulation, phase 2 of the design process.

Chapter 4 This chapter continues the discussion of problem formulation by stressing the need for thoughtful, methodical problem solving. Its introduction of the concepts of design goals and design constraints (i.e., specs), including ergonomic considerations, is of particular importance.

Chapter 5   Section 5.1 "Science: The Foundation of Technical Knowledge" can be used to link introductory courses in physics, chemistry, and mathematics with the engineering profession. Other sections in this chapter focus on the legal protection of intellectual property and on patents as a rich source of technical knowledge.

Chapter 6   This chapter's topics of abstraction (the first segment of phase 3 of the design process) and modeling formats are critical in engineering problem solving and should be included in any basic course.

Chapter 7   Synthesis (the second segment of phase 3 of the design process) and a set of creativity stimulation techniques are included in this core chapter.

Chapter 8   The presentation of ethical issues and product liability laws in this chapter can provide students with a deeper appreciation of the need to perform rigorous analysis in engineering design. As a result, it serves as an optional but quite valuable prelude to Chapters 9 and 10.

Chapter 9   Preventive techniques for hazards analysis and diagnostic techniques for failure analysis are included in this chapter, including HAZOP, HAZAN, fault tree analysis, and failure modes effects and analysis (FMEA).

Chapter 10   This chapter focuses upon various techniques for evaluating and comparing alternative solutions as part of design analysis (phase 4 of the process). In addition, it reviews elements of economic analysis (in particular, cost estimation and the time value of money).

Chapter 11   This final chapter focuses upon implementation, phase 5 of the design process, with particular emphasis given to design for X (where X can represent manufacturability, assembly, quality, reliability, packaging, maintainability, disassembly, and recyclability), materials selection, and fabrication processes.

Please note that both metric and English units appear in the text. We chose to not limit ourselves or the readers to one particular set of units since both sets are commonly used throughout the United States and elsewhere.

In most introductory design courses, instructors must achieve a balance between a qualitative presentation of engineering practices that can be understood by students with little technical knowledge and a more quantitative approach in which substantive analytical techniques are used to develop and evaluate proposed engineering solutions. Such a balance, difficult to achieve when one is dealing with a relatively narrow and specialized topic, becomes even more challenging in a broad introductory course. *Engineering by Design* has been developed and tested to assist instructors in maintaining this balance. More quantitative techniques and examples are included in Chapters 1, 5, 9, 10, and 11, whereas Chapters 2, 3, 4, 6, 7, and 8 are more qualitative.

Instructor's Topic Keys

In order to assist instructors with the delivery of text material, there is a set of topic keys (one key for each chapter in the text). For instructors who are using *Engineering by Design,* these topic keys are available on the official World Wide Web site dedicated to this book:

http://www.awl.com/cseng/Voland

This site also provides links to other design-oriented websites; examples of the work that can be performed by students who are using this text; additional helpful hints for instruction, classroom exercises, and exams; and a chat room for instructors who wish to share their experiences in teaching engineering design.

Classroom meetings cannot be totally unconnected to the textbook, yet they should never be simple oral recitations of the text. The challenge to the instructor is to reinforce the text's presentation by helping students gain new perspectives and a deeper understanding of the material. The topic keys for this book provide a flexible structure through which instructors can meet this challenge.

Each topic key contains the following sections:

- Objectives
- Estimated Class Time
- Overview
- Materials
- Suggested Class Activities
- Reflection
- Preparation Assignment

The purpose of these sections is as follows:

Objectives   Each key begins with a list of the learning objectives.

Estimated Class Time   The number of one-hour meetings needed to cover a given topic is estimated. Since multiple classroom activities are suggested for each topic, the instructor may choose to devote more or less time to a particular topic as he or she deems appropriate.

Overview   A brief overview of the topic, together with some comments about possible issues in its treatment in the classroom, help to focus the coverage of topics.

Materials   This is a list of materials that should be provided to (or by) students during a class meeting.

Suggested Class Activities   This section suggests classroom activities that can be very effective in conveying the course material to the students. These activities range from the traditional lecture approach to various active learning sessions in which students can begin to master the material by applying it to different problems. Some of the categories of activities that may be suggested in a topic key include:

- Debate, in which student teams must prepare for in-class debates
- Exercises, in which students perform design exercises in (or out of) class
- Journal, in which students share their journal (or diary) entries with the class
- Lecture, in which the instructor conveys information directly to the class
- Questions and answers (Q/A), in which students are asked questions to which they must respond in class
- Reports, in which students prepare brief reports on design topics, case histories, and other issues
- Role-play, in which students play roles in case histories or other scenarios
- Roundtable, in which the class is arranged in a circular pattern for discussion

Reflection   At the end of each session, it is useful to review the lesson(s) covered during the meeting, and allow the students to reflect upon what they have experienced.

Preparation Assignment   Of course, students should prepare for each class by reading the appropriate text material, reviewing their lecture notes, and so forth. However, in an interactive learning environment focusing upon complex engineering design issues, students also should perform some preparatory work through various design exercises and other activities. This section gives suggestions for such preparatory exercises and activities.

Contact with the Author

I can be contacted directly via e-mail at gvoland@lynx.neu.edu. I welcome all comments, corrections, and questions.

ACKNOWLEDGMENTS

Any textbook is the result of many people working together to produce the best product possible. This is especially true for *Engineering by Design.*

I first want to recognize my wife and colleague Margaret Voland, who has served as a de facto coauthor of this text since its conception. Margaret wrote the four case problems following Chapter 11, and prepared the Design Project Assignment Template. In addition, she provided invaluable feedback on the remainder of the book. Her work and contributions add immeasurably to the value of this text. Most important to me, her love and support have added immeasurably to my life.

Margaret and I both thank Peter Murby for his valuable technical research on Case Problem 4, and for being such an amiable coworker. We also thank Titus Peachey for his generosity and dedication in quickly responding with his first-hand accounts of humanitarian demining work. We are also grateful to Professor Enoch Durbin of Princeton University for his gracious offer of technical information about his research and economic analysis of vehicle emissions control.

The striking cover of this text, illustrating some of the construction aspects of the Brooklyn Bridge, attests to the skills and sacrifices of the men and women who have improved our world through their work—a small portion of which has been described in the case histories distributed throughout this text. I wholeheartedly recommend that we all continue to study the history of engineering, including its most recent chronicles, and learn from the work of those who went before us.

I thank all those who created the source materials for this text. Hundreds of books and articles were reviewed during the writing of this text, many of which are included among the references at the end of the book. Without the work of those who developed these archival records of engineering cases and practice, this book could not have become the learning tool that I trust it will be.

Dr. Paul King, former Dean of the College of Engineering at Northeastern University in Boston, led many of us in the college to completely revise our freshman engineering program. This revision included the conversion of our traditional engineering design graphics course into a course on engineering design. Earlier versions of this text were class-tested as the notes for this course during several years of development. *Engineering by Design* would not exist without Paul's wholehearted support and commitment to this effort.

Dean Allen Soyster has continued to support our college's program in freshman engineering design by providing students and faculty with enhanced facilities and opportunities for professional growth. Moreover, Associate Dean Richard Scranton has worked continually, year after year, to ensure that our freshman engineering design course is serving the needs of both the students and the faculty.

I also thank my many generous colleagues at Northeastern University who used the manuscript as their course notes. They provided valuable feedback, corrections, and suggestions for improving the text and ancillary materials. Among those deserving special mention are Professors Clayton Dillon, Charles Finn, Susan Freeman, George Kent, and Ronald Willey.

I also am very grateful to the many hundreds of engineering students who have class-tested the manuscript in its earlier forms. Their work provided critical information about the content, format, and supporting materials for the text, and their encouragement and support spurred us onward.

I also would like to recognize my mentor in engineering design, Dr. William J. Crochetiere, former Chairperson of the Department of Engineering Design at Tufts University. Bill not only taught me a great deal about design, but also about the supportive and encouraging ways in which we can help one another to learn.

The staff at Addison Wesley Longman could not have been more supportive and enthusiastic of this project. I wholeheartedly thank Denise Penrose, the original editor of this project who laid the foundation for its development. Denise Olson, Senior Acquisitions Editor, spearheaded the entire project to its successful completion. Production Editor Amy Willcutt kept us all on schedule and coordinated the many different tasks that needed to be completed in a timely manner. Phoebe Ling, Brooke Albright, Morgan Baker, Sarah Corey, and Judith Strakalaitis each contributed in important ways, for which I am grateful. The book's artwork was completed by George Nichols, who did a first-rate and very creative job. The striking cover design was created by Alywn Velásquez and the beautiful text design by Melinda Grosser. As the book's compositor, Scott Silva transformed our manuscript into its impressive final form. Finally, Joan Flaherty pulled all of the elements of the text together in its final months of development, and ensured that we had a product that matched our vision, while maintaining her good humor. All of these professionals shared my belief that this text will help serve the needs of many engineering faculty who are seeking a more effective way to integrate the active study and mastery of design into their courses in a format that is truly accessible and motivational to their students.

I also thank the many anonymous reviewers who provided absolutely critical and thoughtful feedback on the numerous drafts of this text, and who raised the quality of the final version.

I thank my mother Eleanor Voland for her continued love and encouragement.

Above all, I thank the Lord for His support and many blessings in my life.

Gerard Voland
Bridgewater, Massachusetts

DESIGN PROJECT ASSIGNMENT TEMPLATE

With this template, instructors can correlate their students' efforts on a design project with appropriate material from each chapter of this text. Not all of the following assignments need to be performed. The template provides a "menu" from which instructors can select those tasks and topics that they wish to emphasize in a course.

This template also provides the student with an abbreviated summary of the text with key concepts highlighted in **bold** font. Students should use this template to review and correlate these critical elements of the design process as they develop an overall perspective of the material presented throughout the text.

The template is organized according to the five phases of the engineering design process: needs assessment, problem formulation, abstraction and synthesis, analysis, and implementation. Each assignment is accompanied by a reference to the section(s) in the text that correspond to the topic(s) involved.

DESIGN PHASE 1 Needs assessment

The first step of the **engineering design process** is to identify as clearly as possible what **needs** are to be satisfied by a solution based on technology. A design project may originate out of a concern to protect the health and safety of the public, or to improve the quality of life for some people. An existing product or process may need to be redesigned in order to be made more effective or profitable. A company may wish to establish a new product line for commercial benefit, but how well it meets their customers' needs will determine its success. Or an individual may be confronted with a need in his or her own life, and recognize it as an area for a new application of technology. Last, technological developments or scientific discoveries may create opportunities for new engineered products.

Assignment   **Section**

1.1 Become familiar with an area of human or industrial need. The case problems and case problem topics following Chapter 11 will introduce you to a variety of conditions or situations in need of an appropriate application of technology, or your instructor may give you a topic. Investigate the **current situation**. What conclusions can you draw about the present state of affairs? | 2.1

Assignment	Section
1.2 List the **needs** that you perceive in this design problem. Which needs have been addressed by engineers and which needs still require a solution? (These two lists may overlap.)	2.1, 2.2
1.3 From your list of needs, write an **initial problem statement** for your design project. Identify your problem as one of **prediction, explanation, invention**, or a combination of these types, and explain your reasoning.	2.3
1.4 Is your chosen problem related to an existing engineered product(s)? In what way? Has this product failed to perform its function(s) in any way? If so, try to identify any **physical flaws, process errors**, or **errors in perspective and attitude**.	2.4
1.5 Write an initial **design proposal** in which you justify why the design work should be performed, specify whom it will serve, and state where it will be used. State how you would approach the project, working as an individual or on a team, and what your expected schedule is. Itemize your "deliverables" such as a report, visual models, or a working model. Finally, estimate the cost of the entire project.	2.5

DESIGN PHASE 2 Problem formulation

It is important to recognize that a specific problem must be formulated if one is to develop a specific solution. However, in order to avoid being so specific that the problem is stated in terms of a particular set of solutions, the problem statement should focus on the **functions** to be performed by any viable solution. (An exception to this rule is the Revision Method, in which factors other than function become important.) The following assignments will give you experience in using several **heuristics for accurate problem formulation**, which will enable you to refine your initial problem statement.

Assignment	Section
2.1 Produce three **restatements** of your initial problem from Assignment 1.3 (e.g., changing positive terms to negative terms and vice versa; relaxing constraints or boundaries, retaining only critical goals; and specifying desired outputs, required inputs, and how the inputs are to be transformed into the outputs). Have the restatements changed the focus of the problem? Now write an improved problem statement, focusing on the function to be performed by the design solution.	3.2.2
2.2 Describe the **primary source** of the problem statement (corporate engineers, data acquired through private or funded research, consumers, etc.). Can you determine the **cause** of the problem (not merely its symptoms) from the problem statement or the problem source (see Assignment 1.4)?	3.2.3

2.3 Does your problem involve improving an existing product? If so, consider   3.2.4
changing the focus of your project from the functions to be performed by the
product to the needs of the customer vis-à-vis the product, (e.g., ease of use,
minimum maintenance and repair costs, ease of storage; see the **revision
method**).

2.4 Begin a **Duncker diagram** by writing present state (PS) and desired state (DS)   3.2.5
descriptions. Write general solutions, by modifying the PS to satisfy the DS, and
by revising the DS so that the new PS need not be achieved. Write functional
and specific solutions by satisfying in increasing detail the relevant PS and DS.
Review this diagram to obtain new insights into your problem formulation.

2.5 If your problem involves troubleshooting, that is, discovering the cause for   3.3
the failure of a product, system, or process, **Kepner–Tregoe (K–T) problem
analysis** may be helpful for your problem formulation. Create a table in which
you identify the problem **characteristics** (what is the problem? what is it not?),
its **location** (where did it occur? where did it not occur?), its **timing** (when did
it occur? when did it not occur?), and its **magnitude** (how much? how little?).

Engineers always work within specific deadlines, and search for solutions that lie within
rigid constraints. As a matter of practicality, engineers use strategies that assure them
that an acceptable (if not the best) solution to the problem at hand will be found in a timely
manner.

   After an engineering problem statement has been written, the problem is analyzed, or
broken up into its primary components, by formulating a list of **design goals** that must be
achieved by any viable solution. These design goals help the engineer to structure and perform the search for the optimal design within the given constraints.

2.6 There are a number of **general design goals** that are usually (although not   4.3.1
always) associated with engineering design efforts. These include **safety**, **environmental protection**, **public acceptance**, **reliability**, **performance**, **ease of
operation**, **durability**, **minimum maintenance**, **use of standard parts**, and
minimum cost. However, limited resources and conflicting design parameters
may require trade-offs to satisfy the goals. Create a prioritized list of general
design goals for your product, system, or process. (Your list need not have 10
goals, but should have at least 5.) State how your design would satisfy each
listed goal.

2.7 In addition to general design goals, each design must achieve **specific goals**   4.3.2
that pertain to the problem under consideration. Care should be taken not to
state these goals in terms of a specific class of solutions. Specific design goals
may be generated by such considerations as requirements before, during, and
after product use; criteria concerning weight, size, shape, or speed; simulation
of a human capability, such as speech, grasp, vision, or logical thought, and so
on. Give at least 3 specific goals of your design solution.

2.8 **Design specifications** (specs) are quantitative boundaries associated with each   4.4
design goal, within which the search for a solution must be conducted. These
specs usually are numerical maxima or minima, or ranges, associated with the
physical or operational properties of the design; the environmental conditions
affecting the design, or its impact on the environment; the ergonomic (i.e.,
human factors) requirements of the design; and/or the economic or legal
constraints imposed on the design. Through appropriate research, identify
the constraints or boundaries for each of your design goals.

2.9 **Ergonomic data**, which contains statistics on the **size** and **proportions of people**; their abilities in terms of **endurance**, **speed**, **strength**, **accuracy**, **vision**, and **hearing**; their aptitude in **acquiring**, **learning**, and **processing information**, and the like, may contribute critical constraints or specs to your design. Ergonomic considerations may be needed to achieve the safety, productivity, health, and happiness of a designated user population. What ergonomic considerations are you incorporating into your design?

4.5

People often have expectations of a system. This may be due to physical analogies or cultural standards. For example, most of us expect that an upward movement will turn a device on, whereas a downward movement will turn it off. Such expected relationships, called **natural mapping** (see Section 4.5.4), should be thoroughly investigated and incorporated into engineering designs.

The concept of **affordance** is related to natural mapping. Certain materials and/or designs afford the user the opportunity to perform some action. For example, an oversized plate on a door indicates that one should push to open the door, and a handle indicates that one should pull.

Design engineers must acquire broad and deep technical knowledge and develop the ability to apply it as necessary. However, technical knowledge begins with an understanding of **scientific principles**. You should become familiar with a number of scientific and mathematical principles in such fields as mechanics, fluid mechanics, electricity and magnetism, thermodynamics, and computer science.

There are **handbooks** of physical property data, standards and specifications that are very useful for engineers. **Patent records** are also useful sources of technical knowledge.

2.10 In what ways can your design make use of considerations of natural mapping and affordance?

4.5.4

2.11 Identify some of the **scientific principles** underlying your design solution. For example, is it a mechanical system involving the effects of gravity? Does it involve wave phenomena (light waves or sound waves, as in some mine-detection systems [Case Problem 3])? Does it involve voltages or properties of currents (as do fuel cells [Case Problem 2])? Are there changes in state from solid to liquid, liquid to vapor, or solid to vapor (as in paper manufacture [Case Problem 4])? Does it involve fluids in motion (as do airfoil gliders [Case Problem 1])? Discuss the scientific principles underlying your design.

5.1

2.12 Look at a few **handbooks** of physical property data and specifications for engineering design listed in Section 5.2 or similar references. Use references in your design, citing them in a bibliography and/or footnotes in your report.

5.2

2.13 Read the material on **trade secrets**, **trademarks**, **copyrights**, and **patents**. Discuss why individuals or corporations may want to hold a 20-year patent on inventions, and why others may choose not to patent their inventions.

Conduct a **patent search** for devices or processes that are likely to give you useful information for your design work. Start with the *Index to U.S. Patent Classification* to find subject headings for classes and subclasses to search. Use the subject headings to search the *Manual of Classification* for the classes and subclasses that relate to your design. If possible, go to a CASSIS CD-ROM computer database to view all the patents that have been awarded in the selected subclass; alternatively, access the partial patent records that are available through the World Wide Web (www.uspto.gov). Review the

5.3–5.5

Assignment	**Section**

2.13 (continued)   5.3–5.5

> brief descriptions for some of these patents in the *Official Gazette of the United States Patent and Trademark Office.* Finally, review the full patent disclosure (usually archived on microfilm) of the most promising patents.

DESIGN PHASE 3 Abstraction and synthesis

You have, by now, formulated a problem statement, identified general and specific design goals, and acquired background technical knowledge. At this point you need to generate alternative design solutions. Initial concepts of a solution may all fall into only a few categories or types. It is good to expand the number of solution types, through the method of **abstraction**, in order to ensure that you have not overlooked another, better approach or view. Next, **models** will help you to organize data that is important to your problem, structure your thoughts, describe relationships, and analyze the proposed design solutions.

Assignment	**Section**

3.1 To initiate **abstraction,** break up your problem into as many meaningful sub-   6.1
problems as is reasonable. Classify these subproblems under more general
categories, for which the distinctive characteristic of the subproblem is a
special case. In other words, broaden the definition of the subproblem to
give yourself more latitude in thinking (even if it violates some constraints
of the original problem statement). Consider the principles or approaches
that could be used to achieve the objective of each subproblem. Join elements
of the different partial solution categories in some advantageous manner, stay-
ing within the range of each category that would still satisfy the given problem.

3.2 **Models** are more or less accurate and complete representations of proposed   6.2–6.5
design solutions. They may look like and/or function like the engineered
design being modelled. Although models have limited accuracy, they often
demonstrated interrelationships of system components, and test proposed
design changes, relatively quickly and cheaply. Create models of your product
design solution incorporating at least two of the following types:

- **symbolic models** (e.g., equations)
- **analogic models** (e.g., functional equivalents)
- **iconic models** (e.g., scaled visual resemblances)

3.3 A system model is **deterministic** if the model allows a person to know with   6.6, 6.7
certainty the behavior of the system under given conditions. It is **stochastic**
if the system response can be known only probabilistically (with less than
100% certainty). Does your design project include a deterministic or stochastic system model, or both? Explain.

A process model is **prescriptive** if it provides general guidelines on how
a process should be performed in order to achieve a certain goal. It is **descriptive** if it states exactly what process was followed in a design. Does your

Assignment	Section

3.3 (continued) | 6.6, 6.7

> design project contain a prescriptive or descriptive process model, or both? Explain.

Assignment	Section

3.4 Apply **Occam's razor** to your model as needed. Ask what specific contributions or insights are to be provided by the model, and eliminate unnecessary detail that would obscure its purpose or unnecessarily raise its cost. | 6.8

> **Synthesis** is the formation of a whole from a set of building blocks or constituent parts. **Creative thinking** involves the ability to synthesize, or combine, ideas and things into new and meaningful forms. Although a new synthesis may come as a flash of insight, it generally follows a period of careful, and perhaps laborious, preparation. Engineers use a number of proven techniques for stimulating creative thinking. Before studying these techniques, it is helpful to become aware of some common **conceptual blocks** to creative thinking.

3.5 Consider the models of your design solution. Do you think that one or more of the following **blockages to creativity** may be impeding you from designing the best solution? If so, what specific steps can you take to reduce or eliminate the blockage? Some common blockages are | 7.2

- **knowledge blocks:** inadequate scientific knowledge base;
- **perceptual blocks:** stereotyping elements or not recognizing other interpretations of these elements; improperly delimiting the problem or creating imaginary constraints; information overload or not distinguishing what is significant or insignificant in the data;
- **emotional blocks:** fear of failure and need for approval; unwillingness to build upon prescribed paths and methodologies; impatience or attempting to arrive at a solution too quickly;
- **cultural blocks and expectations:** inhibitions as a result of cultural predilections within the corporate environment; limitations as a result of design expectations or preconceptions held by clients;
- **expressive blocks:** misdirection as a result of inappropriate terminology having been used to define the problem or describe the solutions.

> Having eliminated various blockages to creativity, you can now turn to **strategies for generating creative solutions** to an engineering problem. Since no single technique will be effective for everyone, you should experiment with different strategies to find out which ones work better for you.

3.6 With one or more other people, conduct a **brainstorming session**. Since quantity, not quality, of concepts is sought, freely suggest design concepts and associations, without evaluation or concern for practicality. Make a list of the ideas resulting from your brainstorming session. | 7.5.1

3.7 In the world of animals and plants, you can observe many important, and often difficult, tasks being performed in elegant and effective ways. Can you adapt (or gain insight from) a solution already existing in nature to solve the engineering problem under consideration? This is the creativity technique of **bionics**. | 7.5.2

3.8 For making improvements in an existing product or concept, **checklisting** may be helpful. In checklisting one uses trigger words and questions to spark creative thought. Examples are

- **trigger questions:** What doesn't it do? What is similar to it? What is wrong with it?
- **trigger words:** cheapen, rotate, thin
- **trigger categories:** change order (reverse, stratify); change relative position (repel, lower)

Can you apply this technique to your design problem? Explain or give the results.

7.5.3

3.9 At times an engineer is confronted with a problem that is so familiar that it is difficult to conceive of anything but conventional solutions. Alternatively, a problem may be so unusual that it is difficult to relate it to his or her experiences. The creativity technique of **synectics** includes the tasks of

- making the familiar strange, and
- making the strange familiar.

Creatively transfer your familiar problem into a strange context, or your strange problem to a familiar context, and then devise design solutions. You may apply brainstorming, checklisitng, and so on to the new problem context.

7.5.4

3.10 The **method of analogies** recommends progress toward a creative design solution for a given problem by linking it to another problem that resembles it in some way, and that is easier to solve, closer to being solved, or actually solved. An engineer might

- make a **direct analogy** between the given problem and a solved, or an almost solved, problem. Bionics, usually involves direct analogy.
- make a **fantasy analogy** to imagine the problem in an analogous but more convenient form for the purpose of advancing the design process. Synectics involves fantasy analogy.
- make a **symbolic analogy** by using a poetic metaphor, or a literary cliche, to view a given problem in a new way.
- make a **personal analogy** by imagining himself or herself as part of the system, especially under adverse circumstances, in order to gain a new perspective.

Engineers may, in addition, **adapt** a solution for an unrelated problem, or an earlier (and rejected) engineering design for a similar problem, to the current problem, and mold it into a feasible solution.

Make use of one or more of these methods of analogy or adaptation to generate design ideas, or enhance a promising solution under development.

7.5.5

3.11 It is often possible to obtain new insights into your design problem by simply **explaining the problem** to people who not involved in the design effort. Do this and take notes on their comments.

7.5.6

3.12 Your design problem may involve a task that many people find difficult to do (e.g., conserving a particular resource). The **inversion strategy** states that you may be more productive in solving the inverse problem, (e.g. finding many ways of wasting the resource). Then negate (or invert) your ideas in order to solve the original problem.

7.5.7

3.12 (continued)   7.5.7

Inversion also may include changing people's perception of an object or situation. In Case History 7.9 *Jokes for Trash?*, street litter was seen no longer as articles a person was obliged to dispose of properly, but rather the means of amusement for the use of a trash receptacle.

Use inversion to stimulate creative thinking for your design effort.

3.13 An idea diagram follows a line of reasoning opposite to that of abstraction.   7.5.8
In abstraction (Section 6.1), from the problem statement one generates broader categories that contain the initial solution concepts as particular cases. In an **idea diagram**, the problem statement itself is considered the broadest category, then subcategories showing increasing detail are generated in a tree structure in order to generate multiple solutions (see Figure 7.2). Create an idea diagram for your design problem.

Having used abstraction and creativity techniques to develop a set of partial solutions, or components, of the system, and modeling to define and clarify these partial solutions, there remains the task of combining the partial solutions into a whole in order to solve the given problem. The process of combining the partial solutions, or components, is **synthesis**. Synthesis may also be described as reasoning from principles (for example, design goals) to their applications (for example, concrete means for achieving each goal).

3.14 A **morphological chart** allows a person to systematically form different   7.6
engineering designs, each of which is a solution to the same design problem (expressed by a fixed set of design goals). The rows of the morphological chart correspond to different functions, or design goals, which should be achieved by every solution. The columns correspond to different means by which these goals may be achieved (a result, perhaps, of creativity techniques). By selecting one element from each row, while avoiding impossible combinations, a designer may generate a set of alternative solutions to a given problem. Create a morphological chart for your design problem, and from it a set of design solutions.

DESIGN PHASE 4 Analysis

Ethical behavior in engineering can be the difference between success and failure in professional practice, triumph and tragedy concerning a problem, or life and death for a user. Therefore the Accreditation Board for Engineering and Technology (ABET) has called ethical, social, economic, and safety considerations in engineering practice "essential for a successful engineering career" (see Section 8.1).

An engineer needs to be familiar with, and consciously work to comply with, his or her professional society's code of ethics, as well as applicable federal, state, and local regulations, contract law, and torts law. An engineer who does this effectively guards against the loss of his or her professional reputation, and the need to defend oneself in civil or criminal lawsuits (see Section 8.2). He or she will also reap the many benefits associated with honor and appreciation.

4.1 Consider and report what specific **ethical, social, economic, or safety factors** (cited by ABET as necessary professional concepts) have been, or could be, incorporated into your design solution. What difficulties have been, or might be, encountered in incorporating these factors? Can these difficulties be overcome? (Your considerations should cover at least two of the four areas listed.)

4.2 In the case history of the *Challenger,* consider the directive given to the vice president of engineering, to "take off your engineering hat and put on your management hat" in terms of the **NSPE Code of Ethics for Engineers** (See Appendix). Is this directive in conflict with reporting all relevant information (Section II.3.a of the NSPE Code), disclosing potential conflicts of interest (Section II.4.a), or advising a client that a project is believed to be unlikely to succeed (Section III.1.b)? Explain.

Apply the five **fundamental canons** of the NSPE Code of Ethics (see Part I of the code) to your design solutions, and the methods by which they came into being. Have there been infractions?

4.3 Three BART engineers (Max Blankenzee, Robert Bruger, and Holger Hjortsvang) were commended for "courageously adhering to the letter and the spirit of theIEEE code of ethics." Explain the distinction between adhering to the **letter of a code of ethics,** and adhering to **its spirit.**

These engineers contacted their supervisors, a private engineering consultant, and some members of the BART board of directors concerning their assessment of inherent dangers of the system. Have you experienced disagreements with team members or supervisors? What principles helped you resolve the difficulties?

4.4 In Case History 8.5 *A Helicopter Crash,* the company's primary defense was that the system's requirement that autorotation be activated within 1 second of power failure adhered to FAA standards. Analyze this design feature in terms the NSPE's Code of Ethics and inadequate **floor standards** for safety.

4.5 Engineers must **anticipate uses and misuses** of their products, and take precautions against possible injury that may result. Consider the anecdote of "The Dangerous Door." What corrective action would you order for future door shipments?

What injuries might occur from use or misuse of your product, and what steps will you take to prevent them?

Engineering case histories describe how problems were solved, and point out the consequences of the decisions that were made. Case histories that chronicle disasters or engineering failures allow engineers to develop skills in **post-failure diagnostic analysis**. These skills in turn help them to become more adept at **preventive hazards analysis**, or recognizing and avoiding conditions in the design that could lead to disaster (see Section 9.1).

Once failure has occurred, specific sources of failure should be identified as accurately as possible (see Section 9.2). Known hazards in a design solution may be identified through familiarity with industrial and governmental standards, codes, and requirements (see Section 9.3.2). Unknown hazards may be discovered through application of techniques such as HAZOP, HAZAN, fault tree analysis, failure modes and effects analysis, and Ishikawa diagnostic diagrams (see Section 9.4).

Design engineers should try to reduce or prevent the known and unknown hazards through the use of safety features (shields, interlocks, sensors, etc.), a sufficiently large safety factor (strength-to-load ratio), quality assurance programs, redundancy, and appropriate warning labels and instructions (see Section 9.5).

4.6 Consider the railroad warning system at Hixon Level Crossing in England, described in Section 9.1. A 24-second interval from the onset of flashing warning lights to the arrival of a train at the intersection proved to be an insufficient time for a 148-foot road-transporter to cross the tracks. (The carrier needed one full minute to clear the tracks.) Moreover, the transporter crew and its police escort were unaware of the emergency telephone installed near the intersection, and so failed to communicate their need to the railway operators.

    Perform a **failure analysis** on this railroad disaster. Were there **physical flaws**, **errors in process**, or **errors in attitude**? Explain.

    If you have tested a working prototype of your design solution, identify specific sources of failure, which you may have encountered, according to the outline in Section 9.2.4.

9.1, 9.2

4.7 Consider the coal mine waste "tips" that were built up along the side of Merthyr Mountain near Aberfan in South Wales. In the context of this mine-related disaster, describe the **hazard**, **danger**, **damage,** and **risk**. Was the hazard associated with tip 7 primarily **inherent** or **contingent**? Explain.

    Perform research to see what **standards** or requirements apply to the product or system you are developing (e.g., product performance standards, packaging standards, personal exposure standards). Report on your compliance.

9.1, 9.3

In **hazards analysis (HAZAN)** (see Section 9.4.2), an engineer attempts to quantitatively estimate the **frequency** and **severity** of each hazard, and develop an appropriate **response** to each one. Consider the following case histories.

    Case History 9.2 recounts the repeated incidence of corroded airplane engine fuse pins over a 14-year span. Corroded fuse pins eventually resulted in the aircraft wing engine tearing loose. There was the further complication that one broken-away wing engine could dislodge the other wing engine. The corrosion continued even after the hollow cylindrical steel pins were no longer machined on the internal surface, but an insert was used instead. This later corrosion problem was due to removal of bits of anticorrosive primer coating when the inserts were attached.

    Case History 9.3 deals with another type of corrosion problem—the failure of the chain-link Point Pleasant Bridge, after 41 years of use. The design of the eyebars in this bridge made them particularly difficult to inspect. When one eyebar did crack, due to stress corrosion, twisting motion pried loose the cap plate, which was bolted over the pin joining pairs of steel eyebars that made up the links of the bridge chain. Once one link gave way, a chain reaction followed in which other links broke loose, resulting in the bridge's collapse.

    Although there was a high concentration of stresses at the point of failure (between the critical eyebar and a pin), this applied stress was still well below the limit that theoretically could be supported by the steel eyebars. Why then did it crack? Investigation showed that the inner core of the forged steel eyebar had not regained its ductility (i.e., elasticity), because the final cooling of the eyebar was performed too rapidly. This flaw in the manufacturing process gave the eyebars a brittle inner region, which made them susceptible to stress corrosion.

4.8 Consider the hazards associated with the systems described in Case Histories 9.2 and 9.3. Comment on the **frequency** and **severity** of each hazard.

Case Histories 9.2, 9.3

4.9 Assess the hazards associated with your design solution by performing a hazards analysis. Summarize your data by answering the questions, How often? How big? So what? 9.4.2

In **fault tree analysis (FTA)** (see Section 9.4.3) and **hazard and operability studies (HAZOP)** (see Section 9.4.1) one envisions possible causes of undesirable performance by a system.

In FTA, a tree diagram is created by means of top-down reasoning from system failure to the component basic faults that could have led to it. From test data, or historical records, probability or frequency of occurrence may be incorporated into the tree. A designer then may use FTA to make modifications to a system in order to eliminate faults that are most severe in their consequences and/or most frequent in their occurrence.

In HAZOP also, from descriptions of deviations from acceptable behavior of a system, first the consequences of each deviation are considered; second, possible causes that contributed to each deviation are studied; and third, the HAZOP team develops a set of specific actions that could be taken to minimize or eliminate the deviations. HAZOP is often used for identifying flaws within a process; (Case History 9.4 may be considered an example of a flawed process).

4.10 Design a fault tree diagram or perform a hazard and operability study on one of the following case histories: 9.4

      Case History 9.7   *DC-10 Cargo Door Design*
      Case History 9.6   *The* Titanic *Disaster*
      Case History 9.4   *Disasters on the Railroads*

4.11 Postulate an undesirable event for your design solution, and create a fault tree diagram that would lead to this event. Use the symbols in Figure 9.1. 9.4

4.12 When using **failure modes and effects analysis (FMEA)** to troubleshoot a design, an engineer studies each basic component of the design, one at a time, and tries to determine every way in which that component might fail, and the possible consequences of such failures. He or she then tries to take appropriate preventive or corrective action. The engineer advances from analysis of the failure of single components, to combinations of these components, to a failure of the system itself, in a bottom-upward approach to redesigning the product. 9.4.4

    Read Case History 9.10 *The MGM Grand Hotel Fire.* Construct a portion of a **FMEA sheet**, that contains at least four parts or subsystems and five columns (see Table 9.1 as an example), as part of an analysis of the hotel's fire detection, containment, and escape system.

    Create a FMEA sheet for your design solution. Note the way in which a component of the system may fail, and the effect of its failure on the system.

4.13 An **Ishakawa diagnostic diagram** is a graphical display of the components or subsystems of a design, drawn as ribs extending outward from a central spine. It also shows a number of the potential flaws, weaknesses, or hazards associated with each subsystem, drawn as offshoots of each rib. 9.4.5

    Construct an Ishakawa diagram, that has at least three ribs, as a diagnostic tool for studying the hazards associated with your design solution. "Invert" at least three of the weaknesses (the offshoots) to create a set of specific goals for an improved design of the system (see Section 9.4.5, footnote 24).

4.14 State how you will seek to reduce or eliminate the known and unknown hazards you have identified in your design solution. Some means to achieve this are

- **safety features** (shields, interlocks, sensors, etc.),
- a sufficiently large **safety factor** (strength/load),
- **quality assurance program** (testing),
- **redundancy** (back-up systems), and
- appropriate **warning labels and instructions**.

9.5

You now are reaching the final stages of the design process. You used creativity techniques to generate design concepts, and synthesized your knowledge by means of a morphological chart, to arrive at complete design solutions. You have critiqued your solutions with respect to compliance with professional codes of ethics; federal, state and local regulations; contract law and torts law. Further, you have critiqued your designs in terms of preventive hazards analysis. You are ready at this point to select among the design alternatives the **best overall solution**. The outcome of a product is determined largely by the **selection and weighting of the design goals** specified in the engineering design process. In order to evaluate the competing designs, reflect upon the relative importance of each design goal, and the ability of each design to achieve these goals.

4.15 **Rank-order your design goals** as shown in Table 10.1 by assigning to each "row" goal a value 1, 1/2, or 0 to show its relative importance with respect to each "column" goal.

10.2

4.16 Assign **weighting factors** in the range 1 to 100 to your rank-ordered list of design goals, as shown in Tables 10.3 and 10.4.

10.3

4.17 Assign a **rating factor** in the range 0 to 10 to each design solution to show the degree to which that design satisfies each design goal. Refer to Table 10.5.

10.4

4.18 Using the weighting factors and the ratings factors calculated above, create a **decision matrix** to evaluate each alternative design solution. Refer to Table 10.6. Remember that decision factors within 10 percent of each other are a tie.

10.5,
10.6

If certain design goals absolutely must be satisfied to a certain threshold level, you may choose to create a **Kepner–Tregoe decision matrix** (see Table 10.7). The design goals are divided into "musts" and "wants." Only design solutions that satisfy all the "musts" are viable. Decision factors are calculated for viable solutions with respect to the "wants."

4.19 The highest decision factor is only one indicator of a best solution. You should also consider the threat associated with this best solution. Create a **Kepner–Tregoe evaluation matrix of adverse consequences** (see Table 10.8). Identify the risks of each viable design solution. Calculate the threat associated with each risk as the product of the probability of its occurring (scale: 0–1.0) times its relative severity (scale: 1–10). Compute the total threat as the sum of the threats associated with each risk. If the threat for the candidate best solution seems too high, you may consider developing the second-best solution, or redesign the best solution to reduce its inherent risks. (Comment: If an inherent risk in a particular design is of truly disastrous proportions, you may indicate this by choosing a severity scale 1–10,000, for example, where all risks but one are in the 1–10 range of the scale.)

10.6,
10.7

Economic viability is another type of analysis that is necessary in order to design a successful product. An engineer may compare different methods of producing and marketing a product in order to find the method of operation that would optimize profit, efficiency, use of facilities, and aspects of the production process.

Assignment	Section
4.20 Using the **Rule of Thumb method,** estimate the expected retail price of your product. What is the anticipated sales volume? Is it expected to be economically viable? Do you recommend any design changes to improve its competitive status?	10.8

DESIGN PHASE 5 Implementation

You are now in the final phase of the design process. Implementation has to do with the physical realization of a design. However, the requirements of implementation should not be left to the end of the design process. Since materials, manufacturing, assembly, disassembly, recycling, and economic requirements can affect early decisions of the design process, these considerations should be brought to bear on decisions throughout the design effort. The method for doing this is called concurrent engineering. In order to implement a design properly, an engineer needs to be knowledgeable about fabrication materials, materials properties, and fabrication processes. Materials chosen for an engineering design should match the availability, performance (functional), economic, environmental, and processing (manufacturing) requirements of the product or system.

Assignment	Section
5.1 **Concurrent engineering** is simultaneous development of all aspects of a design, by means of teamwork, from the initial concept to its manufacture, maintenance, and disposal, in order to optimize the performance and quality of a product and minimize its cost and production time, while achieving other design goals for its fabrication, maintenance, disassembly, recycling, or disposal, and so on.	11.2

    Concurrent engineering employs the following "design for X" (DFX) considerations. Review each of the eight DFX major categories (i.e., manufacturing, assembly, reliability, quality, packaging, maintainability, disassembly, and recycling) listed below.

- **Design for manufacturing** and **design for assembly**
  - –Employ division of labor.
  - –Use interchangeable parts.
  - –Use assembly line operations.
  - –Use machines where appropriate.
  - –Use modular design and subassemblies (group technology).
  - –Use rapid prototyping.
  - –Minimize number of parts.

5.1 (continued)   11.2
      –Minimize variations of parts.
      –Design multifunctional parts.
      –Avoid separate fasteners.
      –Minimize number of assembly operations.
      –Maximize tolerances for easy assembly.
      –Provide sufficient access for easy assembly.

- **Design for reliability**
      –Select proven components.
      –Incorporate intentional design redundancies.
      –Perform preventive maintenance.
      –Perform corrective maintenance.

- **Design for quality**
      –Perform reactive (or diagnostic) operations.
      –Use proactive (or preventive) strategies (e.g., Taguchi engineering
        method, benchmarking, Quality Function Deployment/"House of
        Quality").

- **Design for packaging**
      –Provide aesthetic appeal and brand-name recognition.
      –Protect the product from spoilage and damage.
      –Provide a range of product sizes and formats.
      –Establish product standards and measures of quality.
      –Prevent misuse of the product.
      –Provide inexpensive and easy-to-open closures.
      –Reduce, reuse, and recycle packaging.

- **Design for maintainability, disassembly, and recyclability**

How have they been implemented in your design? Can more be done with
these methods to enhance the viability of your design? Give at least one concrete example of application of DFX methods to your design for each of the
eight major categories.

5.2 Materials are selected for a given design because they have properties that   11.3
satisfy functional, environmental, manufacturing, and economic require-
ments imposed on the product. Some **material properties** of importance are

- mechanical properties (e.g., strength, stiffness, ductility, toughness)
- electrical properties (e.g., resistance, conductance, dielectric strength)
- thermal properties (e.g., thermal conductivity, specific heat capacity, melting point)
- availability
- cost
- toxicity
- corrosion
- biodegradability
- flammability
- permeability
- texture
- density
- appearance

What are significant material properties of (or parameters for) your finished
product, and of the raw materials you will use? (You may name properties not
included in the list above.) Explain why these properties are important for
your design.

5.3 It is useful to know some basic properties and applications of common **fabri-cation materials**. An engineer should be acquainted with the following common fabrication materials: 11.4

- metals
  −ferrous (iron, steel)
  −nonferrous (aluminum, copper, magnesium, nickel, titanium, zinc)
- polymers
  −thermoplastics (acrylics, nylons, polyethylene, PVC, vinyl, etc.)
  −thermosets (phenal formaldehyde, melamine formaldehyde,
    urea formaldehyde)
  −elastomers (natural rubber, synthetic rubber)
- ceramics
  −brickware
  −whiteware
  −glassware
  −enamels
  −cements
- composites
  −fiberglass
  −reinforced concrete
  −wood

List the materials that will be used to implement your design and note their advantages and disadvantages.

5.4 Appropriately selected and properly executed **fabrication processes** are essential for manufacturing successful products. Fabrication processes can be identified as one of the following eight types: 11.6

- solidification processes (i.e., material is cast into a desirable form while in a molten state)
- deformation processes (e.g., rolling, forging, extruding)
- material removal processes (e.g., grinding, shaving, milling)
- polymer processes (e.g., injection molding, thermoforming)
- particulate processes (e.g., pressing, sintering, hot compaction)
- joining processes (e.g., soldering, riveting, bolting)
- heat and surface treatment processes (e.g., heating, electroplating, coating)
- assembly processes

Identify and evaluate three or more fabrication processes that could be used to produce your design. Present your analysis in tabular format.

CONTENTS

ONE

Engineering Design 1

TWO
Needs Assessment 47

THREE
Structuring the Search for the Problem 77

FOUR

Structuring the Search
for a Solution 103

FIVE

Acquiring, Applying, and Protecting Technical Knowledge 153

SIX
Abstraction and Modeling 211

SEVEN

Synthesis 237

EIGHT
Ethics and Product Liability 275

NINE

Hazards Analysis and Failure Analysis 305

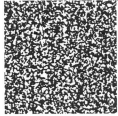

ENGINEERING BY
D E S I G N

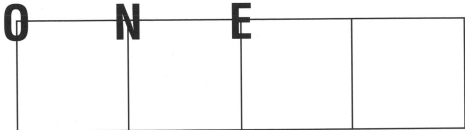

O N E

ENGINEERING DESIGN

I find the great thing in this world is not so much where we stand as in what direction we are moving.   Oliver Wendell Holmes

O B J E C T I V E S

Upon completion of this chapter and all other assigned work, the reader should be able to

- Identify some of the attributes that an engineer should possess in order to be successful.
- Describe the stages of the engineering design process used to develop innovative solutions to technical problems.
- Explain why engineers must be methodical when solving technical problems.
- Discuss the reasons for such current practices in engineering as life-cycle design, design for quality, design for export, design for manufacturing and assembly, and "engineering without walls."
- Explain why modern engineers often work in teams and why it is important to maintain a working environment of mutual trust, accountability, and respect.
- Explain why engineering design—and the study of engineering-requires that we be patient and diligent in our work (even if no apparent progress is visible on the surface) and that we persevere until success finally is achieved.
- Explain the need for technical reports to be well organized and well written, and describe the general format often used for such reports.

O U T L I N E

1.1 Definitions of Engineering

Engineers apply science and technology to develop solutions to practical problems—problems that need to be solved for the benefit of humanity. The Accreditation Board for Engineering and Technology, Inc. (ABET) provides the following definition of engineering as a profession:

> Engineering is the profession in which a knowledge of the mathematical and natural sciences, gained by study, experience, and practice, is applied with judgment to develop ways to utilize, economically, the materials and forces of nature for the benefit of mankind.[1]

Alternatively, our working definition of engineering as an activity is as follows:

> An innovative and methodical application of scientific knowledge and technology to produce a device, system or process, which is intended to satisfy human need(s).

Both definitions emphasize that engineering is not solely "inventing" (although certainly both engineering and inventing may result in a new

---

1. *ABET Accreditation Yearbook,* New York: Accreditation Board for Engineering and Technology, Inc. (1993).

device or process); the two activities differ in rather subtle ways. For example, inventors seek to develop a new (and usually patentable) idea, whereas engineers seek to solve a specific technical problem in the best possible manner—even if this solution is neither new nor patentable. In other words, the objectives of an engineer may differ from those of an inventor. Secondly, an engineer applies scientific knowledge and follows a methodical approach in developing designs, whereas an inventor may do neither.

We must quickly note that many engineers are *also* inventors and that many inventors are both methodical and scientifically knowledgeable. However, engineers should *always* be methodical in their work and able to apply scientific principles as needed.

In a recent survey,[2] representatives from industry were asked to prioritize a list of attributes that an engineer should possess. Their prioritized list follows:

1. **Problem-solving skills**   Engineers must be able to
   - identify and define the problem to be solved
   - develop alternative design solutions
   - implement the solution finally selected.
2. **Effective communication skills**   Engineers must be able to convey ideas effectively in both written and oral form (see Section 1.5 later in this chapter).
3. **Highly ethical and professional behavior**   Engineers must be able to recognize and resolve ethical dilemmas, and behave in a professional manner at all times and under all circumstances.
4. **An open mind and positive attitude**   If engineers are to be successful in solving challenging technical problems, they must be both imaginative and optimistic that their efforts will bear fruit.
5. **Proficiency in math and science**   Engineers must be adept in mathematical techniques and knowledgeable about science.
6. **Technical skills**   Engineers must acquire the appropriate set of technical skills if they are to perform well in their chosen profession.
7. **Motivation to continue learning**   Given that both technology and scientific knowledge are expanding at an incredibly rapid rate, engineers must be willing and able to acquire new skills and knowledge in their areas of expertise.
8. **Knowledge of business strategies and management practices** Engineers must be familiar with such strategies and practices if they are to succeed in industry.
9. **Computer literacy**   Engineers must be familiar with the latest computer technology if they are to use it in effective ways in various engineering applications.
10. **Understanding of world affairs and cultures**   It is critical to understand cultural differences if one is to work in harmony with others from around the world, as tomorrow's engineers will do.

2. Summarized in Landis (1995).

Engineering is a service profession. Virtually all engineered products and processes are responses to people's needs. A significant amount of engineering thought and effort can (and should) be devoted towards improving even such relatively common products as wheelchairs. Each of us in this profession seeks to serve or to assist others by providing them with a safer, happier, and more productive lifestyle.

Engineering can be divided into many distinct (but often overlapping) disciplines in which a particular problem focus or application area is of primary interest. For example, agricultural engineers focus on the food and fiber industries, in which harvesters, milk processors, feed distributors, and other farm machinery are used to produce, process, and distribute food in an efficient and economical manner. Chemical engineers are often involved in the materials processing and production industries, in which fuels, plastics, insecticides, paints, cements, and other substances are developed for society's needs (and where the recycling of wastes and the elimination of environmental pollutants are critical objectives). Civil engineers apply their knowledge of structures, surveying, and hydraulics to design and build dams, tunnels, bridges, roads and other large-scale systems for densely populated areas. Electrical engineers develop better ways to generate, distribute, and utilize electrical energy, communications systems, and computer technology in order to improve people's lives. Industrial engineers integrate materials, equipment, and people to form systems that will generate high-quality products and services in a safe, productive, comfortable, profitable, and efficient manner. Mechanical engineers are involved in the design and development of such products as engines, aircraft, automobiles, pumps, refrigeration units, and solar energy systems. In each of these disciplines, we find engineers striving to improve the quality of their customers' lives.[3]

It is also important to recognize that a problem or application area often overlaps several different engineering disciplines, thereby requiring engineers from various fields to pool their expertise in order to design and develop a viable solution. Engineers seldom work alone and seldom work only with people from their particular discipline. Every engineer should develop the skills necessary to communicate and work with people from other engineering disciplines and professions (see Section 1.6).

1.2 The Engineering Design Process

Mathematicians, physicists, chemists, and other scientists seek unique solutions to the problems that they investigate. In other words, each of these problems has a one-of-a-kind solution. In contrast, engineers focus on

---

3. Of course, many other disciplines and subdisciplines are recognized within engineering including aerospace, biomedical, computer, environmental, manufacturing, nuclear, petroleum, and systems engineering.

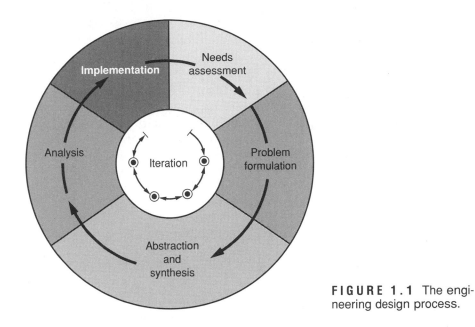

FIGURE 1.1 The engineering design process.

problems for which many practical solutions could be developed; they then seek the best solution from among these many alternatives.

Of course, in order to determine the best solution from among a set of alternatives, engineers must first be able to recognize and develop each of these alternatives—a very formidable task! To perform this task in an effective and efficient manner, engineers often follow a procedure known as the engineering design process.[4] Figure 1.1 shows this process divided into five principal tasks or phases, as follows:

1. **Needs assessment**   The need for a solution is first established.
   - Identify the objective(s) to be achieved by a solution.
   - Identify who will benefit from the solution.
     In what way(s)?
     How do you know?
   - Begin with the end in mind; know where you are going.
2. **Problem formulation**   The "real" problem to be solved is defined in the form of design goals that are to be achieved by any viable solution.
   - Ask if the real problem differs from the problem as initially perceived or presented. In what way(s)?

---

4. Since the early 1960s, many versions of the engineering design process have been prescribed. Some prescriptions of the process are very brief with only four separate stages, whereas others are decomposed into dozens of subtasks to be performed by the designer. However, most variations include the basic tasks of *problem formulation, synthesis, analysis,* and *implementation* (although the labels given to these tasks sometimes vary). We have chosen here to include *needs assessment* and *abstraction* in Fig. 1.1 as distinct elements in the process for we believe that the resulting five-stage process is a truly effective problem-solving algorithm.

What or who was the source of the original problem statement? Did this source bias the statement in some way because of a unique perspective? If so, is the statement then incorrect or incomplete? In what way(s)?

- Structure the search for a solution. Identify as many different pathways leading to possible solutions as you can. Know where you are going and direct your search by pruning those paths that will (probably) not lead to a solution.
- Acquire and apply technical knowledge as appropriate. In order to formulate the problem correctly and completely and to structure the search for a solution, one must make informed—that is, knowledgeable—decisions.
- Identify the design specifications (both explicit and implicit constraints or boundaries within which the solution must lie).
- Identify the resources (time, money, personnel) needed to obtain a solution.
- Prioritize the design goals; continually review this list and modify it as needed during the remainder of the design process. Be aware that your initial prioritization may be incorrect. Be open to change in your goal list. Focus primarily on those goals deemed most important, but recognize that all goals should be achieved by the final solution.

3. **Abstraction and synthesis** Develop abstract (general) concepts or approaches through which the problem could be solved and then generate detailed alternative solutions or designs for the problem.
   - Recall related problems or experiences previously solved, pertinent theories, and fundamental approaches (if any exist) to resolving this type of problem.
   - Expand your thinking as to what is possible and what is not possible.
   - Seek to develop solutions that are fully acceptable to all involved. What approaches can be taken to solve the problem? Which of these approaches is most valid? Why?
   - Reconsider the problem statement: Is it still valid or must it be modified?
   - Be creative; use established and appropriate techniques for generating as many detailed solutions as possible.
   - Combine ideas for achieving each of the individual design goals into total solutions. Seek to make the whole (i.e., the complete design) greater than the sum of the parts (i.e., the individual ideas or subsolutions).
   - Once again, expand your thinking as to what is possible and what is not possible. Be adaptable.
   - Again reconsider the problem statement: Is it still valid or must it be modified? Does your list of goals need to be modified? If so, in what way?

4. **Analysis** Compare and evaluate alternative designs.
   - Choose a basis for comparing your alternative design solutions by establishing objective evaluation criteria.
   - Be critical of your work. Try to see your designs objectively and recognize each of their weaknesses or shortcomings, as well as their strengths.
   - Consider fabrication/implementation requirements for each solution—for example, raw materials and standard parts ("off-the-shelf" components) to be used; manufacturing processes needed to shape the raw materials into final form; the impact that production, distribution, operation, and disposal of the fabricated design may have upon the environment, etc. Compare and contrast the requirements for each proposed design.
   - Are each of the proposed solutions ethical in concept and operation (safe, environmentally responsible, etc.)?
   - Eliminate alternatives that do not satisfy critical design goals (i.e., those goals that *must* be satisfied for the problem itself to be solved).
   - Anticipate and avoid failure by eliminating weaknesses in your designs; focus upon others and their needs and expectations. Are there any inherent hazards in your designs? Can these hazards be eliminated or minimized?
   - Does each design alternative satisfy appropriate ergonomic requirements (human–machine system design goals and specs)? If not, why not? Improve and refine each of your proposed designs, if possible.
   - Construct prototypes of the most promising designs (if possible) and test/evaluate/refine these solutions.
   - Select the best alternative from among those designs that remain as viable solutions to the problem.
   - Revise and refine this best design as appropriate; eliminate or minimize weaknesses and shortcomings of the design. Can this best design be improved by combining it with elements from any or all of the other (rejected) alternatives?
5. **Implementation** Develop the final solution and distribute it to your intended clients/customers/users.
   - After successfully fabricating, testing, and evaluating a design prototype (if such testing is possible), proceed with full production.
   - Distribute to user population and obtain feedback for the next-generation design

A sixth step in the process, reflection, may be included. During this step, one contemplates the lessons learned and the knowledge acquired as a result of the just-completed design effort. Most of us quickly shift our attention from one completed task to the next unresolved task without stopping to consider the ways in which we have grown from the experience. A period of formal reflection can be immensely helpful in clarifying those aspects of the experience that can be used to perform future tasks in a more effective manner. Reflection is particularly valuable if all members of the design team can share their final thoughts about a project once it has been completed, thereby encouraging each person to identify and assess the benefits of the experience.

Engineering design is a naturally iterative process. As indicated in Figure 1.1, iteration can occur between any and all of the stages in the design process. One does not necessarily complete the entire design cycle (traveling from needs assessment through implementation) before returning to one of the earlier stages of the process to correct and modify earlier results.

For example, the engineer may acquire a deeper understanding of the problem as he or she evaluates a set of alternative solutions during stage 4 in the process. Recognizing that the real problem to be solved is far different from the one originally described in the problem statement (during stage 1), the engineer revises both the problem statement and the goal list to reflect this new insight correctly. With this now appropriate target (a correct problem statement and goal list) at which to aim, the engineer is more likely to be successful.

Iteration also may occur as a result of tests on various design prototypes, resulting in a gradual refinement of the basic solution concept.

Review Case History 1.1 *Incomplete Design The Early Quest for Manned Flight* and Case History 1.2 *Methodical Design The Wright Brothers' Success in Manned Flight* for illustrations of the need to apply the engineering design process carefully and in its entirety.

1.3 Nth-Generation Designs

Engineering design does not end with an optimal solution. There is *no such thing as a perfect solution to an engineering problem* because of the compromises that one usually must make in order to resolve conflicts among the design goals. For example, maximizing the durability of a design while minimizing its production costs can be very difficult to achieve fully, in which case the engineer strives to obtain an acceptable balance between these two goals.

As a result, the search for a better Nth-generation design solution to a problem may continue endlessly (e.g., once a product is marketed, feedback about the design is collected from the users, and the product then undergoes redesign to better meet the users' needs). Thus, the design process is repeated again and again as new and better solutions are developed.

The keyboard for both the typewriter and today's computer is an example of a successful design that has undergone—and continues to undergo—numerous iterations as engineers strive to improve its value as a communication interface. Review Case History 1.3 *Nth-Generation: Design Refining the Typewriter (1867–Present)* for details.

1.4 Current Practices in Engineering Design

Just as engineering solutions undergo change from one design generation to the next, so, too, does the engineering profession itself change as the years go by. Engineers change the world and in turn are affected by the very changes that they wrought. As a result, engineering is an extremely dynamic profession, continuously changing in response to a changing world.

Many current engineering practices were neither recognized nor followed only twenty years ago; among the most important of these practices are:[5]

- **Life-cycle design**   Engineers increasingly focus on the entire life cycle of a design—from conception through its manufacture and use to its final disposal—in order to ensure that the design will be successful.

  Manufacturing decisions can affect the economic viability and the functionality of a design. For example, attractive packaging for a consumer product might require that the retail price be raised beyond profitability, leading to commercial failure. Or a poorly chosen fabrication process might reduce the strength of a product, leading to functional failure.

  Most consumers dislike maintaining, servicing, or repairing their products, and designs that do not minimize or eliminate the need for such maintenance and repair are more likely to become commercial failures.

  Furthermore, materials and components should be recycled to protect the environment and to generate cost savings for the manufacturer (both in terms of materials and reduced service facilities). One reflection of this effort for greater recyclability is that 76 percent of the average automobile is now recycled, according to the American Automobile Manufacturers Association.

- **Design for manufacture and assembly**   Design engineers should work to ensure that any proposed solution can be properly manufactured. Although this need to consider the manufacturability of a design may appear obvious, product concepts were once simply "thrown over the wall" from the design engineers to those in manufacturing. The manufacturing engineers then were faced with the challenge of fabricating the designs no matter how difficult or expensive that might be to accomplish.

  Today, design and manufacturing engineers work together to produce products that are innovative, yet cost effective and manufacturable. Components are combined when possible; if unnecessary, they

---

5. Machlis, Wingo, Murray, Hogan, Baker, Schofield, Teague, Smock, Lynch, and Gottschalk (1995).

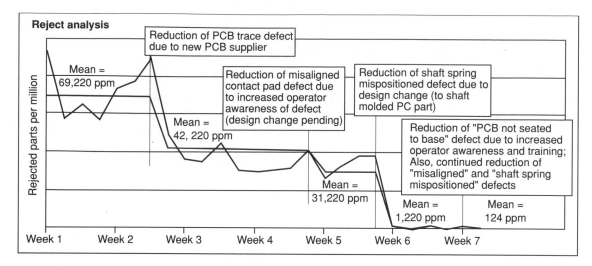

FIGURE 1.2 Quality improvement methods used by Cherry Electric of Waukegan, Illinois, reduced the average number of defective parts in a power switch assembly from 69,220 ppm (parts per million) to 124 ppm. *Source:* From Machlis et al.; reprinted with permission from *Design News Magazines,* copyright © 1995.

are eliminated. Materials and fabrication processes are selected to maximize company resources while serving the needs of the consumer. Such efforts can result in substantial savings; for example, design engineers at General Motors reduced the unit cost for the 1995 Chevrolet Cavalier and 1995 Pontiac Sunbird by 13 percent by eliminating, replacing, or redesigning various components.

- **Design for quality** The principles of quality assurance (see Chapter 11) are applied to a design in order to ensure low failure rates coupled with high performance levels. These principles help the engineer to first identify and then eliminate manufacturing and design defects in a product. The result is a design that is safer for the consumer to use and more cost effective for the manufacturer to produce.

  Figure 1.2 illustrates the potential benefits of quality improvement efforts. Within a matter of weeks, Cherry Electric of Waukegan, Illinois (a supplier to automobile manufacturers), reduced the average rate at which defective parts appeared in a power switch assembly from 69,220 ppm (parts per million) to 124 ppm.

- **Faster design cycles** The need to produce engineering solutions quickly and effectively is being met through the use of computer technology and concurrent engineering. Computer-aided design (CAD), computer-aided manufacturing (CAM), finite element analysis (FEA), and microprocessor controls have allowed engineers to reduce significantly the time required to develop new designs. Concurrent engineering—in which different phases of engineering design, development, and manufacture are performed simultaneously—has further reduced this time by increasing effective communication between all those involved in the production of a design. (We will return to the topic of concurrent engineering in Chapter 11.)

- **Engineering without walls** Engineering departments within different companies often work collaboratively to achieve a common goal. Companies are discovering that they sometimes can produce better designs more quickly and in a more cost-effective way by sharing expenses and pooling their resources.

  Virtual corporations are being formed in which engineers at remote sites work together via the Internet. By sharing their expertise, these engineers are better able to ensure that the needs of all those affected by a product (consumers, manufacturers, distributors, etc.) will be satisfied by the final design solution. In addition, they are better able to develop designs that will use the shared resources of their companies to full advantage.

  Such communication across industrial walls is increasing as firms come to realize that they sometimes should work together as allies.

- **Design for export** Increasingly, products are developed for the international marketplace. Global product standards have been developed in many industries as companies seek to broaden their markets by working directly with foreign customers. Case History 1.4 *Single-Use Cameras* describes an example of such a design.

Recognizing such trends in engineering design as they emerge is critical, for these trends usually reflect better ways to perform engineering tasks. However, although new tools may develop and manufacturing methods may change, engineers should continue to be methodical, knowledgeable, collaborative, patient, and diligent in their work. Furthermore, the design process continues to provide a map for successful engineering work.

1.5 Writing Technical Reports

1.5.1 ▓ The Need to Communicate Effectively

As in other professions, engineers must be able to convey ideas effectively in formal reports.[6] On April 2, 1995, *the Wall Street Journal* published a survey in which the ten most critical business writing problems were identified as:

1. Poor organization
2. Spelling and capitalization
3. Grammar and punctuation

6. Partially based upon "Guide for Written Communication" by Edward G. Wertheim (Northeastern University, College of Business Administration, 1996).

4. Misused words
5. Redundancy
6. Hedging (i.e., being noncommittal and unwilling to clearly state one's position on an issue)
7. Lengthy paragraphs
8. Lengthy sentences
9. Passive language (e.g., "it is recommended that . . ." instead of "I recommend . . .")
10. Inappropriate tone

Engineering design proposals, technical reports, and other documents should avoid these weaknesses to be of a truly professional calibre and to reflect careful thought and preparation. A well-written and organized presentation can lead to implementation of a design concept, whereas a poorly written report can have the opposite effect. Of course, a design concept must have intrinsic merit and value; a well-written document will not be able to overcome an essentially weak concept.

The following general guidelines may be helpful in preparing a technical report.

- **Purpose**   First ask, What is the purpose of this report?

  Write a one-sentence statement of the report's purpose, focusing upon the problem to be solved, the task to be performed, or the response sought from your intended audience. What is the main idea that must be conveyed? Include a statement of this purpose—in some appropriate form—in your introduction.

  Remember that the introduction should generate enough interest in the reader so that he or she will continue reading the remainder of the report. It should briefly describe the situation under consideration ("Numerous tasks must be performed in hospital emergency rooms, such as . . ."), the difficulties or problems to be solved ("Often these tasks either conflict with one another or they are impeded by the physical layout of the facility"), and the proposed solution ("Our proposed layout includes partitioning the facility according to each of the tasks that must be performed . . ."). These descriptions must be coherent and brief, perhaps stated in as little as one or two paragraphs, if appropriate.

- **Audience**   Be certain that you are satisfying the needs of the audience.

  Consider the audience for your report: What do they need to know and understand in order to respond properly to your work?

  After the first draft of the report has been written, ask if you have made the presentation truly accessible to readers. Have you obscured the essence of the report by relying upon technical language and terminology that may be unfamiliar to the reader? The comedian Milton Berle would tell his television comedy writers to make a joke more "lappy" if he believed it might go over the heads of the audience—that is, he wanted to ensure that the joke would land directly in the lap of each audience member and evoke laughter in response. A technical presentation similarly must be "lappy": The reader must understand the significance of the work being reported, and the report recommendations must be stated clearly in terms of the actions to be taken by the reader.

- **Organization**   Next, begin with a general overview of the presentation and its organization. Prepare an outline of the main points to be covered and the underlying arguments for each statement or conclusion in the report.

  Based on the initial outline, does it appear that the report will be organized properly for the benefit of the reader? WIll the presentation flow from one section to another (and within each section) without abrupt changes or discontinuities that might confuse the reader?

  After the first draft of the report has been written, once again consider the organization of the presentation but at a more detailed level. Are the paragraphs linked in a meaningful and logical manner? Have you provided appropriate transitions between paragraphs and between sections? Do you use signposts such as subheadings and bold or italic fonts to help guide the reader through the material? Should paragraphs be added? Should some be deleted or modified? Is each paragraph self-contained (i.e., do later sentences in a paragraph support or expand upon the initial sentence)? Is each paragraph of an appropriate length? Are tables/diagrams/figures self-explanatory? Do they serve a useful purpose? Are they effectively integrated into the report? Could other information (currently in text form) be summarized in a table or figure that is missing from the report? How should such a table or figure be designed? Are titles and captions properly included in each table or figure?

  Has each of the conclusions and other statements contained in the report been justified by supporting discussion, data, and documentation? Are there generalizations in the report that are too vague or meaningless?

- **Grammar, syntax, and punctuation**   Have you used correct grammar, syntax, and punctuation in your report?

  Spelling is easily checked electronically via a word processor. However, not all spelling errors will be corrected by a computer (*form* instead of *from*, *affect* instead of *effect, there* instead of *their,* and so forth). Manually check all spellings and words in your final draft. Syntax and grammar can sometimes be more difficult to correct, yet it must be done! Ask a colleague to review your draft for clarity and correctness, if possible.

- **Rewrite**   All reports require a substantial amount of editing and rewriting. Winston Churchill, renowned for his eloquent and meaningful speeches, believed that a speech was not ready for presentation until it had been rewritten (at least) seven times!

  Is the message lost in unnecessarily dense, technical, or long-winded verbiage? Are the main points easily recognizable or are they lost among less important concerns?

  Have you justified each of your statements? Have you addressed all issues of importance? Are all sections in the report truly worth the reader's time or should some be eliminated?

Has each idea been developed fully and expressed in clear, easily understood language? Have you included all supporting evidence for your conclusions and recommendations? Finally, is the entire paper coherent?

1.5.2 ■ Technical Reports

Technical reports should conform to the traditional format for such documents. A technical report typically includes the following sections:

- **Title page**
- **Contents**
- **List of figures**
- **List of tables**
- **Abstract (or Summary)**

    The abstract or summary provides the reader with a very brief overview of the most important elements of the report; for example, the problem targeted for solution, the final design developed, conclusions, and recommendations for subsequent action. The abstract must be very concise and direct, and is therefore usually no longer than two or three paragraphs. Abstracts sometimes appear as the only published record of a report, thereby providing a public source of information for others who may be interested in acquiring the entire report.

- **Introduction**

    The introduction is an opportunity to describe the system or situation under investigation, provide a succinct statement of the problem to be solved, and summarize the needs of the potential clients/users in terms of the corresponding goals that are sought in an optimal design solution.

- **Relevant background information**

    The writer should provide the reader with detailed descriptions of:
    - the user population(s) to be served
    - the environment in which the design solution will be used
    - the design specifications or constraints within which a solution must be developed
    - prior art in the field (e.g., patented and/or marketed designs) and other aspects of the work to be performed

    When appropriate, this information should be summarized in tabular or graphical form for both clarity and brevity.

- **Methodology**

    The process or procedure that was followed in developing the final design solution should be described in sufficient detail so that the reader will be able to appreciate the care with which the work was performed. All assumptions and decision points, together with the rationale for each decision, should be included. Again, use charts, diagrams, and tables as appropriate.

- **Alternative solutions developed**

    Summarize each of the alternative solutions that were developed. Provide sufficient detail so that the reader will be able to understand the composition of each design (what it is), together with its functioning (how it works). This section could be included as an appendix to the report, unless otherwise specified by your instructor or supervisor.

- **Final design solution**

    The final design solution should be described in sufficient detail so that the reader will be able to recognize both its strengths and its weaknesses. Detail drawings of the final design may be included as an appendix, unless otherwise specified. Also include a summary of the evaluation and comparison effort that led to your choice of the final design (e.g., you might summarize your evaluation in the form of a decision matrix; see Chapter 10). Be certain that all of the decisions that you made in developing and selecting your final design solution are clearly stated in the report. Justify each of these decisions with supporting arguments and data. Also, a feedback table summarizing the reactions of various groups (e.g., potential users, components or materials suppliers, manufacturing personnel, marketing specialists, financial analysts) to the proposed design should be included in this section. Finally, include details of any economic evaluations of the designs that were performed.

- **Conclusions**

    Comment upon whether the final solution satisfies the original design goals that were sought. In particular, respond to the questions: will the needs of the user(s) be satisfied by the design? Should the design be implemented? Why or why not? Justify your statements with pertinent data, logical reasoning, and references to appropriate discussions elsewhere in the report.

    Also, identify the risks associated with the implementation of your solution (and with the failure to implement). Do not be defensive about your work; instead, be honest in your evaluation and conclusions!

- **Recommendations**

    Suggest further work that could or should be performed on this design, its development, and its implementation. Consider the question: If you had one more year to work on this project, what would you do?

- **Bibliography**

    Following standard style guidelines, list all references and other resources (e.g., people) used in the development of the design and the report.

- **Appendices**

    If appropriate, include additional technical data, journal articles, patent information, marketing studies, correspondence, and other materials that could not be included in the formal body of the report without disrupting its continuity or extending its length to unreasonable limits.

Variations on this format can be used, as appropriate for the problem under investigation and for the purpose of the report. For example, a progress report would not be expected to be as comprehensive as a final report since it describes work in progress and would follow a somewhat different format in recognition of this fact.

1.5.3 ▨ Cover Letters

A cover letter should be attached to a formal technical report, identify the significant results of the work, and summarize the contents of the report in two or three paragraphs. It often is signed and dated by all team members.

In writing a cover letter, strive for clarity and meaningful content. Consider the purpose of the letter and the response sought from the reader, and write accordingly. Be concise, but informative. Value the reader's time and effort.

The sample letter in Figure 1.3 contains seven elements found in virtually all business letters. These elements include (a) the return address (sometimes included in a company's letterhead stationery) of the sender, (b) the date, (c) the name, title, and address of the recipient, (d) a salutation, (e) the body or main substance, (f) a close, and (g) the signature, name, and title of the sender. The notation "Enc." below the signature and flush with the left margin indicates an enclosure with the letter.

The body of the sample letter contains three sections often included in such communications: (1) an introductory or background paragraph in which the purpose of the letter is identified, (2) a section describing the methodology that was followed, and (3) a concluding section in which the results and their interpretation are provided. Most cover letters contain these elements in some form in order to provide the reader with the answers to such basic questions as, Why am I receiving this letter? How was the work performed (or the results obtained)? What are these results and what do they mean?

The sample letter is in block style; paragraphs are not indented (i.e., they remain flush with the left margin). Block style is considered intermediate in its level of formality. Full block, in which all elements (including the return address, date, and signature) are flush with the left margin, is most formal. Semi-block, in which each paragraph is indented, is least formal.

1.6 Teamwork

Modern engineering design usually depends upon people working together effectively in teams. Technical problems of current interest can be quite complex and multidimensional, often requiring broad interdisciplinary solutions. Working in formal design teams, engineers from various disciplines can more effectively devote and coordinate their diverse skills and knowledge to develop innovative solutions quickly and efficiently.

All members of a design team first must agree upon a single problem formulation and a set of initial expectations so that everyone will be moving in

123 Academic Road
Boston, MA 02115   (a)

December 4, 1998   (b)

Dr. Xavier Cellent
Professor, Mechanical Engineering
367 Snell Engineering Center   (c)
Northeastern University
Boston, MA 02115

Dear Professor Cellent: (d)

I am happy to report the final results of assignment number 10 "Stability of a Ladder," together with my interpretation of these fingings. As you may recall, this problem focused upon the question of whether a 12-ft ladder leaning against a wall at an angle of 35 degrees with a 160-lb. man standing four feet from the top of the ladder would remain stable.   (1)   **Purpose/ introduction/ background**

Using three equations of motion (two for the x and y force components acting on the ladder, together with a third that describes the torques acting on the system), I developed a C program that allows the user to specify a set of input conditions. These conditions include the angle of inclination of the ladder, the coefficients of friction for the wall surface and the floors, the weight and location of the person standing on the ladder, and the length and weight of the ladder itself.   (2) (e)   **Methodology**

With the data provided in the original assignment, I determined that the ladder would indeed remain in place. However, if the angle of inclination is increased to 38 degrees, the ladder will begin to slip. I am enclosing a hard copy of my program and my results, together with an electronic copy of the program on diskette.   (3)   **Results/ interpretation**

I look forward to your review of these results

Sincerely,   (f)

Isaac M. Smart

Isaac M. Smart   (g)

Enc.

FIGURE 1.3 Sample cover letter.

the same general direction (see Chapters 3 and 4). The overall workload for developing a final design should be shared equitably among all members of the team. Moreover, each member must accept personal responsibility for the completion of those specific tasks assigned to him or her. A design proposal (discussed in Chapter 2) is one mechanism that can be used to assign tasks to individual team members and monitor progress toward the completion of those tasks.

There also must be mutual accountability; that is, all members of the team must share responsibility for the completion of all tasks and the development of the final design. Leadership and decision-making responsibilities should be shared—either formally or informally—among all members in order to increase each person's motivation and commitment to the project and to enhance the quality of the final design.

Teamwork is built upon mutual trust and respect among all members of the group. The working environment must be one in which everyone feels comfortable about sharing an idea or opinion. Every individual's contributions must be seen as important to the entire group effort. Nevertheless, conflicts will occur: each person must realize that his or her ideas may be justly criticized (and perhaps ultimately rejected) by the team. However, all criticisms must be constructive in nature and respectful of the person whose work is being criticized. Teamwork requires that all members remain able and willing to collaborate and help one another to succeed.

1.7 The Value of Perseverance in Design

As is true of most things in life, engineering design is mastered only through practice. Such practice requires that we be patient and diligent in our work and that we persevere until success finally is achieved. Of course, all engineers must work within deadlines and other constraints, so there is always a practical limit to the amount of effort and time that can be devoted to a particular project. Sometimes projects must be abandoned because of unexpected financial and/or technical difficulties that cannot be resolved within the given time limit. Nevertheless, perseverance is important in engineering design: Projects seldom end in success without significant levels of dedication and commitment being exhibited by the engineering staff.

As seen in the work of the Wright brothers (Case History 1.2), engineers should acquire a robust (i.e., deep and broad) understanding of a design problem in all of its many dimensions if they are to produce satisfactory solutions. These dimensions include the specific needs of the customer, the environment in which the design will be used, scientific principles which are pertinent to the problem, any previous solutions that may have been developed, and so forth.

Clearly a substantial investment of time and effort is needed to acquire a robust understanding of a problem. During the period in which this investment is being made, progress in developing a design solution may appear slow or even absent. The engineer, in frustration, then might decide to skip certain tasks in order to accelerate progress on the project—failing to recognize the negative impact that such action may have on the final design.[7]

---

7. The engineer might even abandon the entire project.

Both engineering design and the study of engineering require diligent effort, sometimes with little or no apparent progress. When confronted by a frustrating situation in which your work may not seem to be bearing much fruit, remember the behavior of the moro (Chinese bamboo) plant.

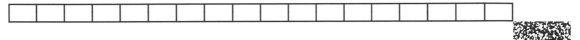

The Moro Plant: Patience and a Strong Foundation Leads to Success

After planting moro seeds, a gardener must then water, feed, and care for this plant year after year with no apparent results. The first sprouts finally appear above ground *five years* after the initial planting. However, these shoots then grow at the phenomenal rate of two or more feet per day for the next six weeks until they have reached a final height of 90 feet![8]

The initial five-year period of growth allows an extensive root structure to develop, thereby providing the huge shoots with the underlying support that they require.

Engineering design work sometimes will be like the development of the moro plant: No apparent progress will be visible on the surface, but a strong foundation will be developing underneath—a foundation upon which final success will depend.

SUMMARY

- An engineer should possess a number of attributes to be successful: These attributes include:
    - Problem-solving skills
    - Effective communication skills
    - Highly ethical and professional behavior
    - An open mind and positive attitude
    - Proficiency in math and science
    - Technical skills
    - Motivation to continue learning

8. Weldon (1983).

- Knowledge of business strategies and management practices
- Computer literacy
- Understanding of world affairs and cultures
- Five stages of the engineering design process are used to develop innovative solutions to technical problems:
    - Needs assessment
    - Problem formulation
    - Abstraction and synthesis
    - Analysis
    - Implementation

    Reflection is an unofficial but valuable sixth stage during which one contemplates the lessons learned during the just-completed design effort.
- Nth-generation design efforts allow one to iteratively refine and improve a technical solution to a problem.
- Current engineering practices include:
    - Life-cycle design
    - Design for quality
    - Design for export
    - Design for manufacturing and assembly
    - Engineering without walls
    - Faster design cycles
- Engineers usually work in teams, which requires mutual trust, accountability, and respect among all members.
- Both engineering design and the study of engineering require that we be patient and diligent in our work and that we persevere until success finally is achieved.
- Technical reports must be well organized and well written and serve the needs of the intended audience.

PROBLEMS

1.1 Engineering has improved the quality of life in many ways. Prepare a brief report on one specific engineering contribution (i.e., a product, system, or process such as an irrigation method, a sewage system or a particular type of harvester) to
    a. energy conservation
    b. environmental protection
    c. food production

d. communication

e. transportation

f. shelter

g. health

In completing this assignment, please do the following:

    i.   Explain the need for a solution to the problem.

    ii. Describe the design that was developed as a solution to the problem (use sketches as needed).

    iii. Identify the specific engineering disciplines (e.g., chemical, civil, electrical, industrial, mechanical) that were involved in the contribution that you select, together with any major individual contributors.

    iv. Use a tabular format to present the chronological development of this contribution, including any earlier solutions that preceded the final design.

    v. List all references in complete form.

1.2 Read Case Histories 1.1 and 1.2.

  a. How does lowering the elevators result in an airplane diving downward through the air? Sketch your explanation in terms of deflected air flow, velocities, and changes in air pressure.

  b. Explain (with sketches) why a combination of rotations about the roll and pitch axes results in an airplane banking through the air (with the rudder then used for slight corrections in this turn.)

1.3 Calculate the horsepower that the Wright–Taylor engine would have needed to provide if Smeaton's coefficient of air pressure were indeed equal to 0.005 (see Case History 1.2).

1.4 a. Describe (in a few words) ten types of engineered devices and systems (e.g., computers, highway systems, security systems, food processing systems, fire extinguishers).

  b. Next, evaluate one of the components in each of these systems that depends upon the decisions made by a design engineer. In particular, compose a list of possible questions that might be asked by the engineer about each of these components as the system designs were developed. For example, a fire extinguisher should be reliable so that it will work whenever needed; among the design elements that the engineer then must consider in order to ensure such reliability is the valve to be used in the device. The engineer might need to ask (and answer) such questions as:

- Will this valve clog or otherwise fail under certain extreme conditions?
- What are these extreme conditions (High temperatures? High or low viscosities for the chemical compound(s) used in the extinguisher? High moisture levels? High dust levels? Others?)
- Why would the valve fail under such extremes?
- What can be done to prevent such failure?

c. Finally, try to imagine what types of knowledge are needed for the design engineer to answer each of the questions that you generated (e.g., in the case of a fire extinguisher, he or she would need to have some knowledge of fluid behavior, chemistry, valve mechanisms, etc.). Compose a list of these different types of knowledge.

1.5 Many prominent individuals have been trained as engineers, including:[9]

- Former presidents of the United States (George Washington, Herbert Hoover, and Jimmy Carter)
- Astronauts (Neil Armstrong)
- Motion picture directors (Alfred Hitchcock)
- Business leaders (William Hewlett, cofounder of Hewlett-Packard)
- Sports figures (Tom Landry, former head coach of the Dallas Cowboys professional football team)
- Military leaders (Sheila Widnall, Secretary of the U.S. Air Force)

Identify five other prominent individuals who were trained as engineers. Also identify the engineering discipline of each person that you identify. (Hint: You may want to check computerized newspaper databases that may be available at your local library. Most important, ask your librarian to suggest appropriate references.)

1.6 a. Select a product that has undergone several redesigns over the years. Research the earlier (historical) versions of this product and create a table listing the various changes that were made in each design generation of the product. List the changes in chronological order, noting the year in which each change occurred.

b. Next, prioritize these changes, listing them in order of importance (as opposed to chronological order).

c. Finally, explain your reasons for the specific way in which you prioritized the changes made in the product, that is, your reasons for stating that some changes are more important that others.

1.7 Prepare a brief report on the development of keyless typewriters, first developed in 1984 by the Japanese firm Matsushita.[10] Relate this design to the development of other keyless or touch-sensitive systems through which one can enter data.

1.8 Prepare a brief report on the development of commercial videocassette recorders for mass production. Include discussions of the Betamax and VHS systems, focusing upon their relative technical merits and commercial success.

1.9 Prepare a brief report on the development of the transistor, including the bipolar (junction) design and the Field Effect Transistor (FET). Discuss the impact of this device in electronic systems.

1.10 Compare five alternative designs for one of the following products:

---

9. Based in part on Landis (1995).
10. d'Estaing and Young (1993).

- Toasters
- Garage door openers
- Running shoes
- Exercise equipment
- CD players
- Vacuum cleaners
- Bicycle helmets
- Bookbags

Discuss the relative merits (in terms of performance, cost, durability, aesthetics, ease of operation, maintenance, and other factors) of these designs. (Hint: *Consumer Reports* or similar publications in which commercial products are evaluated and compared may be helpful in this assignment.)

CASE HISTORY **1.1**

Incomplete Design: The Early Quest for Manned Flight

INTRODUCTION

The historical quest for manned flight illustrates why one should follow the 5-step engineering design process (Fig. 1.1) in order to methodically develop solutions to challenging technical problems.[11]

Many people essentially applied an abbreviated version of the design process in their attempts to develop devices (e.g., balloons, airships, gliders, airplanes) for manned flight (see Fig. 1.4).[12] As each new generation of design solutions was developed, the problem statement became more focused; that is, the design goal gradually changed from simply "flight" to "free, sustainable, and controllable flight." However, the development of kites, balloons, and airships was haphazard, relying upon trial-and-error and the inspiration of one inventor or another.

In contrast, the Wright brothers more closely followed the structured problem-solving procedure that we now call the engineering design process to develop and refine their glider and airplane concepts (Fig. 1.5). As a result, Wilbur and Orville Wright behaved more like today's engineers than yesteryear's inventors: They were not only creative but truly methodical and analytical in their work, adding the steps of synthesis and analysis that were missing in the earlier efforts of others. Their painstaking approach bore fruit on December 17, 1903, when they completed the first successful manned, controlled, and sustained flight in a heavier-than-air, gasoline-powered biplane. On that day, the age of the airplane began, thanks to the engineering work of the Wrights.

11. See Travers (1994), d'Estaing (1985), *The How It Works Encyclopedia of Great Inventors and Discoveries* (1978), Breedon (1971), Macaulay (1988), and Singer et al. (1958).

12. Of course, none of the early pioneers of manned flight (including the Wright brothers) was actually aware of the prescriptive design process of Fig. 1.1; however, the patterns of their efforts correspond to the principal elements of this process, as described in Fig. 1.4 and 1.5.

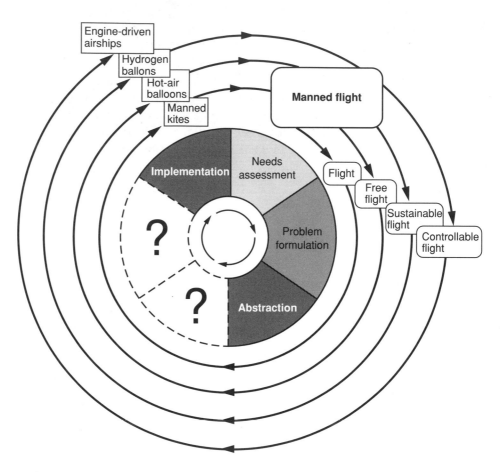

FIGURE 1.4 Incomplete design process led to limited success.

LIGHTER-THAN-AIR CRAFT: KITES, BALLOONS, AND AIRSHIPS

People have long sought to soar birdlike through the air. As early as the thirteenth century, the Chinese conducted successful reconnaissance missions with manned kites fixed to the ground. Leonardo da Vinci designed ornithopters (aircraft with flapping wings), helicopters, and parachutes during the fifteenth century, but his work remained largely unrecognized for 300 years. As the centuries passed, many brave and inventive efforts eventually led to successful manned flight.

The first truly free flight of an unoccupied hot-air balloon occurred after the brothers Joseph-Michel and Jacques-Etienne Montgolfer noted that hot air from a fire caused bags to rise. Their balloon (powered by hot air from a small stove burning wool and straw) rose to 1500 feet on June 4, 1783.

Of course, a hot-air balloon could only maintain its flight until the hot air within it cooled. To overcome this shortcoming, balloons were filled with lighter-than-air hydrogen. Hydrogen had been isolated as a gas by

Henry Cavendish in 1766 and the physicist Jacques A. C. Charles (famed for Charles' law in thermodynamics) launched the first hydrogen-filled balloon (known as a "Chaliere") on August 17, 1783—less than three months after the Montgolfers' initial hot-air balloon flight.

The following month, the Montgolfers launched the first balloon flight that included passengers: A lamb, a rooster, and a duck were transported into the air on September 19, 1783.

FIGURE 1.5 Complete design process led to the Wright brothers' success.

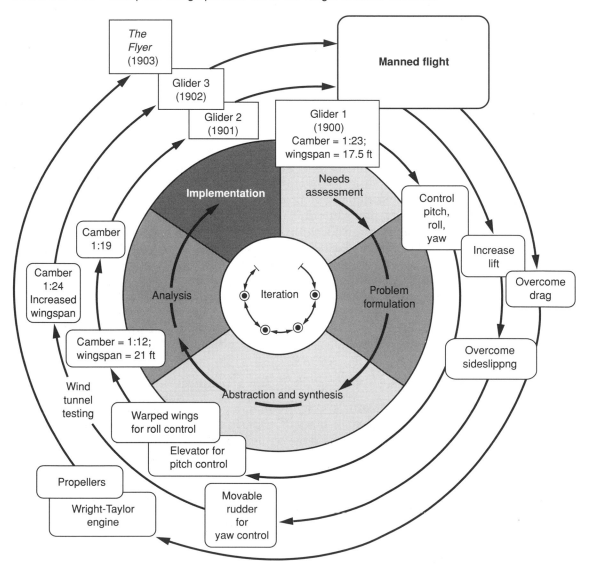

On October 15, 1783, Jean-Francois Pilatre de Rozier flew in a tethered hot-air balloon (i.e., one that was moored to the ground), rising to a height of 84 feet and remaining aloft for four and a half minutes. (Recall that the Chinese may have performed such tethered flights with kites several centuries earlier.)

Finally, on November 21, 1783, the first manned free flight in a hot-air balloon was made by Pilatre de Rozier and the Marquis d'Aalandes in Paris. They traveled 7.5 miles in 26 minutes, attaining a maximum altitude of 3000 feet. The balloon age of manned flight had arrived.

(Tragically, Pilatre de Rozier was also the first casualty in such manned flight. His balloon caught fire during an attempt to cross the English Channel and crashed on June 15, 1785.)

Balloons subsequently were used in successful flights across the English Channel (first accomplished on January 7, 1785, by F. Blanchard and J. Jeffries) and the Alps (on September 2–3, 1849, by F. Arban), together with unsuccessful efforts to cross the Adriatic Sea (on October 7, 1804, by F. Zambecccari and his colleagues) and the Atlantic Ocean. Balloons were also used by F. Tournachon to capture the first aerial photograph of Paris in 1858 and J. Glaisher and H. Coxwell to ascend to more than 20,000 feet in 1862. In 1870 both passengers and mail were first transported by balloon.

Hydrogen balloons became engine-driven airships or dirigibles in 1852 when Henri Giffard added a three-horsepower steam-driven propellor and flew 17 miles. Unfortunately, even with the addition of a propellor, airships remained difficult for pilots to control and direct in windy weather. Small internal combustion and electrical engines replaced steam engines on some airships during the 1880s and 1890s.

Dirigibles were further improved by Ferdinand von Zeppelin, who used large metal frames, basing his work on that of David Schwarz. In 1897 Schwarz had designed and built an airship with a rigid aluminum frame. Unfortunately, his ship crashed but von Zeppelin obtained the patent rights of Schwarz's designs from his widow. Airships were very popular for many years, transporting passengers across the Atlantic Ocean both quickly and safely. However, their popularity almost instantly evaporated when the Hindenburg exploded and crashed above New Jersey in 1937, killing 36 people.[13]

HEAVIER-THAN-AIR CRAFT: GLIDERS AND AIRPLANES

Balloons and airships are lighter-than-air vehicles, whereas gliders and airplanes are both heavier-than-air craft. In the late nineteenth century, attention slowly turned to the development of gliders and airplanes for manned free flight.

---

13. Modern airships use helium to provide lift since this nonflammable gas is only 1/7 as dense as air. In addition to a very large supply of helium, the airship's envelope contains adjustable ballonets (small compartments of air). These ballonets are used to adjust the airship's altitude by either increasing or decreasing the relative amount of air within the ship's envelope, thereby altering the airship's weight. (More air means that the upward lift provided by the helium must overcome more weight; similarly, less air corresponds to less weight.)

Although gliders had been introduced in the early 1800s (Sir George Cayley[14] constructed one in 1804), not until 1856 was the first manned glider flight completed by Jean-Marie Le Bris.

Otto Lilienthal's (1848–1896) engineering factory manufactured many products of his own invention, such as light steam motors, pulleys, and marine signals. He had been fascinated by the flight of birds since childhood and, by studying the stork and other birds, he recognized both the need for an arched wing in gliders and the role that rising air currents played in soaring. His 1889 book, *Der Vogel flug als Grundlage der Fliegekunst,* on the flight of birds and his 1894 essays about flying machines became the very foundations of modern aeronautics.

Lilienthal completed numerous glider flights during the 1890s based upon many models and experiments. He held onto these gliders with his arms and moved his body to maintain balance. His gliders weighed as much as 44 pounds with wing spans of up to 23 feet. Then, in 1896, he developed a new glider design: a biplane with flapping wings powered by a motor. Unfortunately, he was killed in a crash before he could complete the development of his motorized glider. However, the work of this aviation pioneer was continued by his brother Gustav.

Like Lilienthal, P. S. Pilcher was investigating methods to achieve powered flight but his untimely death in a glider accident in 1899 ended this effort.

Others such as Felix du Temple (1874), Alexander Mozhaiski (1884), and Clement Ader (1890) had attached steam engines to their aircraft, but they each enjoyed only limited success in achieving manned flight. Of particular note is Ader's *Eole,* recognized as the first engine-powered manned airplane to achieve a takeoff (on October 9, 1890). The *Eole* rose several inches above the ground for a flight totalling about 160 feet.

Then in 1896 Samuel Langley's steam-powered monoplane *Aerodrome 5,* with a wingspan of 16 feet, flew for 90 seconds and travelled about three-quarters of a mile; however, his craft was unmanned.

Langley served as secretary of the Smithsonian Institution. In 1898 he was provided with government funding by President William McKinley and the War Department to develop an aircraft for possible use in the Spanish-American War. After the design, construction, and successful testing of a gasoline-powered model in 1901, Langley was ready to test a manned, gasoline-powered aircraft. Unfortunately, his *Aerodrome* with a 53-horsepower gasoline engine sank into the Potomac River after being launched from a catapult on October 7, 1903.

Langley's efforts, including his work entitled *Experiments in Aerodynamics,* provided additional evidence that successful manned flight in a controlled and sustainable manner was about to occur. Humanity's triumph

14. Cayley is credited with introducing the concept of fixed wings (as opposed to the flapping wings of ornithopter designs) for gliders in 1799. Cayley also conceived of a helicopter that could convert to a fixed wing form; his 1843 concept for such a convertiplane was finally realized in 1955 with the McDonnell XV-I. In recognition of his contributions to aerodynamics, Cayley has been called the father of aerial navigation.

over the air would be achieved by two tireless, committed, and methodical mechanical geniuses: the Wright brothers.

The early developers of kites, balloons, and airships approached their quest for manned flight in a decidedly haphazard fashion; as a result, only limited and sporadic progress was made during a time span of more than one hundred years. In contrast, the Wright brothers were about to achieve free, sustainable, and controllable flight in less than five years by following the structured problem-solving procedure that we now call the engineering design process.

CASE HISTORY 1.2

Methodical Design: The Wright Brothers' Success in Manned Flight

The Wright brothers' quest for both controllable and sustainable manned flight is a classic case study in the methodical application of scientific principles to solve a practical problem; in other words, it is a classic example of engineering design in action (see Fig. 1.6).[15]

Wilbur Wright (1867–1912) and Orville Wright (1871–1948) designed, manufactured and sold bicycles (Fig. 1.7). They had been interested in flight since their boyhood when they were given a whirligig (i.e., a model helicopter) by their father in 1878. (The toy helicopter was powered by rubber bands and had been invented by Alphonse Penard some years earlier.) However, their professional dedication to aircraft development began only in 1896 upon hearing of Otto Lilienthal's death (another indirect but nevertheless significant contribution of Lilienthal to manned flight).

The Wright brothers did not simply construct a series of flying machines and make adjustments intuitively until they found the correct configuration of design characteristics. Rather, their design and development work was based upon scientific principles of fluid dynamics. The Wrights commenced upon a series of thoughtful, meticulous, and methodical experiments, first with a biplane kite and then with various gliders of their own design. These experiments were coupled with their detailed observations of birds' flights.

15. See Travers (1994), Copp and Zanella (1993), d'Estaing (1985), *The How It Works Encyclopedia of Great Inventors and Discoveries* (1978), Breedon (1971), and Singer et al. (1958).

FIGURE 1.6 The historic flight by Orville Wright on December 17, 1903 at 10:35 A.M. Wilbur can be seen running near the right wingtip. *Source:* Courtesy of the United States Library of Congress, LC-USZ62-13459.

MATHEMATICAL MODELS AND SCIENTIFIC PRINCIPLES

In order to understand the work of the Wright brothers, let us begin with a brief review of certain scientific principles for movement through a fluid medium, in this case, air.

An airplane must overcome air resistance as it flies, just as a moving object submerged within a fluid must overcome resistance as it pushes against (and redistributes) the fluid. This resistance or reaction force R acts in a direction that is perpendicular to the surface itself and can be expressed in terms of two components: the lift L, which is perpendicular to the fluid flow direction and the drag D, which is parallel to the flow. (Case Problem 1 presents a more detailed explanation of the reaction force's origin in terms of Newton's second and third laws.[16]) For successful flight through an airstream, one needs to both overcome drag with an adequate propulsive thrust and use lift to overcome gravity in an effective and safe way. These were the fundamental challenges that confronted the Wrights.

In order to calculate the amount of lift L applied to a wing by the airstream through which it moves, one needs to first understand the effects of several factors upon which L is dependent. In the early 1800s George Cayley, recognized that any change in lift was proportional to changes in certain other variables; for example,

---

16. Case problems are included in the special section at the end of the text.

FIGURE 1.7 Wilbur and Orville Wright. *Source:* Corbis-Bettmann.

$$L \alpha \; \theta \tag{1.1a}$$

and

$$L \alpha \; A \tag{1.1b}$$

where θ is the angle of attack (i.e., the angle of the wing with respect to the airstream), and A denotes the wing's surface area.

Some years earlier, John Smeaton had experimentally determined that the air pressure p was proportional to the square of the velocity (V^2) according to

$$p = kV^2 \tag{1.2}$$

where the proportionality constant k (known as Smeaton's coefficient of air pressure) varies with the angle of attack. In addition, both Smeaton and Cayley noted that the lift L varied proportionately with the upward curvature or, more precisely, the camber of the wing:

$$L \alpha \; \text{camber}$$

Camber refers to the ratio of the wing's maximum height to its chord length, which is the linear distance between the leading and trailing edges.

Isaac Newton was the first to propose that the reaction force R applied to a surface immersed within a frictionless fluid was given by[17]

17. Leonhard Euler demonstrated that R in fact varies with the $\sin \theta$, not $\sin^2 \theta$, at low speeds.

$$R = \rho(V^2)A \sin^2\theta \qquad\qquad (1.3)$$

where ρ represents the density of the fluid.

After all, Newton's second law relates the force F to the mass M and corresponding acceleration a of an object to which the force is applied:

$$F = ma \qquad\qquad (1.4)$$

Thus, a larger (reaction) force R will be required to move material of greater density (i.e., greater mass per unit volume), as stated in Newton's expression for R. Also notice the correlation between Newton's expression for R and the observations of Cayley and Smeaton; such self-consistency is most gratifying since L is simply the perpendicular component of R relative to the fluid flow direction; hence, the dependence of R and L on A, V, and θ certainly should be quite similar.

Although the Wrights did not necessarily apply the above equations *directly* to their development of the airplane, they did base their work on very similar relationships used by Lilienthal, Langley, and Octave Chanute; that is, for the lift L:

$$L = k\, V^2\, A\, C_L \qquad\qquad (1.5)$$

where C_L is a coefficient of lift that incorporates the dependence of L upon both θ and the camber of the wing. (Notice the appearance of Smeaton's coefficient k in this equation; the actual value of this coefficient was to be of critical importance to the Wrights, as we will discover shortly.) For the drag D, the equations were

$$D = D_I + D_P \qquad\qquad (1.6)$$

$$D_I = \text{induced drag} \qquad\qquad (1.7)$$
$$\quad = kV^2\, A\, C_D$$

and

$$D_P = \text{parasitic drag} \qquad\qquad (1.8)$$
$$\quad = kV^2\, A_F\, C_D$$

where C_D is a coefficient of drag similar to C_L, and where D_P is an additional effective drag experienced by the airplane because of its frontal surface area A_F.[18]

The Wright brothers experimentally calculated the ratio (C_D/C_L) of the drag-to-lift coefficients and therefore used a slightly modified version of the above expressions:

$$D_I = kV^2\, A\, C_L\, (C_D/C_L) \qquad\qquad (1.9)$$

$$D_P = kV^2\, A_F\, C_L\, (C_D/C_L) \qquad\qquad (1.10)$$

---

18. The Wrights realized that the parasitic drag on their aircraft could be reduced if the pilot were to lie flat on his stomach (rather than be seated in a more comfortable upright position), thereby decreasing his frontal surface area and the airstream's corresponding resistance to the plane's forward motion.

Finally, they used a simple power equation[19] to determine the engine horse-power P that would be needed to provide sufficient thrust to overcome the total drag D and so that the craft would travel with some desired velocity V; that is,

$$P = D * V \tag{1.11}$$

With these rather rudimentary mathematical models of flight behavior, the Wrights were able to develop their aircraft designs. Whenever failure struck (as it did repeatedly), they returned to careful experimentation and observation in order to correctly adjust the values of constants and variables appearing in the above equations. This methodical and interative transition between scientific theory and its practical application—through which both the theory itself and its application are successively developed and refined—is the essence of good engineering practice.

THE DEVELOPMENT YEARS: 1900–1903

Having summarized some of the underlying scientific relationships used to model air flight, let us now review the major milestones in the Wright brothers' development of their aircrafts from 1900 to 1903.

In 1900 the Wrights designed and constructed a 52-pound glider with a wingspan of 17.5 feet and a camber of only 1/23 (recall that lift is essentially proportional to camber). They flew this first glider as a kite near Kitty Hawk, North Carolina, at least partly to assuage their father's anxiety about the hazards of their work. (After all, many people had already perished in similar attempts to achieve manned flight.) Using Equation 1.5 and various data, the Wrights were able to estimate the amount of lift that would be applied to their glider. According to these calculations, the glider should have been able to lift 192 pounds (the combined weights of the plane and one human pilot—Wilbur and Orville each weighed 140 pounds) in a 25 mile-per-hour (mph) airstream with a 3° angle of attack θ. Unfortunately, it could lift such a weight only in 35 mph winds or with θ equal to 35°—both of which were impractical values.

It was decided that the craft would need to be flown if they were to truly understand its inherent weaknesses and strengths. Wilbur first flew the aircraft as a glider by maintaining relatively steep angles of attack and restricting it to flights in low-velocity winds.

PITCH CONTROL: PREVENTING STALLS WITH AN ELEVATOR

The Wrights recognized that a human pilot would need to control a heavier-than-air craft about all three axes of rotation: roll, pitch, and yaw. In particu-

---

19. Power is defined as work per unit time t, where work W is the product of the force needed to move an object and the distance Δx through which the object is moved. We then have
$V = \Delta x/\Delta t$
D = drag force that must be overcome by the net thrust provided by the engine (and which must be less than or equal to this thrust)
and thus
$P = (\Delta \bullet \Delta x)/\Delta t = \text{Work/time}$

lar, controlling rotation about the pitch axis was a primary objective since a lack of such control had resulted in Otto Lilienthal's death some years earlier.

Lilienthal's glider had stalled in midair and then crashed. Such stalls are due to the movement of a wing's center of lift CL (or center of pressure). This center is very similar to the concept of center of gravity in which one can treat the effect of gravity upon an object as if it were applied to a single point on the object. A wing's center of lift is simply that point at which the net effect of all air pressure forces can be assumed to act.

Balance (and level flight) is achieved when the centers of gravity and lift coincide with one another. However, Langley had noted that the distance between the center of lift CL and the wing's front or leading edge decreased with increasing angles of attack. An unexpected wind could then cause the angle θ of attack to increase, resulting in the CL moving forward of the center of gravity with a corresponding increase in both lift and θ. This cyclic process would then continue until θ becomes so large (approximately 15° or more) that the airflow above the wing becomes turbulent and a stall occurs. (Lift is due to the lower pressure and higher velocity of the airstream above the wing; lift vanishes once this airstream is no longer smooth.).

The Wright brothers installed a flat horizontal surface on the front of their glider to act as an elevator.[20] By properly raising or lowering this elevator, the pilot could control rotation about the pitch axis rather than simply allowing the angle of attack to be determined by the relative movement of the CL.

ROLL AXIS CONTROL VIA WARPED WINGS

Next, the Wrights discovered a way to control rotation about the roll axis. Rather than simply shifting their bodies to maintain balance about this axis (as had Lilienthal), the Wrights began to use wings that could be warped or twisted to varying degrees by the pilot through a system of pulleys and cables. Such warping allowed one to adjust the relative angles of attack of each wing and thus the corresponding lift forces applied to these wings. This method to control rotation about the roll axis occurred to Wilbur after he observed buzzards twisting their wing tips to maintain balance about the roll axis.

Unfortunately, successful rotation about the roll axis can lead to a corresponding rotation about the yaw axis and a hazardous phenomenon known as sideslipping, a phenomenon that soon became apparent to the Wrights during their next series of flights in 1901.

1901: INSUFFICIENT LIFT, SMEATON'S COEFFICIENT, AND SIDESLIPPING

Wilbur and Orville returned to the Kitty Hawk region in 1901 bearing a new glider, one with a wingspan of 22 feet and a camber ratio of 1:12. Even with the expanded wingspan and camber, the amount of lift generated was

---

20. George Cayley (1799), W. S. Henson (1841), and Otto Lilienthal (1895) had earlier incorporated rear elevators for longitudinal control in their designs, whereas both Sir Hiram Maxim (1894) and the Wright brothers used a front elevator to achieve such control.

still inadequate. The Wrights had to find the reason—or abandon their great adventure.

They flew each wing of the glider separately as a kite in order to carefully observe and evaluate its behavior. Wilbur noticed that in winds of high velocity, the wing was pushed downward because the *CL* moved behind the center of gravity and towards the trailing edge of the wing. This reversal at high speeds of the *CL*'s generally forward movement was totally unexpected.

To counteract this effect, the Wrights decreased the camber to 1:19. The lift did indeed increase with this change, but it continued to remain below expected values. Disappointed but undeterred, Wilbur and Orville decided that Lilienthal's aeronautical data (which had provided much of the quantitative basis for both their design work and that of many other researchers) would need to be verified by experiment.

They constructed a small wind tunnel to test more than 200 glider wing designs. (Their tests were so accurate that modern instrumentation has improved only slightly upon their calculations.) Almost one full year was devoted to this experimentation, leading to their final conclusion that none of Lilienthal's values could be the primary source of error in their calculations. Since all other possibilities had now been eliminated, only Smeaton's coefficient *k* remained a viable candidate for this error source.

Fortunately, Wilbur and Orville discovered that data from their 1901 glider could be used to determine a more accurate value[21] (about 0.0033 as opposed to Smeaton's original estimate of 0.005) for Smeaton's coefficient. The new lower value for *k* corresponded to both lower lift and drag—consistent with the Wrights' flight tests.

The Wrights also noticed another serious problem known as sideslipping; that is, as the glider banked into a turn, it slipped sideways and lost altitude. Such sideslipping is caused by the increase in drag that accompanies any increase in lift. During a turn, the lift acting on one wing is intentionally increased by enlarging its angle of attack, resulting in the plane's desired rotation about the roll axis. However, there is then an unintentional increase in drag on the lifted wing, resulting in its deceleration relative to the other wing. This imbalance in drag on the two wings causes the plane to rotate about the yaw axis. The corresponding change in angle of the wings' leading edges with respect to the airstream can then lead to a decrease in lift so that the plane sideslips downward.

1902: SUFFICIENT LIFT AND YAW CONTROL

The Wrights returned once again to Kitty Hawk in 1902 with a new and improved glider. Based upon their newly determined and more accurate value of the Smeaton coefficient, their glider now had a larger wing surface area and a reduced camber (1:24) for greater lift.

---

21. With modern techniques and instrumentation, we now know that Smeaton's coefficient equals 0.0026.

To overcome the problem of sideslipping, they first installed two fixed vertical rudders.[22] Although these rudders eliminated sideslipping during shallow turns by decreasing rotation about the yaw axis, they were unable to prevent sideslipping during sharply inclined turns. After all, these fixed rudders were passive: They could not adjust automatically to overcome large and undesirable yaw rotations. Orville replaced these fixed rudders with a single movable one, thereby allowing the pilot to *actively* control the aircraft's rotation about the yaw axis in order to eliminate any sideslipping.

The Wrights had now achieved control about all three axes of rotation—a landmark achievement in manned flight!

1903: A NEW LIGHTWEIGHT ENGINE LEADS TO SUCCESS

After more than 1000 successful and controlled glides in their new aircraft, they knew that the next step would be to add an appropriate propulsion system that could overcome drag.

A lightweight, self-contained engine would be necessary. A steam engine would require water and a boiler; in contrast, an internal combustion engine would be more self-contained. Unfortunately, all existing internal combustion engines were too heavy; they would add so much weight to the aircraft that it would not achieve take-off.

At this point, Charles Taylor became an important contributor. Taylor was the mechanic who operated the Wrights' bicycle shop whenever they were in Kitty Hawk. Orville provided Taylor with the necessary specifications for the engine and then asked Taylor to build it. Six weeks later, Taylor (collaborating with Orville) had completed his assignment.

Orville calculated that the gasoline engine would need to provide 8.4 horsepower and weigh 200 pounds or less. He used the following values for his calculations:

Lift L required = Total weight of aircraft + pilot + engine (in lbs)

$\qquad = 285 + 140 + 200 = 625$ lbs

A = Surface area of wings = 500 ft^2

A_F = Frontal area (parasitic drag) = 20 ft^2

C_L = Coefficient of lift

$\qquad = 0.311$ (for a maximum angle of attack = 2.5°)

$\qquad = 0.706$ (for a maximum angle of attack = 7.5°)

C_D/C_L = Drag-to-lift ratio

$\qquad = 0.138$ (for a maximum angle of attack = 2.5°)

$\qquad = 0.108$ (for a maximum angle of attack = 7.5°)

k = Smeaton coefficient of air pressure = 0.0033

22. Both George Cayley (1799) and W. S. Henson (1841) proposed the use of rudders for directional control.

Now recall that[23]
$$P = D * V \qquad (1.11)$$
= horsepower needed to overcome drag D

Therefore, one needs to calculate both the total drag D and the velocity V. The velocity can be obtained from the following expression:

$$L = k V^2 A C_L \qquad (1.5)$$
$$V = [L/(k A C_L)]^{1/2}$$
$$[625/(0.0033)(500)(.706)]^{1/2} \sim 23 \text{ mph}$$
$$= [625/(0.0033)(500)(.311)]^{1/2} \sim 35 \text{ mph}$$

Given the velocity, one can then calculate the total drag D

$$D = D_I + D_P \qquad (1.6)$$

For an angle of attack only equal to 2.5°, we then obtain for the induced drag:

$$D_I = kV^2 A C_L (C_D/C_L) \qquad (1.9)$$
$$= (0.0033)(35)^2(500)(0.311)(0.138)$$
$$= 86.7 \text{ lbs}$$

Similarly, for parasitic drag

$$D_P = kV^2 A_F C_L (C_D/C_L) \qquad (1.10)$$
$$= (0.0033)(35)^2(20)(0.311)(0.138)$$
$$= 3.5 \text{ lbs}$$

Therefore,

$$D = 90.2 \text{ lbs}$$

and

$$P = D * V$$
$$= (90 \text{lbs}) * (35 \text{ mph})$$
$$= 3,150 \text{ lb-mile/hr} * (1 \text{ HP}/33,000 \text{ ft-lbs/min})(1 \text{ hr}/60 \text{ min}) * (5280 \text{ ft/mile})$$
$$= 8.4 \text{ HP (horsepower)}$$

The Wright–Taylor four-cylinder engine could generate 12 horsepower and weighed only 180 pounds. The Wrights now had their power source. One item remained: propellers to convert the generated power into forward thrust.

Wilbur noted that a propeller was essentially a wing that has been twisted. He and Orville returned to their wind tunnel experimental data and developed a theoretical model of propeller design. With this model, they were able to predict the thrust that would be provided by a given propeller.

They added two propellers of their own design and construction at the rear of their new gasoline-powered biplane, *The Flyer*. Rear mounting of these propellers minimized any additional turbulence due to the propellers themselves. In addition, they eliminated the problem of torque from the pro-

23. Calculations based upon Copp and Zanella (1993).

pellors pulling the airplane sideways by simply having the propellors rotate in opposite directions (i.e., contrarotating propellors).

Their labors were not in vain. On December 17, 1903, near Kitty Hawk, North Carolina, Orville and Wilbur Wright completed the first successful manned, controlled, and sustained flight in a heavier-than-air, gasoline-powered biplane with two propellers (Fig. 1.6). Orville's first flight lasted for 12 seconds, covering a distance of only 131 feet and attaining a maximum altitude of about 10 feet. That same day, Wilbur flew 852 feet in 59 seconds.

Within two years, the Wright brothers were travelling 24 miles (albeit in a circular path) in 38 minutes, 3 seconds during a single flight (again the result of diligent, thorough experimentation and methodical engineering design). In 1908 Wilbur completed a 120-kilometer (74.5 miles) flight in less than two hours while visiting France. That same year, their latest aircraft was the first to carry a passenger during tests for the United States Army. Unfortunately, during these tests, a passenger—Lt. Thomas E. Selfridge—also became the first airplane fatality when he perished in a subsequent flight; Orville suffered a broken leg and cracked ribs in that same flight. However, due to news reports of this fatal accident, the Wrights and their airplanes finally received wide public recognition throughout the United States. Further publicity followed in 1909, when hundreds of thousands of people watched Wilbur Wright as he flew twenty miles above the Hudson River in New York, completing a round trip between Governor's Island and Grant's Tomb.

Although others would later produce aircraft of far more advanced designs and capabilities (e.g., replacing warped wings with ailerons for controlling roll[24]), the methodical labor of the Wrights had fulfilled humanity's long-held dream of achieving both controllable and sustainable flight.[25]

RAPID PROGRESS

Progress in manned flight was to continue at a rapid rate following the initial triumphs of the Wright brothers. In 1905 a factory dedicated to the manufacture of airplanes was founded in France by the brothers G. and C. Voisin, who then constructed two biplane box-kite gliders for Ernest Archdeacon and Louis Bleriot. These gliders led to the Voisins' later development of the popular box-kite biplane design that included separate mainplane and tailplane units. Similarly, Bleriot is credited with developing the tractor[26] monoplane, which continues to be the standard structural design for most modern airplanes.

---

24. The use of ailerons for lateral control was suggested by M. P. W. Boulton (1868) and later by both R. Esnault-Pelterie (1904) and A. Santos-Dumont (1906).

25. Elements of the Wright brothers' design can be found in modern light aircraft. For example, the 1986 *Beech Starship I* includes rear-mounted propellers and front-mounted tailplanes, features that were designed by a computer to ensure optimal performance (see Blackburn and Holister, 1987).

26. The propellor is located in the front of the airplane in tractor designs, whereas rear-mounted propellors are used in pusher airplanes.

The first European airplane flight soon followed when Alberto Santos-Dumont flew 65 yards at Bagatelle, France, on October 23, 1906.

Flying an aircraft manufactured by G. Voisin, Henri Farman completed the first long-distance, noncircular flight (17 miles) in 1908 at Issey-les-Moulineaux, France. The following year, Bleriot became the first person to cross the English Channel in an airplane (the *Bleriot XI*); this first crossing required a total of 37 minutes.

In contrast to simply flying above water, on March 28, 1910, Henri Fabre's seaplane became the first to takeoff from water as it soared into the air (with floats in place of its wheels) at Martigues, France. That same year, the first airplane take-off from the deck of a ship occurred when Eugene Ely accomplished the feat.

In 1912 more airplane firsts occurred: Harriet Quimby became the first woman to cross the English Channel, A. Berry made the first successful parachute-jump from an airplane, and Armand Deperdussin's craft was the first to fly more than 100 miles per hour—quite a progression of aircraft technology in the nine short years since Orville Wright had flown those first 131 feet through the air.

Daily international air travel between Paris and London began in 1919 with passenger flights above the English channel. That same year, the first nonstop transatlantic crossing was completed by John Alcock and Arthur Brown (Charles Lindbergh later made the first nonstop *solo* flight across the Atlantic Ocean in 1927).

Floyd Bennett and Richard Byrd flew over the North Pole in 1926 and Charles Lindbergh completed his 34-hour solo flight across the Atlantic Ocean (New York to Paris) the following year. Frank Whittle patented his jet engine design for aircraft propulsion in 1930. Then in 1931, Hugh Herndon and Clyde Pangborn flew across the Pacific Ocean. Amelia Earhart became the first woman to complete a solo flight across the Atlantic in 1932.

In 1933 Wiley Post completed the first solo flight around the world in 7 days, 8 hours. That same year, Boeing introduced its prototype of the modern airliner, the 10-passenger Model 247. Lockheed developed its 21-seat Douglas DC-3 Dakota airliner along similar lines, introducing it to the public in 1935. (More than 13,000 DC3's were built between 1935 and 1946, the vast majority of which were intended for military service during World War II. Surplus DC3s were used for civilian passenger air travel after the war ended, leading to the popularity of such travel by the public.)

Below is a brief chronological listing of the major events in the development of manned flight:

1200s	Manned kites used in China
1783	Free flight in a hot-air ballooon
1780s	Hydrogen balloons
1852	Airships (engine-driven balloons)
1870s–1890s	Gliders and steam-engine airplanes
1896	Half-mile, unmanned airplane flight

1903	First manned, sustained, and controlled airplane flight
1908	Long-distance, noncircular flight
1909	English Channel traversed by plane
1910	Seaplane take-off from water
1912	Plane travels 100 mph
1919	Flight across Atlantic Ocean
1926	Flight over North Pole
1927	Solo flight across Atlantic Ocean
1931	Flight across Pacific Ocean
1933	Solo flight around the world

Again, take a moment to reflect upon the enormous rate of progress described above. After centuries of effort, humankind had substantially conquered the air within the lifetime of a single generation.

SUMMARY: FUNDAMENTALS OF MANNED FLIGHT

Although today's aircraft are based upon a far more vast understanding of aerodynamics, the fundamental elements of airplane flight were captured in the Wrights' *Flyer*.[27] These elements include the concepts of drag, lift, and controlled rotation about three axes.

Drag A resistive drag force acting against forward motion of the plane as it moves through the airstream will always accompany lift. Lift and drag are the two components forming the total resistance or reaction force experienced by the plane as it travels through the air (see Case Problem 1). Thus, to achieve successful flight through an airstream, a plane must overcome this drag force with a propulsive thrust from its own engine(s).

Lift The amount of lift can be increased by enlarging the angle of attack between the wing and the airstream. Unfortunately, if this angle becomes too large the airstream will no longer be able to flow smoothly over the wing; the pressure differential is then lost and the plane enters a stall (loss of lift).

Axial control A plane can rotate about any of three distinct orthogonal axes: roll, pitch, and yaw (see Fig. 1.8a). These axes intersect at the plane's center of gravity lying between its wings.

Rotation about the roll axis is achieved through the use of ailerons (small flaps) on the outer rear edges of the wings (Figure 1.8b). The pilot can adjust the relative amounts of lift acting on each wing by raising and lowering these ailerons asymmetrically (i.e., one is raised as the other is lowered), thereby producing some desirable rolling action.

27. Also see Blackburn and Holister (1987), Macaulay (1988), and Copp and Zanella (1993).

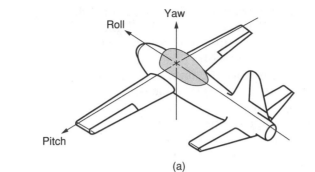

(a)

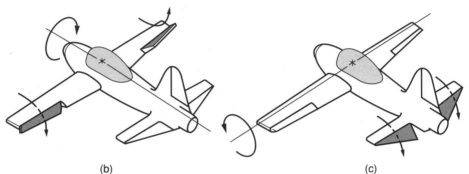

(b)   (c)

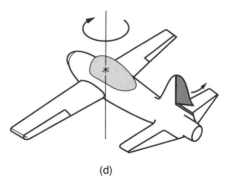

(d)

FIGURE 1.8 Axial airplane control: (a) Three principal axes: roll, pitch and yaw; (b) Rotation about the roll axis is achieved with ailerons; (c) Rotation about the pitch axis is achieved with elevators; (d) Rotation about the yaw axis is achieved with rudders. *Source:* Illustrations adapted from Blackburn and Holistor, 1987.

Elevators (flaps on the tailplane) are used to increase or decrease the wings' angle of attack and the corresponding lift applied to each wing, thereby causing a rotation of the plane about its pitch axis. The elevators are raised to lower the tail and vice versa (Fig. 1.8c).

Finally, a rudder allows the pilot to precisely adjust turns about the yaw axis (Fig. 1.8d). Such turns are usually made through a combination of rotations about the roll and pitch axes, with the rudder then used for slight corrections as needed as the plane banks through the air.

A steering wheel/joystick is rotated to produce rolls and moved forward or backward for altitude adjustments (pitch rotation), whereas the rudder (and thus rotations about the yaw axis) is controlled via pedals.

Cruising, take-offs, and landings require additional control. For example, maximum lift is needed during take-offs and landings. For this reason, some wings include trailing edge flaps that can be both lowered to increase the wing's curvature and extended to increase the wing's area, leading to greater lift (and greater drag). In contrast, spoilers (flaps located along the top of a wing) are used to decrease lift while increasing drag. The pilot is then able to properly manipulate the lift and drag forces acting on the airplane during its flight.

The swept wing is used to minimize drag and conserve fuel (since fuel must be consumed to provide the thrust to overcome drag forces). However, a swept wing not only reduces drag (a beneficial reduction) but also lift (the cost associated with a reduced drag). Flaps are included on such wings to provide increased lift as needed. (Swept wings also are often larger than straight wings in order to achieve sufficient lift.)

Swing wings are used on some high-speed aircraft. During take-offs and landings, these wings assume a straight wingform to generate increased lift, whereas the wings sweep backward in flight in order to reduce drag.

Delta wings are used on supersonic aircraft because these wings remain safely inside the shock wave that forms about the plane as it flies faster than the speed of sound.

The *Flyer* was not the result of simple trial-and-error efforts. Instead, it was the result of truly methodical engineering design in which the Wright brothers tried to overcome all foreseeable difficulties in their quest for controllable and sustainable manned flight. Each new challenge was met with imagination, the development of more accurate and more detailed theoretical models, and carefully designed experimentation based upon these theoretical models. The Wrights successfully incorporated synthesis and analysis in their work, completing the engineering design process as applied to manned flight.

Charles Taylor summarized a critical aspect of the Wrights' work—and that of all other successful engineers—when he made the following observation:

> (Wilbur and Orville) sure knew their physics. I guess that's why they always knew what they were doing and hardly ever guessed at anything.[28]

28. Copp and Zanella (1993).

Engineers *must* understand science if they are to apply it in practical forms. The Wrights provided us with an outstanding example of creativity coupled with detailed scientific knowledge, methodical work habits, exhaustive analysis, and critical thinking in service to humanity.

They also demonstrated the value of teamwork in engineering design, working together very effectively. Each brother complemented the talents and skills of the other and provided a sounding board off which ideas could be bounced. Today, most engineers work in teams and assist one another to be successful.

CASE HISTORY 1.3

*N*th-Generation Design: Refining the Typewriter

Early eighteenth- and nineteenth-century typing machines were designed to print raised characters that could be read by the blind.[29] The British engineer Henry Mill (although a prototype of his 1714 design was apparently never constructed), William A. Burt (whose 1829 machine featured a manually rotated circular carriage), Xavier Progin (who in 1833 marketed a circular device for the blind), Charles Thurber (who developed an automatic means for rotating the circular carriage on his typewriter in 1845), and others developed very clever devices as the years went by, but the invention of the modern typewriter did not occur until 1867.[30]

Christopher Latham Sholes, together with Carlos Gidden and Samuel W. Soule, developed a carriage that would shift automatically to the left as each character was typed. In addition, the keys of their typewriter all struck the same point on the platen as they were typed; that is, pianoforte movement (Progin's 1833 machine featured a similar movement). In fact, Sholes referred to the device as the "literary piano"—a very descriptive name. They were issued a patent for this work in 1868.

ERGONOMIC AND MECHANICAL GOALS

Sholes recognized that their machine, although satisfactory, was imperfect. One of its principal flaws was that quickly typed keys tended to collide or

29. A brief case history in Chapter 2 focuses upon Pellegrini Turri and his 1808 creation of a typing device for Countess Caroline Frantoni, the blind woman whom he loved.
30. Based upon Travers (1994) and d'Estaing (1985).

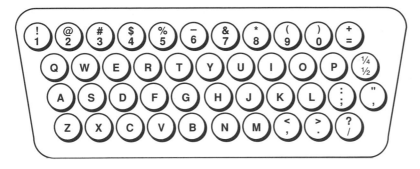

FIGURE 1.9 The QWERTY keyboard.

jam. He decided to separate the most frequently used keys on the keyboard (corresponding to the most frequently used letters of the English alphabet). Of course, a simply random placement of these keys about the keyboard was unacceptable since the user might then be required to strike the keys with awkward and difficult motions. Sholes needed to satisfy two goals: a mechanical one (separation of the keys from one another) and an ergonomic one (placement of these keys in a configuration that would aid rather than hinder the typist). His solution was the QWERTY layout, so named because of the first six letters along the top left of the keyboard. Developed over a five-year period involving more than 30 different models, the QWERTY layout became the standard typewriter keyboard configuration (Fig. 1.9).

Other more efficient keyboard layouts have been developed in recent years (e.g., the Dvorak simplified keyboard shown in Fig. 1.10). One might then ask why the QWERTY configuration remains the standard for today's computer keyboards. The answer: *entrenchment!* Sometimes it is very difficult to change people's habits even when such change offers substantial benefits. The QWERTY layout is most familiar to typists, programmers, and other users of keyboards. After all, typing classes usually focus upon the QWERTY system; as a result, most professional typists expect to use this layout. Industry has been somewhat reluctant to adopt new, more efficient keyboard layouts because of the retraining that would be necessary for their typists to become adept with these layouts. Such retraining would necessitate a temporary loss in productivity and efficiency (although long-term benefits of the new keyboard layouts could be substantial). Because of these factors, most keyboard manufacturers are very cautious about introducing any changes in their products that might alienate their customers and harm their sales. Although other keyboard layouts may indeed be more efficient than the QWERTY system, these new layouts have been introduced to the market in very small doses, thereby allowing people to become familiar with them while limiting risks to the manufacturers. (Of course, many computer software systems now allow the user to alter the keyboard layout electronically, replacing the QWERTY system with one more suited to the needs of

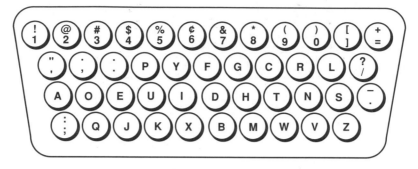

FIGURE 1.10   The Dvorak simplified keyboard.

the user. However, keyboards themselves continue to be manufactured with the original QWERTY layout.)

In 1873 the Remington Arms Company bought the patent rights to the improved design from Sholes and his business partner, James Densmore, for $12,000. The Remington Model I was released to the public and soon thereafter Mark Twain submitted to a publisher the first manuscript typed on this machine.

FURTHER IMPROVEMENTS IN NTH-GENERATION DESIGNS

Each later generation of the basic typewriter design continued to include improvements developed by others. These improvements included

- the shift-key mechanism (1878)
- the first portable machine (the "Blick"), developed by George C. Blickensderfer (1889)
- the front-stroke design (1890)
- electrically powered systems, first developed and manufactured by Thaddeus Cahill in a 1901 form that was a commercial failure because of its exorbitant cost: $3925 per machine; the much more successful IBM Electromatic, developed by R. G. Thomson over an 11-year period, was released in 1933
- the portable electric typewriter (1956)
- a stationary carriage with IBM spherical ball elements for changing font styles (1961)
- the IBM 72BM, an electronic typewriter with a magnetic tape that provided it with a memory (1965)
- rotating print wheels (1978)
- a portable electronic typewriter (Olivetti's Praxis in 1980)

A successful engineering solution is only satisfactory; improvements can—and should—always be made through Nth-generation design efforts.

CASE HISTORY 1.4

Single-Use Cameras

One example of a product that was redesigned for export is Eastman Kodak's Fun Saver Panoramic 35-mm single-use camera. (The consumer returns the camera, together with the film inside, for developing; a new camera then must be purchased for further picture taking.) In 1993 Kodak researchers began to work with people in Europe, Japan, and the United States in order to identify those features that should be incorporated into the next-generation design of this camera.[31]

It was found that Japanese consumers wanted a design that would be thinner, lighter, and require less time to recharge its flash mechanism. Germans sought greater recyclability of the parts, whereas other respondents desired rounded corners that would make the camera more comfortable to carry and use.

Because of this focus upon the world market, the camera was redesigned to fit around the film pack, thereby minimizing its size. Rounded corners, a plastic ellipse that prevents a user from inadvertently blocking the picture, an automatic recharger so that the flash will always be available to light a scene, and components that either can be reused or recycled also were introduced. Furthermore, the needs of photofinishers were flected in the use of a standard (35-mm, 400-speed) film and a cover that could be opened easily.

Marketplaces increasingly are becoming international in scope. As a result, products often are designed for both domestic use and export.

31. Machlis et al. (1995).

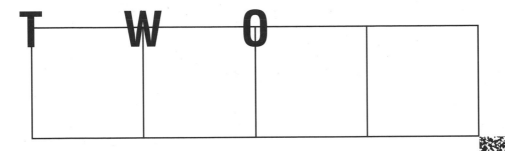

NEEDS ASSESSMENT

Those who bring sunshine to the lives of others cannot keep it from themselves. Sir James Barrie

Upon completion of this chapter and all other assigned work, the reader should be able to

- Recognize the need to justify an engineering problem-solving effort.
- Discuss the ways in which the need for a design may develop.
- Classify problems in distinct categories of prediction, explanation, and invention, or some combination thereof.
- Recognize the hazards associated with focusing upon oneself and not upon the needs of others.
- Prepare a design proposal that justifies the need to develop a technical solution to a problem.

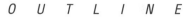
2.1   Establishing Need

Engineers develop solutions to practical problems—problems that need to be solved for the benefit of humanity. In fact, the initial step of the engineering design process is to identify the specific need that is to be satisfied by a solution based upon technology.

    The need for a new or redesigned product may have resulted from any number of situations, including

- Concern for the health, safety, and quality of life of the public.
- Recognition that an existing product must be redesigned in order to[1] (a) eliminate shortcomings in the original design or fabrication by

1. See, for example, Walton (1991).

incorporating new technology and manufacturing methods; (b) better serve the changing needs of the user population; (c) increase the commercial viability of the product by responding to competition in the marketplace; and/or (d) reduce costs.

- Personal experience of the design engineer in performing a particular task (or in observing others while they perform this task), resulting in the conclusion that a new or different technical solution would improve such performance by making it easier, more accurate, more efficient, less wasteful of material resources, and so forth.

Opportunities for new and improved products and systems are endless. We only need to consider the unmet needs of our neighbors and the world around us.

2.1.1 ■ Safety and Quality of Life

As noted above, an innovative design often is needed to protect the health and safety of the public or to improve the quality of life for some people.

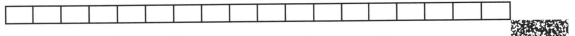

A Typewriter for the Blind

The typewriter is an example of a design that was developed for a loved one in order to enhance the quality of her life.[2]

As we noted in Chapter 1, most early typewriters were designed to print raised characters that could be read by the blind. In 1808, Pellegrini Turri created one of the first true typewriters for Countess Caroline Frantoni, the woman he loved. Countess Frantoni was blind and Turro's machine allowed her to correspond with him. Unfortunately, no other information about the Turro typewriter now exists, other than 16 letters that were exchanged by the Countess and Turro between 1808 and 1810. These letters, now stored in Reggio, Italy, are evidence of the technological innovations that *can* be generated through love and affection in response to someone's real need(s).

Case History 2.1 *Kidney Dialysis Treatment* provides another example of a design that focuses upon health and safety issues.

2. See d'Estaing (1985).

2.1.2 ■ Improving an Existing Product or System

An existing product or process may be in need of modification in order to be truly effective and/or commercially viable. An example of such product redesign and development is taken from the medical world in the following anecdote.

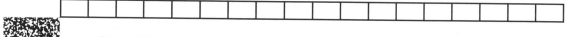

The DC Heart Defibrillator

Valuable engineering designs often are improvements of earlier solutions with distinct weaknesses or shortcomings. In fact, most engineered products and patented designs fall into this category.[3]

A heart defibrillator is used to overcome ventricular fibrillation in which a person's heart has suddenly lost its natural rhythm. The heart's sinoatrial node synchronizes the contraction of the heart's upper chambers (the atria) with its lower chambers (the ventricles). When this synchronization is lost, the heart can no longer pump blood throughout the body, leading to death. A defibrillator applies an electric shock to the heart to return synchronization to the system.

Early alternating current (AC) heart defibrillators (first used experimentally in 1948) had several shortcomings, including the risk of a complete hospital-wide power loss. In 1959 Dr. Karl William Edmark, a cardiovascular surgeon, began to develop the direct current (DC) heart defibrillator, which could provide an electric charge of significantly lower voltage within a very short time interval, thereby providing the benefits of an AC defibrillator without its associated dangers. In 1961 Edmark's defibrillator first saved a life: that of a 12-year-old girl during open-heart surgery.

There is no such thing as a perfect engineering solution to a problem. Improvements can always be made to a design.

Other examples of improved designs are in Case History 2.2 *Credit Cards for the Blind* and Case History 2.3 *Cleaning up Oil Spills*.

2.1.3 ■ Commercial Incentives

A new product line may be driven by the commercial needs of a company or an individual. However, such products must respond to their customers' needs (i.e., needs in the marketplace) or they are doomed to failure.

3. See Woog (1991).

FIGURE 2.1   A modern version of the *Kwik-Lok* closure.

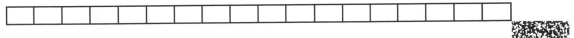

The *Kwik-Lok* Closure

In 1952 Floyd Paxton was seeking a new commercial product for his company to manufacture.[4] The company produced equipment used for nailing wooden crates; however, sales were expected to decrease as cardboard boxes became more popular as an alternative to wooden crates. The company "needed" a new product line that would allow it to remain commercially viable.

A local produce firm (Pacific Fruit and Produce) had begun to store apples in plastic bags, using rubber bands for closure. The owner of Pacific Fruit and Produce mentioned to Paxton that a new type of closure would be helpful. Other closures (such as wire twists and tape) either broke too easily or did not allow numerous reopenings and closings of the bags, and rubber bands were cumbersome to use.

Paxton took a piece of plexiglass and whittled the prototype of the *Kwik-Lok* closure. He next designed and constructed a manual machine for their manufacture, followed by a high-speed die-punching machine. Food producers embraced the new product, which became ubiquitous in food stores everywhere (Fig. 2.1).

After more than forty years, the *Kwik-Lok* continues to be an extremely popular closure for plastic bags.

For another example of a design driven by commercial incentive, see Case History 2.4   *Durable Pants for Miners*.

2.1.4 ▦ Personal Experiences

Sometimes the need for a new application of technology becomes apparent through events in one's own life, as seen in the following example.

---

4. See *Woog* (1991).

The Quick-Release Ski Binding

In 1937 the champion skier Hjalmar Hvam fractured his leg after jumping a crevasse because he could not release his foot from his ski. In response, he conceived the design for the world's first quick-release ski binding.[5] A spring held a Y-shaped pivoting metal toepiece against the skier's boot unless a lateral force was applied from the right or left—in which case the toepiece swiveled to release the boot from the binding. (A strap continued to hold the ski to the heel in order to avoid loss of the ski.)

Hvam patented his design in 1939. Untold numbers of skiers were saved from injury because one skier recognized a need and developed a simple yet creative solution.

Sometimes the need for a new application of technology becomes apparent through events in one's own life.

2.1.5 ■ Opportunities from Scientific Advances

Advances in technology and/or a scientific discovery may create the opportunity for a new engineered product, as illustrated by the next example.

How Color Printing Led to Air Conditioning

Human beings are usually comfortable when in temperatures between 20°C and 25°C (with a corresponding humidity somewhere between 35 and 70 percent).

Many have attempted to provide a means for ensuring such comfort through various types of air cooling. In ancient times, air was cooled (via evaporation) as it passed through wet mats hung over doorways and windows. By the beginning of the twentieth century, air cooling systems of limited effectiveness had been installed in Carnegie Hall and the New York Stock Exchange.

Then, in 1902, Willis H. Carrier (1876–1950), a mechanical engineer, was striving to eliminate the negative effect that humidity had upon color printing.[6] Carrier realized that cold air could absorb humidity from warm air. He designed a cold water spray system through which the relative humidity of the air within a room could be controlled by simply regulating the temperature of the spray. (He received a patent for this design in 1904.)

Carrier continued his work in this area, eventually developing the science of *psychrometry* (the measurement of water vapor in air). Then, in 1911, he laid the foundations for a complete air conditioning system.

---

5. See Woog (1991).
6. Based upon Travers (1994) and *The Way Things Work* (1967).

Air conditioning was subsequently installed in many public facilities: movie theatres and department stores (beginning in 1919), high-rise buildings (1929), railroad trains (1931), automobiles (beginning with the Packard in 1939), and Greyhound buses (1940). Perhaps most significant, H. H. Schutz and J. Q. Sherman developed the individual room air conditioner during the 1930s.

Ammonia was used as the refrigerant in early air conditioning systems. Unfortunately, ammonia is both corrosive and unpleasant. In modern times, ammonia has been replaced by freon, a nontoxic and nonflammable fluorocarbon discovered by Thomas Midgley, Jr. in 1930. However, being a fluorocarbon, freon can be quite hazardous to the earth's ozone layer if it is allowed to escape to the atmosphere. Hence, the need exists for further improvement in air conditioning systems.

Engineering solutions often result from scientific advances. Engineers should be prepared to take advantage of such advances whenever an opportunity arises by (a) searching for better ways to perform tasks, (b) learning from experience, and (c) continuously updating and upgrading their skills and knowledge.

2.2 Service to Humanity

It is ironic that engineers are sometimes caricatured as reclusive, socially awkward individuals who are happier working with machines than with other human beings. Nothing could be further from the truth! Since engineers usually focus upon society's most basic physical needs (e.g., food, shelter, heat, health care, communication), they *must* work closely with many people, from potential customers to fellow engineers, marketing specialists, financial advisors, field and factory workers, stylists, technicians, managers, and others who will contribute to the development of a successful design solution. Engineering is a people-oriented and people-rich profession. Engineers are focused upon the needs of others whom they serve. Case History 2.5 *Garrett Morgan's "Safety Hood" and Traffic Signal* provides a memorable example of this fact.

2.3 Types of Problems: Prediction, Explanation, Invention

It is sometimes helpful to consider the type of problems that one might need to solve. Parnes (1967) defined three basic types of problems:[7]

7. Also see, for example, Higgins, Maitland, Perkins, Richardson, and Piper (1989).

- **Problems of prediction** in which one seeks to calculate a result or predict a system's behavior by applying equations, physical laws, tools of data analysis, and so on (e.g., determining the reentry speed and direction of a falling space satellite.)
- **Problems of explanation** in which one seeks the cause(s) for a phenomenon or observed behavior (e.g., searching for the cause of a satellite prematurely falling to earth).
- **Problems of invention** in which one seeks to develop a new and effective solution to a problem (e.g., developing a way of preventing satellites from prematurely falling to earth).

If a problem can be classified as one of these three types, the engineer can then begin to identify the data, physical laws, and other information that will be needed to develop a solution.

Furthermore, the engineer should consider if the problem under consideration is solely one of prediction, explanation, or invention—or if it is, in fact, a *combination* of these problem types. For example, one may seek to prevent satellites from falling (a problem of invention), but he or she must first determine the cause(s) for such flight failures (a problem of explanation).

2.4 Focusing upon Others: A Key to Success

We will refer to an inability of an engineering solution to perform its intended function(s) as a design *failure,* whereas an *error* is the underlying cause for such failure. We contend that the likelihood of a design error will be reduced if the engineer *focuses upon others* (in particular, the expected users of the design). Whenever we focus solely upon ourselves (or upon those with whom we work) and ignore the needs of those who will be directly affected by our work, we become more likely to overlook important aspects of a problem that can be the difference between success and disaster.

For example, most of us find formal public speaking to be very stressful; such anxiety is a reflection of our natural fear of failure. We become overly concerned that we will be embarassed by a verbal mistake, temporary loss of memory, or by an overall poor performance. However, if we instead focus upon serving the needs of our audience and strive to present relevant information in an understandable and useful form, our initial nervousness often will be replaced by a calm, deliberate, and effective speaking style.

Physicians must always be aware of the consequences of their behavior. If a decison is made carelessly or if an action is taken with indifference, the patient can die. Similarly, engineers must always be aware of the potential hazards associated with their work. Obviously, if a bridge collapses or an explosive gas leaks from a chemical reactor, people may perish. However, most engineering systems can be hazardous in more subtle ways. For example, if a stray electrical signal causes the cargo door of a jumbo jet to open unexpectedly while the plane was in flight, the sudden depressurization of the fuselage might result in a crash. If at all possible, such an event must be foreseen and prevented through effective engineering design. But such potential hazards will not be recognized if the engineering team fails to focus upon the needs and safety of their customers. Engineers—like physicians—must be aware of the harmful consequences of actions taken and not taken.

Whenever an engineered system or product fails, the underlying cause of the failure (i.e., the error) must be identified and eliminated. The cause of a failure may lie at any or all of three different levels:

1. **Concrete level** A specific physical source is identified as a contributory cause of the failure. For example, the failure of the o-ring seal between the interlocking joints of the booster rockets in the *Challenger* space shuttle allowed escaping gases to be ignited by the rocket flames in the 1986 disaster.

2. **Process level** *Invalid assumptions, faulty reasoning,* or the *flawed execution* of a procedure leads to the failure. In the case of the *Challenger* tragedy, one invalid assumption was that the o-ring gaskets and putty would be sufficient to ensure that the interlocking joints would remain sealed.

3. **Values/attitudes/perspective level** A flawed value system contributes to the failure. Morton–Thiokol (M–T) was the booster rocket manufacturer for the *Challenger* shuttle. M–T executives overrode the objections of fourteen engineers when they approved the final launch of the shuttle system. These executives apparently did not want to disappoint NASA officials for fear that such disappointment would adversely affect upcoming contract negotiations between Morton–Thiokol and the space agency. A focus upon self (company concerns) led to poor decision making (the approval of the launch despite persuasive evidence that the gasket system might fail).

    The values and attitudes exhibited by NASA officials also can be criticized. NASA policy had been that a launch would be approved only if it could be demonstrated that the system was safe. Instead, officials reversed this policy, declaring that the shuttle launch would be postponed only if the o-ring system was proven to be unsafe. This subtle change in policy led to the tragic loss of seven lives.

The fourteen engineers who recommended that the launch of the *Challenger* be postponed acted responsibly. Engineering codes of ethics (see Chapter 8) provide professional guidelines for responsible behavior. Responsibility can be interpreted as the *ability to respond* to other people's needs. Whenever we are indifferent, overconfident, callous, impulsive, or otherwise concerned more with ourselves than with others, we increase the likelihood of failure in our work. The final case histories in this chapter illustrate the need to focus upon others in order to avoid failure and avert disaster; these include Case History 2.6 *Lake Peigneur Goes down the Drain;* Case History 2.7 *The Flixborough Explosion;* Case History 2.8 *Maintenance Procedure Leads to the Crash of a DC-10;* Case History 2.9 *Bhopal: Simultaneous Failures Lead to Disaster;* and Case History 2.10 *Rushing to Fill the Teton Dam Leads to Its Collapse.*

Unfortunately, focusing upon others does not guarantee that we will never experience failure. However, focusing upon oneself virtually ensures that we will confront disaster in one form or another. *The true measure of a good engineer is the ability to respond to others' needs in an effective, professional, and appropriate manner.*

2.5 The Design Proposal

Often, the engineer must prepare a formal design proposal for the work planned, demonstrating to management that the expected results will indeed justify a substantial[8] investment of effort, time, and money in the project. Justifying the need to search for a solution to the problem will help the engineer to formulate the problem correctly (step 2 of the design process, reviewed in the next three chapters).

The design proposal provides the engineer with the opportunity to (a) justify the need for a technical solution to a problem and (b) to express this need in precise and accurate terms. Design proposals may be formal (i.e., written and submitted to management, the client firm, or some external funding agency for support) or informal (i.e., prepared for use by the engineering team itself in order to develop an understanding of the work to be performed and the reasons for this work).

Sometimes the reason for developing a solution to the problem may seem obvious. At other times, the justification for solving a problem may come directly from management. Whatever the reason or the source, the need to develop a solution to a problem must be justified by the engineer in concrete and specific terms.

A design proposal should answer the following questions about the work to be performed: Why? Who? Where? How? When? What? How much (will it cost)? These questions can be used to structure the proposal by subdividing it in the following way:

1. **Objective (Why?)** The precise objective to be achieved by the proposed effort should be described in a clear and concise form. Do not assume that the need for a solution is self-evident.

    Focus upon function; that is, what is the core task(s) that must be performed by any viable solution to the problem? (In Chapter 3, we

---

8. Any investment should be viewed as substantial because it prevents some resources from being directed towards other efforts.

will return to this need to focus upon function in the generation of a problem statement.)

2. **Background (Who? Where?)**   Next, the necessary contextual background of the problem should be provided so that the reviewers of the proposal will be able to understand *who* will be served by a solution to the problem (i.e., the potential clients or users of the design) together with the environment in which the design is expected to be used (*where*). Existing solutions and prior work on the problem should be described and evaluated.

   This portion of the proposal should demonstrate that the engineer has a truly robust understanding of the problem to be solved.

3. **Methodology (How? When?)**   In this section of the proposal, the engineer should describe the approach that will be used to design and develop the desired solution.

   Each task to be performed should be identified, together with the person(s) who will be completing these tasks. A schedule for the expected completion of these tasks should also be included. A Gantt chart similar to Figure 2.2 can be used to denote the specific tasks to be performed, their relative interdependencies (i.e., some tasks can be performed concurrently, whereas others must be performed sequentially because of their inherent interdependencies), and their expected completion dates.

   This portion of the proposal should persuade managers and any other reviewers that the engineers have structured the problem-solving effort in a carefully planned format that will likely lead to success.

4. **Expected results (What?)**   Next, the expected results of the effort should be described. Will the deliverables of the work be abstract (e.g., a set of recommendations presented in the form of a report) or concrete (e.g., a

FIGURE 2.2   Example of a Gantt chart (in textual form).

Proposed Scheduls

Phases	1	2	3	4	5	6	7	8	9	10	11	12	13	14	15	16	17	18	19	20	21	22	23	24
Preliminary system	X	X	F	X																				
Design layouts	F	X	X	X																				
Subsystems - dev				X	X	X	X	X	X	X	X	F	X	X	F									
Testing and analysis													X	X	F	X	X	F						
Full system - evaluation															F	X	X	F	X	X	X			
Implementation/distribution					X	X			X	F							X	F			X	X	X	F

Note: F = work performed in field; X = work performed at company offices.

manufactured product)? In addition, the expected benefits and potential risks of the effort should be delineated. Presumably these benefits outweigh the risks (otherwise the design effort would not have been proposed); one should explain why this is expected to be the case.

5. **Costs (How much will it cost?)**   Finally, the expected costs of the proposed effort should be given. Labor, materials, facilities, and any other factors that will affect the economic feasibility of the project should be included and justified.

Written with clarity and accuracy, the design proposal should be concise yet complete. The engineer should seek clarification about any element of the proposed work (such as the anticipated costs for the project) that is not fully understood.

The design proposal is a plan: It helps the engineers (and management) to develop an appreciation for why a problem should be solved, what tasks must be performed to develop a solution to the problem, how these tasks might be accomplished, and the investment that must be made in the effort. Its initial value is to serve as a mechanism through which the design project can be justified and structured.

After the project has begun, the engineers may discover that their original plan should be modified in order to deal with aspects of the design effort (e.g., new information, technical difficulties, logistical problems) that were not foreseen when the proposal was prepared. After all, design is an iterative process in which corrections and refinements to a project plan can be expected. Although incomplete as a description of the eventual work that will be performed, the design proposal nevertheless can serve as a guide for the engineers as they complete this work.[9]

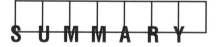

SUMMARY

- Engineering designs are intended to satisfy human needs.
- Technical problems can be classified into three basic categories: problems of prediction, problems of explanation, and problems of invention (or some combination thereof).
- Success is more likely if one focuses on others and not on oneself.
- Design proposals provide the engineer with the opportunity to justify the expenditure of time, money, and effort for a problem-solving project, and to generate an initial plan for completing the project.

9. Martin (1993) discusses proposal preparation in detail.

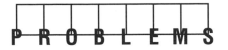
PROBLEMS

2.1 Identify three types of societal needs that engineering has satisfied. Discuss the degree to which these needs have been satisfied. If the need(s) have not been fully satisfied by the engineering solution(s), discuss the possible reasons for this less-than-optimal result.

2.2 Prepare a brief report describing two examples of engineering failures that are not discussed in this text. Discuss the possible underlying causes for these failures. Were these failures inevitable? If not, how might these failures have been avoided?

2.3 Prepare a brief report describing the development of an engineering solution that was driven by one or more of the following factors:

- Desire to enhance safety and quality of someone's life
- Need to improve an existing product or system
- Commercial or economic need
- Personal need
- Desire to apply a scientific advance

Describe the historical development of this solution beginning with the recognition of a need. Evaluate the degree to which this need was satisfied by the engineered solution, and discuss the critical aspects of the design that either helped or hindered its success.

2.4 Describe an example of each of the following problem types: prediction, explanation, and invention. (Do not list the same examples given in this chapter.)

2.5 Prepare a brief report on one of the following systems. Discuss the needs that were to be satisfied by this solution. Which of these needs have been fully satisfied and which have been only partially satisfied?

- Jackhammers
- Smoke detectors
- Elevators
- Expert systems
- Bookbags
- Cellular telephones
- Sewage treatment facilties
- Fax machines

CASE HISTORY 2.1

Lowering Costs and Increasing Availability in Kidney Dialysis Treatment

AN EARLY CHALLENGE: THE COLLAPSE OF ARTERIES AND VEINS

The body perishes if its blood is not cleansed, either naturally by the kidneys or artificially via dialysis by filtering the blood through a machine. In 1939 Dr. Willem Kolff developed a dialysis machine for treating patients suffering from uremia (renal or kidney failure). Unfortunately, this machine, even with subsequent improvements, had many weaknesses. One major shortcoming: Each week a patient would need to be attached repeatedly to the machine, requiring multiple incisions and leading to the eventual collapse of the person's arteries and veins.[10]

THE SCRIBNER CANNULA: A SIMPLE YET ELEGANT SOLUTION

This less-than-optimal situation was corrected in 1960 when Dr. Belding Scribner (a nephrologist), together with Dr. Albert Babb (a nuclear engineer), Wayne Quinton, and Dr. Loren Winterscheid, designed the Scribner cannula. When not undergoing cleansing via dialysis, the patient's blood is directed by this shunt through a U-shaped tube that is permanently attached to the person's arm. By simply removing the shunt, the patient then could be attached easily to the dialysis machine. Although the shunt system required refinement by the team, the concept itself was ultimately successful: Multiple incisions leading to artery collapse were no longer necessary. A simple yet elegant solution could now be used to save lives.

PORTABILITY LEADS TO AVAILABILITY

Next, the team developed the first dialysis machine that could simultaneously serve several patients. In addition, the machine was so simple that it could be operated by the patients themselves.

Unfortunately, these improved machines still could not satisfy the demand for dialysis; treatment was limited and expensive. Patients were chosen for treatment by anonymous panels, meaning that some people were saved and some were not. In response to this intolerable situation, Scribner, Babb, and Quinton designed the world's first portable home dialysis machine, thereby dramatically decreasing costs and increasing availability. Caroline Helm, a high school student who had been rejected by a panel for dialysis treatment, became the first person to use the new portable home system in 1965. Another life-saving improvement in dialysis treatment had been developed through the creative collaboration of these specialists.

10. See Woog (1991).

Engineering designs often are developed to protect the health and safety of the public or to improve a person's quality of life.

CASE HISTORY 2.2

Credit Cards for the Blind

Credit cards have been difficult for the blind to use.[11] In response to this need, Accubraille, the company that first introduced braille stickers for automatic teller machines (ATMs), developed braille-encoded cards. Not only do these braille cards help to satisfy the needs of the blind, but they also assist companies who issue credit cards to comply with the Americans with Disabilities Act.

Braille may seem to be an obvious addition to credit cards and other products used by the blind; however, even such "obvious" design improvements are not made until someone recognizes and responds to a person's need. Significant innovations in a design often can be achieved by focusing upon the needs of specific user groups.

CASE HISTORY 2.3

Cleaning up Oil Spills

Following the 1989 Exxon *Valdez* oil spill in Alaska, Congress passed the Oil Pollution Act of 1990. By this act, any company that stores or transports petroleum must prepare plans for cleaning up any oil spill that may occur on land or water.[12]

Oil spills occur more often than one might expect; for example, 16,000 separate spills were identified by the Environmental Protection Agency during 1992. Moreover, each spill can be very detrimental to living organisms and their environment. (It has been estimated that only one quart of oil is

11. See Block (1994)
12. See Penenberg (1993).

needed to pollute 750,000 gallons of water.) Furthermore, cleaning up an oil spill can be expensive: Exxon spent $2.5 billion for the *Valdez* clean-up.

The traditional method of mechanical skimming has not been very effective. Only 20 percent (or less) of an oil spill can be cleaned via this process, resulting in significant damage to the environment and the loss of most of the spilled oil. Fortunately, many companies are developing new methods for cleaning up oil spills.

One new method involves the use of Elastol, a substance manufactured by General Technology Applications, Inc, and based upon a chewing gum ingredient called polyisobutylene. When applied to a spill, Elastol causes the oil to collect into a thin skin, which is then removed with a rotating drum. The oil and the polyisobutylene in the collected skin are then separated. It has been estimated that 97 percent of the oil can be recovered via this process.

Another method uses Eco Spheres (also known as Heller's beads after their inventor Adam Heller, a professor of chemical engineering at the University of Texas in Austin). Eco Spheres are miniscule glass beads, coated with titanium dioxide, that are distributed over an oil spill. Sunlight interacts with the titanium dioxide in a photocatalytic process to convert the oil on each bead into carbon dioxide, water, and a sandy residue that can then be collected.

Engineers are continually developing new processes for performing difficult tasks in more effective ways.

CASE HISTORY 2.4

Durable Pants for Miners

The California Gold Rush of 1849 indirectly led to the development of Levi jeans and clothing.[13] Soon after the first gold strike, Levi Strauss (1829–1902) traveled to San Francisco to make his fortune, not by discovering gold but by selling various types of cloth goods to the many miners who were also seeking their fortunes. By the time Strauss arrived in the city, his wares consisted only of canvas duck (heavy cotton fabric for tents and wagon covers); he had sold his other goods during the long journey from New York. The miners were uninterested in the canvas duck; however, one suggested that what was really needed were durable pants.

Strauss acted upon this suggestion. He had a tailor use the canvas duck to make wear-resistant pants that proved to be extremely popular among the

13. See Sacharow (1982).

miners. Once his original stock of canvas was consumed, Strauss sought other durable materials from which to make pants. His relatives sent a broad range of cloths, including *serge de Nimes* or denim. This French cotton cloth was chosen by Strauss as the material for his popular pants. He later dyed the material, added copper rivets at appropriate locations to increase the strength of the pants, and introduced the double arc of orange thread on the back pockets to distinguish his pants from those of his competitors. (In 1873 Strauss and Nevada tailor Jacob W. Davis received a patent for the idea of using rivets to increase the strength of their clothing products.)

Levi Strauss achieved commercial success by responding to a very specific need within the mining community in a creative yet effective way. His company eventually became the world's largest manufacturer of apparel.

CASE HISTORY 2.5

Garrett Morgan's "Safety Hood" and Traffic Signal

Engineers often serve others through very concrete and direct actions. One striking example is the truly heroic behavior of Garrett Morgan (Fig. 2.3).

Morgan invented the gas mask (first known as a "safety hood") in 1912.[14] Modern gas masks (or respirators) provide protection against smoke, toxic fumes, dust, and other airborne hazards. Morgan's mask was originally designed for use by firefighters when rescuing people from burning structures.

The subsequent popularity of safety masks was primarily due to Morgan's courageous actions following the 1916 collapse of a tunnel below Lake Erie at the Cleveland Water Works. Twenty-four workers were trapped 228 feet below the lake. Squads totalling eleven men mounted two separate rescue attempts, but gas explosions killed nine of these rescuers.

Morgan was then called to the scene. He, with his brother Frank and two volunteers, donned gas masks and entered the tunnel. They successfully rescued the 20 surviving workers.

Morgan formed the National Safety Device Company to manufacture and distribute his gas masks. Fire departments throughout the United States placed orders for this new safety device. The gas mask also became a piece of standard equipment in other fields such as mining and chemistry.

Modern gas masks protect their wearers from various fumes and gases through filtration, physical absorption, and/or chemical absorption. They can be effective if (a) toxic fumes are in relatively low concentrations and

14. See Travers (1994) and *The Way Things Work* (1967).

FIGURE 2.3 Garrett Morgan invented the gas mask and the four-way traffic control light. *Source:* AP/World Wide Photos.

(b) the air is sufficiently oxygen-rich; otherwise, an independent oxygen supply must be used.

The face piece is attached securely to the wearer so that air only can enter through a cannister that acts as a filter. This cannister may contain a mechanical filter made of cellulose, glass, or other fibrous material that is used to remove droplets from the air prior to inhaling. In addition, gas molecules can be extracted from the air via physical adsorption on surface-active materials (e.g., active charcoal). In this process, molecular or capillary action attracts the gas to the surface of the material, as seen in dehydrators and refrigerative systems. If necessary, fumes also can be removed via chemical absorption. For example, alkalies have been used to eliminate acid fumes in the air through chemical reaction.

Today numerous respirators are available for application in a broad range of environments and occupations in which toxic fumes may be present. An elegance and simplicity is seen in the design of these life-saving "gas masks" in which both mechanical and chemical processes are used to purify air for the wearer.

Garrett Morgan continued to design new and useful products. In 1923 he received a patent for the first automatic traffic control signal. Earlier traffic control devices had only two signals: stop and go. In contrast, Morgan's signal could also warn drivers to proceed with caution when necessary. Morgan formed another company (G. A. Morgan Safety System) to market his invention and the Morgan signal became the standard traffic control device. Eventually, General Electric purchased his patent rights to the four-way electric traffic light signal for $40,000.

We all would do well to remember Garrett Morgan's intrepid behavior in risking his life to save others. His actions demonstrate why an engineer should always design and develop a solution as if it will be used by both family and friends.

CASE HISTORY 2.6

Lake Peigneur Goes down the Drain

The 1980 disappearance of Louisiana's Lake Peigneur provides a stark lesson in both the importance of communication and the fact that any engineering project must be designed and implemented with due consideration of its environment.[15]

Lake Peigneur was relatively large, measuring about one mile by two miles. Many people enjoyed fishing on the lake, which was adjacent to a tourist attraction known as the Rip Van Winkle Live Oak Gardens. In 1980 the Texaco Oil Company was searching for oil beneath the lake. Their drilling operation had reached a depth of about 1250 feet (at a point where the water's depth was no more than six feet) when the drill unexpectedly stopped. The drill then began to move in large vertical leaps (up to 10 feet high) and the drilling rig itself started to list. Workers quickly abandoned the rig just before it disappeared below the surface of the lake. A large whirlpool then started to form, eventually consuming several nearby barges, a tugboat, and some shrimp boats. The whirlpool continued to grow until it had sucked in 65 acres of the Rip Van Winkle Live Oak Gardens. All fishermen and the tugboat crew made it safely to shore before the entire lake drained into the drilling hole. A total of 3.5 billion gallons of water disappeared from the surface of the earth in only seven hours.

The reason for this strange event was as follows: About 1300 feet below the lake there was a branch of the Diamond Crystal salt mine. Salt mines can be quite large. This single branch had a ceiling that was 80 feet high, a width that was equivalent to a four-lane highway, and a length of about 0.75 miles from the mine's central shaft. Texaco's drill punctured the ceiling of the cavern and the lake waters flowed into the salt mine. Fifty-one miners barely escaped the mine before it filled with water. (Lake Peigneur began to gradually refill once the mine was flooded.)

15. See Gold (1981) and Perrow (1984).

The presence of the mine below the critical drilling point was indicated on only one of several charts that Texaco was using to guide their drilling operation; the other charts indicated that the mine was elsewhere. Texaco never bothered to contact directly or work with the Diamond Crystal mining firm to resolve this conflict in data. However, Diamond Crystal was given an opportunity to object to the issuance of a dredging permit to Texaco when they were first informed of the drilling operation by the U.S. Army Corps of Engineers. Even after this notification of a drilling operation that could affect their salt mine, Diamond Crystal failed to contact Texaco.

This case demonstrates that the operating environment of an engineering system can drastically affect its performance. The total lack of communication between Texaco and Diamond Crystal led to disaster for both the drilling and mining efforts, hundreds of millions of dollars in damages, and a total of seven lawsuits. Fortunately, no one was killed by this failure in both communication and environmental systems design.

SUMMARY OF FAILURES AND TYPES OF ERRORS

- Lack of communication
- Lack of knowledge regarding operating environment (error in design due to erroneous data and invalid assumptions)

The expected operating environment must be considered fully during the design and implementation of an engineering system. Otherwise, one risks unexpected (and potentially disastrous) interdependencies and interactions between the system and its environment.

In addition, failure to communicate and share relevant data with others can lead to disaster for all. The parties involved in this incident worked in isolation from one another. Such isolation is one example of focusing upon oneself rather than others—behavior that can and does lead to failure.

CASE HISTORY 2.7

The Flixborough Explosion

Most of us intuitively understand that ignorance can lead to errors. Those who wish to succeed in a given engineering discipline must strive to acquire as much pertinent technical knowledge as possible. While we are in school, our instructors guide us through various topics, methodologies, and techniques that are fundamental to professional engineering practice. A major challenge that confronts each of us after we leave the classroom is to recognize what we do not, but should, know and understand.

The failure to recognize the need for specific expertise apparently led to the June 1, 1974, disaster that subsequently influenced operations throughout the chemical industry. On that date, the Nypro (UK) Limited Works plant in Flixborough, England, exploded, killing 28 men, injuring 89 other people, and causing extensive property damage to more than 1800 homes and 167 businesses.[16] The plant manufactured caprolactam (a substance used to create nylon) from cyclohexanone and cyclohexanol, which were formed by first mixing cyclohexane and a catalyst with oxygen. During this multistage process, liquid reactants overflowed from one of six large reactors to the next. Each reactor was approximately 16 feet in height, 12 feet in diameter, and contained about 20 tons of reactants.

Earlier water had been poured over one reactor in order to condense some leaking cyclohexane vapor. The water contained nitrates that caused a crack to develop in the reactor. The crack was discovered and a temporary bypass pipe was used in place of the reactor while it underwent repair, thereby allowing production to continue. This 2400-pound, 20-inch diameter temporary pipe was supported by scaffolding and included two bends. Bellows were used at each end of the pipe to provide some flexibility in the connections.

An expert in piping design should have been involved in the construction and installation of this temporary bypass connection. Unfortunately, those who were given the responsibility for maintaining the plant's level of production did not realize that they lacked the necessary expertise. Two months after the temporary pipe was installed, it twisted under slight additional pressure within the system, causing two holes to develop.[17] Within 50 seconds, between 30 and 50 tons of cyclohexane escaped to form a vapor cloud over the plant. This cloud was then ignited (possibly by a nearby furnace), destroying the plant's control room (in which most of the deaths occurred), a nearby office block, and other structures in the resulting explosion.

SUMMARY OF FAILURES AND TYPES OF ERRORS

- Failure to form a properly constituted and knowledgeable engineering team, resulting in inadequate knowledge and expertise (error in design due to conceptual misunderstandings, invalid assumptions, and faulty reasoning)
- Failure to protect populated areas (error in design due to invalid assumptions and faulty reasoning)

As we noted, it is difficult to recognize what we do not know. Because of this difficulty, it is important that any engineering operation have a broad

---

16. See Kletz (1988, 1991), Simon (1994), and *Petroleum Review* (1974).

17. In an alternative theory (known as the "eight-inch hypothesis") of the Flixborough disaster, it is suggested that the initial failure may have occurred in an eight-inch diameter pipe and not in the 20-inch diameter temporary bypass. However, the official Court of Inquiry concluded that the spontaneous failure of the temporary pipe did indeed lead to the explosion.

range of relevant expertise distributed throughout its staff. If all personnel have similar training, knowledge, and experience (e.g., all engineers are members of the same discipline), they may not realize the need for additional outside expertise when that need arises. As Kletz (1988) notes, there were many chemical engineers at the Nypro plant site in Flixborough, but not a single professional mechanical engineer,[18] who, although he or she may not have been an expert in piping, would probably have known that such an expert was needed for the temporary connection. Similarly, there is often a need for chemical engineers (and others) in operations that are primarily mechanical in nature. Engineering is increasingly interdisciplinary as new and more complex systems are developed to satisfy humanity's needs. It is therefore incumbent upon those who operate such systems to include a broad cross-section of relevant interdisciplinary experience among their engineering staff if safety, efficiency, quality, and productivity are to be achieved. The engineers' narrow view of the technical aspects of the task to be performed led to their violation of a principal rule in engineering: Never perform any task for which you are not qualified. Focusing upon self in any number of ways can be deadly.

In addition, the layout of any hazardous operation should seek to minimize human suffering and property damage in the event of an accident. Occupied structures that must be located near a hazard (e.g., explosions, floods, fire, toxic gas) should be designed to protect the people within their walls. In response to the Flixborough tragedy, many chemical process control rooms are now constructed to withstand an explosion. If possible, structures should be located safely away from hazardous sites.

Plants can be inherently safe (in which hazards are minimized or eliminated) or extrinsically safe (in which additional equipment is installed to increase safety).[19] Flixborough led to developments in both of these approaches toward safer operations. Companies throughout the chemical industry installed additional protective equipment and developed specific operating procedures to increase safety (i.e., plants were made more extrinsically safe). Companies also recognized that inventories of hazardous materials should be reduced to a minimum whenever possible, thereby lessening the danger in the event of an accident (i.e., intrinsically safe operations).

In addition, safer reaction processes have been developed for manufacturing certain chemical products; this approach to increased safety is known as substitution. Once rebuilt, the Nypro plant produced cyclohexanol via the hydrogenation of phenol rather than through the oxidation of cyclohexane (unfortunately, the manufacture of phenol also can be quite dangerous). Moreover, the concept of substitution can be extended beyond the realm of chemical reaction processes to other engineering operations: If there exists a safer yet practical method for accomplishing a goal, then use it! (Unfortunately, most of us do not seek safer ways to accomplish a task until an accident occurs.)

18. The one mechanical engineer on site had left the facility by the time of the temporary construction and his replacement had not yet arrived.
19. Kletz (1985, 1988).

CASE HISTORY 2.8

Maintenance Procedure Leads to the Crash of a DC-10

On May 25, 1979, American Airlines Flight 191 (a DC-10 jumbo jetliner) crashed at Chicago's O'Hare Airport; 273 people died in this crash, including two persons on the ground. The plane crashed after an engine and its pylon (the structure connecting the engine to the wing) tore loose from beneath the left wing.[20]

The subsequent investigation determined that a particular maintenance procedure had caused a crack to form in a flange, which in turn led to the engine's loss. In this procedure, the upper and lower bearings on the pylon are to be inspected and replaced if necessary. However, this apparently simple task actually became a formidable act to accomplish because of the enormous size and weight (13,477 pounds) of the combined engine/pylon assembly. Recognizing the effort that would be required, American Airlines decided that this maintenance should be performed whenever an engine is scheduled to be dismounted from a wing to undergo other regular servicing. American Airlines then developed a time-saving method in which the entire engine/pylon assembly is lifted from the wing by a forklift for servicing in a single step. (United Airlines had been using an overhead crane successfully to remove engine assemblies from their DC-10s.) McDonnell–Douglas (manufacturer of the DC-10) had reservations about the proposed forklift approach; they noted that the forklift should support the engine assembly at its exact center of gravity. In response, engineers at American Airlines accurately determined the precise location of this center of gravity via theoretical calculations. Unfortunately, forklift operators could only estimate (by visual inspection) the location of this center during the maintenance procedure. The flange of the aft bulkhead developed a 10.5-inch crack during servicing; fatigue growth of this crack during 430 hours of subsequent flight time eventually led to the loss of the entire engine assembly. Following the disaster, cracks were discovered in eight other DC-10s, also apparently due to the forklift operation.

The National Transportation Safety Board (NTSB) stated that the difficulties posed by required maintenance tasks on the DC-10 should have been foreseen by McDonnell–Douglas and that the flanges should have been designed in a way that would minimize possible damage.

Yet the plane should have been able to fly with only two of its three engines intact because the possible loss of an engine had been anticipated by the designers of the DC-10. Why, then, did it crash?

When the engine/pylon assembly broke away from the left wing, it severed the hydraulic cables (including those for the backup system) that con-

---

20. See Bisplinghoff (1986), Perrow (1984), and Mark (1994a).

trolled the leading edge flaps on this wing. These flaps, or "slats," which are normally extended at the time of takeoff to provide more lift for the plane, then retracted, sending the plane first into a stall and then a final roll. If the pilot had known that the flaps on one wing were retracted, he might have made adjustments in engine power that would have allowed the plane to land safely. Unfortunately, he did not know, because part of the lost engine assembly was the generator that powered the electrical stall and flap position warning systems. Hence, various elements of the plane's design that were meant to ensure safety were actually overcome by the loss of the engine assembly and the subsequent severance of numerous cable lines.

Since the 1979 Chicago accident was blamed by the NTSB on a maintenance procedure (which was indeed the direct cause of the engine loss but only one of the underlying factors for the plane's crash), McDonnell–Douglas did not have to modify the wing flap design. After all, the aircraft manufacturer had calculated that the probability of a simultaneous loss of engine power and slat damage during a take-off was less than one in 1,000,000,000. However, a similar accident had already occurred in Pakistan in 1977, two years before the Chicago crash. Then, in 1981, two more incidents occurred: First, on January 31, three leading edge wing slats on a DC-10 climbing on take-off from Dulles Airport were struck by an engine cowl and fan casing that broke loose following the fracturing of a fan blade. Fortunately, no significant damage resulted from this incident. Second, on September 22, cables to the leading edge slats were again cut as a DC-10 was taking off in Miami; the take-off was quickly and safely terminated.

After four incidents, McDonnell–Douglas reconsidered the likelihood that such accidents might occur and installed locking safety mechanisms that prevent dangerous slat retractions during take-offs.

SUMMARY OF FAILURES AND TYPES OF ERRORS

- Failure to develop safe maintenance and inspection procedures (error in original design due to misunderstanding of maintenance problem, invalid assumptions, and faulty reasoning; error in implementation (maintenance) due to invalid assumptions and misunderstanding of operator limitations)
- Failure to recognize inherent hazard (slat retractions with no warning signals transmitted) (error in design due to invalid assumptions and faulty reasoning)
- Failure to react to data indicating severity of hazard and failure to redesign when necessary (error in (re)design and implementation due to improper priorities, invalid assumptions, and faulty reasoning; possible error of intent)

It is imperative to react quickly and positively to data that indicate the presence of a significant hazard. Maintenance and inspection procedures can reduce the danger, but often only engineering redesign of a system is able to eliminate the hazard.

A disturbing focus upon self was exhibited by (a) the design engineers at McDonnell–Douglas when they failed to consider the truly Herculean efforts required for proper maintenance of the engine/pylon assembly, (b) the engineers at American Airlines who failed to recognize the limitations of forklift operators in performing maintenance work with sufficient precision, and (c) McDonnell–Douglas in failing to respond quickly to the hazards associated with the leading edge flaps. Whenever we fail to consider the needs of others in our engineering design work, the likelihood of failure looms.

CASE HISTORY 2.9

Bhopal: Simultaneous Failures Lead to Disaster

On December 3, 1984, a set of simultaneous failures at the Union Carbide plant in Bhopal, India, led to a massive loss of human life.[21] The official death toll was 1754, with approximately 200,000 injuries; however, other estimates placed the number of fatalities between 3,000 and 10,000, and the number of injured as high as 300,000. In addition, the long-term economic, social, and psychological damages to the community and to the company's employees were staggering.[22]

The plant manufactured carbaryl (the active ingredient in the pesticide known as Sevin) by first combining alpha-naphthol with methylamine to form methyl isocyanate (MIC), which then reacted with phosgene to form the final product. The MIC produced on-site was stored in two underground tanks, each with a diameter of 8 feet and a length of 40 feet. In addition, any MIC that failed to meet the company's specifications was stored in a third tank for reprocessing and refinement.

MIC is highly toxic; its vapor can cause blindness, lung damage, and other physical traumas as it burns flesh and attacks the nervous system. MIC is also highly reactive: It must not come in contact with water, zinc, copper, tin, iron, or other substances, or a runaway chemical chain reaction may occur. For this reason, it is often surrounded by a protective shield of dry nitrogen and stored at temperatures of $0°$ C or less.

Such an extremely dangerous substance clearly demands the use of a carefully designed, sophisticated, and well-maintained safety system in order to ensure workers' health. However, plant personnel (especially management) need to recognize that any safety system, no matter how sophisticated and

21. See Kletz (1988), Martin and Schinzinger (1989), Ingalls (1994), and Shrivastava (1992).
22. Shrivastava (1992) provides a particularly detailed description of the extensive social and psychological impact of the accident.

well-designed, is useless if it is turned off. At the Bhopal plant, many of the safety mechanisms were either inoperative or incorrectly set at the time of the accident.

In addition, state-of-the-art automatic safety devices were not used throughout the Bhopal plant. For example, workers were expected to detect MIC leaks with their eyes and noses.[23] Plant safety further deteriorated because of inadequate training of new workers and laxity in the use of protective clothing and equipment.

Apparently, the accident occurred after one worker neglected to insert slip blinds (backup metal safety disks) that are meant to act as secondary seals when filters are removed in the presence of a valve leak.[24] He had been instructed to flush some pipes with water and clean the filters between the storage tanks and the processing unit. He isolated the storage tanks from those parts of the system that were to be cleaned by closing certain valves but he failed to insert the safety disks. Water and chloroform, blocked by two clogged exit valves, then backed up and leaked through an isolation valve into the storage tank via a jumper line. This jumper line had been installed in May 1984 in order to allow venting of MIC gas from the tanks to a scrubber unit while repairs were being made to either the relief valve vent header or the process vent header (pipes within the system). On the day before the accident, the jumper line was opened because of repairs to be made to the process vent header, thereby allowing water to eventually enter tank E610 on December 3. This in turn led to the runaway reaction at such high temperatures and pressures that more than 25 tons of an MIC vapor/liquid mixture were released to the atmosphere, eventually covering a populated area of 15 square miles near the plant.

One might be tempted to conclude that human error in failing to insert the slip blinds was the cause of the Bhopal disaster. In fact, this was simply one element in a series of failures that led to the tragedy.

The initial failure was in allowing large amounts (more than 100 tons) of a deadly intermediate product (MIC) to be stored on site. Such products should be manufactured as needed so that large and unused inventories do not exist. The approximately 15,000-gallon capacities of the Bhopal MIC storage tanks were more than three times that of storage tanks at other plants. Subsequent to the Bhopal disaster, intermediate product inventories were reduced dramatically by many chemical manufacturing companies (including Union Carbide, which indicated a 74 percent reduction in their inventories of 36 toxic chemicals during 1985).

Furthermore, each tank was never to be filled to more than 60 percent of its capacity, thereby allowing the gas to be diluted in an emergency. The tank in which the runaway reaction occurred had been filled to between 75 percent and 87 percent of its capacity (the second failure).

---

23. Martin and Schinzinger (1989).

24. It remains uncertain how water first entered the MIC storage tank, thereby triggering the runaway chemical chain reaction leading to the tragedy. An alternative theory focuses upon the possibility of sabotage. See Kalelkar (1988).

The third failure was in allowing the MIC tank to become contaminated by whatever means that might occur. (Tank E610 was also contaminated by high levels of chloroform.) Given the hazards of runaway reactions, appropriate protective measures should have been in place and enforced so that such contamination could not occur.

Once the runaway reactions began, the temperature of the storage tank should have been controlled by the refrigeration system. Unfortunately, this refrigeration unit had been deactivated five months earlier so that the temperature of the MIC was relatively high—between 15° C and 20° C (the fourth failure). Plant personnel may have believed that there was less need for such safety equipment since the plant was temporarily shut down.[25]

The MIC vapor should have been absorbed by the vent gas scrubbing system as it was released through the tank's relief valve; however, the vent gas scrubbers were on stand-by status and could not become fully operable in time to limit the extent of the disaster (the fifth failure). The plant design also included a flare system to burn any vapor that escaped the scrubbers. Once again, this safety feature was undergoing repair and therefore inoperable (the sixth failure).

The set point (the trigger value for activation) of the alarm that warned of high temperatures was too high, so the alarm was never activated (the seventh failure).

The various protective subsystems (refrigeration, scrubbing, flare, and relief valve) may not have been able to contain the runaway reactions and the subsequent release of a vapor/liquid mixture even if they had been operative; that is, these subsystems may have been underdesigned.[26]

Finally, a heavily populated shanty town had been allowed to develop adjacent to the plant. The staggering tolls of death and long-term injuries would have been much lower if people had been prevented from living so near to the plant either via government action or through purchase of the adjacent land by the company itself, as suggested by Kletz.[27]

Was this disaster totally unexpected? Consider the following fact: 28 major MIC leaks (of a total 97) occurred at the Union Carbide plant in Institute, West Virginia, during the five years preceding the Bhopal disaster because of various equipment malfunctions and human errors.[28] One month before the tragedy in Bhopal, 14,000 pounds of an MIC and chloroform mixture leaked into the atmosphere from the Institute plant.

The exact and simultaneous combination of failures that occurred in Bhopal may have been unexpected. Nevertheless, the possibility of some of these failures certainly should have been recognized (and then minimized) in

25. Kletz (1988).

26. Ibid.

27. One might believe that a disaster of Bhopal's magnitude only happens once in many years. Unfortunately, this is only partly true: a single tragedy of Bhopal's magnitude is indeed rare; however, disasters of somewhat smaller magnitude occur with alarming frequency. For example, during 1984 alone and in addition to Bhopal, a fire at a liquefied petroleum gas processing plant killed 550 people in San Juanico near Mexico City, and 508 people perished when a petrol pipe burst and ignited in Cabatao, Brazil. [Kletz (1988)].

28. See, for example, Shrivastava (1992).

light of the accidents at Institute, West Virginia. As engineers, we must work to prevent the loss of human life by learning from our failures.

SUMMARY OF FAILURES AND TYPES OF ERRORS

- Failure to protect populated areas (error in design due to invalid assumptions and faulty reasoning)
- Failure to prevent contamination (error in design due to invalid assumptions and faulty reasoning)
- Storage of large amounts of toxic material (error in design and implementation due to faulty reasoning)
- Overfilling tanks (error in implementation: maintenance and use)
- Shutting down hazard prevention subsystems (error in implementation: maintenance and use)

The Bhopal tragedy is a stark reminder that we must focus upon the needs and safety of others in our engineering work. Safety systems must be well designed, maintained, and fully operative in case of an accident. In addition, hazards (e.g., large inventories of toxic substances) should be eliminated or at least minimized. Possible catalysts for dangerous events (e.g., contamination that could lead to a runaway reaction) must be prevented. And human beings should be distanced from any hazardous site.

Disasters often require a set of unusual circumstances (e.g., a series of system failures) to occur simultaneously. If any one of the various failures at Bhopal had not occurred, the disaster probably would have been much less extensive and harmful.

Since Bhopal, efforts have been made to enhance safety throughout the chemical industry via improvements in automatic warning systems, tank design and construction, site layout, production operations and monitoring, personnel training, storage operations, maintenance procedures, emergency response systems, management priorities, and many other areas.

CASE HISTORY 2.10

Rushing to Fill the Teton Dam Leads to Its Collapse

Another example of overconfidence in an engineering design leading to disaster is the collapse of the Teton Dam in Idaho on June 5, 1976.[29]

29. See Boffey (1977), Martin and Schinzinger (1989), and Shaw (1977).

Overconfidence in the fundamental integrity of the dam's design was reflected in the following decisions:

- The absence of continuous and adequate monitoring for leaks as the dam was quickly filled with water. The leak that led to the reservoir's collapse was detected only five hours before the disaster occurred.
- The initial filling of the dam while its outlet system remained unfinished. Normally, if a leak is detected, the main outlet can be opened to release water and prevent a total collapse. In the case of the Teton Dam, such a release of water could not be performed because the outlet system was inoperable.
- The casual transmission of data that indicated water may have begun seeping from the dam. This data was finally received by the dam's designers six weeks after it had been collected and one day after the dam had collapsed.

This overconfidence in design led to $400,000,000 in damages and, most important, the loss of 11 people.

SUMMARY OF FAILURES AND TYPES OF ERRORS

- Inadequate monitoring and inspection (error in implementation due to overconfidence)
- Inoperable hazard prevention system (error in implementation: operations and maintenance)
- Failure to communicate data in a prompt and appropriate manner (error in implementation via communication)

This case contains several important lessons in engineering design and implementation. First, we must never be overconfident about our designs. Overconfidence and impulsive behavior simply reflect a focus upon oneself rather than an appropriate concern for others.

Furthermore, careful monitoring and testing should be performed whenever a system is about to begin its service. Critical subsystems (particularly those relating to safety) must be operable before the system is activated. Finally, immediate corrective action should follow the collection of any data that indicate the existence of hazardous conditions.

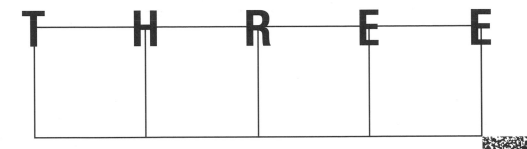

THREE

STRUCTURING THE SEARCH
FOR THE PROBLEM

The mere formulation of a problem is far more often essential than its solution, which may be merely a matter of mathematical or experimental skill. Albert Einstein

Be sure you are right, then go ahead. Davy Crockett

3.1 Focus on Function

Recognizing that a specific problem should be formulated if one is to develop a specific solution is important. Unfortunately, a problem statement sometimes becomes too specific; that is, the engineer describes the problem in terms of a particular solution, limiting creativity and often leading to simple modifications in an existing solution rather than a breakthrough design. Instead a problem statement should focus upon the function(s) to be performed by any viable solution.

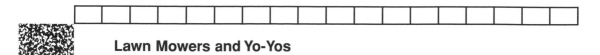

Lawn Mowers and Yo-Yos

The following anecdote emphasizes the importance of the problem statement in engineering design and the need to maintain one's focus on function.[1]

1. Although undocumented, this anecdote was provided by a reliable source.

During the mid-1960s, an engineering meeting was held at a gardening equipment firm. The engineers were informed that the lawn mowers manufactured by the company were losing market share due to increased competition from other manufacturers. In order to overcome this loss, the engineers were told to solve the following problem: Design a new lawn mower that will be the most popular product in the field.

The design engineers commenced work on this problem. After considerable time and effort, they generated a number of "new lawn mower" designs (e.g., a lawn mower with additional safety features (guards, automatic shut-off switches, etc.), one with multiple blades for more effective cutting, and another that could be folded into a compact shape for easy storage). Unfortunately, none of these designs were so revolutionary that it would take over the market and become the resounding success sought by the company's management.

At this point, one of the senior engineers suggested that they return to the original problem statement and rewrite it with a focus on function, as follows: *Design an effective means of maintaining lawns.*

The engineering design team then began to brainstorm[2] new ideas for achieving this modified objective. In brainstorming, one seeks to generate as many ideas as possible in a relatively short time, without stopping to evaluate any of these ideas during the brainstorming session. The assumption is that if one can generate 50 ideas in two hours (for example), at least some of those ideas will be worth pursuing. A number of new approaches for maintaining lawns were suggested, including a chemical additive for lawns that would prevent grass from growing to heights greater than two inches.

One engineer then commented that he had been watching his three-year-old son playing with a yo-yo on the lawn. The child was joyful as he swung the yo-yo in a circular horizontal arc above his head. The engineer suggested that perhaps a high-speed spinning cord could be used to cut grass. This idea led to the design of a breakthrough product that did indeed become extremely popular in the marketplace.

Can you guess what type of garden-care product was developed?

Do not define the problem in terms of an existing product; otherwise, you may simply generate variations of this product and fail to recognize a valuable new approach. Instead, focus upon those *functions* that are desired in a solution to the problem and formulate the problem statement in terms of these functions.

3.2   Formulating the Real Problem

Engineering is an iterative process: One should be prepared to reconsider assumptions, decisions, and conclusions reached during the earlier stages of the design process if any new results indicate the need to do so. For example, the initial problem statement may be too vague, ill-conceived, or simply incorrect. One must determine the object of a search before beginning to search; an incorrect problem statement is very unlikely to lead to the optimal solution of a problem.

---

2. Brainstorming is a creative technique that is reviewed in Chapter 7.

3.2.1 ■ The Dangers of Misdirected Search

A problem statement can be incorrect, leading to a misdirected search for solutions (as opposed to defining the problem in terms of a specific solution instead of a functional objective, thereby unwittingly limiting the search for solutions). For example, consider the following design anecdote.

Re-entry of Space Capsules

During the early days of the space program, it was understood that upon re-entry into the earth's atmosphere and due to frictional heating, the outside of a space capsule would rise to a temperature much higher than any known material could withstand. Thus, a research directive (i.e., a problem statement) was issued: *Develop a material that is able to withstand the extremely high temperatures of re-entry.*

When the *Apollo* moon landings occurred during the late 1960s and early 1970s, such heat-resistant materials had still not been developed. Yet our astronauts returned safely to earth! How could this be?

The answer is that the problem formulation eventually had been redirected towards the *true* objective, which was to protect the astronauts, not to develop a heat-resistant material.

This reformulation of the problem statement was the breakthrough step that led to the final solution. Researchers noted that some meteors reach the earth's surface without completely disintegrating. These meteors are not completely destroyed because their surfaces vaporize when they become molten, so that only some of their material is lost. Vaporization acts to cool a surface: This process is known as ablative cooling. Space capsules were then protected with heat-shielding material that vaporizes at high temperatures. The heat due to friction is thereby dissipated in the form of vapors.

Case History 3.1   *The Tylenol Case* also illustrates the hazards of misdirected search.

The problem statement is the most critical step in the engineering design process. Clearly, if you begin to journey to an incorrect site, you are very unlikely to correct your travel plans until you have arrived at the site and discovered your error.

We must strive to determine the real problem to be solved. In real-life engineering—as opposed to textbook problem solving—you may be asked to solve ill or incorrectly defined problems, you may be provided with insufficient or incorrect information, and you will be given a deadline for developing the "best" solution.

3. Fogler and LeBlanc (1995).

How, then, does one determine the real problem to be solved? Numerous heuristics have been proposed for properly formulating a problem. (A heuristic is often described as a rule of thumb, which simply means that it is a systematic approach that seems promising because it led to success in the past. However, unlike a law of physics, it does *not* guarantee success because it has not been verified through numerous problem-solving efforts and trials.[4]) The engineering design process shown in Figure 1.1 is, in fact, a heuristic.

Some heuristics for accurate problem formulation are as follows.

3.2.2 ■ The Statement–Restatement Technique

The statement–restatement technique[5] has four objectives, as discussed below.[6] One seeks to achieve these objectives by stating and restating the problem in different and innovative ways. The assumption is that by restating the problem in different forms, one will develop a deeper and more accurate understanding of the actual problem that needs to be solved.

1. **Determine the real problem (in contrast to the stated problem).** How would this be achieved? Fogler and LeBlanc (1995) recommend the use of various restatement triggers, such as:

   - *Varying the emphasis* placed on certain words and phrases in the problem statement. Next ask yourself if the focus of the problem itself has changed. If so, in what way? Is this a better focus?

   - *Substituting explicit definitions* of certain terms in the problem statement for these terms. Does this result in a different and more precise statement? If so, in what way? Why?

   - *Changing positive terms to negatives* and vice versa; for example, try to identify ways in which energy is being wasted in a plant rather than seeking ways to save energy. Has the focus of the problem statement changed? How? Why?

   - *Replacing persuasive and/or implied words* in the problem statement (such as "obviously" and "clearly") with the reasoning behind the use of such words. Is this reasoning valid? What is the evidence for such reasoning? If the reasoning is invalid, should the problem statement be modified? In what way?

4. An example of a heuristic is the guideline used by Conrad Pile, a merchant in Tennessee's Wolf River Valley during the early 1800s. Pile reasoned that a customer was probably industrious—and thus a good credit risk—if he had patches on the front of his trousers, in contrast to someone with patches on the back of trousers (the assumption being that anyone with patches on the rear must be lying down much of the time). This simple rule of thumb helped Pile to become a successful merchant and trader (Lee 1985).

5. See Parnes (1967) and Fogler and LeBlanc (1995).

6. We have modified the terminology used in the statement–restatement technique to be more consistent with engineering design problem solving.

- *Expressing words in graphical or mathematical form[7] and vice versa.* Has this improved your understanding of the problem to be solved? How? Why?

Table 3.1 illustrates the use of such triggers to generate seven solution paths that could be investigated.

2. **Determine the actual constraints or boundaries (in contrast to the given or inferred boundaries).** One way to achieve this objective is to relax any constraints that are contained within the problem statement. Design constraints usually should be quantitative rather than qualitative (e.g., "less than 100 lbs" as opposed to "lightweight"); however, both quantitative and qualitative boundaries can be relaxed. One can relax a quantitative boundary by simply adjusting the numbers (e.g, using "less than 200 lbs" in place of "less than 100 lbs"). Similarly, a qualitative boundary can be relaxed by simply replacing certain key words, (e.g., replacing the word "lightweight" with "not burdensome to move or lift").

   After relaxing a constraint, ask yourself if the problem itself has been modified in a significant way. If not, then work within the more relaxed constraints. Sometimes engineers work within perceived problem boundaries that are imaginary, thereby limiting the solutions that can be considered and making the problem-solving task more difficult. If the problem has changed, determine the cause for this change. One relaxed constraint may have altered the entire problem focus.

3. **Identify meaningful goals (in contrast to a set of given or inferred goals).** Sometimes one defines a problem in terms of particular goals that must be achieved by any viable solution. In engineering design, these goals are qualitative (e.g., "minimum cost", "safety") as opposed to constraints that are usually quantitative (e.g., "less than $10,000"). (Design goals will be discussed more fully in Chapter 4.) Are all the stated goals equally important? Usually, the answer to this question is no. Try to prioritize the goals and then focus upon the most critical ones as you rewrite the problem statement.

4. **Identify relationships between inputs, outputs, and any unknowns.** What is the desired output(s) or benefit(s) of the design? What are the inputs to the design (e.g., raw materials, people, equipment, money)? How will the inputs be transformed into the desired output(s)? What is unpredictable in the process? Why? What additional data needs to be collected?

   Restate the problem after answering these questions. The problem statement should then include what is known, what is unknown, and what is sought in a solution.

---

7. Graphical and mathematical models are more fully discussed in Chapter 6.

TABLE 3.1 Reformulating the problem via the statement–restatement technique.

Initial problem statement: Increase the number of commuters who use public transportation.

Varying the emphasis upon certain words and phrases
- **Increase** the number of commuters who use public transportation.
    Possible solution path 1: Increase the number of customers by decreasing the price (sell monthly passes at reduced price?).
- Increase the number of **commuters** who use public transportation.
    Possible solution path 2: Advertise the benefits (savings, safety, etc.) of public transportation
- Increase the number of commuters who use **public transportation**.
    Possible solution path 3: Provide or reserve highway lanes for buses.

Substituting explicit definitions for key words
- Increase the number of **people travelling to work each day** who use **trains/buses.**
    Possible solution path 4: Encourage employers to reward employees who use public transportation.
    Possible solution path 5: Provide office areas (desks, computers, etc.) in trains/buses for those who would like to work while commuting.

Changing positive terms to negatives and vice versa
- **Reduce** the number of commuters who use public transportation.
    Possible solution path 6: Investigate the reasons for people failing to use public transportation (e.g., high costs, discomfort, inconvenience) and try to eliminate these negative factors or minimize their impact.

Replacing persuasive and/or implied words and investigating the underlying reasoning; expressing words in graphical or mathematical forming
- The initial statement, Increase the number of commuters who use public transportation, assumes that such an increase obviously will be beneficial because the number of people using private transportation then will be reduced (i.e., this is the underlying reasoning for the problem statement). We might describe this situation in mathematical form as follows:

    Public commuters + private commuters = constant (total)

    Thus, if we increase the number of people in one category, the number in the other group must also decrease.
    Possible solution path 7: The above reasoning may be faulty. For example, new commuters may be entering the system each day, thereby adding to the total number. Perhaps instead we should try to reduce the total number of commuters (those who use both private and public modes of transportation), for example, by increasing the number of people who telecommute from home.

The statement–restatement technique assumes that one begins with a fuzzy or ambiguous formulation of the problem, which is very often the case. By focusing on the above objectives and activities, the initial problem statement can often be refined and corrected until one has a correctly formulated expression of the problem to be solved.

3.2.3 ▓ Determine the Source and the Cause

This heuristic simply states that one should consider the source of the problem statement.[8] Does or can that source (a person, a journal article, data) explain how the problem statement was developed? Furthermore, does the problem statement focus upon the cause of the problem or merely its symptoms? The engineer needs to focus upon the source or cause of the problem, just as a physician needs to treat the cause of an infection and not simply its symptoms.

Case History 3.2 *Blowing in the Wind: Boston's Hancock Tower* emphasizes the need to determine the true cause of an undesirable situation.

3.2.4 ▓ The Revision Method

Often the engineer is not confronted with a totally new problem, but instead must improve an existing product. The manufacturer may have invested substantial amounts of human resources, money, and equipment to design, develop, manufacture, distribute, and promote this product. The product cannot be abandoned without losing all of this investment. In addition, the product may be a current marketing success but it is expected to face increased competition from similar products in the near future. (This is particularly true if patent protection on the design is about to expire; see Chapter 5.) Hence, improvements in the design must be made. The revision method can be used when one is searching for a fresh perspective on this task or as a creativity technique for generating new ideas (see Chapter 7).

The method simply assumes that the focus of the design effort occasionally should revert to the product or solution (rather than the specific function to be achieved by the solution) *if* one has exhausted all efforts in reformulating the problem or *if* one needs to stimulate creative thinking in order to generate new design concepts. For example, the benefits of focusing upon function were discussed in the lawn mower design anecdote. However, if we were seeking specific ways to improve an existing lawn mower design, it might be wise to change the problem statement from *Design an effective means of maintaining lawns* to *Design an effective means of maintaining lawn mowers,* in which case we would now be searching for ways to minimize the lawn mower's maintenance and increase its durability. Such a focus might then lead us to identify the sources of damage to a lawn mower. For

---

8. See Fogler and LeBlanc (1995).

example, stones and rocks can cause damage to the blades and undercarriage of a mower. In order to overcome these hazards, an improved design might use a set of strategically placed rollers that would allow the mower to move vertically when traversing a rock, thereby limiting damage to the machine. By changing to a focus upon the product, additional needs of the customer (i.e., ease of use, minimum maintenance and repair costs, ease of storage) can be recognized and a revision of the existing design can be developed.

3.2.5 ▨ Present State and Desired State via Duncker Diagrams

Another strategy[9] for properly formulating a problem is to specify the present or problem state (PS), and the desired state (DS) or the solution state of the process or system under development. The engineer then modifies either the PS statement, the DS statement, or both until there is a satisfactory correlation between the two. For example, a student currently enrolled in both an engineering design and a physics course might begin with the following statements:

PS: I need to study physics.
DS: I want to earn an A in engineering design.

Unfortunately, the PS does not seem to have anything in common with the DS, in this example; that is, the PS appears to be irrelevant to the DS. However, upon reflection and reformulation, the student might rephrase these statements as follows:

PS: I need to study physics because I have an exam next week.
DS: I want to earn an A in engineering design.

A gap remains between the two statements, so further revision is needed:

PS: I need to study physics because I have an exam next week, but the only extra time that I can devote to physics is already scheduled for my term project in engineering design.
DS: I want to earn acceptable grades in both engineering design and physics.

We now begin to see the relationship between the PS and the DS, but further refinement is still needed:

PS: I am not sufficiently prepared for my upcoming physics exam and I also need to work on my term project in engineering design.
DS: I want to earn acceptable grades on both my engineering design term project and my physics exam.

9. See Higgins, Maitland, Perkins, Richardson, and Piper (1989) and Fogler and LeBlanc (1995).

Finally, a direct and obvious correlation between the PS and the DS exists. The student may now decide to investigate possible solution paths leading from the PS to the DS, such as *I must become more efficient; I will speak with my professors, and seek tutoring help; I will decrease the number of hours each week that I spend watching television and devote this time to my academic work; I will reformulate my term project so that less time is needed to complete it.* The student might also investigate some combination of these actions. (Of course, not all of these solution paths may be valid.)

Also notice that the final versions of the PS and DS statements match in terms of specificity (project, exam) and explicit requirements (acceptable grades).

Duncker Diagrams: General, Functional, and Specific Solutions Duncker diagrams[10] are a graphical tool that can be used to develop a set of matching PS and DS statements. These diagrams focus on developing solutions at three levels: general, functional, and specific.

General solutions are of two types: (1) those that require some action be taken in order to achieve the desired state, and (2) those that transform the DS until it matches the PS, thereby eliminating the need to achieve the original DS but perhaps necessitating some change in the PS in order to make it acceptable. Sometimes NO action whatsoever is taken because we discover that the present state is actually preferable to all alternative states: This is sometimes called the *null solution*.

Functional solutions are then generated without consideration given to feasibility; one simply considers any and all possibilities for solving the problem (i.e., these are "what if" solutions). Finally, these functional solutions are transformed (if possible) into specific solutions that are indeed feasible. Figure 3.1 presents an example of a Duncker diagram for the problem of public transportation used earlier in Table 3.1. (This type of diagram is similar to idea diagrams used to stimulate creative thinking and which we review in Chapter 7).

Consider another fictional situation in which Duncker diagramming might be helpful:[11]

When first marketed, Family-Comp computers were very popular for home use. Although admittedly offering only limited capabilities in such tasks as word processing and computation, these machines were very inexpensive.

However, sales began to decline and it was determined that the computers were becoming less popular as people realized that they had limited need

---

10. Duncker (1945).

11. This example is totally fictional; however, its structure is similar to one given by Fogler and LeBlanc (1995), which involves a breakfast cereal product.

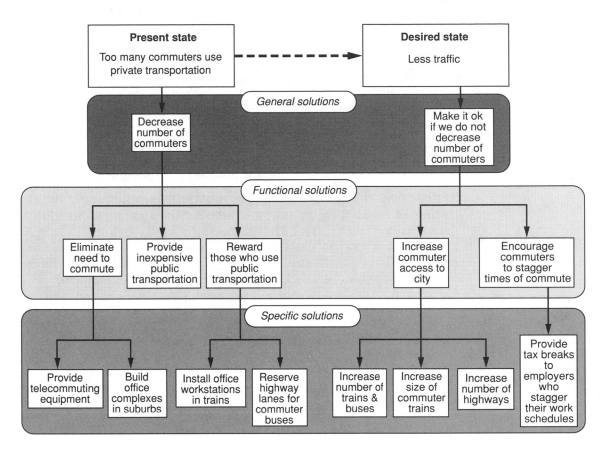

FIGURE 3.1 Duncker diagram for public transportation problem.

for such a machine when performing most household chores. Management then decided that a new strategy needed to be devised if their Family-Comp computers were to regain their former popularity in the marketplace.

We might then ask: What are some strategies that could be considered? For every benefit there is usually a cost. For example, perhaps the computers would become more desirable if additional capabilities were added to each machine; however, this strategy would necessitate additional company investments in design and development, and it would increase the retail price of each computer. A second alternative would be to (once again) persuade the customers that they need a Family-Comp computer for household tasks, perhaps through a massive promotional campaign emphasizing various unusual or new applications of the computer in the home. However, such an effort—even if imaginative—might not succeed. What, then, can be done to resolve this dilemma?

A Duncker diagram for this situation can be used to generate possible solution paths (Figure 3.2). Notice how this diagram links the three levels of solutions, together with the transformation from the present state to the desired state. It allows us to organize our thoughts and refine various solution paths. Each of the resulting specific solutions should then be carefully refined and evaluated.

3.2.6 ■ Other Heuristics

Some other heuristics or guidelines for problem formulation include:

- Restate the problem and explain it to another person (not someone on the design team); this may lead to a deeper understanding or another perspective of the task(s) to be performed.
- Are there similar problems that have been solved? If so, how does your problem differ (if at all) from these other problems? Why? Could some other existing solution be used or adapted to solve your problem?

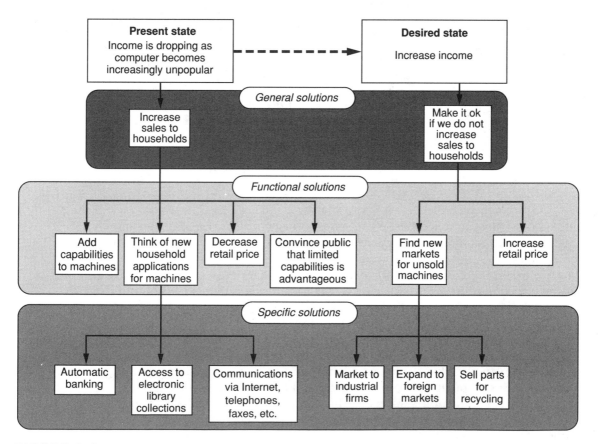

FIGURE 3.2 Duncker Diagram for fictional case study of Family-Comp computers.

The heuristics for problem formulation are similar to those used for modeling (see Chapter 6). In both cases, we seek to clarify both the problem itself and our search for its possible solution(s).

3.3 Kepner–Tregoe Situation Analysis

Charles Kepner and Benjamin Tregoe (1981) have developed a four-step problem-solving method that can be very effective for engineering design.[12] These steps are as follows:

- **Situation analysis (SA)**   The most urgent or critical aspects of a situation are identified first. Kepner–Tregoe situation analysis allows us to view the problem from different perspectives and at varying levels of detail. It is useful in focusing our efforts to determine what we know, which task(s) should be performed, and in what order these tasks should be completed.
- **Problem analysis (PA)**   The cause of the problem or subproblem must be determined (corresponding to a focus upon the *past* since we seek the preexisting cause of a current subproblem).
- **Decision analysis (DA)**   The action(s) needed to correct the subproblem and/or eliminate its cause must be determined (a focus upon developing a correction to the subproblem in the *present* time).
- **Potential problem analysis (PPA)**   The action(s) needed to prevent recurrences of the subproblem—and/or the development of new problems—must be determined (corresponding to a focus on the *future*).[13]

In other words, we should seek the cause(s) of the subproblems, decide on the steps to follow in developing viable solutions, and anticipate any negative impact that may result from implementation of the final solution(s). In this chapter we will review Kepner–Tregoe situation analysis and problem analysis; in Chapter 10 we will examine decision analysis and potential problem analysis

See Case History 3.3   *Apollo 13: Preventing a Tragedy* for an example of how effective the Kepner–Tregoe method can be when it must be applied in an abbreviated form because of time constraints.

3.3.1 ▓ Situation Analysis

In Kepner–Tregoe analysis, the various aspects of a situation are first evaluated through the use of three criteria: timing, trend, and impact. *Timing* simply means that one should determine the relative urgency of each aspect of a problem (i.e., which subproblem is most urgent and requires our immedi-

---

12. See Kepner and Tregoe (1981) and Fogler and LeBlanc (1995).
13. We discuss new problems that can result from a design solution in Chapter 9.

ate attention? Which are less urgent?). *Trend* is used to indicate the expected growth pattern of the subproblem: does the difficulty seem to be increasing or decreasing, and at what rate? Finally, *impact* refers to the expected negative consequences of the subproblem: How severe is the problem?

One of three values or rankings are then assigned to each subproblem for each of the above criteria; these rankings are H (high), M (moderate), and L (low).

The following multidimensional problem illustrates this first step in Kepner–Tregoe analysis.

The Water Tank Disaster

The following news story is based on the Nassau edition of *Newsday,* the Long Island, N.Y., newspaper (April 24, 1981) and OPLOW, American Water Works Association, vol. 7, no. 6, June 1981, p. 3.

Inadequate safety precautions and an accident inside an empty water tank caused the deaths of two workmen in New Jersey on April 23. At 4 P.M., a scaffold inside the tank collapsed and caused the two men painting the tank to fall to the bottom. Stranded there, they were overcome by paint fumes and eventually lost consciousness. John Bakalopoulos, 34, of Brooklyn, N.Y., and Leslie Salomon, 31, also of Brooklyn, were not wearing oxygen masks. The Suffolk County Water Authority's contract for the painting job specified that workmen wear "air hoods," masks connected to air compressors. The masks were available, but Bakalopoulos and Salomon had decided not to wear them because they were unwieldy. Instead, Bakalopoulos wore a thin gause mask designed to filter out dust and paint particles. Salomon wore no mask.

Peter Koustas, the safety man who was handling the compressor and paint feed outside the tank, asked a nearby resident to call firemen [sic] as soon as he realized the scaffold had collapsed. Then he rushed into the tank with no oxygen mask, and he, too, was overcome by the fumes and lost consciousness.

The men lay unconscious for hours as rescue efforts of more than 100 policemen, firemen, and volunteers were hampered by bad weather. Intense fog, rain, and high winds made climbing the tank difficult and restricted the use of machinery. Several men collapsed from fatigue.

Inside the tank, conditions were worse. Because of the heavy fumes, rescuers used only hand-held, battery-powered lights, fearing that sparks from electric lights might cause an explosion. Lt. Larry Viverito, 38, a Centereach, N.Y,. volunteer fireman, was overcome by fumes 65 ft (20 m) above the floor of the tank. Fellow rescuers had to pull him out.

Rescuer John Flynn, a veteran mountain climber, said he hoped he would never have to go through anything like that night again. For five hours he set up block-and-tackle pulleys, tied knots, adjusted straps on stretchers, and attached safety lines and double safety lines. The interior of the tank was as blindingly white as an Alpine blizzard—completely and nauseatingly disorienting. Fans that had been set up to pull fresh air into the tank caused deafening noise.

When Flynn first reached the tank floor, he stepped into the wet paint and began to slide toward the uncovered 4-ft (1.2 m) opening to the feeder pipe in the center of the floor. Flynn was able to stop sliding, but John Bakalopoulos wasn't as fortunate.

As rescuers watched helplessly, Bakalopoulos, still out of reach, stirred, rolled over, and in the slippery paint slid into the feeder pipe. He plunged 110 ft (34 m) to the bottom.

Bakalopoulos was dead on arrival at the University Hospital in Stony Brook, N.Y., Peter Koustas, rescued at 1:45 A.M. and suffering from hypothermia, died the following morning when his heart failed and he could not be revived. Only Leslie Salomon survived.

(Source: The above anecdote is reprinted with permission from M. Martin and R. Schinzinger, *Ethics in Engineering* 2nd. ed., New York: McGraw-Hill, 1989, pp. 116–118.)

There are many aspects of this problem that could be considered in a design evaluation if one seeks to prevent similar accidents. In fact, there are so many dimensions to this situation that it may be difficult to determine the most important subproblems on which to focus—or even to determine the "real" problem that needs to be solved. Kepner–Tregoe situation analysis may help, as shown in Table 3.2.

TABLE 3.2 Kepner–Tregoe situation analysis of the water tank problem.

Concerns	Subconcerns	Priorities		
		Timing	Trend	Impact
Painters	1. Scaffolding collapse	M	L	M
	2. Paint fumes	H	H	H
	3. Open feedpipe	H	L	H
	4. Air hoods not worn	M	M	M
	5. Lack of communication	M	L	M
	6. Inadequate training	M	M	H
Rescuers	1. Limited access to tank	M	M	M
	2. Weather conditions			
	a. Climbing hazardous	M	M	M
	b. Machinery limited	M	M	M
	c. Fatigue	M	M	H
	3. Paint Fumes			
	a. Limited lighting	M	L	M
	b. Unconsciousness	H	H	H
	4. Other tank conditions			
	a. Blinding	M	L	M
	b. Noise (fans)	L	L	L
	c. Open feeder pipe	H	L	H
	d. Slippery paint	M	L	M
	5. Training/preparedness	M	L	M

H = High level of concern/urgency; M = Moderate level of concern/urgency;

Although there may be additional concerns that could be identified (such as rescue expenses and the subsequent use of the water tank), let us assume that Table 3.2 includes the major elements of the problem. A review of the priorities given to each subconcern indicates that "paint fumes" received high levels of concern in all three categories (timing, trend, and impact) for both paint crew members and their rescuers. Therefore, we should initially focus on this most urgent aspect of the situation.

This first step in Kepner–Tregoe analysis further requires that we classify each aspect of a situation into one of three categories, corresponding to the next step (problem analysis, decision analysis, or potential problem analysis) to be performed in resolving the problem. In the case of the water tank problem, since we already know the cause of the paint fumes (the paint itself), Kepner–Tregoe problem analysis is unnecessary; we would move directly to decision analysis (see Chapter 10) and strive to eliminate the need for painting the tank.

3.3.2 ■ Problem Analysis

Kepner–Tregoe problem analysis seeks to distinguish between the following elements:

- **What** is the problem (and what is it not?)
- **When** did the problem occur (and when did it not occur)?
- **Where** did the problem occur (and where did it not occur)?
- What is the **extent** of the problem (number, degree, etc.)?

TABLE 3.3 Principal elements of Kepner-Tregoe problem analysis.

Consideration	Is	Is not	Distinction	Cause of distinction
Identity	What is the problem?	What is NOT the problem?	What is different between "is" and "is not"?	Why?
Location	Where is problem found?	Where is problem NOT found?	What is the difference in locations?	Why?
Timing	When did problem occur?	When did problem NOT occur?	What is different in timings?	Why?
Magnitude	Not much? How many? How extensive?	How little? How localized?	What distinguishes the affected and nonaffected elements?	Why?

This approach seeks to ask the right questions in order to determine the cause of the problem or subproblem. It focuses upon identifying a problem's *characteristics, timing, location,* and *magnitude.* One is also expected to consider the negatives (e.g., what is *not* the problem, when did it *not* occur) since this may lead to the cause of the problem. Table 3.3 summarizes Kepner–Tregoe problem analysis.

Specific forms of the questions (What? Where? When? Who? How? Why?) can be asked during Kepner–Tregoe problem analysis and are summarized in Table 3.4.

TABLE 3.4 Questions to ask during Kepner–Tregoe problem analysis.
Source: Adapted and modifed from Fogler and LeBlanc, 1995.

	Is	Is not
What	What is known?	What is NOT known?
	What was observed?	What was NOT observed?
	What are the constraints?	What are NOT constraints?
	What is important?	What is NOT important?
	What are the goals/objectives?	What are NOT goals?
	What can be expected?	What is NOT expected?
When	When did the problem occur?	When did the problem NOT occur?
	When must solution be implemented?	When is solution NOT needed?
	When did changes occur?	When did changes NOT occur?
	When were instruments calibrated?	When were instruments NOT calibrated?
Who	Who can provide more information?	Who can NOT provide information?
	Who is the customer?	Who is NOT the customer?
	Who performed (each) task?	Who did NOT perform (each) task?
	Who is source of information?	Who is NOT source of information?
	Who is affected by problem?	Who is NOT affected by problem?
Where	Where did problem occur?	Where did problem NOT occur?
	Where are input sources located?	Where are input sources NOT located?
	Where is equipment located?	Where is equipment NOT located?
	Where are products shipped?	Where are products NOT shipped?
	Where is customer located?	Where is customer NOT located?
Why	Why is problem important?	Why is problem NOT important?
	Why does solution work?	Why does solution NOT work?
	Why is there a problem?	Why is there NOT a problem?
How	How is problem related to other problems?	How is problem NOT related to other problems?
	How can a task be performed?	How can a task NOT be performed?
	How did problem develop?	How did problem NOT develop?

As an example of Kepner–Tregoe problem analysis, consider the following anecdote.

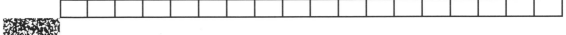

The Airplane Rash

In 1980 some Eastern Airlines flight attendants aboard a new type of airplane began to develop a red rash that lasted for about 24 hours.[14] The rash was limited to the attendants' arms, faces, and hands. Furthermore, it only appeared on routes that traveled over large bodies of water. In addition, only some—not all—attendants were affected on any given flight, but the same number of attendants contracted the rash on every flight.

When those flight attendants who had contracted the rash recovered and then flew on other (older) planes over the same water routes, no rashes appeared.

The attendents became anxious about this mysterious illness, and numerous physicians were asked to determine the cause of the problem—to no avail. In addition, the airplane cabins were examined by industrial hygienists, but again nothing was found to be amiss.

Given the above information, let's use Kepner–Tregoe problem analysis to determine the cause of the rash, constructing Table 3.5.

Based upon the above analysis, we conclude

- that the source of the rash must be related to materials found only on the newer planes,
- that these new materials come in contact with the hands, arms, and face of some (but not all) flight attendants, and
- that such selective contact is related to crew procedures that only occur during flights over large bodies of water.

Such an analysis leads us to the actual cause of the rash: life vests or preservers. These water safety devices were made of new materials, were found only on the newer planes and were demonstrated by selected flight attendants on routes over bodies of water.

Kepner–Tregoe problem analysis can be useful in our efforts to determine the cause(s) of a problem. As noted earlier, we will return to Kepner–Tregoe analysis in Chapter 10.

14. Fogler and LeBlanc (1995) and *Chemtech,* vol. 13, no. 11, (1983), p. 655.

TABLE 3.5 Kepner–Tregoe problem analysis of the airplane rash.

	Is	Is not	Distinction	Possible cause
What	Rash	Other symptoms	Skin	External contact
When	Flights over water	Flights over land	Different crew procedures	Materials
Where	New planes	Old planes	Materials, design	Materials, design
Extent	Some attendants	Other attendants	Different crew procedures	Materials
	Face, arms, hands	Other areas of body	Exposed skin	External contact

SUMMARY

- Formulation of a problem statement is a most critical step in an engineering design project since it will determine the direction in which the effort proceeds; the misformulation of a problem may result in a final design that is of little value or that may even be hazardous.
- Problem statements should focus upon the function(s) to be achieved by the desired design solutions.
- A number of techniques and strategies can be used to define the real problem to be solved; these include the statement–restatement technique, the source/cause approach, the revision method, and the present state–desired state (PS–DS) strategy via Duncker diagrams.
- Kepner–Tregoe situation analysis can be performed to evaluate various aspects of a situation in terms of three important criteria (timing, trend, and impact), thereby helping one to identify those issues that are most critical.

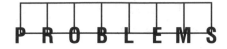

3.1 Apply one of the techniques for correctly defining a problem discussed in Section 3.2 to a problem of your choice. Present your results with clarity and identify any significant breakthroughs or corrections that were made as the result of applying the technique.

3.2 Apply one technique for correctly defining a problem to the following design problems:
a. Space capsule
b. Lawn mower
c. Boston's John Hancock Tower (see Case History 3.2)
(Do not use the same technique that was applied to the anecdotal problems in this chapter.)

3.3 Perform a literature search (books, journals, magazines, articles) and describe an example of engineering design in which a problem was initially defined incorrectly. (Hint: Investigate such references as *When Technology Fails* (edited by N. Schlager, 1994) and technical journals such as *Chemtech*.) What were the consequences of this misdefined problem statement?

3.4 Apply one or more of the techniques discussed in this chapter to correctly define the problem discovered in Problem 3.3. Show all work.

3.5 Apply Kepner–Tregoe situation analysis and problem analysis to a design case study selected by your instructor.

3.6 Select a problem (technical or nontechnical) of your choice and use the Kepner–Tregoe methods to analyze it. Present your results in tabular formats. (Include the source of the problem if it is taken from the literature.)

3.7 Toll booths on automotive turnpikes can result in a number of difficulties, such as unexpected delays, stress, unexpected lane changing and merging, accidents, and so forth. Apply Kepner–Tregoe situation analysis to determine the most critical issues associated with toll booths and their underlying causes. Present your results in tabular form.

CASE HISTORY 3.1

The Tylenol Case

Seven people were fatally poisoned in 1982 when they swallowed Tylenol cold remedy capsules that had been laced with cyanide.[15] This tragedy led to the widespread use of tamper-evident (TE) packaging for many products. Tamper-evident packaging allows the consumer to recognize if any tampering to the product has occurred. This packaging includes:

- PVC shrink neckbands (plastic seals around the lip or cap of a container)
- blister cards (plastic-coated cards that seal and display the product), and
- drop rings (plastic rings around the neck of a bottle from which a strip is torn, leaving a ring that must be slide upward to unlock the cap. The drop ring design is an example of a child-resistant closure that is relatively easy for adults to open.)

Johnson & Johnson (the manufacturer of Tylenol) sought to develop protective packaging that would make any tampering immediately evident to the consumer and thereby eliminate the potentially deadly consequences of such tampering. Their final package design included three different tamper-evident features. Johnson & Johnson felt confident that these features would protect their customers if every customer remained alert to any telltale signs of danger provided by the packaging.

In 1986 Johnson & Johnson discovered that they were wrong. A woman in New York died from cyanide poisoning after swallowing a Tylenol capsule. The company then realized that TE packaging was insufficient protection; the capsules themselves would need to become tamper-proof!

Note that the original problem—tampering with the packaging is not clearly evident to the consumer—was misformulated. The actual problem to be solved was prevent tampering with capsules. This failure to formulate the problem correctly then misdirected the search for a solution towards product packaging rather than the product itself.

However, there did not seem to be any way in which to completely prevent tampering with capsules. Confronted with this dilemma, Johnson & Johnson acted admirably in order to protect the public: They stopped manufacturing encapsulated products! Solid caplets, shaped like capsules but presumably tamper-proof, became the preferred form for such products.

---

15. Refer to such articles as *"Tylenol Case Raises Questions About Capsules"* (1986) and "T-E Packaging: Watchful Consumers Make It Work" (1993).

Sadly, not all capsule products throughout the pharmaceutical industry were removed from the marketplace. In 1991 poisoned Sudafed capsules took the lives of two people in Washington.

CASE HISTORY 3.2

Blowing in the Wind: Boston's John Hancock Tower

Completed in 1972, the John Hancock Tower in Boston is 60 stories in height (790 feet) with a steel frame and covered with 4.5' × 11' floor-to-ceiling reflective panels of double-glazed glass. Although its architectural design was aesthetically pleasing, its structural design contained hidden flaws.[16]

Beginning in November 1972, the glass panels began to fall from the building's facade. Eventually so many of these nearly 500-pound panels had to be replaced with plywood sheets that the structure became known as the "Plywood Palace." The building remained unoccupied for four years until the source of the problem could be identified and corrected.

Wind tunnels tests were conducted on a model of the building and its environment while additional tests were run on the building itself. The tests indicated that the wind loads were higher near the lower portion of the structure than near the upper floors—exactly the opposite of the designers' expectations (the glass panels were thinner at the lower floors, reflecting these expectations).

Finally, the source of the failure was determined. Laboratory tests and inspection of some damaged panels at the tower revealed that the outer glass sheet or light usually cracked before the inner light (see Figure 3.3). Since the wind loading on any single panel should have been evenly distributed to the inner and outer lights, it seemed surprising that the outer sheets exhibited a greater tendency to crack first. Chips of glass in the lead solder provided the clue to the mystery: The solder connecting the reflective coating to the inside of the outer light and to the lead spacer was too rigid. As a result, the outer light in a panel received the brunt of the loads, causing it to crack. Eventually, all 10,344 panes of glass were replaced with single sheets of tempered glass.

16. See Ross (1984) and Levy and Salvadori (1992).

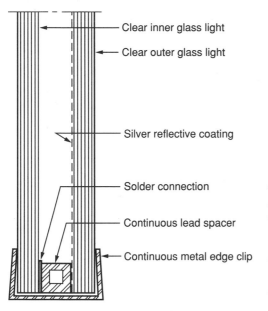

Clear inner glass light

Clear outer glass light

Silver reflective coating

Solder connection

Continuous lead spacer

Continuous metal edge clip

FIGURE 3.3 Epoxy between the spacers and the outer window sheet was too strong, failing to allow distribution and damping of wind-induced vibrations. *Source: Why Buildings Fall Down* by Matthys Levy and Mario Salvadori, copyright © 1992 by Mathys Levy and Mario Salvadori. Reprinted by permission of W. W. Norton & Company, Inc.

A Kepner–Tregoe problem analysis might have generated Table 3.6, which would then have directed attention to the reflective coating and solder used on the outer lights.

TABLE 3.6 Kepner–Tregoe problem analysis of the Hancock Tower's windows.

	Is	Is not	Distinction	Possible cause
What	Shattered windows	Other damage	Glass	Materials, design
When	Outer lights shatter first	All lights shatter simultaneously	Reflective coating on outer lights	Loading, materials (coating, solder, etc.)
Where	Lower floors (mostly)	Upper floors (fewer instances)	Height, thinner lights at lower levels	Distribution of wind loads
Extent	Complete loss	Frame and other	Glass	Materials

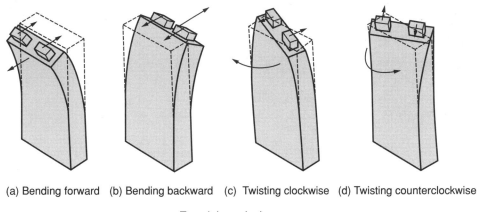

(a) Bending forward   (b) Bending backward   (c) Twisting clockwise   (d) Twisting counterclockwise

Tuned dynamic dampers

FIGURE 3.4 Mass damping system used in the John Hancock building.
Source: From Ross (1992)

The wind tests had uncovered another potentially more significant flaw in the building: Apparently it was too flexible along its longer edge.[17] Two tuned dynamic dampers had been installed to offset the building's motions along its shorter dimension. Each damper consisted of a 300-ton lead mass connected with springs to the structure and riding on a thin layer of oil on the 58th floor of the building. As the tower moved in one direction or the other, the inertia of these huge masses resisted this motion, effectively damping the movement as shown in Figure 3.4. Unfortunately, the buidling's potential movement along its longer dimension had been under-estimated because the designers had failed to consider the P-delta effect in which the structure's weight adds to the effect of the wind, effectively increasing the bending motion in a given direction. Diagonal steel bracing weighing 1650 tons was added to the building in order to offset this possibly hazardous condition.

Although very costly and potentially hazardous, the building's flawed glass facade led to the discovery of a potentially more dangerous shortcoming. Sometimes the symptoms of one problem in a design will provide us with the opportunity to discover far more critical flaws.

17. Some structuralists felt that this additional concern was insignificant and unwarranted.

CASE HISTORY 3.3

Apollo 13: Preventing a Tragedy

Apollo 13 was 205,000 miles from the earth when the crew noticed a sudden drop in the electrical voltage of Main B Buss, one of their two power-generating systems. They also heard a loud banging sound. The voltage then rose to normal. Even as Duty Commander John L. Swigert, Jr., reported these observations to NASA control in Houston, the voltage in the Main B system dropped to zero and the voltage in Main A began to fall.[18]

The astronauts were in deadly danger. Without power, *Apollo 13* would become their tomb in space. The engineers in Houston, realizing that action would need to be taken immediately in order to prevent a tragedy, began to evaluate what was known (e.g., voltage drops, a loud noise) as they collected further information from the crew.

Then, only thirteen minutes after the first voltage drop, Commander Swigert reported that their number 2 cryogenic oxygen tank was reading empty and that the ship appeared to be venting gas into space. Moreover, the ship's other tank also was losing oxygen.

Using an abbreviated version of Kepner–Tregoe problem analysis, the Houston engineers quickly deduced that a rupture in the number 2 oxygen tank would explain all of the observed phenomena. A loud noise, such as that first heard by the astronauts, would accompany a rupture in the tank. Furthermore, the gas being vented into space could be the lost oxygen from the tank (as well as that from the damaged number 1 tank). Finally, since *Apollo 13*'s power-generating systems depended upon oxygen in order to operate, a decrease in the oxygen supply would explain the observed loss of electrical power.

Once the situation was understood, appropriate actions could be taken to preserve the remaining oxygen and conserve electrical power, thereby allowing the crew of *Apollo 13* to return safely to earth.

Eventually, the reason that the oxygen tank ruptured also was discovered. Prior to launch, a ground crew had connected a heater inside the number 2 oxygen tank to a 65-volt power supply (instead of *Apollo 13*'s 28-volt supply), thereby fusing the heater's automatic cutoff switch to the ON position. When the astronauts then activated the heater, it remained on, eventually overheating the oxygen until the tank exploded into space.

Kepner–Tregoe analysis allows one to quickly evaluate a set of observations and identify the most likely underlying cause for a problem. As such, it is a trouble-shooting or diagnositc tool of significant value.

18. Refer to Kepner and Tregoe (1981).

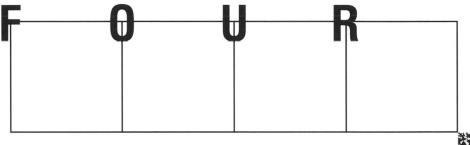

FOUR

STRUCTURING THE SEARCH FOR A SOLUTION

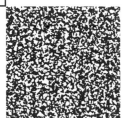

If you don't know where you're going, you'll probably end up someplace else.   Yogi Berra

O B J E C T I V E S

Upon completion of this chapter and all other assigned work, the reader should be able to

- Explain the need to properly structure the search for a solution to a problem.
- Design each task in a problem-solving effort so that it is most fruitful; that is, so that it will provide the most information or guidance.
- Use various attributes of the final solution state to guide earlier decisions made along the solution path.
- Define the most common or general engineering design goals.
- Differentiate between general and specific design goals.
- Differentiate between design goals and design specifications.
- Eliminate paths that do not satisfy the desired design goals and/or specifications.
- Use anthropometric, anatomical, physiological, and other types of ergonomic data to formulate design goals and constraints for human–machine systems.

4.1 The Need to Structure Your Search for a Solution

Before we turn our attention (particularly in Chapters 6 and 7) to various techniques that can be used to generate alternative design solutions, we must first recognize that an exhaustive search for possible solutions is neither practical nor wise.

In most problem-solving situations, one should not seek to generate and evaluate every possible solution to a problem since limitations must be placed on the amounts of time, effort, and money that can be invested in a project. Engineers always work within specific deadlines and search for solutions that lie within rigid constraints. As a result, they must design and develop solutions in an efficient manner. The only way to accomplish this goal is to (a) evaluate both the current problem state and the desired final solution state, and (b) develop a strategy for successfully traversing the path from the problem state to the solution state.

Problem formulation and specification of the desired attributes of the solution state were discussed in Chapter 3. We now turn our attention to the task of developing a strategy that will lead us directly from the problem state to the desired solution state. Such a strategy must *guarantee* success; that is, it must assure us that an acceptable (if not the best) solution to our problem will be found in a timely manner.

4.2 Designing the Search Strategy

In order to structure the path to be followed to a final design solution, we need to consider and compare aspects of the problem and solution states. Specifically, we need to

- Eliminate paths that do not satisfy the desired design goals and/or constraints, since these paths do not lead to viable solutions.
- Design our tasks so that they are most fruitful; that is, so that they will provide the most information or guidance in our problem-solving effort.
- Use various attributes of the final solution state to guide our choices in earlier decisions made along the solution path.

4.2.1 ▒ Eliminating Impossible Solution Paths

We begin with the need to eliminate impossible or unpromising paths between the problem state and the solution state. Consider the following puzzle of determining the weights of coins as an illustration of this need.[1]

1. Rubinstein (1985).

Twelve Coins and a Scale: The Easy Version

Among twelve coins, eleven are "normal" and one is heavier (in weight). We are given the following task: Determine which coin is the heavy one by using a balance scale no more than three times.

This problem is similar to one in engineering design since we must work within real constraints (i.e., we can use the balance scale only thrice) and we must find the "true" solution to the problem (i.e., we must determine the odd coin with certainty and not through chance, such as accidentally discovering the identify of the heavy coin through some unplanned event). What is a possible path or strategy that guarantees success?

One possibility is the sequence of weighings shown in Figure 4.1. We first place six coins on each side of the balance; the scale must then tip downward in one direction or the other (since the heavy coin must be among the twelve coins on the scale). Assuming that it tips downward to the right, we then know that the heavy coin is among the six on the right side of the scale. The number of candidate coins (among which the oddball lies) has been reduced from twelve to six by this first weighing.

We then weigh these remaining six candidates by placing three of them on each side of the scale. Once again, the scale must tip downward, indicating which set of three coins must include the heavy one. Only three coins now remain to be evaluated.

On the final weighing we place two of the remaining three candidates on the scale. If it remains balanced, the third candidate (off the scale) is the oddball; otherwise, the heavy coin is the one indicated by the imbalance of the scale.

No matter what happens during each of the three weighings, the above strategy guarantees that we will identify the odd coin. We eliminated impossible solutions (coins that could not be the heavy one) with each weighing until only one possibility remained.

Of course, other strategies also could be used; for example, we could have placed only eight coins on the scale during the first weighing (four on each side), with four coins off the scale. Given this initial configuration of coins, lay out a tree diagram (similar to the one in Fig. 4.1) of the coin configurations that should be used during the second and third weighings, along with the possible outcomes of each weighing, to guarantee success in identifying the odd coin.

4.2.2 ▨ Extracting the Most Useful Information

Next, we need to structure our problem-solving tasks so that they provide the most useful information as guidance in our subsequent decision making. Let's continue with the coin problem.

Twelve Coins and a Scale: The Difficult Version, Part I

The earlier version of the coin identification problem could be called easy because of the initial information with which we were provided; that is, we knew that the odd coin was heavier than the other eleven. Now let's consider the more difficult version of this

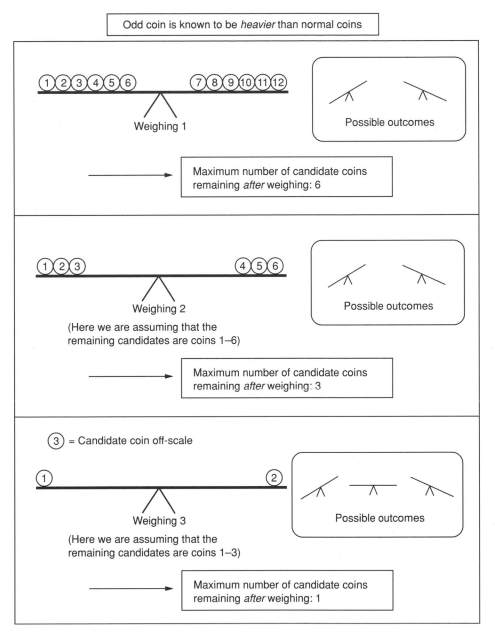

Odd coin is known to be *heavier* than normal coins

Weighing 1

Possible outcomes

Maximum number of candidate coins remaining *after* weighing: 6

Weighing 2

(Here we are assuming that the remaining candidates are coins 1–6)

Possible outcomes

Maximum number of candidate coins remaining *after* weighing: 3

③ = Candidate coin off-scale

Weighing 3

(Here we are assuming that the remaining candidates are coins 1–3)

Possible outcomes

Maximum number of candidate coins remaining *after* weighing: 1

FIGURE 4.1 The easy version of the twelve-coin problem.

problem in which we must both identify the odd coin, which could be heavier or lighter than the other eleven, *and* determine if it is indeed heavier or lighter than a normal coin.

We begin our work by recognizing the constraints in which we must work: only three weighings are allowed! Thus, we must extract as much information as possible from each weighing. This is an important fact to recognize: When you set up an experiment or any task that is designed to provide information, structure the task so that it will provide the most information and the most useful information possible. For example, our first weighing might consist of eight coins on the balance (four on each side) with the other four coins off the scale. If the scale remains balanced, then we know that the odd coin lies among the four that are off the scale. However, if the scale becomes unbalanced, we know that the odd coin lies among the eight on the scale. Thus, the number of candidate coins will be no larger than eight after this first weighing. In addition, we will know that certain coins are definitely normal (eight on the scale or four off the scale), knowledge that we will be able to use in designing the coin configurations for the second and third weighings. In other words, placing eight coins on the scale will allow us to extract much useful information from the first weighing.

Now: Try to solve this problem before continuing. Remember: You need to identify both the odd coin and whether it is heavier or lighter than the other eleven normal coins.

4.2.3 ■ Evaluating the Final Solution State

Finally, we should carefully evaluate the final solution state before deciding on the intermediate steps to follow along the path to a solution.

Twelve Coins and a Scale: The Difficult Version, Part II

One could try—through trial and error—to find configurations for the second and third weighings that would lead to success. Unfortunately, this trial-and-error approach to problem solving corresponds to an exhaustive search of all possible solution paths, with the search ending only if the correct path is found. This is not an efficient strategy as you may have discovered if you tried to solve this problem.

A better strategy is as follows: Consider the third and final weighing *before* you design any coin configuration(s) for the second weighing. We know from the easy version of this problem in which the odd coin was initially identified as heavier than the normal coins that the third weighing can be used distinguish between three candidate coins. In fact, we must have the number of candidates limited to three or less by this third and final weighing! Otherwise, we will not be able to determine the odd coin in this final weighing.

Thus, we must somehow eliminate all but three coins as possibilities with the second weighing. Since as many as eight coins may remain as candidates from the first weighing, this means that the second configuration of coins on the scale must guarantee elimination of five candidates. How might this be accomplished?

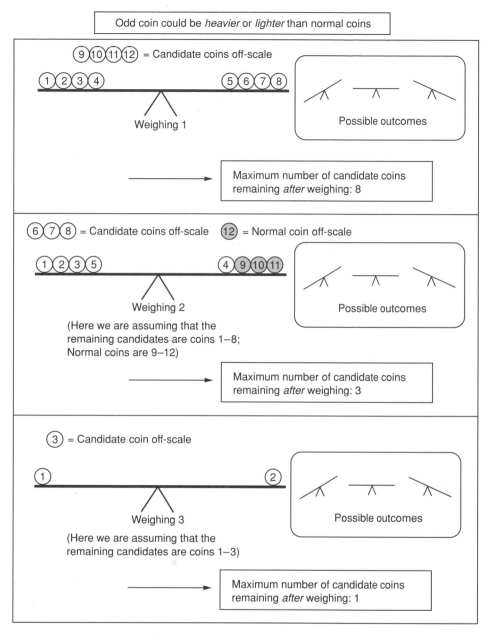

Odd coin could be *heavier* or *lighter* than normal coins

⑨⑩⑪⑫ = Candidate coins off-scale

①②③④   ⑤⑥⑦⑧

Weighing 1

Possible outcomes

Maximum number of candidate coins
remaining *after* weighing: 8

⑥⑦⑧ = Candidate coins off-scale   ⑫ = Normal coin off-scale

①②③⑤   ④⑨⑩⑪

Weighing 2

(Here we are assuming that the
remaining candidates are coins 1–8;
Normal coins are 9–12)

Possible outcomes

Maximum number of candidate coins
remaining *after* weighing: 3

③ = Candidate coin off-scale

①   ②

Weighing 3

(Here we are assuming that the
remaining candidates are coins 1–3)

Possible outcomes

Maximum number of candidate coins
remaining *after* weighing: 1

FIGURE 4.2   The difficult version of the twelve-coin problem.

Consider the solution shown in Figure 4.2. This configuration guarantees that both
the identity of the odd coin and whether it is heavier or lighter than a normal coin will
be determined within three weighings.

The coin puzzle illustrates an important lesson about engineering problem solving: Structure the search for a solution so that you are guaranteed success in a timely fashion. The need to structure one's search becomes even more important in an engineering project in which (unlike the coin puzzle) there usually are many viable design solutions that could be developed for a single problem. Structuring the search can help engineers to find quickly as many of these multiple solutions as possible, thereby increasing the likelihood that the presumably best design will be selected.

One effective strategy for structuring the search is (a) design your data collection and information gathering efforts so that the most and most useful information can be extracted to formulate the initial situation or problem state to be corrected; (b) next, evaluate the final solution state by identifying the desirable elements (i.e., functional capabilities or characterisitics) that should be part of any solution; (c) finally, select intermediate steps along the path from the problem state to the final solution state and develop the desired solution by following this path.

Notice how this same strategy was used to solve the coin puzzle. In fact, this prescribed search strategy is similar to the engineering design process itself. Phase 1 (needs assessment) of the design process generally corresponds to step (a) in the search strategy. During phase 2 (problem formulation), a set of design goals are identified, corresponding to those elements that are desired in any solution to the problem (i.e., the intermediate step [b] of the search strategy). The final three phases of the design process then correspond to step (c) of the search strategy.

We next focus upon the task of identifying design goals during phase 2—in essence, anticipating the desirable characteristics of the final solution state (step b).

4.3 Subdividing the Problem into Design Goals

After an engineering problem statement has been formulated, it should next be decomposed into a set of design goals that must be achieved by any viable solution to the problem. Formulation of design goals is a form of analysis, in that we are breaking the problem into its primary constituent components. The benefit of such a goals list is that it can be used as a sort of map, guiding us in our search for the best solution (i.e., it helps us to structure and perform the search).

To formulate a set of appropriate design goals, the engineer should gather as much detailed information as possible about the expected users, the environment in which the solution is to operate, and other relevant factors. This data collection might include further review of existing products, technical literature, and patents. In addition, knowledgeable individuals in the area of application should be identified and contacted for further information.

The environment in which the design is to operate should be described. The physical factors that may affect the design must be determined (e.g., temperature, humidity, dust levels, vibrations). Also, it may be necessary to place specific requirements upon the solution in accordance with the specific abilities or needs of a particular user group (e.g., the elderly).

4.3.1 ▨ General Goals

A number of common or general design goals are usually (although not always) associated with engineering problem-solving efforts. These goals include:

- Safety
- Environmental protection
- Public acceptance
- Reliability
- Performance
- Ease of operation (operating conditions)
- Durability
- Minimum maintenance
- Use of standard parts
- Minimum cost

Safety   A paramount goal of engineering is to produce truly safe products, systems, and processes. A design solution should not threaten the safety of those who will produce, distribute, and/or operate it. The following example illustrates the series of designs through which automobile windshields became progressively safer.

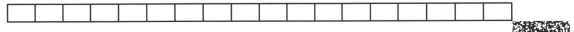

Laminated Glass for Automobile Windshields

The first automobile windshields were simply sheets of plate glass that could shatter and cause severe injuries to the driver and passengers of a car.[2] To reduce this risk, John C. Wood created laminated glass windshields in 1905 by gluing a layer of celluloid between two sheets of glass. However, there were certain weaknesses in this solution, including (a) the celluloid gradually yellowed, thereby decreasing visibility, (b) high manufacturing costs, and (c) curved laminated glass required a frame.

2. See Travers (1994).

St. Gobain, the French glassmakers, developed toughened glass in 1928 by first superheating the glass and then cooling it. This temperature-hardened glass had several advantages, including breakage into small pieces upon impact and no requirement of a frame. However, hardened glass is also quite rigid and refuses to yield adequately when struck, thereby increasing the likelihood that a driver or passenger will be injured in a collision. As a result, engineers continued their efforts to improve laminated glass windshields.

Later windshield enhancements included tungsten wires embedded within the laminated glass for defogging and deicing, wraparound construction, tinted glass, and differential zoning that ensured large uncracked portions of the windshield would remain directly in the driver's line of sight. Each of these developments was meant to increase the overall safety of the automobile.

Design engineers should always consider the safety of the user. However, many products (such as the automobile windshield) have this very important goal as their *primary* focus.

Environmental Protection   In engineering design, safety extends beyond our concern for those who will produce, distribute, or operate a product to the environment itself. Design solutions to technical problems should never harm our environment—from their production through their operating lives to their final disposal. Engineers must strive to anticipate and eliminate any hazards to the environment in their designs. This concern for the environment (sometimes dubbed "green engineering") has achieved the status of a general design goal, common to virtually all engineering problem-solving efforts. The following example illustrates the increasing importance of this goal in engineering design.

Environmentally Safe Batteries

Battery manufacturers are striving to develop long-lasting and environmentally safe power sources for automobiles, computers, and other electronic systems.[3] Batteries should be reliable, relatively inexpensive, lightweight, and preferably rechargeable; they also should not harm the environment during their use or disposal.

Lead-acid batteries used in automobiles are heavy and cumbersome; in addition, the acid used as the electrolyte in such batteries may leak, giving off hazardous fumes or causing corrosion. (Charged particles travel through the electrolyte to form an electric current.) Rechargeable nickel-cadmium batteries (used in flashlights and other

3. See Wald (1994).

devices) contain liquid sodium-hydroxide electrolytes that also may leak. Nickel metal-hydride batteries (found in some laptop computers) use relatively scarce materials, thereby limiting their appeal for inexpensive mass production.

However, one design does show promise: the lithium-polymer battery. In this design, a lithium alloy (acting as the anode) attracts positively charged ions while a carbon-oxide cathode attracts negatively charged ions. The current flowing between these two materials travels through a solid polymer electrolyte, thereby minimizing the hazard of leakage. These batteries are lightweight, fabricated in the form of lithium-polymer-oxide sandwiches that can be as thin as 0.02 inch. Hence, they may provide an environmentally safe, rechargeable, compact, lightweight, and inexpensive power source.

Lithium-polymer batteries already are used in hearing aids and other devices. Research is now focused upon developing larger and more powerful versions of these energy sources.

As engineers, we must always consider the environmental impact of the products and processes that we design. Case History 4.1 *Improper Waste Disposal at Love Canal and Times Beach* and Case History 4.2 *The Collapse of the Baldwin Dam* describe some consequences of failing to consider the interdependencies between a design and its environment.

Public Acceptance An estimated 80 percent of new products are commercial failures.[4] Such failures often result from one or more misunderstandings or preconceptions held by the design engineer about the customer's expectations for the product and its function(s). Thus, to increase the likelihood of product acceptance, the engineer should try to identify any such customer expectations that may exist about a proposed design.

In order to ascertain if expectations exist about a product, potential customers should be identified as precisely as possible in terms of such factors as their mental and physical abilities, economic means, and—most important—specific needs that are to be satisfied by the design solution. Although marketing surveys may be conducted by the engineering firm, the engineer also should endeavor to learn more about the customer group to be served. *Do not work in a vacuum; know and respect your customer!*

Once expectations about a product have been identified, these expectations should then be carefully evaluated to ascertain if they truly are in the best interests of the customer. Fulfilling expectations may fail to satisfy the real needs of the customer (needs which may not even be recognized by the customer). In such situations, customers must be persuaded that their preconceptions and preferences about a product are inappropriate and that the final design will satisfy their needs more effectively than they had expected.

Although it can be difficult to ascertain what expectations are actually held by the customer about a new design, it is critical to identify and address such expectations. Failure to do so will almost ensure that the design will be rejected in the marketplace.

4. Swasy (1990).

Shopping Cart Security Systems

Although 350,000 supermarket shopping carts are stolen annually at an estimated cost in excess of $30 million, most supermarket chains have not embraced the various designs that have been developed to prevent such theft.[5] These designs include (a) a fifth wheel that is activated and locked in position if the cart is pushed over uneven terrain (e.g., a bumpy road or cracked sidewalk). This fifth wheel, once activated, restricts the cart to purely circular motion, rendering it useless to the thief; (b) a vending system in which the user deposits a coin in order to unlock a cart in the supermarket; this deposit can be retrieved when the cart is returned to the store.

Supermarket operators have been reluctant to adopt these security devices because of their concern that customers could become insulted or frustrated by such systems. Engineers need to take such concerns about reactions to a product into account when developing a new design.

Concerns about public acceptance can prevent adoption of an otherwise effective design. Case History 4.3 *Perrier Water,* and Case History 4.4 *The Pentium Computer Chip* further illustrate the importance of this general design goal.

Reliability   The design solution should perform each and every time it is used; at least, this is the ideal level of reliability sought by the engineer. For example, a reliable camera is one that operates every time it is used.

Performance   An engineering design should perform well in the given environmental and operating conditions. A camera can be reliable (i.e., it operates every time it is used) but nevertheless perform poorly (it only produces blurred photographs). In order to enhance performance, the engineering team should anticipate (a) the reactions of the design to its environment (temperature, humidity, intensity of loads or applied forces, and so forth) and (b) its ability to operate according to specifications (e.g., duration of operation, power requirements).

Ease of Operation   The design should be easy to understand and easy to operate. (This goal sometimes is referred to as "operating conditions".) Once again, it is imperative that the user be identified; user populations with differing abilities and needs will require different levels of ease, as illustrated in the following example.

5. See Jefferson (1991)

Childproof Caps for Older Adults

The design of child-resistant caps for containers of prescription and nonprescription medicines must achieve two extremes in ease of operation because they are directed at two very different populations. These caps should be relatively easy for adults (particularly the elderly) to open, whereas they should be very difficult for children to open.[6]

In 1971, 216 children died after accidentally swallowing medicines and certain other household products. The following year, the Consumer Product Safety Commission (CPSC) introduced regulations that required child-resistant caps for containers of such products.

These regulations were somewhat successful: Although only 42 children died from such poisonings in 1992, hospital emergency rooms treated 130,000 children for accidental poisonings during 1993. In many of these cases, adults had removed or replaced child-resistant caps on containers when they became frustrated with the difficulties in opening these caps, thereby inadvertently exposing children to danger. (Even many young adults have found child-resistant caps difficult to open.)

The CPSC modified their original regulations for the first time in 1995. An age test that includes older adults as well as children must now be applied to cap designs. The test is as follows: 90 percent of 100 adults between 50 and 70 years old must be able to open the package within five minutes; these adults must then be able to open the package again within one minute. In addition, 80 percent or more of 100 children under the age of five years should not be able to open the package within ten minutes.

The redesigned caps are expected to require less force to open. Such caps may include tabs that must be pushed inward as the cap is turned, in contrast to older designs that required the user to simultaneously push and turn the entire cap. Such designs (already used for Scope mouthwash and Aleve pain reliever) should help to satisfy the needs of the elderly while protecting the young.

Case History 4.5 *Two Routes to Simplicity: Camcorder Redesigns* also demonstrates that ease of operation can be critical to the success of a design.

Sometimes flexibility in operating procedures can lead to disaster. Engineers must recognize the burden that options may place upon the operator of a system. In particular, if disastrous consequences may result from operator error in choosing from among a set of options, then either (a) all but one option must be eliminated (i.e., the operator is thereby restricted and can perform only the correct action) or (b) inappropriate selections, if allowed to be made by the operator, must be prevented from causing harm. The following case demonstrates the importance of this fact.

---

6. See Davidson (1995).

The DC-8 Crash

The DC-8 aircraft allowed pilots to raise ground spoilers (metal flaps on the plane's wings) after touchdown, thereby decreasing lift, by pulling a control lever. Pilots were also given the option to lift this lever before touchdown, after which the spoilers would be raised automatically once the plane touched the ground. Unfortunately, 109 people were killed when a pilot *pulled* the lever *before* touchdown, thereby reducing lift and crashing the plane.[7]

From a design perspective, one could argue that the automatic option to arm the spoilers prior to landing simply eliminated the need for pilots to remember to pull the lever after touchdown. However, this option actually transferred the need for critical recall and appropriate action by pilots from "after touchdown" to "before touchdown." It required pilots to remember which of two actions should be performed while the plane was still in flight, thereby enhancing the chances for disaster if an error was made.

After two (or more) planes crashed because of pilots pulling the control lever before touchdown, McDonnell–Douglas (manufacturer of the DC-8) installed locking mechanisms that prevented the ground spoilers from being raised while the plane was in flight.

Durability Any product must be designed to resist wear and operate properly for a specified amount of time. The durability of a system's components may vary, but the design should include provisions for minimizing wear of certain key components, such as a provision for lubricating moving parts or use of specific wear-resistant materials.

Use of Standard Parts Many standard or "off-the-shelf" parts are available in the marketplace from manufacturers. These should be incorporated into a design whenever possible, since it is usually less costly (in terms of money, effort, and time) to purchase these ready-made components than to fabricate them from raw materials. In addition, an investment in expensive tooling equipment and specialized personnel may be necessary for such custom fabrication. Such expenses may result in the design being rejected as infeasible.

The design engineer should be familiar with as many different types of components (and their manufacturers) as possible. For example, relays, solenoids, motors, gearing, piping, leadscrews, washers, and hundreds of other standard components are available in the marketplace. Catalogs are available from most manufacturing firms that describe the components available for purchase. (Many of these catalogs will be sent free of charge to those who request a copy.)

In addition, handbooks should be consulted for guidelines on the use of such components. For example, standard values for such component char-

7. See Eddy, Potter, and Page (1976) and Kletz (1991).

acteristics as size, electrical resistance, and viscosity often can be found in handbooks. These recommended values help us to select components that will perform properly under a given set of operating conditions. Failure to use such values may invite disaster, as shown in Case History 4.6 *O-Rings and the* Challenger *Disaster.*

Familiarity with a wide variety of components and their proper use facilitates the generation of creative yet feasible design solutions since it broadens one's technical knowledge base. Handbooks and other standardized data can guide engineers along the safe and tested path when using off-the-shelf components. Remember: Learn from others what is already known and then apply it in new and wonderful ways.

Minimum Cost   Our list of general engineering design goals also includes the need to minimize costs whenever possible, for otherwise (a) a proposed design may be deemed economically infeasible and never produced, (b) unexpected costs may cause a design to become economically infeasible after production has begun, leading to its commercial failure, or (c) costs may be reduced in an arbitrary fashion, leading to design or manufacturing flaws. Costs include those for the design and development of the solution, its production, its distribution, advertising and promotion, and so forth.

The challenge in reducing costs is to do so while maintaining or increasing the quality of the design. An example of a product in which this challenge was met is described in Case History 4.7 *Fax Machines.*

Inappropriate cost cutting can be very expensive. Consider the following example.

The Hubble Space Telescope Failure

The $1.5 billion Hubble Space Telescope is an example of a design that suffered because of budget constraints.[8] After this telescope—weighing 12 tons and measuring 43 feet—was launched into orbit in April 1990, scientists discovered that the performance of its primary mirror (94.5 inches in diameter) was producing blurred images; astronauts would have to journey into space to correct the problem.

The Hubble mirrors, which work in conjunction as a system, had never been tested before the launch. In 1977 NASA had awarded a $64.28 million contract to the Perkin-Elmer Corporation for grinding this mirror. The Eastman Kodak Company had submitted a bid to perform this task for $99.79 million. Kodak—unlike Perkin-Elmer—proposed that the entire mirror system be thoroughly tested before launch; such testing was reflected in the higher cost of their (losing) bid.

8. See Broad (1991).

NASA chose the route, and the bid, that appeared to be less costly because the agency needed to invest its limited funds in the developing space shuttle program. Unfortunately, this 1977 decision to minimize costs in an inappropriate way led to a loss of public confidence in the NASA program when the Hubble system failed in 1990.

All costs need to be identified and estimated with care, and a cushion should be included in the overall budget to cover various unanticipated expenses. Unexpected or unrecognized costs can be fatal to a product, as seen in the following case.

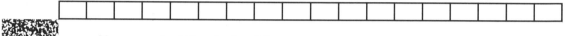

Unexpected Costs for Diapers

Customers were enthusiastic about UltraSofts diapers when they were introduced in 1989 by the Weyerhauser firm. After all, UltraSofts had a pleasing clothlike external cover (unlike other brands with plastic covers) and superabsorbent pulp material that was actually woven into the pad itself.

Marketing tests indicated that parents preferred UltraSofts to other leading brands by a two-to-one margin. Despite the apparent desirability of this new product, it became a commercial failure because of unexpected costs and financial limitations.

Some of the diaper's components (such as its lining) had been fabricated on a pilot basis by suppliers; however, these suppliers were unwilling to increase production without long-term contracts with Weyerhauser, contracts that could not be awarded during the early production period.

Perhaps more important, production problems developed when the superabsorbent material was sprayed into the pad; some of this material missed the pad entirely, instead coating—and ultimately corroding—the machinery's electronic components. Other production problems (e.g., transformers overheating) also became apparent, leading to a 22 percent increase in the wholesale price of the diapers to retailers. This led to a reduction in profits for the retailers and made it more difficult for UltraSofts to compete for shelf space with the leading brands.

(Recognizing that the superabsorbent material could have many uses, Weyerhauser looked for other marketing opportunies. For example, this material can be used to wrap fish and other fresh foods; in Japan, it has been used in place of sandbags to absorb water in underground tunnel systems.)

Minimum Maintenance and Ease of Maintenance Two other general engineering goals, related to durability, are to minimize necessary maintenance work and to ensure that such work is as easy to perform as possible. Maintenance

9. See Swasy (1990).

is required of virtually all engineered products, and it can be expensive in terms of materials and labor.

Many people are reluctant to perform maintenance tasks diligently, leading to failures, worn parts, or an unnecessarily brief operating life for the engineered system. Moreover, maintenance often includes the regular inspection and replacement of parts, reminding us that systems should be designed to allow for proper and easy inspection. For a successful, competitive, and safe product, maintenance should be minimized (and, if possible, eliminated) and made easy.

The following case illustrates the impact that improper maintenance can have on a design.

The Mianus River Bridge Collapse

On June 28, 1983, a 100-foot section of eastbound Interstate Highway I-95 across the Mianus River Bridge in Greenwich, Connecticut, suddenly fell to the river far below (Fig. 4.3a).[10] Three people were killed. Subsequent investigation indicated that three factors may have contributed to the disaster:

1. the skewing of the bridge along a diagonal between its east and west abutments, possibly decreasing the structural integrity of the structure;
2. the use of only two nine-foot-deep plate girders to form the bridge span, thereby providing no back-up support if one of these girders failed; and
3. the effect that improper maintenance may have had on the bridge.

The span that collapsed was supported from an adjoining span by a pin-and-hanger assembly as shown in Figure 4.3b. Apparently, corrosion due to rust had weakened the 25-year-old assembly until it could no longer support the live loads of the traffic. This corrosion may have been accelerated when drains along the bridge were paved shut and gutters were not replaced, allowing water mixed with road salt to flow over the assembly. The developing corrosion went undetected by inspectors because the bridge design failed to provide easy access for such inspection. The findings of a jury trial in 1986 indicated that poor maintenance had probably led to the tragedy.

The partial collapse of the Mianus River Bridge resulted in enhanced bridge maintenance and inspection procedures, together with the subsequent strengthening of those bridges with pin-and-hanger assemblies.

Maintenance and inspection must be priorities in the design of any engineered system.

10. See Robison (1994c).

FIGURE 4.3a On June 28, 1983, a 100-foot section of Interstate Highway I-95 across the Mianus River Bridge in Greenwich, Connecticut, suddenly fell to the river far below. Three people were killed. *Source:* Corbis-Bettmann.

Satisfying Multiple Goals A product may fail because of its inability to satisfy several general engineering design goals simultaneously (e.g., minimum cost together with minimum maintenance), as demonstrated in the following case.

FIGURE 4.3b Pin-and-hanger assembly. *Source: Why Buildings Fall Down* by Matthys Levy and Mario Salvadori, copyright © 1992 by Mathys Levy and Mario Salvadori. Reprinted by permission of W. W. Norton & Company, Inc.

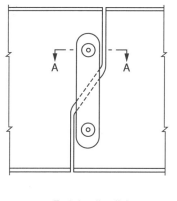

End detail at link

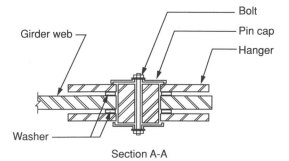

Section A-A

Coca-Cola's BreakMate

Coca-Cola's BreakMate system was a miniature soda fountain dispenser designed for use in the one million offices throughout the United States with 45 or fewer employees.[11] These offices were deemed too small to sustain a standard soda vending machine, and the BreakMate was intended to provide coca-cola an alternative commercial route to these millions of potential customers. In addition, the machine could have served as the forerunner of a household soda fountain. By successfully marketing such home fountains, Coca-Cola would have made considerable progress in addressing the environmental difficulties associated with cans and bottles.

In the BreakMate system, water was chilled, carbonated, and mixed with syrup. The soda was then dispensed in cups.

Problems began when the cost of each machine grew substantially beyond the early estimates. These increases were largely due to the rise of the deutsche mark relative to the U.S. dollar since this system was manufactured in Germany. This rise made it difficult for the BreakMate to be profitable when distributed to those offices with fewer than 10 employees, which accounted for 75 percent of the targeted market. In addition, the sales volume of the cola syrup itself was less than expected, reducing profits for the machine's distributors. Thus, the design faced a combination of high costs and reduced sales.

Furthermore, it was reported that the system was difficult to maintain. In some offices, the drip trays that collected overflow were not cleaned when necessary; this led in turn to such problems as insect infestations, which further reduced demand for the product.

Although 25,000 units had been distributed by 1993 at a cost of $30 million to Coca-Cola, the BreakMate was viewed as a failure. It appears that office and home soda fountains will remain limited in number until a truly cost-effective and virtually maintenance-free design can be developed.

Please note that the general engineering design goals outlined here are not all-inclusive; other common goals can be identified. However, this outline certainly contains many of the design goals that most frequently must be achieved by any viable solution to a technical problem.

11. See Emshwiller and McCarthy (1993).

4.3.2 ■ Specific Goals

In addition to the preceding general engineering goals, each design must achieve specific goals that pertain to the particular problem under consideration. As an example, consider the problem of restraining the drivers and passengers of an automobile at the moment of impact during a crash. One might be tempted to define the goals in terms of a specific solution; for example, improved seat belts or air bags. This would be a mistake since it would limit the imagination of the design engineer. Instead, the problem might be divided into time intervals: before impact, during impact, and after impact. Such a division would then suggest certain specific goals, such as[12]

- Before impact, the design must not severely restrict the movement of the user or cause discomfort.
- During impact, the design must operate effectively and automatically since the user cannot be expected to initiate or control its operation under such conditions.
- After impact, the design must not restrict the user from escaping the vehicle quickly.

Many other specific goals (regarding weight, size, shape, speed, and other factors) may be important as well. Consider the following example.

Duplicating Human Capabilities: The Voice Synthesizer

A specific goal that often is sought in a engineering design is the simulation of some particular human capability (such as speech, grasp, vision, or logical thought). Many of today's industrial robots have replaced people in the performance of difficult, tedious, and/or dangerous tasks. The voice synthesizer is a device that has been developed to imitate—and in certain situations replace—human voices.[13]

An 1830 device known as the Euphonia used bellows, gears, and levers to roughly duplicate the human voice as it greeted people with "How do you do?" and recited the alphabet; it was developed by Joseph Faber. More sophisticated devices— primarily designed to serve the blind as aids in reading—were made possible with the computer. The MITalk System (developed at the Massachusetts Institute of Technology during the 1960s) was able to translate text into synthesized speech. Beginning in 1976, the Kurzweil reading machine (created by Raymond Kurzweil) became extremely popular among blind users due to its unlimited vocabulary and amazing reading capabilities.

Of course, computer chips are now commonly used in automobiles, games, telephone answering machines, and other devices to produce synthesized human speech.

---

12. Voland (1987).
13. Based upon Travers (1994).

Such artificial speech can provide instructions, warnings, and other information in a friendly and effective manner.

Both general and specific design goals should be achieved by the final solution to an engineering problem, as shown in the following example.

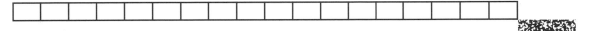

Reseeding Forests

Reseeding of forestland in the form of tree farms began in 1941 when the first such farm was dedicated by Governor Arthur B. Langlie in Montesano, Washington. Tree farms represent the important recognition that our forests are not an inexhaustable resource.[14]

However, such reseeding must be performed with both speed and relative ease if it is to be effective and economically feasible. Although speed of operation often appears as an important factor in many engineering problems, it is not so universal that it could be given the status of a general goal; rather it is a specific goal unique to particular problems.

The challenge, then, was to find some way to perform reseeding with sufficient speed and relative ease. One method to accomplish this task was developed during the 1970s by Philip Hahn, who designed the "double-barreled shotgun" approach for reseeding Douglas fir trees. One barrel consists of a dibble with which one can punch holes in the ground; the second barrel then drops a seedling (with its roots in compacted soil) into the hole. This tool effectively doubled the speed with which reseeding could be accomplished.

Specific goals help us to define the particular problem that is to be solved and maintain our focus upon the functions that are desired in a design solution. Case History 4.10 *The Development of Band-Aids* demonstrates the importance of specific goals in the development of a successful design.

4.3.3 ■ Continuously Re-evaluate the Goal List

If the goals that must be achieved by a design solution are not carefully defined, then the development of the so-called best solution will be very difficult or even impossible to achieve. During the entire design process, these project goals should be continuously re-evaluated. Initial goals may be

14. See Woog (1991).

divided into more specific tasks. Additional goals will likely be recognized as the engineer becomes more familiar with the functions that a viable solution must perform. The list of design goals will remain in a state of flux until the final solution to the problem has been developed, tested, and implemented.

4.4 Working within Constraints: Design Specs

The next task is to identify the constraints or specifications (specs) associated with each design goal. These specs represent the quantitative boundaries within which our search for a solution must be conducted. Once again, we seek to structure the search so that it can be performed in an efficient manner while ensuring that the final design will be truly effective as a solution to the problem.

For some goals, it is understood that they should be achieved completely (e.g., the goal of safety); that is, the corresponding specs are simply 100 percent achievement of these goals. Other goals (such as lightweight or minimum cost) must be defined in quantitative terms so that one has specific targets at which to aim when trying to satisfy the goals and so that one can measure the degree to which the goals have been achieved. For example, the corresponding specs for lightweight and minimum cost might be less than 25 pounds and less than $1,000, respectively.

Specs may be of several types:

- *Physical,* including space allocation or dimensional requirements, weight limits, material characteristics, energy or power requirements, etc.
- *Functional or operational,* including acceptable vibrational ranges, operating times, and so forth.
- *Environmental,* such as moisture limits, dust levels, intensity of light, temperature ranges, noise limits, potential effects upon people or other systems that share the same environment, etc.
- *Economic,* such as limits on production costs, depreciation of equipment, operating costs, service or maintenance requirements, and the existence of any competitive solutions in the marketplace.
- *Legal,* such as governmental safety requirements, environmental or pollution control codes, and production standards.
- *Human factors/ergonomics,* including the strength, intelligence, and anatomical dimensions of the user (see Section 4.5 for a fuller discussion of the use of ergonomics in design).

Constraints can force us to use system characteristics creatively and effectively in the development of an engineering design, as shown in the next example.

Pulse Oximetry: Measuring the Oxygen Content in Blood

Working within constraints can help to inspire engineers to develop solutions in which these constraints are used to advantage. For example, the need to measure the amount of oxygen in a hospital patient's blood noninvasively, continuously, and without pain led to the development of pulse oximeters.[15] (Notice that these goals must be achieved completely; that is, the specs correspond to 100 percent achievement of each goal.)

First developed in Japan, pulse oximetry was later commercialized in 1981 by Nellcor, Inc., a medical equipment firm in California. In this technique, a sensor is attached to the patient's finger, toe, nose, or earlobe. A detector then measures the amount of red light from a diode that is transmitted through the skin. More light will be absorbed by dark red blood that is relatively low in oxygen, whereas more light will be transmitted through blood that is rich in oxygen and bright red. Thus, a blood variable (color) that depends upon oxygen level is used to real advantage.

Now used in hospital operating rooms and intensive care units, pulse oximeters have reduced the fatality rate of patients under anesthesia. In addition, fetal oximeters are now under development that could prevent oxygen deprivation that leads to brain damage in babies during delivery.

The design engineer is responsible for identifying—sometimes through exhaustive research—all of the constraints or boundaries that must be satisfied by a solution. Otherwise, a design may be developed that is illegal, hazardous, or infeasible.

4.5 Ergonomic Constraints in Design

Ergonomics (sometimes referred to as "human factors engineering") focuses upon the variations that exist among different human populations and the effects of these variations on product design decisions. Ergonomic data often represent critical design constraints or specs that must be achieved if the safety, productivity, health, and happiness of the designated user population for a particular product, system, or process is to be enhanced.

People can differ broadly in height, weight, intelligence, age, physical abilities, and many other ways. Such variations in the client population for which an engineering solution is developed must be carefully reflected in the design itself. After all, if an automobile seat were not adjustable but only designed to accommodate the "average" man in terms of height and weight,

15. See Pollack (1991).

it then would fail to satisfy the needs of most other people. User characteristics that may be important to the success of an engineering design include

- visual acuity
- hearing discrimination
- hand–eye coordination
- reaction time
- sensitivity to temperature, dust, and humidity
- reading skills

The following incident illustrates why human factors should be included whenever a solution to a problem is prepared.[16]

The Sinking Boat

A group of officials from the foreign service of a western government were scheduled to visit their embassy in a southeast Asian country. Plans were made by a member of the embassy for their visit; these plans included a chartered canal-boat tour of the capital city's business district. Each day this boat carried ten people along the canal routes of the city without incident.

Unfortunately, when the boat left the dock carrying the five visitors along with four other foreign service workers and the driver, it began to sink. All those aboard the boat safely swam to shore.

Why did the boat sink? It was assumed that the boat could transport ten people since it did so on a daily basis. However, the western visitors and most of the others aboard the boat were substantially heavier than the natives; their combined weight was beyond the capacity of the boat to support. One human factor—weight—led to this accident.

During the past 40 years, much effort has been expended to collect and evaluate data describing the variations among human beings. Some of this data is anthropometric; that is, it focuses upon the variations in size and proportion among people. Other data describe the variations in people's relative physical abilities (e.g., endurance, speed, strength, accuracy, vision, and hearing) based upon physiological and anatomical knowledge. Similarly, psychology can provide understanding about our abilities to acquire, learn, and process information. All of these data can be used to formulate important ergonomic constraints for an engineering design to ensure that it will satisfy the needs of its intended user population.

16. Kepner and Tregoe (1981).

4.5.1 ▨ Anthropometry

As noted, anthropometry focuses upon the variations in the sizes and proportions of people. Much of these data have been tabulated in the form of percentiles; that is, percentages of people who are at or below a particular value. For example, the 50th percentile for men's heights corresponds to the medium value: 50 percent of men in the sample population lie at or below this height, and 50 percent lie above this height. Similarly, the 95th percentile would correspond to a height above which only five percent of the sample population were found.

The sample population for anthropometric data necessarily is based upon specific sample sets of people who have been measured and tested in various ways. These sample sets usually are assumed to be representative of the population as a whole. However, the engineer should be careful to note the sample set(s) upon which an anthropometric range is based in order to ensure that there is a correspondence between the sample and the intended user population of a particular design. For example, data based upon samples of law enforcement personnel may be skewed or biased in some manner since police officers often need to satisfy both minimum and maximum requirements of height, weight, strength, education, and other attributes; it would be inappropriate to describe the general public with such data.

Figure 4.4 illustrates the form in which anthropometric data is sometimes presented, indicating a series of measurements at the 97.5 (large), 50 (median), and 2.5 (small) percentiles. A product designed for people at the 50th percentile may be unsafe or difficult to use by those people who lie at the extremes of the population (the very tall, the very small, etc.). However, a product designed for use by everyone may not be satisfactory for anyone because it is too imprecise, ill defined, or unspecified. The question then arises: For what population range does one design?

The answer depends upon many factors: the degree of specificity with which the user population can be defined (e.g., the elderly, children, athletes, those working in a particular trade or profession), the expected application or purpose of the product, cost constraints, and so forth. In general, however, one usually focuses upon the following ranges:[17]

Civilian populations: 2.5–97.5 percentile (95 percent of population)
Military populations: 5–95 percentile (90 percent of population)

In other words, a product intended for a military population is usually designed so that all people lying between the fifth percentile and the 95th percentile will be able to use it. The product then is not expected to be used by those lying outside this range (e.g., in terms of height, very short or very tall people would be excluded from the target population).

Obviously, anthropometric data should match the expected user population(s) of a product.[18] For example, if one is designing a baby carriage, then

17. Diffrient, Tilley, and Bardagjy (1974)
18. Middendorf (1990).

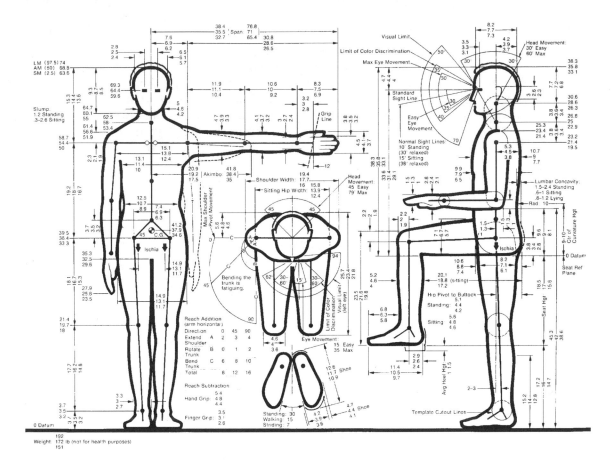

FIGURE 4.4 Anthropometric data (distances in inches) for men at the 97.5, 50, and 2.5 percentiles. *Source:* Diffrient, Tilley, and Bardagjy, *Humanscale 1/2/3 Manual* Cambridge, MA: The MIT Press, copyright © 1974. Reprinted with permission.

one should consider data that describes variations among infants in the intended age range rather than children of other ages.

In addition, data should be current for changes do occur among human populations as time goes by (e.g., the average height of a population can increase by approximately 0.3 inches every 10 years, resulting in a significant difference over a 50-year period).

It is also important to recognize that human populations often do not vary in a uniform manner. Body heights and lengths do vary uniformly since they depend upon bone structures (Fig. 4.5). However, weights, widths, and girths may not vary uniformly throughout a population since these measurements depend upon age, amount of fat, musculature, and other factors that may diverge widely (Fig. 4.6). Although variations in height occur symmetrically about the 50th percentile, variations in weight are asymmetrically

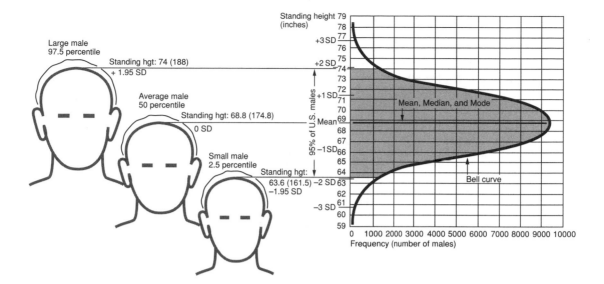

FIGURE 4.5 Uniform variations in male heights about the 50th percentile. *Source:* Diffrient, Tilley, and Bardagjy, *Humanscale 1/2/3 Manual,* Cambridge, MA: The MIT Press, copyright © 1974. Reprinted with permission.

distributed about this point. (For an example of anthropometry in design, see Case History 4.9 *The Ergonomically Designed Lounge Chair.*)

4.5.2 ▨ Human Capabilities and Limitations

When designing a product, system, or process, one must recognize that there are certain abilities and limitations commonly associated with human behavior.[19] These include the time necessary to respond to certain stimuli, the accuracy with which particular tasks can be completed, and the ability to recognize visual, auditory, and tactile stimuli.

All systems involving feedback require a time delay or response time during which (a) the input signal(s) are processed, (b) the appropriate response is determined, and (c) the action is initiated.[20] The human body behaves in this same manner when responding to stimuli. For example, it requires about 300 milliseconds (ms) to respond to a visual stimulus but only 200 ms to respond to auditory stimuli.

The time delay also depends upon the type of response that is desired.[21] A manual response requires only 75 percent of the response time needed to produce a verbal response. Moving an object 45 centimeters (cm) may

19. See Baumeister *and* Marks (1967); Middendorf (1990).
20. Voland (1986).
21. Middendorf (1990).

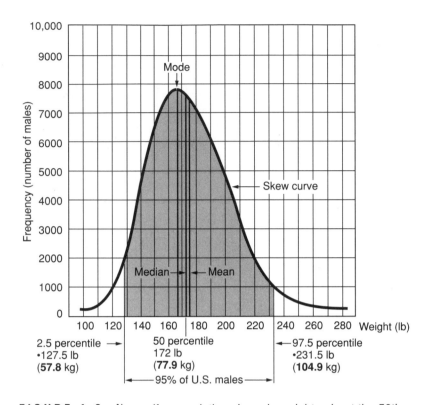

FIGURE 4.6 Non-uniform variations in male weights about the 50th percentile. *Source:* Diffrient, Tilley, and Bardagjy, *Humanscale 1/2/3 Manual,* Cambridge, MA: The MIT Press, copyright © 1974. Reprinted with permission.

require 125 ms, whereas moving it only one-third of that distance (15 cm) requires almost as much time (100 ms).[22]

Most of the 206 bones in the human body are controlled by one of three types of joints: ball, hinge, and pivot. A ball joint (e.g., shoulders) provides the greatest flexibility in movement by allowing a maximum number of degrees of freedom; however, it also requires the most effort in order to achieve accuracy in these movements. In contrast, hinge joints (e.g, fingers) provide the least flexibility in movement but require little effort for accurate positioning. Pivot joints (e.g., elbows) provide moderate levels of flexibility and require moderate effort for accurate movements.[23]

This means that the design engineer must consider the degree of accuracy required to complete a task, together with the joint(s) that will be used for this task. A product should be designed so that the required level of accu-

22. Most of the time is used in accelerating and decelerating one's hand and then in properly positioning the object.
23. Middendorf (1990).

racy is achieved with the minimum amount of effort. For example, moving an object laterally requires use of the elbow's pivot joint whereas the shoulder's ball joint is needed to move the same object directly forward or backward. Hence, the lateral movement will require considerably less effort to achieve the same degree of accuracy.

Users often need to respond in some manner to information that is provided by a product or system. Such information feedback may be in the form of visual or auditory stimuli (or both), such as a blinking light or a buzzer. When designing such feedback mechanisms, the engineer must recognize both the visual and auditory limitations exhibited by most people. For example, visual limitations include the ability to distinguish colors only over a relatively narrow band of the electromagnetic spectrum (see Fig. 4.7). Moreover, the ability to distinguish objects in one's field of vision is very dependent upon the relative intensity of light (Fig. 4.8).

Visual signals (lights, signs, information displays) are often less expensive to produce than auditory signals. Visual signals can also be varied by shape, position, and color. In addition, visual signals can be passive (a warning label) or active (a warning light that is activated in the event of an emergency). Finally, visual signals can be quite varied in the amount of detailed information that they provide (e.g., a static sign versus a changing message that prompts the user to initiate some action).

However, visual signals are not pervasive: They require that the user is attentive to the signal. In contrast, sound is pervasive and can be used to alert a person even if his or her attention is directed elsewhere. In addition, human beings respond faster to auditory signals than to visual stimuli. Unexpected sounds can induce fear or other forms of alertness very quickly. Sometimes an auditory signal will be used to attract the attention of a user to visual information.

FIGURE 4.7 Wavelength comparisons: (a) Visible spectrum (in millicrons); (b) Electromagnetic spectrum (in meters). *Source:* Adapted from Edel, 1967.

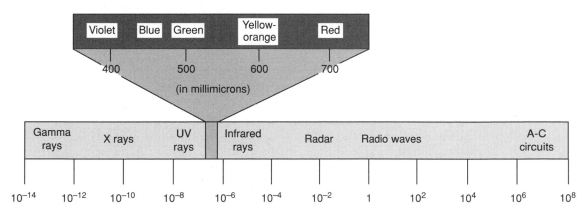

Milliamberts

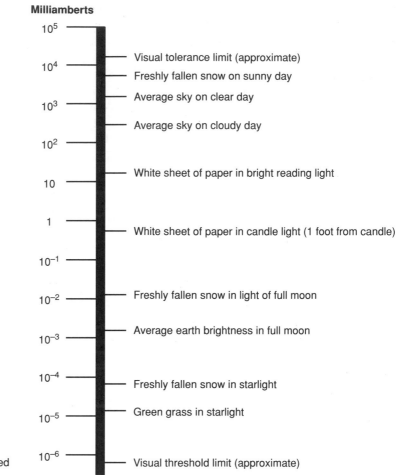

10^5

10^4 — Visual tolerance limit (approximate)
— Freshly fallen snow on sunny day
— Average sky on clear day
10^3
— Average sky on cloudy day
10^2

— White sheet of paper in bright reading light
10

1 — White sheet of paper in candle light (1 foot from candle)

10^{-1}

10^{-2} — Freshly fallen snow in light of full moon

10^{-3} — Average earth brightness in full moon

10^{-4} — Freshly fallen snow in starlight

10^{-5} — Green grass in starlight

10^{-6} — Visual threshold limit (approximate)

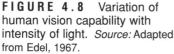

FIGURE 4.8 Variation of human vision capability with intensity of light. *Source:* Adapted from Edel, 1967.

If necessary, auditory signals can be varied in tone, duration, volume, and other ways in order to provide selective feedback information to the user. Products (such as computers and automobiles) often use distinct clicking sounds, buzzers, bells, or other auditory signals to alert a person of a change in condition or of the need to perform a specific action.

However, one must consider background noise (and possible hearing loss by the user) when designing a product that will include auditory signals. Figure 4.9 presents the range of some common sounds that can be distinguished by human beings.

Distinctive tactile information can be provided by an object through its shape, texture, size, hardness, or relative motion. Although visual and auditory stimuli are used more frequently to convey information, tactile design can be very effective. For example, specific shapes and locations for aircraft control knobs are specified by the U.S. Air Force to allow pilots to

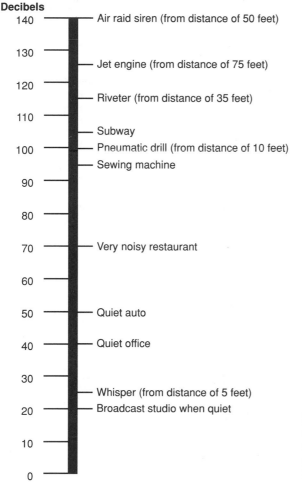

Decibels

dB	Sound
140	Air raid siren (from distance of 50 feet)
130	
	Jet engine (from distance of 75 feet)
120	
	Riveter (from distance of 35 feet)
110	
	Subway
100	Pneumatic drill (from distance of 10 feet)
	Sewing machine
90	
80	
70	Very noisy restaurant
60	
50	Quiet auto
40	Quiet office
30	
	Whisper (from distance of 5 feet)
20	Broadcast studio when quiet
10	
0	

FIGURE 4.9 Comparison of common sounds in terms of decibel levels. *Source:* Adapted from Edel 1967, and McCormick 1964.

distinguish each knob by touch.[24] Visual inspection is not necessary to locate the appropriate control and may not be possible in an emergency.

4.5.3 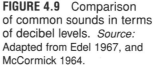 Ergonomic Impact on Input Data and Displays

Data must be entered and displayed without ambiguity if a person is to use any mechanized system properly. Although this objective for the design of input, display, and control systems may seem to be obvious, too often it is forgotten in actual practice. Even advanced computer controlled systems require accurate input data in order to perform their tasks correctly. Automatic safety devices and other sophisticated equipment often can reduce the likelihood that human error will lead to disaster; however, they cannot be expected to overcome errors that are embedded within their input data.

24. See Woodson and Conover (1964); McCormick (1976); Middendorf (1990).

An Input Data Error Forces an Aircraft to Land in a Jungle

An airplane exhausted its fuel supply after the pilot incorrectly set the inertial navigation system to 270° instead of the correct value of 027°. Twelve people died when the plane was forced to land in the Brazilian jungle. A simple transposition of digits had led to disaster.[25]

Whenever possible, systems should be designed with additional safeguards to verify that raw data (particularly those which are entered manually) are indeed correct and accurate. Human–system interfaces should echo any data that is manually provided back to the user so that he or she may verify that these values are indeed correct. Moreover, such echoing should be in a form that requires the user to carefully—and not cursorily—inspect the data. For example, the system could require the user to input critical data more than once in order to increase the likelihood that it is entered correctly. In addition, numerical values (270°) could be echoed in textual form (two hundred seventy degrees), thereby increasing the likelihood that the user will recognize an error than if the values were simply echoed in numerical form. Furthermore, any inconsistency among various input data that is detected by the system should be brought to the attention of the user. Such interface design features can reduce the impact that human errors in data input may have in critical situations.

4.5.4 ▓ Natural Design and Feedback

People often have natural expectations when interacting with a system. For example, most of us expect that an upward movement will turn a device on or increase its power, whereas downward actions are associated with turning a machine off or decreasing its power. Engineers should try to develop designs in which these natural and expected relationships are used in human–machine interactions, thereby reducing the likelihood that the user will operate a device incorrectly. Norman (1988) refers to this type of design—in which the engineer uses physical analogies and cultural standards to relate operator expectations to corresponding actions—as natural mapping.

He also describes the concept of affordance, in which certain materials and/or design objects "afford" the user the opportunity to perform some action or task. For example, glass is transparent and breakable, and wood is

---

25. See Kletz (1991) and Learmont (1990).

opaque and porous; hence, glass telephone booths may be shattered by vandals and wooden ones may be covered with graffiti.

Many of today's products contain such subtle yet visible clues to guide the user. For example, an oversized plate on a door indicates that one should push on it to open the door whereas a small bar or handle suggests that one should pull. Other products are designed intentionally to prevent the user from making a mistake: There is only one way to insert a 3.5-inch computer diskette into a disk drive.

Unfortunately, many products continue to be designed poorly (ask the average user of a video cassette recorder if it is easy to operate and program). These products often lack visible clues that could guide the user to successful operation. Too little information may be given for the user to take full advantage of a multipurpose device. In other systems, the user may be inundated with information, much of which is unnecessary, so that the devices become difficult to operate, particularly in emergency or stressful situations.

Controls also should provide feedback to the user whenever an action has been completed. Such feedback might consist of a simple clicking sound whenever a switch is thrown, or it might involve lights or some other display mechanism. Feedback reassures the user that a control has indeed been activated and that the desired response of the system can be expected.

For example, recall the different sounds and messages that are used in telephone systems to prompt the caller to take some additional action (e.g., leaving a voice message at the sound of the beep, entering a number for a calling card, or remaining on hold until someone is available). Without these feedback guides, using a telephone would be much more difficult, error-prone, and time-consuming.

Middendorf (1990) suggests that the engineer follow a procedure similar to the following one when developing human–machine interfaces based upon ergonomic factors:

1. Identify the expected interactions between the user and the product during its operation.
2. Identify the operations that will require monitoring or control by the user. Recognize the limitations of the expected user(s) in performing such tasks.
3. Evaluate the expected environment in which the product will be used. Consider such factors as intensity of light, availability of space, background noise, temperature range, dust levels, humidity values, and vibration levels.
4. Finally, identify those operations that can be automated or made easier to perform manually. Consider the amount of training and experience required by the user to perform the task(s) properly.

We should strive to incorporate ergonomic considerations in our designs in order to optimize the interactions between the user and the product or system which is to be developed.

- One must properly structure the search for a solution to a problem by formulating the desired characteristics or design goals, recognizing the constraints within which any solution must lie, and eliminating those paths that will not satisfy these goals and constraints.
- Design goals are qualitative in nature, whereas design specifications are quantitative.
- There are both general and specific engineering design goals; general goals (e.g., minimum cost, safety, reliability, performance, and ease of operation) are sought in most engineering design solutions, whereas specific goals (e.g., lightweight, water-resistant, transparent) are used to help define the characteristics and desired functions of a particular design.
- Anthropometric, anatomical, physiological, and other types of ergonomic data and principles can be used to formulate certain types of design specs and to optimize interactions between the design and its user(s).

PROBLEMS

4.1 Select a design solution/product of your choice and identify
   a. the general goals that were probably sought by the engineering team in this solution,
   b. the specific goals that were probably sought by the engineering team in this solution,
   c. the goals that are not achieved by this solution but that probably should be achieved by the next-generation design.
4.2 What is the difference between the problem statement and the list of goals to be achieved by a design solution? Be as specific as possible in stating this difference.
4.3 Explain why a design engineer must be thoroughly familiar with the goals to be achieved by a solution and why this list of goals will probably be continually updated during the entire design process.
4.4 Consider the design goals that should be considered when designing an automobile seat, taking into account the various user populations that may use the seat. Design and sketch the optimal seat for the driver of an automobile and, based upon ergonomic considerations, identify

the most important aspects of your design on your sketch.

4.5 It has been estimated that 680,000 people in the United States use wheelchairs. Use ergonomic data describing wheelchair user populations (from library reference books) to design a drinking fountain that will accommodate both wheelchair and non-wheelchair users. Sketch and note all significant aspects of your design. Attach documentation of the ergonomic data that you used in developing your design.

4.6 Select two alternative designs of a common household product. Sketch these two designs. Include a detailed comparison of the two designs based upon ergonomic considerations. Determine which design is superior and in what ways it is superior. Remember that the designs may have distinct strengths as well as weaknesses relative to one another; one may not necessarily be superior to the other in all ways. Include the results of your analysis and the ergonomic factors that were considered on your sketch in tabular form.

CASE HISTORY 4.1

Improper Waste Disposal at Love Canal and Times Beach

Technology serves humanity in innumerable ways by improving the quality—and even the length—of our lives. However, technology also often produces by-products that can be deadly if not properly recycled or disposed. Because of this fact, engineers should include appropriate mechanisms for the safe disposal of waste products in their designs of new solutions to technical problems. Safe, efficient waste disposal and recycling should be one of the primary goals for almost all engineering design efforts.[26]

Love Canal   In 1953 the Niagara Falls Board of Education purchased a block of land known as Love Canal from the Hooker Chemical Corporation for a token payment of one dollar. Unfortunately, the board was not aware that Hooker had been burying highly toxic chemicals on this land since 1942 and that more than 20,000 tons of such waste lay beneath the surface. Hooker continued to withhold information from local authorities as first a school and then a bustling community was built on the site during the ensuing years. Finally, in 1976, heavy rains caused many basements to flood and the long-buried toxic chemicals began to rise to the surface with the water. Many residents became ill; the rates of cancer, miscarriages, birth defects,

26. See Regenstein (1993, 1994a, 1994b).

FIGURE 4.10 Soil samples are collected by the United States Environmental Protection Agency in Times Beach, Missouri. *Source:* Corbis-Bettmann.

and other health problems became unusually high. By 1978 several homeowners (led by 27-year-old housewife Lois Gibbs) were demanding that the federal, state, and local governments act to protect the residents of Love Canal. Tests then conducted by the state of New York indicated that more than eighty chemical compounds (many considered to be carcinogenic) were present in the air, water, and soil; the level of contamination in the air alone was up to 5000 times beyond that which was deemed acceptable for human health. In 1980 the community was declared a federal disaster area and most residents were relocated to other neighborhoods. (Sixty families chose to remain in Love Canal; about 1000 decided to move to new locations.)

Times Beach During 1971 mixtures of reclaimed oils and waste chemicals were spread on roadways as a dust suppressant throughout Missouri. Unfortunately, some of these chemicals were hazardous. In particular, the mixture used in the town of Times Beach, Missouri, contained dioxin (tetrachlorodibenzoparadioxin or TCDD), an extremely toxic chemical compound. Dioxin has been deemed by the U.S. Environmental Protection Agency (EPA) to be carcinogenic, mutagenic, teratogenic, and fetotoxic. It is found in some herbicides and also generated as a waste by-product when paper is bleached white and when certain plastics are burned.

Although illnesses among both the human and animal populations of Times Beach appeared almost immediately following the 1971 distribution of the deadly dust suppressant mixture, it was not until 1982 that the town's contamination became intolerable (Fig. 4.10). Flooding by the Meramac River led to the town being covered by dioxin-contaminated soil and water. Times Beach was no longer fit for human habitation; it was purchased by the EPA for $33 million and its inhabitants were relocated.

Unhappily, the extent of the hazard in Missouri was not limited to Times Beach; 100 other regions also contaminated by dioxin had been located throughout the state by 1993.

Environmental disasters such as Love Canal and Times Beach focused the national spotlight on the problem of hazardous wastes. In 1980 the EPA estimated that between 32,000 and 50,000 sites had been contaminated by hazardous wastes. President Jimmy Carter signed the Environmental Emergency Response Act (popularly known as the "Superfund") into law, thereby beginning the federal government's official response to environmental polluters and the hazards that they create.

Much work remains to be done. The truly daunting task of cleanup has only begun: Only 149 of the 1,256 Superfund hazardous waste sites designated as priority locations for immediate action by the federal government had in fact been cleaned by 1993. Nevertheless, at least the problem of hazardous wastes and their disposal is now recognized by many and progress in coping with this problem has been made. For example, by 1993 approximately half of all wastes generated by the chemical industry were being recycled.

CASE HISTORY 4.2

The Collapse of the Baldwin Dam

Sometimes we are working within constraints that we do not recognize; this can result in unexpected side effects of our work that are lethal, as illustrated by the collapse of Los Angeles' Baldwin Dam on Saturday, December 14, 1963, resulting in the deaths of five people and extensive damage to more than 1000 homes and 100 apartment buildings.[27]

At 3:38 P.M., water erupted from the dam until a 75-foot wide, 90-foot high, V-shaped break had formed in the dam's wall (Fig. 4.11). Flood waters were eight feet high as they swirled through the streets below the dam.[28]

27. See Hamilton and Meehan (1971), Holmes (1994), Jessup (1964), Martin and Schinzinger (1989), Marx (1977), Sowers and Sowers (1970), and Voland (1986).

28. The author vividly remembers this dam collapse from his childhood: "My mother, Eleanor Voland, courageously maneuvered our car through the streets below the dam as water (together with a piano, furniture, mud, and other debris) rushed and rose all around us. Far above, water could be seen pouring through the huge gap in what had been the dam's wall. We finally reached safety when several young men lifted our car—and those of many others—over a set of raised railroad tracks that acted as a waterbreak. These tracks directed the tide of water towards the Ballona Creek drainage runoff, which then carried it to the ocean. I remain very grateful to those anonymous young men who helped save so many."

FIGURE 4.11 The V-shaped gap in the Baldwin Hill dam wall can be seen in this aerial photograph. *Source:* Corbis-Bettmann.

The dam was opened in 1951. The designer recognized that the chosen site would be near minor earthquake faults and that the dam's rock foundation (composed of siltstone and sandstone) would be subject to erosion. To counteract these factors that could lead to a shifting of the earth and erosion of the foundation, a set of protective liners and drainage systems were installed. Beginning with the topmost layer and working downward, the liners of the foundation were as follows: (a) three inches of asphalt, (b) compacted soil with a thickness that varied between 5 and 10 feet, (c) two inches of sand and gumite, (d) four inches of pea gravel to collect and drain any water seeping through the upper layers, and (e) a final bottom layer of asphalt. These liners were meant to prevent water seepage into the dam's siltstone/sandstone foundation and any subsequent erosion. In addition, the dam was thoroughly inspected on a regular basis.

Why then did the Baldwin dam—designed to withstand earthquakes—suddenly fail after 12 years of reliable service? The answer lies in the numerous oil wells that dotted the surrounding Baldwin Hills at that time. As oil and gas were removed, the soft ground below the hillsides began to settle, gradually causing damage to the reservoir's liners. Water seepage and erosion of the reservoir foundation's soil continued until the dam collapsed.

These oil wells had been productive for many years but they were beginning to run dry by 1963. One method of forcing oil from nearly depleted reserves, known as secondary recovery, is to inject water underground at very high pressure. Unfortunately, when this method was used in the Baldwin Hills/Inglewood oil fields to extend the wells' productivity, it appears to have resulted in further damage to the dam's liners. Figure 4.12 shows the locations of injected water pockets throughout the area. Figure 4.13 compares the cumulative time histories of ground ruptures with the amounts of under-

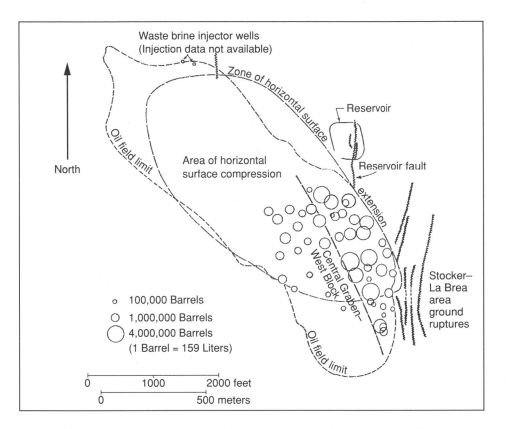

FIGURE 4.12 Injected water bubbles throughout the Inglewood oil fields. *Source:* Reprinted with permission from Hamilton and Meehan, "Ground Rupture in Baldwin Hills," *Science,* vol. 172, 23 April. Copyright © 1971 America Association of Science.

ground water injected throughout the east block section, providing persuasive evidence of the cause-and-effect relationship that appears to have existed between injections of water underground in the east block region of the Inglewood oil fields and the appearance of surface faults near the reservoir. In particular, notice the steady increase in the number of injection wells with relatively high pressure gradients between 1957 and 1963.

SUMMARY OF FAILURES AND TYPES OF ERRORS

- Failure to anticipate the secondary effects of actions taken; that is, oil drilling and water injection (error in design due to invalid assumptions and faulty reasoning about design constraints in the form of

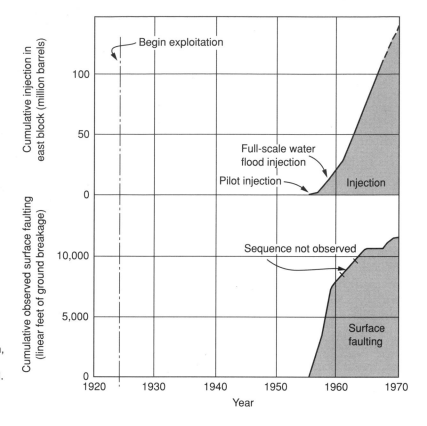

FIGURE 4.13
Cumulative time histories of ground ruptures and water injection in the east block region of Baldwin Hills. *Source:* Reprinted with permission from Hamilton and Meehan, "Ground Rupture in Baldwin Hills," *Science,* vol. 172, 23 April. Copyright © 1971 America Association of Science.

environmental-system interactions, and error in implementation through inadequate communication and deterioration)

Most of us realize that environmental factors such as corrosion, moisture, dust, vibration, and heat can have a very significant impact upon our engineered systems. These factors form the basis of many of the design boundaries or specs within which we must work. However, it is important to remember that not only can its environment affect a system but that the system itself (e.g., an oil field operation) can have a very real and negative impact upon its environment. To ignore this fact is to invite disaster, as shown by the oil-drilling and water-injection systems that changed the hillside environment of the Baldwin Hills Dam until its protective foundation liners could no longer prevent erosion.

CASE HISTORY 4.3

Perrier Water

Once achieved, product acceptance should remain a primary goal of any company, lest it (and the company) be lost forever.

In 1990 the French firm Source Perrier found itself enmeshed in a potentially disastrous public relations crisis regarding the purity of its world-famous sparkling mineral water.[29] Scientists at the Mecklenburg County Environmental Protection Department in Charlotte, North Carolina, had accidentally discovered that their latest supply of Perrier water (used in some of their laboratory tests) contained benzene, a carcinogenic or cancer-causing substance. It was determined that the benzene appeared in concentrations varying between 12.9 and 19.9 parts per billion, thereby exceeding the U.S. Food and Drug Administration's mandated limit of only five parts per billion.

Eventually, 160 million bottles of Perrier water were recalled worldwide (an estimated $200 million action), and the source of the contamination was identified and corrected, but not before the company had committed some costly public relations errors.

Among these errors was an early announcement that the contamination was limited to North America and that its source was cleaning fluid applied to bottling machinery. However, within days the true source of contamination was announced: Benzene was one of the substances naturally present in the spring water itself. Perrier used charcoal filters to remove these impurities, but, unfortunately, these filters had not been replaced when necessary, thereby allowing small amounts of benzene to remain in the water.

Since the levels of benzene were so low, the company seemed to view the entire incident as unfortunate and expensive but of minor importance. Nevertheless, most customers had assumed that Perrier's spring water was naturally pure; any indications to the contrary would require appropriate reassurances by the firm that their water was indeed beneficial for one's health.

The news that Perrier water required purification, the factual corrections made to the company's earlier announcement about the source of contamination, and other developments together had a very negative impact on the image and sales strength of the product. Although Perrier survived the incident and continues to remain strong in the spring water market, its market share decreased from 13 percent to 9 percent in America and fell from 49 percent to below 30 percent in Great Britain; both operating profits and the company's share price fell.

A candid and accurate response to a potentially disruptive incident—with due recognition of the public's concerns—can ease a company's efforts to regain their customers' trust.

29. See "When the Bubble Burst," *Chemtech,* (February 1992) pp. 74–76.

CASE HISTORY 4.4

The Pentium Computer Chip

Another instance in which the public's confidence in a product was severely shaken involved the Pentium microprocessor computer chip manufactured by the Intel Corporation.[30] In November 1994, Intel admitted that errors could be made by the Pentium chip during certain specific mathematical operations. However, the company refused to automatically replace the flawed chips unless a customer could demonstrate that he or she needed to perform the mathematical calculations in which errors could occur.

Intel was perceived as insensitive to their customers' needs and unresponsive to their concerns. Within a few weeks, the strongly negative reactions from the public, retailers, computer manufacturers, and others to the limited replacement policy had grown into a public relations disaster. Intel's stock value dropped when IBM announced that it would no longer ship their new personal computers with the Pentium chip. Finally, Intel agreed to replace any Pentium chip upon request.

In January 1995, Intel announced that it would take a one-time, pretax charge of $475 million to cover the replacement costs of Pentium chips. The company later reduced the price of the chip and settled 11 class action product liability lawsuits.

As of April 1995, most customers had not sought replacement of their Pentium chips.

One final note of interest: Reidy (1994) contrasts Intel's initial response to customer concerns about its Pentium chip to that of L.L. Bean. In 1994 L.L. Bean sold Maine balsams as Christmas trees via mail-order, packing each tree in a moisture-proof wax container. Unfortunately, some of these trees were not being delivered on time. The company decided to automatically reimburse any customer who requested a refund, recognizing the importance of maintaining their customers' trust and confidence.

It is absolutely critical to retain the public's trust and confidence in a manufacturer and its product line; without such trust, a design can quickly fail in the marketplace.

---

30. Refer, for example, to such articles as those by Day (1994), Lewis (1994), Husted (1994), Jones and Sullivan (1994), Clark (1994), Millan (1994), Markoff (1994b, 1994c), Reidy (1994), Fisher (1995), and Hill (1995).

CASE HISTORY 4.5

Two Routes to Simplicity: Camcorder Redesigns

Two Japanese manufacturers of camcorders followed different routes to achieve greater simplicity in their products.[31] Both companies were convinced that sales would increase if their products became easier to use.

Sharp Corporation eliminated features that were difficult for the average customer to operate by completely redesigning its ViewCam. For example, the viewfinder in the camcorder was replaced by a liquid crystal display screen on which the scene being filmed is shown. Furthermore, this display screen is mechanically separate from the lens, thereby allowing the operator to comfortably view the screen while pivoting the lens to various positions.

In contrast, the Matsushita Electric Industrial Company simply eliminated such features as anti-jitter circuitry and the automatic zoom in its Pattoru camcorder (marketed under the Panasonic brand name). The result is a simple and inexpensive product that appeals to those who dislike more complex electronic devices.

Positive initial sales of both the ViewCam and the Pattoru seem to indicate that the public is indeed willing to embrace products that sacrifice some capabilities but are easier to operate.

CASE HISTORY 4.6

O-Rings and the *Challenger* Disaster

In 1986 a special presidential commission investigated the explosion of the *Challenger* space shuttle.[32] The explosion was found to have resulted from a simple design defect: The O-rings in the *Challenger* joint-and-seal assembly (Fig. 4.14) failed to stop hot (6000° F) gases in a solid-fuel booster rocket from burning through the wall of the orbiter's liquid-hydrogen tank, thereby releasing hydrogen and causing a disastrous chain reaction. The oxygen tank

---

31. See Hamilton (1993).

32. Based upon *Kamm* (1991) and the *Report to the President by the United States Presidential Commission on the Space Shuttle Challenger Accident* (1986).

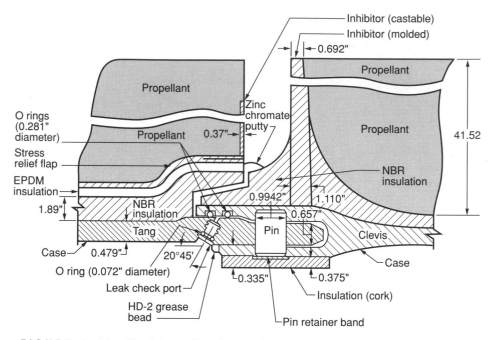

FIGURE 4.14 The joint-and-seal assembly to the *Challenger* space shuttle. Source: From the *Report to the President by the United States Presidential Commission on the Space Shuttle* Challenger *Accident,* 1986.

wall adjacent to the hydrogen tank was supported in part by the pressure within its neighbor; this wall collapsed as the hydrogen tank emptied and its pressure decreased. The flames of the rocket engines then ignited 160 tons of escaping liquid hydrogen and liquid oxygen.

In order to understand this design flaw, we need to first consider the function of an O-ring. An O-ring is a thin rubber torus or doughnut that is placed in a groove between parts that are to be sealed (Fig. 4.15a). When pressure is applied to the O-ring by a fluid (liquid or gas) entering the groove, it is pushed to one side and effectively prevents the fluid from further passage through the gap between parts (Fig. 4.15b). Of course, it is imperative that the fluid pressure be applied to the entire one side of the O-ring so that the ring is pushed forward and upward to seal the gap; otherwise, the O-ring will be pressed downward by the fluid and the seal is broken (Fig. 4.15c). If the groove is not sufficiently wide, the O-ring will not be able to move forward and upward to complete the seal.

The gap between body parts (i.e., tang and clevis) in the *Challenger* had to be 0.004 inch or larger; otherwise, the O-ring would be pressed downward to fill the groove, thereby allowing the gas to travel over it and escape. However, the rocket body parts to be sealed were twelve feet in diameter.

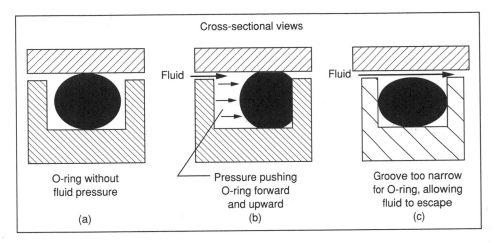

Cross-sectional views

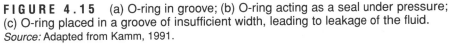

O-ring without
fluid pressure

(a)

Pressure pushing
O-ring forward
and upward

(b)

Groove too narrow
for O-ring, allowing
fluid to escape

(c)

FIGURE 4.15 (a) O-ring in groove; (b) O-ring acting as a seal under pressure;
(c) O-ring placed in a groove of insufficient width, leading to leakage of the fluid.
Source: Adapted from Kamm, 1991.

Kamm (1991) notes that because of the large size of these parts and expected imperfections in their manufacture, the gap could be as small as 0.000 inch at some locations.

Kamm also emphasizes that standard O-ring handbooks indicate that any O-ring with a 0.275-inch diameter will require a corresponding groove width between 0.375 inch and 0.380 inch in order to function properly. However, in the *Challenger* design the groove was only 0.310 inch in width in order to contain a nominal O-ring of 0.280 inch diameter. In other words, the groove apparently was too small for the O-ring to complete the seal!

We will focus upon other aspects of the *Challenger* disaster in Chapter 8 during our discussion of engineering ethics.

This tragic event emphasizes

- the need for engineers to understand how each component functions (or fails to function) within a system,
- the value of handbooks and manufacturer catalogs as sources of design data, and
- the significance of manufacturing tolerances (i.e., acceptable variations in the size or shape of a part) in engineering design.

CASE HISTORY 4.7

Fax Machines

Facsimile (fax) machines are used to electronically transmit and receive text, drawings, photographs, and other types of information across telephone lines. Each image is converted by an optical scanner into millions of electronic signals (pixels) that are transmitted to a receiver where the original pattern is then reconstructed.[33]

These devices, now ubiquitous in industries throughout the world, are not of recent vintage. The concept of the fax machine dates back to 1842 when first invented by Alexander Bain. The later development of scanning devices (by Frederick Bakewell in 1848) and a synchronized transmission and reception system (by Ludovic d'Arlincourt in 1869) eventually led to the first successful electronic transmission of photographs from Cleveland to New York in 1924.

For the next sixty years, the fax machine remained a device that was used primarily by newspapers and other businesses to send photographs from site to site. Then, beginning in the mid 1980s, fax machines became an increasingly popular form of communication, as reflected in the following sales figures:

1986	187,500 machines
1990	1,400,000 machines
1993	2,500,000+ machines

What, then, led to this phenomenal growth in popularity? One reason is the significant reduction in retail price achieved between the end of the 1970s (when a machine could cost as much as $10,000) and 1989 (when a model with limited features could be purchased for as little as $500). In addition, the time required to transmit a single page of text decreased from almost six minutes in 1979 to about three seconds in 1989, thanks to advances in microprocessor and image sensor technologies. During the 1990s, fax machines underwent further development as plain paper replaced both thermal sheets (which were coated with a dye that became visible whenever toner was applied) and photosensitive paper, thereby eliminating paper curling and the need for the user to invest in specialized materials. Finally, the quality of reproduction continues to improve.

A substantial reduction in price, coupled with improved performance characteristics, can transform a product with very limited appeal into one that serves the needs of millions.

33. See Behr (1995), Travers (1994), and "Japan Does It Again with Fax Machines" (1989).

CASE HISTORY 4.8

The Development of Band-Aids

The development of the Johnson & Johnson (J & J) product known as the Band-Aid brand adhesive bandage is an example of specific goals that guided the designer to the final solution of a problem.[34]

In 1920 Earle E. Dickson worked as a cotton buyer in the J & J purchasing department. One day, he used gauze and adhesive to bandage his wife Josephine's fingers (she frequently cut or burned herself while working in the kitchen). Upon reflection, he decided to design a bandage that would

- retain its sterility,
- be easy to apply directly to a wound, and
- remain in place after its application to the wound.

(These were specific design goals unique to the particular problem under consideration.)

After more thought, Dickson placed strips of surgical tape on top of a table and affixed gauze to the center of each strip's exposed adhesive surface, taking care that some of the adhesive remained uncovered by the gauze. Finally, he covered the gauze with crinoline (cotton interlining) to complete his design. Whenver an injury occurred, a portion of these bandage strips could be cut with scissors to the desired length, the crinoline removed, and the bandage applied directly to the wound.

Dickson described his design to the management at J & J. The company president, James W. Johnson, embraced the design as a new product and Band-Aids were introduced to the world. Dickson eventually became a member of the J & J Board of Directors and a company vice president.

Band-Aids originally were produced in 18-inch by 3-inch bandage strips that then were cut by the customer to whatever size was needed. Then in 1924 the popular 3-inch by 3/4-inch strips were introduced. Further developments included the introduction in 1928 of aeration holes to hasten the healing process and in 1939 sterilization of the entire Band-Aid.

Dickson's original design managed to achieve each of his three specific goals: the use of crinoline ensured that the bandage would retain its sterility, the gauze/adhesive strips allowed the user to apply the preformed bandage with relative ease, and the adhesive of the surgical tape ensured that the bandage would remain in place after its application to the wound. Formulating the problem in terms of the specific goals that must be achieved is a critical step towards the solution.

34. See Sacharow (1982).

CASE HISTORY 4.9

The Ergonomically Designed Lounge Chair

The design of a lounge chair can be a challenging assignment.[35] It must conform to the needs of the user, providing sufficient yet comfortable support for the body.

Chair design must include consideration of the following factors:

- The function of the chair (e.g., a desk chair will be used for different tasks than a lounge chair and must reflect this difference).
- The number and type of people who will use the chair (e.g., will the chair be used only by men, only by women, only by the elderly, and so forth). If there is a variety of people who will be using the chair, it should be adjustable to accommodate their differences in size, proportions, strength, and other attributes.
- Related to the function is the amount of time during which the chair will be in continuous use (e.g., an uncomfortable chair cannot be used for long periods of time).
- The environment of use (i.e., will the chair be used in conjunction with a table or other furniture? Must the user be able to reach control devices, such as the dashboard elements in an automobile? How much space is available for the chair?)

Next, one must consider the variations that are possible in the chair design itself. These include:

The seat In a lounge chair, 75 percent of the body weight is supported by the seat with only 8 percent supported by the backrest and the remaining 17 percent supported by the floor. One must specify

- *Seat length* Length should be large enough to provide adequate support under the thighs, yet small enough so that the (short) user does not need to slide forward and away from the backrest.
- *Seat width* This must be large enough to support the ischia (two bony tuberosities or 'sitting bones' at the base of the pelvis) and the buttocks.
- *Seat height* This should be small enough so that one's feet and legs do not dangle; however, if it is too small, it becomes difficult for the user to rise from the chair. Height should also be consistent with any tables or other objects with which the user must interact. It also should accommodate the range of expected users.

---

35. Based upon material from Diffrient et al. (1974).

- *Seat angle* 0° for dining tables and similar functions; 0° to 5° for table and desk chairs.
- *Seat padding* Discomfort results if padding is too hard; however, unduly soft padding also becomes uncomfortable because the body weight is transferred from the ischia to the nearby tissue.
- *Front edge* Hard edges can compress one's flesh against the thigh bones, leading to discomfort, swelling feet and possibly venous thrombosis. In addition, a clearance of approximately three inches is needed to allow one's legs to move backward when rising.
- *Seat covering* Coarse coverings can be uncomfortable, yet slippery coverings may be unsafe. Furthermore, one must consider absorption of moisture (e.g., perspiration), porosity, cleaning, and maintenance of the covering.

Backrest The backrest supports several parts of the body, including
- *Lumbar* the backrest should follow the natural curvature of the lumbar region. Lumbar support is needed in nearly vertical chairs.
- *Thoracic* Lounge chairs need to support both the thoracic and lumbar areas.
- *Sacrum* A sacrum support distributes the load over a wider region, stabilizing the pelvis and increasing comfort.

Headrest You are probably aware that headrests are used in automobiles as a precaution against neck injuries due to whiplash. However, for backrests placed at angles greater than 30° relative to the vertical direction, a headrest is needed to prevent one from resting his or her head on the backrest by sliding forward in the chair, thereby resulting in poor posture. Angle, position and size should be compatible with the expected range of users.

Armrests Armrests support the user's arms and also are used to rise from chairs. Length, width, height, padding, and covering all must be considered.

Angles The backrest-to-seat angle usually varies from 95° to 120°. However, a value of 130° is most relaxing for a lounge chair, although this relatively large angle may inhibit conversation, reading, and other activities.

Backrest angles usually range from 10° to 45° relative to the vertical directions. If more than 30°, a headrest should be provided.

One may also need to consider other aspects of chair design (e.g., the chair's mobility). Figure 4.16 summarizes many of the factors involved in the design of a lounge chair.

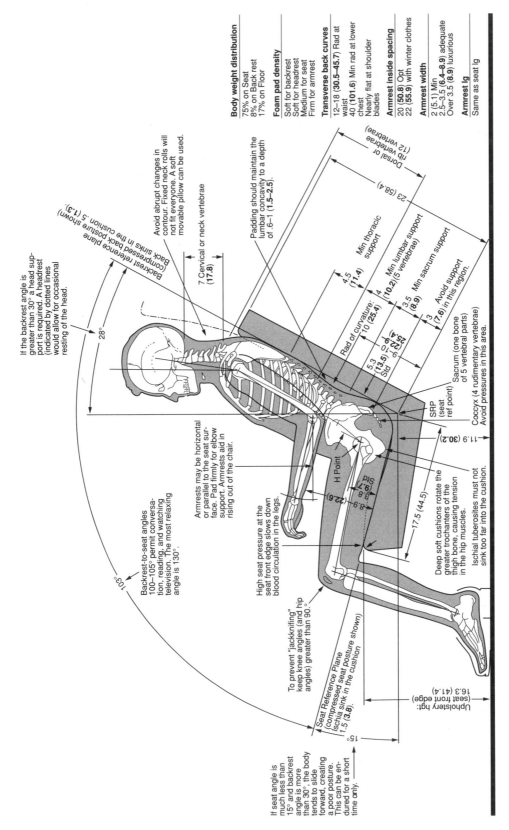

FIGURE 4.16 Some ergonomic factors in the design of a lounge chair (50th percentile male is shown seated). *Source:* Diffrient, Tilley, and Bardagjy, *Humanscale 1/2/3 Manual*, Cambridge, MA: The MIT Press, copyright © 1974. Reprinted with permission.

Body weight distribution
75% on Seat
8% on Back rest
17% on Floor

Foam pad density
Soft for backrest
Soft for headrest
Medium for seat
Firm for armrest

Transverse back curves
12–18 (30.5–45.7) Rad at waist
40 (101.6) Min rad at lower chest
Nearly flat at shoulder blades

Armrest inside spacing
20 (50.8) Opt
22 (55.9) with winter clothes

Armrest width
2 (5.1) Min
2.5–3.5 (6.4–8.9) adequate
Over 3.5 (8.9) luxurious

Armrest lg
Same as seat lg

Dorsal or rib vertebrae (12 vertebrae)

23 (58.4)

Min thoracic support

4.5 (11.4)

Min lumbar support
4 (10.2) (5 vertebrae)

Rad of curvature:
10 (25.4)

3.5 Min sacrum support
(8.9)

5.3 (13.5) Std
9–10
22.9–
25.4

3 Avoid support
(7.6) in this region.

Sacrum (one bone of 5 vertebral parts)

SRP (seat ref point)

Coccyx (4 rudimentary vertebrae)
Avoid pressures in this region.

11.9 (30.2)

H Point

8.9 Std
22.6 (8.9)

17.5 (44.5)

Deep soft cushions rotate the greater trochanters of the thigh bone, causing tension in the hip muscles.

Ischial tuberosites must not sink too far into the cushion.

Upholstery hgt: (seat front edge)
16.3 (41.4)

Seat Reference Plane (compressed seat posture shown)
Ischia sink in the cushion
1.5 (3.8).

15°

If seat angle is much less than 15° and backrest angle is more than 30°, the body tends to slide forward, creating a poor posture. This can be endured for a short time only.

To prevent "jackknifing" keep knee angles (and hip angles) greater than 90°.

High seat pressure at the seat front edge slows down blood circulation in the legs.

Armrests may be horizontal or parallel to the seat surface. Pad firmly for elbow support. Armrests aid in rising out of the chair.

Backrest-to-seat angles 100–105° permit conversation, reading, and watching television. The most relaxing angle is 130°.

105°

If the backrest angle is greater than 30° a head support is required. A headrest (indicated by dotted lines would allow for occasional resting of the head.

28°

Backrest reference plane (compressed back posture shown) Back sinks in the cushion .5 (1.3).

7 Cervical or neck vertebrae (17.8)

Avoid abrupt changes in contour. Fixed neck rolls will not fit everyone. A soft movable pillow can be used.

Padding should maintain the lumbar concavity to a depth of .6–1 (1.5–2.5).

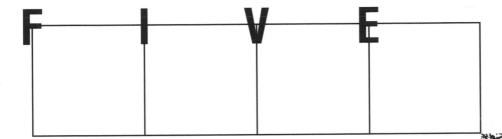

FIVE

ACQUIRING, APPLYING,
AND PROTECTING
TECHNICAL KNOWLEDGE

*Knowledge is the only instrument of production that is not subject to
diminishing returns*   J. M. Clark

O B J E C T I V E S

Upon completion of this chapter and all other assigned work,
the reader should be able to

- Explain the importance of scientific and technical knowledge in
  the practice of engineering.
- Recognize the need to be familiar with many different sources
  of technical information, such as engineering manuals, manufacturers' catalogs, books, magazines, journals, and patents.
- Distinguish between three distinct types of property: real,
  personal, and intellectual.
- Discuss the advantages and disadvantages of the following
  mechanisms for protecting intellectual property: trade secrets,
  trademarks, copyrights, and patents.
- Distinguish between utility, design, and plant patents.
- Define the principal criteria (novelty, nonobviousness, and
  usefulness) used to evaluate utility patent applications.
- Recognize that engineers and inventors should maintain careful records in anticipation of patent applications.
- Recognize that patent protection is based upon the claims
  contained in the patent disclosure.
- Find encouragement in the fact that many creative people without technical training have been awarded patents.

5.1 Science: The Foundation of Technical Knowledge

Design engineers must acquire robust (i.e., broad and deep) technical knowledge and develop the ability to apply it whenever necessary. State-of-the-art designs require a deep understanding of one or more specialized engineering domains, whereas a broad view of how our own areas of expertise relate to other fields (both within and outside of engineering) will provide us with opportunities to apply our knowledge in novel ways (see, for example, Case History 5.1 *A Special Mix of Experience Leads to Photocopying* at the end of this chapter). These opportunities sometimes involve immediate corrections or redesigns of earlier solutions in which a "bug" or flaw has become apparent (see Case History 5.2 *Technical Knowledge and Quick Thinking Save a Life*).

Moreover, designers often are ignorant of—or choose to neglect—the historical records and the lessons contained therein about engineering failures of the past. As Petroski (1994) correctly notes, successful precedents are often emphasized in engineering design whereas failures are either ignored or minimized. This is understandable since we all seek to emulate the successes of our predecessors. Nevertheless, current jeopardy to historically recognized types of hazards can be reduced only if we first become aware of these dangers by carefully reviewing the technical literature and then eliminating (or at least minimizing) them through innovative engineering design. In other words, we must follow the path from hazards awareness and analysis to hazards avoidance, as described in Chapter 9 (see also Case History 5.3 *The Tacoma Narrows Bridge Destroys Itself*).

Technical knowledge begins with an understanding of scientific principles. For example, jet planes and rockets are based upon Newton's first law:

> To every action there is always opposed an equal reaction: or, the mutual actions of two bodies upon each other are always equal, and directed to contrary parts.[1]

In a jet plane or rocket, the vehicle is pushed forward as heated exhaust is ejected from a rear nozzle. (Rockets can operate in airless space since they have their own oxygen supply for fuel combustion; in contrast, jet engines require an external air intake.[2])

Today's solar cell is another example of scientific theory converted into a practical device.[3] The photovoltaic effect, discovered by Edmond Becquerel in 1839, led to the development of selenium photovoltaic cells

---

1. Isaac Newton, as quoted in Sears, Zemansky, and Young (1974).
2. See, for example, Yenne (1993).
3. Yenne (1993).

nearly fifty years later by Charles Fritts. In 1954, more efficient photoelectric cells made of silicon were created, followed by silicon solar cells for space satellites.

The coin-testing mechanisms used in common vending machines also depend upon the creative application of scientific principles.

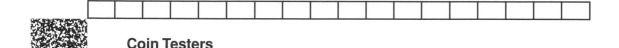

Coin Testers

These devices perform sequential tests on such basic physical characteristics of a coin as its diameter, thickness, weight, iron content, and mass.[4] For example, in some machines a coin, once inserted, must follow a path designed to accommodate coins of a specific diameter. Larger coins are blocked and must be ejected by the customer through activation of a release button, whereas coins that are too small in diameter will simply slip into the return chute. The weight of a coin can be tested by counter-weighted balance arms; if the coin is sufficiently heavy, it will be able to rotate the counterweight, thereby allowing the coin to continue on its journey (rejected coins once again are delivered to the return chute). A magnet can be used to test the iron content of the coin; an iron "slug", for example, will become attached to the magnet, forcing the customer to press the release button that will activate a wiper to sweep the slug from the magnet and into the return chute. The magnet also will slow any coin with an unacceptably high iron content. Such coins will not have sufficient speed in their final descent to bounce over a rejector pin; instead, they will be directed into the return chute. In summary, the expected properties of a coin are used to evaluate any candidate that is inserted into the machine, and each test can be based upon scientific principles of behavior due to these properties. As a result, coin-testers are wonderful examples of applied science.[5]

We will next consider a small portion of the common technological applications of scientific principles in the world around us, so that we can

---

4. *The Way Things Work* (1967) and Sutton and Anderson (1981).

5. Similar authenticity tests can be performed on the inherent characteristics of paper money. These tests may include: a gross density examination in which a photocell is used to measure the amount of light that can pass through the paper; an interferometric test that checks the lines in the paper via optical wave patterns (i.e., a light passing through both the paper and a specific moire grid pattern in the machine creates a wave pattern that will lead to rejection or acceptance of the paper bill); an examination of the bill's length; and a test of the magnetic ink on the bill.

begin to appreciate how important knowledge of such principles is to effective engineering design.

5.1.1 ■ Inclined Planes, Levers, and Counterweights

The use of *inclined planes* in performing certain tasks[6] is based upon the definition of work in physics:

$$\text{Work} = \text{Force} \times \text{Distance} \tag{5.1}$$

According to this equation, the force or effort that must be invested to perform this work can be reduced if the distance over which the force is applied can be increased. Therefore, it is easier to push an object along an inclined plane or sloping ramp to a given height than to vertically raise the object to that height (assuming that frictional effects are negligible). Inclined planes were used to construct ancient engineering projects with greater ease, such as the Egyptian pyramids and roadways. In addition, the principle of the inclined plane is used in wedge devices (such as axes, chisels, shovels, and plows), and in spiral screw systems (such as augers, drills, and the Archimedean water screw used to lift water for irrigation systems).

Another device based upon force × distance ratios is the *lever*. In so-called first-class levers, a rigid rod can pivot about a fixed point (the fulcrum) located between two forces, where one force is the load to be moved and the other force is the effort being applied to perform this movement. The balance equation for a lever is then

$$(\text{Effort}) \times (D_{\text{effort}}) = (\text{Load}) \times (D_{\text{load}}) \tag{5.2a}$$

D_{effort} represents the distance between the applied effort and the fulcrum, whereas D_{load} is the distance between the load and the fulcrum. Thus, the ratio of the forces is equal to the inverse ratio of the distances:

$$\text{Effort/Load} = D_{\text{load}}/D_{\text{effort}} \tag{5.2b}$$

A large load can be raised with only a small effort if the relative distances D_{load} and D_{effort} are selected appropriately. In Figure 5.1, the imbalance (a) between the small weight W and the larger weight $2W$ can be eliminated by moving the larger weight nearer to the fulcrum (b). Balances, crowbars, pliers, and other tools are first-class levers in which the ratios of forces and distances are used to reduce the effort needed to perform a given task.

In second-class levers, the load lies between the fulcrum and the effort force (Fig. 5.2a); examples of such levers include wheelbarrows and nutcrackers. Finally, in third-class levers such as tweezers and fishing rods, the effort force lies between the fulcrum and the load (Fig. 5.2b).

Such diverse systems as elevators, blocks and tackle, and tower cranes all use *counterweights* and the fundamental principles of work and torques to accomplish their tasks.

---

6. See, for example, Macaulay (1988), *The Way Things Work* (1967), Weber (1992), and Yenne (1993).

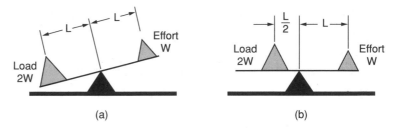

(a)   (b)

FIGURE 5.1 First-class levers: (a) Imbalanced about the fulcrum (pivot point); (b) Balanced.

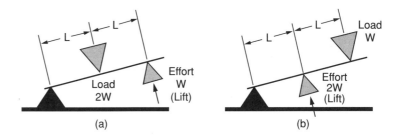

(a)   (b)

FIGURE 5.2 (a) Second-class levers; (b) Third-class levers.

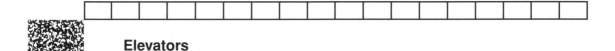

Elevators

Consider the elevator system shown in Figure 5.3: in (a), a motor is used to raise the entire weight W of the elevator car, whereas in (b) a counterweight C is used to offset some of the car's weight so that the motor is only required to lift the difference $W - C$ between the two masses. The counterweight system in a modern elevator can be used to offset the weight of the cab itself and up to one half of the passenger load,[7] thereby minimizing the load actually placed on the hoisting motor.

---

7. Sutton and Anderson (1981).

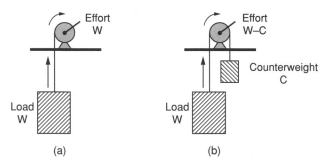

FIGURE 5.3 Elevator system: (a) Without counterweight; (b) With counterweight.

Shadoofs

The shadoof is a manually operated counterweight system used since antiquity to ease the task of lifting water. It consists of a horizontal beam that pivots about a central support; a temporary water container is located at one end of the beam with a counterweight placed at the other end. A person must expend more energy to lower the water container into a lake or river because of the counterweight; however, this counterweight then offsets the weight of the water once the container is filled, thereby making it much easier to lift and pivot the water to a permanent storage vessel near the shadoof.

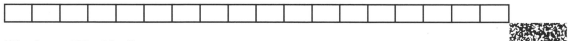

Block and Tackle Systems

A similar use of counterbalances is used in block and tackle systems (Fig. 5.4). In Figure 5.4a, the load $2W$ is attached to an effort W by a cord running over a single pulley. (The pulley dates to the third century B.C.; it was developed by Archimedes.) The radial distance R (i.e., the pulley's radius) at which both forces act about the center of rotation is known as the lever arm of this system; the effort cannot lift the load since its torque is overcome by that of the load.

In Figure 5.4b, a second moving pulley has been added. The load $2W$ is now supported by the two falls of cord directed upward from this pulley. A "pull" or torsion equal to W (i.e., one-half of the load $2W$) acts in each of these falls, and thus the effort W

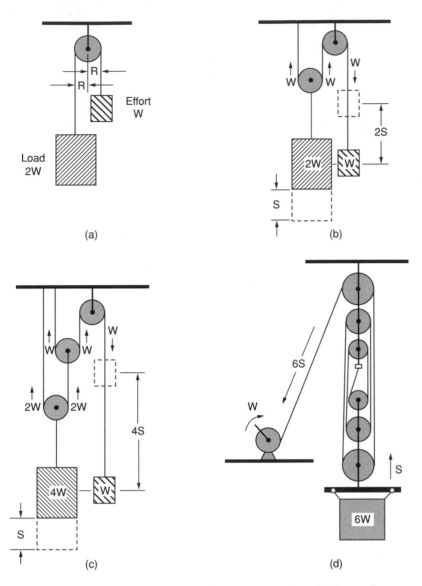

FIGURE 5.4 Pulley systems: (a) Single fixed pulley; (b) Two-pulley system: one moving and one fixed; (c) Three-pulley system: two moving and one fixed; (d) Six-pulley (block and tackle) system. *Source:* Adapted from *The Way Things Work,* 1967.

must only balance one-half of the load. Notice that the effort W must move through a distance of 2S if it is to perform work equal to $S \times 2W$, that is, the amount of work required to move the load 2W through a distance S (where, for simplicity, we are ignoring frictional losses in the pulley system). Thus, a mechanical advantage is achieved with this combination of pulleys: An applied force (effort) can be used to move a weight larger than itself, albeit through a smaller distance than the force itself must move.

Figure 5.4c presents a system of three pulleys, two of which move and one of which is fixed. In this case, an effort W can be used to move a load equal to $4W$ if the effort travels a distance four times as large as the distance S through which the load is moved. Finally, a multiple block and tackle system is shown in Figure 5.4d, in which three moving pulleys are attached to three fixed pulleys. A force W, applied by the motor, can be used to move a load equal to $8W$ if the distance through which the load moves is only one-eighth the distance through which the cord moves. This is an example of the following expression:

$$\text{Load} = (2)^P \times (\text{Effort}) \qquad (5.3)$$

where P denotes the number of moving pulleys within the system. Thus, for the three moving pulleys shown in Figure 5.4d,

$$\text{Load} = (2) \times (2) \times (2) \times \text{Effort}$$
$$= 8 \times \text{Effort} \qquad (5.4)$$

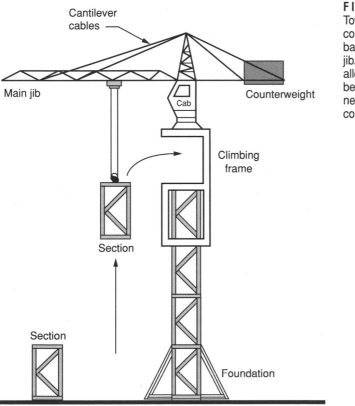

Tower cranes use counterweights to balance the main jib. A climbing frame allows sections to be added as needed during a construction project.

Cantilever cables

Main jib

Counterweight

Cab

Climbing frame

Section

Section

Foundation

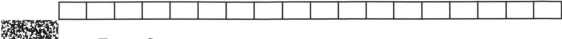

Tower Cranes

Tower cranes (Fig. 5.5) also use counterweights to balance the applied loads that they must lift. These cranes are very cleverly designed to build themselves to any desired height. Preformed sections are lifted into a climbing frame which then hydraulically raises itself atop the new section—an impressive example of creativity combined with practicality.

In summary, a design engineer should be aware of the principles of inclined planes, levers, and counterweights since numerous engineered devices depend upon these principles.

5.1.2 ▩ Wave Phenomena

Many other engineering devices use specific types of wave phenomena to perform certain tasks.[8] For example, radar guns use microwaves to measure the speed of fastball pitches thrown during baseball games. The difference in frequencies between the incident waves striking the ball and the reflected waves bouncing off the ball correspond to the speed at which the ball is traveling.

Such diverse engineered optical systems as eyeglasses, microscopes, and telescopes all depend upon the principle of refraction. According to this principle, a light wave (and sound waves) will bend as it passes from one transparent medium to another because of a change in the average speed of the wave. For example, the speed of light equals c (approximately 186,000 miles per second) when traveling in a vacuum, but decreases to 75 percent of this value when traveling through water, 67 percent of c through glass, and 41 percent of c through diamond.

We refer to the average speed of the wave since photons of light are being absorbed by the molecular electrons of the material through which they are passing, and new photons are being emitted from these excited electrons. The absorbed and emitted electrons share the same frequency so that it seems to be a continuous wave of light traveling through the medium. This process of absorption and reemission takes time to occur, so that the average speed of the light wave is less than c although the instantaneous speed of each photon of light still equals c.

The index n of refraction for light traveling through a given medium is defined as the ratio of the speed c of light in vacuum to the average speed of light in that medium:

$$n = c/(\text{average speed in medium}) \tag{5.5}$$

8. See, for example, Hewitt (1974), Sears et al. (1974), Sutton and Anderson (1981), Weber (1992), and Yenne (1993).

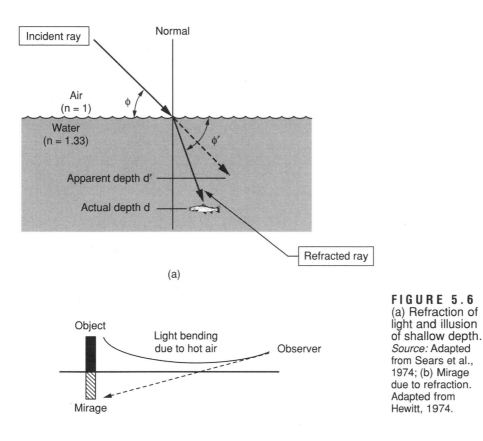

(a)

FIGURE 5.6
(a) Refraction of light and illusion of shallow depth. *Source:* Adapted from Sears et al., 1974; (b) Mirage due to refraction. Adapted from Hewitt, 1974.

As shown in Figure 5.6a, when light crosses a boundary from air ($n = 1$) to water ($n = 1.33$), it slows and bends toward the normal direction relative to the boundary between media. Conversely, an increase in speed when crossing between media causes the light to bend away from the normal. Such bending occurs because one side of the light wave is crossing the boundary and changing its speed before the other side of the wave can do so.

Because of refraction, a fish will appear to lie at a more shallow location d' along the line of sight than its actual depth location d (Fig. 5.6a). Snell's law of refraction allows us to calculate this change in apparent depth from the indices of refraction:

$$n \sin f = n' \sin f' \tag{5.6}$$

where n and n' are indices of refraction in the two media, and where f and f' are the corresponding angles of the incident and refracted rays of light. One can then obtain the following ratio between the apparent depth d' and the actual depth d:

$$d'/d = -n'/n \tag{5.7}$$

Thus, a fish will appear to lie at approximately 75 percent of its actual depth in water.

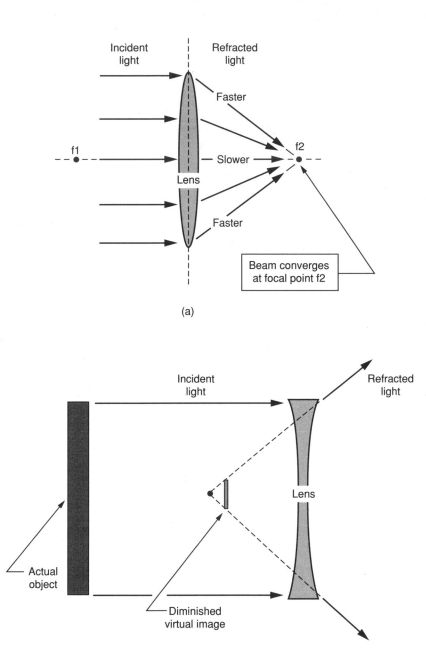

FIGURE 5.7 Refraction of light (a) by a converging lens (convex) and (b) by a diverging lens (concave). *Source:* Adapted from Sears et al., 1974, and Hewitt, 1974.

Mirages will appear in very hot air due to atmospheric refraction: The hot air is less dense, allowing the light to travel faster and bend upward. As a result, an object will appear to be reflected below the surface (Fig. 5.6b) when, in fact, it is refracted. Similarly, highways sometimes appear to be wet when the sky is refracted in hot air.

Refraction is used to advantage in optical systems. A converging or convex lens (one that is thicker at the center than near the edges) refracts incident light so that it converges at a focal point of the lens along its principal axis (see Fig. 5.7a). A magnifying glass uses such a convex lens to produce enlarged virtual images of objects that lie within the focal distance of the lens (an illusion in which the brain assumes that the light rays have traveled from the large object directly along straight lines to the eye[9]).

In contrast, a diverging or concave lens (which is thicker near the edges than at its center) causes the incident light to refract away from the lens, as seen in Figure 5.7b. As a result, diverging lenses produce diminished virtual images of objects; finders in cameras, for example, use diverging lenses to produce reduced images of the scenes that will be captured on film.

Many design engineers often apply the principles of wave phenomena to their work, as the following examples illustrate.

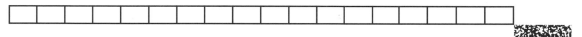

Eyeglasses

Eyeglasses were invented by Alessandro della Spina (a Dominican friar) and the physicist Salvino degli Armati in 1280 in Florence, Italy. Armati wore these first convex eyeglasses to correct the loss in his vision that resulted from the experiments with refraction that he had been conducting. In eyeglass design, such convex lenses are used to correct hyperopic (farsighted) vision whereas concave lenses are used to correct myopic (nearsighted) vision.

The normal eyeball is nearly spherical and about one inch in diameter (Fig. 5.8a). An incident image is focused by the cornea (a transparent membrane) onto the retina (a collection of nerve fibers on the inner part of the eye). Myopic eyes are too long (or the lens is too spherical), so that the image is focused in front of the retina (Fig. 5.8b); in contrast, hyperopic eyes are too short (or the lens is too flat), resulting in an image focused behind the retina (Fig. 5.8c). With the aid of properly designed diverging and converging lenses, images can be focused directly upon the retina.

The near point of vision is the smallest distance between the eye and an object that can be seen clearly. The ciliary muscle in the eye tenses in order to achieve such a near-focus; this process is known as "accommodation." The ability of the eye to perform such accommodation often decreases with age, as seen in Table 5.1.

A converging lens can be used to correct a loss of near-focusing ability. Consider, for example, the Gaussian form of the thin lens formula (applicable to a thin lens with two refracting surfaces):

$$(1/d) + (1/d') = 1/f \qquad (5.8)$$

9. See, for example, Macaulay (1988).

(a) Normal eye

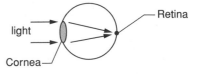

(b) Myopic eye (nearsighted)

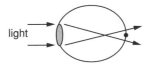

(c) Hyperopic eye (farsighted)

FIGURE 5.8
Common eye config-
urations: (a) Normal
eye; (b) Myopic eye;
(c) Hyperopic eye.

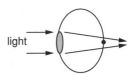

where *d* denotes the object's distance from the lens, *d'* is the image distance, and *f* equals the focal length of the lens along its principal axis. If someone with hyperopic (farsighted) vision cannot focus clearly upon any object that is nearer than 100 cm, and we want to correct this situation to 25 cm (i.e., *d* = −100 cm and *d'* = 25 cm), then we need a converging lens with a focal length *f* equal to 33 cm, according to Equation 5.8.

TABLE 5.1   Near point of vision as a function of age.

Age (yr)	Near Point (cm)
10	7
20	10
30	14
40	22
50	40
60	200

Source: Based upon Sears, Zemansky, and Young, 1974.

Refracting Telescopes

Two converging lenses are used in astronomical refracting telescopes, invented by Hans Lippershy in Holland in 1608 and first used by Galileo Galilei for astronomical observations in 1609. One lens (the objective lens) has a long focal length and produces a real but inverted image inside the focal length of the other (eyepiece) lens. The eyepiece then magnifies this image. The magnifying power of a telescope is given by the ratio of the focal lengths (f.l.) of the objective and eyepiece lenses:

$$\text{Magnifying power} = \text{objective lens f.l.}/\text{eyepiece lens f.l.} \tag{5.9}$$

Three lenses are used in terrestrial telescopes, in which the middle lens inverts the image produced by the objective lens so that it is right-side up when viewed by the observer through the eyepiece.

Acoustical Design of Concert Halls

Acoustical engineering is another area of design in which one must understand wave phenomena.[10] Theaters and music halls are designed to control the reverberation of sound waves (i.e., echoes) in specific ways. The sounds produced by the original source (e.g., a singer, actor, or orchestra) and by echoes should reinforce one another rather than act in competition. In a live amphitheater, the echoes will continue long after the source of the sound has ended. Reverberations in large concert halls, for example, may last two seconds or longer—perfect for choral or orchestral music. Unfortunately, such live rooms may be inappropriate for lectures or plays in which the speaker's words may be lost amidst the echoes, or for musical performances in which many softer sounds may be overcome by louder reverberations; such quiet performances need reverberation times of one second or less. Hence, the acoustical engineer tries to design the room for the particular types of performances that will be given therein.

Sound reverberations can be controlled by both the shape and the furnishings of a room. If a concert hall is irregularly shaped (rather than simply rectangular in form), the sound can be diffused more evenly throughout the room. Moreover, ornamental decorations can be used to properly diffuse the sound. Symphony Hall in Boston was

---

10. Our brief discussion of acoustics is based primarily upon Flatow (1988).

designed by the Harvard physicist Wallace Clement Sabine, who is known as the father of modern acoustic design. His design includes sound-diffusing wall niches filled with Greek and Roman statuary, creating an acoustical environment that is superb for concert performances.

Since the wavelengths of musical notes can vary greatly in size, so too must the diffusing objects vary in size. Private boxes in modern theaters have convex surface paneling for scattering the sound throughout the room. The walls of these theaters often have a rough texture, again designed to improve the acoustics of the rooms. Chandeliers, carpeting, upholstery, insulation, and curtains all add to the acoustical character of a theater by either dampening sounds or enhancing their diffusion. For example, the Kennedy Center in Washington, D.C., has interior carpeting (a short pile of 70 percent wool and 30 percent nylon) that is designed to minimize sound absorption, whereas the exterior carpeting (a deep pile of 100 percent wool) is intended to absorb distractive outside noises. An acoustical engineer must consider all aspects of wave phenomena when designing an environment in which people will interact primarily through sound.

5.1.3 ▦ Piezoelectricity

Compared to the mechanical watches of the past, today's quartz watches can be extremely accurate yet remain relatively inexpensive. These modern timepieces take advantage of the piezoelectric effect exhibited by some crystals and ceramics. Quartz crystals in particular contain positively charged silicon ions together with negatively charged oxygen ions. When placed under pressure, these oppositely charged ions move toward opposite sides of the crystal, resulting in an electric charge.

The piezoelectric effect is simply one of many physical phenomena that can be applied in creative engineering design.

Quartz Watches

In 1929 a quartz resonator was successfully used as part of a clock system by Warren Alvin Marrison.[11] Seiko introduced the first quartz watches for commercial production in 1969. In quartz watches, a battery applies an electric charge to the crystal, causing it to vibrate and produce pulses at its resonant frequency. (In 1988 Seiko and Jean d'Eve developed a design in which miniature dynamos are used in place of batteries.) An integrated circuit (IC) chip is then used to reduce this high-frequency output to a

11. Crane (1993), d'Estaing and Young (1993), and Macaulay (1988).

single pulse per second, which in turn is used to drive the motor and gear train in ana-log watches.

The two major challenges of the quartz watch system is to reduce the high resonant frequency of the crystal oscillator to a usable range of about one pulse per second and to adjust this final frequency so that the watch will maintain proper time-keeping (e.g., keeping proper time to within one minute per year). The first challenge—frequency reduction—is met not only through the use of the IC chip but also by forming the crystal into the shape of a miniature tuning fork. This form, in which the arms of the tuning fork vibrate rather than the opposite faces of the original crystal, reduces the resonant frequency to about 30 kilohertz. The second challenge, accuracy, is met by laser cutting the crystal into the tuning fork shape so that its resonant fre-quency is initially too high. The frequency is then reduced to below the desired stan-dard by adding minute deposits of gold to the fork arms. Some of this gold mass is then evaporated with the aid of a laser, allowing the frequency to rise until it equals the desired value.

5.1.4 ▩ The Joule Effect

James Prescott Joule proposed that a resistant conductor would glow with luminous energy if an electric current is applied to it: This phenomenon is known as the Joule effect.[12] The generation of such luminous energy reflects the effect of heating upon the conductor. Joule's Law states that the heat Q that is developed in a resistant conductor as a steady current I passes through it during a time interval t is given by the expression

$$Q = 0.24\ V \bullet I \bullet t \qquad (5.10)$$

where V is the potential difference between the terminals. (Units: if t is in seconds, V is in volts, and I is in amperes, then Q will be in units of calo-ries.) Furthermore, the first law of thermodynamics is a statement of the gen-eral principle that the total energy in the universe (heat energy as well as all forms of potential and kinetic energy) is constant. More specifically, this law states that when work W is expended in generating some amount of heat Q (or, conversely, if heat is spent in performing work), one can use the follow-ing conversion formula to calculate the relative amounts of W and Q involved in the process:

One joule unit of work = 4.184 calories of heat

This conversion relation is known as the mechanical equivalent of heat. Thus, the work performed by the current as it passes through the resistant conduc-tor is converted into heat, creating a luminous glow in the conductor.[13]

Thomas Alva Edison and England's Sir Joseph Wilson Swan (among oth-ers) independently created incandescent electric light designs based on the Joule effect.

---

12. Wherry (1995) and Yenne (1993).
13. See, for example, Heil and Bennett (1939) and Bueche and Wallach (1994).

Once again, knowledge of a physical phenomenon (the Joule effect) led to the development of a device that improved the quality of people's lives.

Incandescent Electric Lights

The first incandescent electric light based on the Joule effect using a carbon conductor or filament was created in 1838 in France, but it proved to be impractical because the filaments were quickly incinerated. The challenge, then, in converting the Joule effect from pure theory into practice was to create a long-lasting vacuum in which the filament could glow without consuming itself.

Although Sir Joseph Wilson Swan developed an electric light bulb using a carbon filament in 1878, he could not maintain a vacuum within it; he also failed to patent his design. In contrast, Thomas Alva Edison developed a bulb in which a long-lasting vacuum could be maintained, thereby preventing immediate oxidation of the filament. Like Swan, he used a filament made of carbon that could glow for many hours. In 1880 Edison was awarded patent #223,898 for his design. (Swan attempted to market his electric light design in the United States but was prevented from doing so by the patent protection provided to Edison.) Later, in 1883, these two men became partners in the Edison & Swan Electric Company in order to produce the electricity needed for large-scale incandescent lighting.

Although a long-lasting vacuum had been developed, durable carbon filaments remained difficult to manufacture in large quantities and at low cost. Until this final challenge could be overcome, the electric light would be of limited use. In 1881 Lewis Howard Latimer, an African American and the son of former slaves who became one of the world's foremost experts in electric power and lighting, successfully developed a method that could produce durable, inexpensive carbon filaments. Thanks to Latimer's manufacturing process, the electric light became commercially viable for use in homes and offices. (Latimer's process is reviewed in Chapter 11.) A more long-lasting filament made of tungsten was later created by William David Coolidge in 1910.

5.1.5 ■ Sublimation

Sublimation refers to a substance undergoing a change in phase directly from solid to vapor, that is, without passing through an intermediate liquid state. Sublimation of a given substance will occur only under specific conditions of temperature and pressure. The process of freeze-drying coffee depends upon sublimation and the fact that a substance can be vaporized at lower temperatures if the atmospheric pressure is reduced.

Without knowledge of sublimation, freeze-dried coffee would not exist.

Freeze-Dried Coffee

In the manufacture of freeze-dried coffee,[14] first a coffee liquor is prepared by percolating coffee and partially evaporating its water content to form a concentrated mixture. This liquor is then frozen and formed into granules. Finally, these frozen granules are vacuum-dried via sublimation; that is, the remaining water is vaporized in a vacuum at a relatively low temperature. With this two-stage (freezing plus drying) method, the coffee never needs to be heated to the higher temperatures (500° F or so) used in making powdered instant coffee—temperatures that may harm the coffee's flavor.

5.1.6 ▨ Magnetism

A magnetic field will induce an electric current in a conductor that moves through the field.[15] This effect is used in turbines to generate electricity.[16] An electromagnet is connected to the shaft of the turbine; as it spins, it induces an electrical current in the conducting wire surrounding the shaft. Of course, some initial input energy is needed to spin the blades or vanes of the shaft. This input energy is obtained directly from moving water, wind, or solar power, or indirectly through the combustion of common fuels (e.g., oil, coal, natural gas, or fissionable materials such as uranium) to produce steam.

Magnetism also has been used for metal detection. If a metal object passes through a magnetic field, an electric current together with a secondary magnetic field will be induced in the object. The presence of this metal object is then revealed by the distortions in the applied magnetic field caused by the induced secondary field.

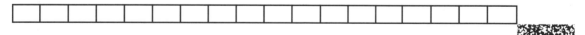

Electric Motors

Magnetic fields also play a critical role in electric motors. For example, in some motors a field magnet is placed near the armature shaft. An electric current is applied to a wire wound about the armature in order to induce a magnetic field.[17] The repulsion

14. Sutton and Anderson (1981).

15. See, for example, Sutton and Anderson (1981) and Yenne (1993).

16. This induced electromotive force, which results in a temporary current in the conductor, was discovered by Michael Faraday in 1831; see Heil and Bennett (1939).

17. This is known as the Biot-Savart law. (It also has been called Laplace's law and Ampere's law; see, for example, Heil and Bennett (1939).

between the like poles of this induced magnetic field in the armature and those of the mounted field magnet (together with the attraction between unlike poles) will generate a torque on the shaft, causing it to rotate. As the shaft completes a 180° rotation, the unlike poles of the armature and the field magnet begin to align with one another. A commutator then reverses the electric current, thereby reversing the poles of the armature. The shaft continues to rotate another 180° until it returns to its original position, at which point the commutator once again reverses the current. The shaft will continue to rotate as this sequence of events is repeated.[18]

A deep understanding of magnetic and electric phenomena is reflected in many of today's engineering designs.

5.1.7 ■ Pressures

Anemometers or air speed indicators provide one example of the application of pressure differences to solve a technical problem. The speed of an airplane can be determined by converting air pressure measurements outside the plane to a set of equivalent velocity readings.

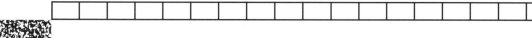

Airspeed Indicators (Anemometers)

A diaphragm in the conventional airspeed indicator of a plane is connected to both a Pitot tube (an open-ended cylinder) and a static tube (sealed at its front end but open to the atmosphere through holes along its length). The Pitot tube is directed parallel to the airflow. As the plane accelerates, the air rushing into the Pitot tube will lead to an increase in air pressure. Any difference between the air pressure in the Pitot tube and the atmospheric pressure in the static tube will cause the diaphragm to expand or contract, leading to a corresponding measurement of the air speed.[19] (More recent efforts have focused upon the development of laser-based airspeed indicators.)

Once again, scientific knowledge was converted to an effective engineered device.

---

18. Nikola Tesla, inventor of the induction motor (U.S. patent #382,280) in 1888, demonstrated that rotating magnetic fields could be used to eliminate commutators in motors; see *The National Inventors Hall of Fame* (1984).
19. d'Estaing and Young (1993) and Sutton and Anderson (1981).

5.1.8 ▨ Binary Notation

Finally, we should recognize that scientific knowledge includes mathematical modeling formats. (Chapter 6 focuses on many types of modeling schemes.) For example, binary notation (in which one can write a given value as a series of zeroes and ones) was developed in 1679 by Gottfried Wilhelm von Leibniz. In this notation system, a "1" denotes the presence of a contributing number and a "0" denotes its absence. An expression of the form "1101" indicates the presence of one digit equal to eight, one digit equal to four, zero digits equal to two, and one digit equal to one; thus, the total value of "1101" equals (8 + 4 + 0 + 1) or 13.[20]

Without knowledge of the binary notation created by von Leibniz so long ago, the designers of the Colossus and other electronic computers would not have been successful.

Computers and Cryptography

The development of modern electronic computers was dependent upon the binary notation system from the seventeenth century. The absence or presence of electrical impulses in a computer simply corresponded to the zeroes and ones in a given number set. If a circuit was open in the computer, no impulse was transmitted, corresponding to a 0 in the binary system; conversely, a closed circuit produced an impulse that corresponded to a 1. As a result, numerical values could be represented by a set of electrical impulses corresponding to the binary code.

The first fully electronic computer was the Colossus Mark I, constructed between 1941 and 1943 by Alan Turing, M. H. A. Neuman, Thomas H. Flowers, and others to help the Allies decipher the secret military communication codes of the Axis powers during World War II. (The IBM Mark 7, developed in 1937 by Howard Aiken, used electromechanical switches rather than electronic ones.) The Nazis transmitted coded messages in the following way: The original (input) message would be aligned with a so-called key stream (a particular arrangement of letters) and converted into the corresponding letters from this stream. This coded message would then be converted one more time by aligning the key stream with an output string of letters. Hence, the original word "ship" might be transformed first into "dgyt" by the key stream, and then finally converted into the output message "kcbn." The receiver of the coded message could simply reverse this process, given the correct key stream.

---

20. Flatow (1992) and Yenne (1993).

Mathematician and cryptographer William Friedman had noted that strings of English text, when aligned, would produce matches between letters about seven percent of the time, whereas random strings of letters would match only four percent of the time. The difference was due to the fact that letters in the English language do not appear randomly; some occur more frequently than others. The cryptographers working for the Allies were very successful in deciphering Axis messages throughout the war by using this frequency differential. However, the key stream (known as the "Ultra Code") used for the most secret of the Nazi transmissions could produce an apparently random set of 16,000,000,000,000,000,000 (16 quintillion) combinations of letters. The task of deciphering any message with that many possible coded combinations was too much for human cryptographers—but not for an electronic one.

In order to crack the Ultra Code, the Allied cryptographers first constructed a Teletype tape-reading decoding machine based upon their knowledge of the German Teletype code-generating machine. Unfortunately, tape messages had to be read by this decoding machine numerous times before the message could be partially deciphered; each reading damaged the tape until it eventually was destroyed. A better technique was needed.

Thomas Flowers suggested that the work he had performed before the war to convert part of England's telephone system from mechanical switches to electronic ones might be the answer to the Allies' problem. The telephone switching project had been named Colossus, and the cryptographers replaced part of their tape-decoding machine with an internal electronic memory based upon the Colossus system. Since many of the fifteen hundred vacuum tubes in this new machine (dubbed the Colossus Mark I) would quickly burn out if they were turned on and off, Flowers further suggested that the machine simply never be turned off, thereby extending the operating life of these tubes.

Thanks to the success of the Colossus machines in deciphering Axis messages, the Allies were able to modify their plans for such operations as the D-Day landing in Normandy, successfully overcoming Nazi preparations. The war effort was helped inestimably by the development of these Colossus computers.

The existence of the Colossus machines remained shrouded in so much secrecy after the war that for many years most people believed that the ENIAC (Electronic Numerical Integrator And Computer), developed by John Mauchly and John Eckert at the University of Pennsylvania between 1942 and 1946, was the first electronic computer. In recent years, however, the Colossus has finally received the recognition that it deserves as the world's first fully electronic computer.

5.1.9 Summary

The above examples (summarized in Table 5.2) illustrate the breadth and depth of science appearing in engineering designs. Of course, there are innumerable other instances of scientific principles used in engineering problem solving that we could cite; some of these are listed in Table 5.3.

We must conclude that effective engineering design is heavily dependent upon scientific knowledge. The more one knows about science and its applications in technology, the more likely it will be that he or she will be successful as a design engineer.

TABLE 5.2 Summary of examples.

Scientific Principles / Theories	Examples of Engineering Applications
Counterweights and balances	Elevators, shadoofs, block and tackle systems, tower cranes
Inclined planes	Wedge devices (axes, chisels, shovels, plows), spiral screw systems
Joule effect	Incandescent electric lights
Levers (first, second, and third class)	Balances, crowbars, pliers, wheelbarrows, nutcrackers, tweezers, fishing rods
Magnetism	Turbines, motors, metal detectors
Newton's first law	Jet planes, rockets
Photovoltaic effect	Photovoltaic, photoelectric, and solar cells
Piezoelectric effect	Quartz watches
Pressure differences (Pitot tube)	Airspeed indicators
Sublimation	Freeze-dried coffee
Unique physical characteristics (materials)	Coin testers, paper money testers
Wave phenomena	Radar guns, eyeglasses, microscopes, telescopes, acoustical designs
Binary notation	Computers

5.2 Sources of Technical Knowledge

We can only apply knowledge with which we are familiar. As we noted at the beginning of the preceding section, design engineers must acquire robust (broad and deep) technical knowledge and develop the ability to apply it whenever necessary. The successful design engineer seeks to increase his or her technical knowledge because

- such knowledge often provides innovative ideas and problem-solving strategies; and
- a thorough understanding of scientific principles and physical phenomena is required for proper analysis of new design concepts.

There are many sources of technical knowledge, such as other technically knowledgeable individuals, engineering journals, textbooks, magazines, reference manuals, newspapers, manufacturer catalogs, and so forth.

TABLE 5.3 Examples of physical phenomena used in technical applications.

Adsorption   Attraction of a gas or liquid to the surface of a material through either molecular or capillary action. *Example applications:* Dehydrator (silica gel), refrigerative systems.

Archimedes' principle   A body immersed in liquid will experience a buoyant force equal in magnitude to the weight of the fluid that is displaced. *Example application:* Hydrometer.

Eotvos effect (Coriolis): The weight of a body changes if it is moving along an east/west direction on the surface of the earth: It becomes lighter when moving eastward but heavier when moving westward (an example of the Coriolis effect with gravitational considerations). *Example application:* Pendulum motion (harmonic when moving in a north/south direction, anharmonic when moving in a east/west direction).

Magnetostriction (e.g., Joule, Villari, Wertheim, and Wiedemann effects)   If a ferromagnetic material (e.g., iron, cobalt, nickel) is placed in a magnetic field, it will experience a change in its dimensions and vice versa (i.e., a change in dimensions will result in a corresponding change in magnetization). *Example applications:* Sonar, fathometer (device to measure depths in water and locate fish), milk homogenization, strain gages.

Photoemissive effect   Electrons are emitted from a metallic surface when it is exposed to radiation. *Example application:* Motion picture sound systems (light striking the cells of the sound strip can be varied by modifying the strip's density, resulting in signals that are then transformed to sound).

Purkinje effect   When the level of illumination is low, the human eye becomes less sensitive to light of longer wavelength. *Example applications:* Vehicle taillights (half-violet and half-red for maximum visibility in both night and daylight situations), color selections for paint and clothing.

Work hardening   The strength of a material can be increased by deforming it with stresses that lie beyond its elastic limit (i.e., larger stresses will then be required to further deform the material). *Example application:* Metal-working processes (e.g., swaging, coining, and cold rolling).

Source: Based on Hix and Alley, 1965.

Moreover, the Internet provides access to numerous technical databases that can be helpful in both formulating and solving an engineering problem. When working on a design problem, one should try to consult as many different sources for pertinent information as possible.

Examples of some important sources of information about standards, specifications, and manufacturers' catalogs are the following:

- *Index of Federal Specifications and Standards,* U.S. General Services Administration

- *Index of Standards and Specifications,* U.S. Department of Defense
- *MacRae's Blue Book* (a collection of manufacturers' catalogs)
- *The Process Industries Catalog* (chemical engineering catalog), Van Nostrand Reinhold
- *Sweet's Catalog File,* New York: McGraw-Hill (a collection of manufacturers' catalogs with particular emphasis upon materials and machine tools)
- *Thomas Register of American Manufacturers* (addresses, classified by product)
- *VSMF Design Engineering System* (monthly update on microfilm of vendors' catalogs of electronic and mechanical components)

Similarly, the following are examples of the many sources of technical data that can be useful to the design engineer (publication dates are not included because many of these references are frequently re-issued in updated editions):

- Ballentyne, D. W. G., and Lovett, D. R. (Eds.), *A Dictionary of Named Effects and Laws in Chemistry, Physics and Mathematics,* London: Chapman and Hall
- Beaton, C. F., and Hewitt, G. F. (Eds.), *Physical Property Data for the Design Engineer,* New York: Hemisphere Publishing
- Blake, L. S. (Ed.), *Civil Engineer's Reference Book,* Boston, MA: Butterworth
- *Dictionary of Engineering,* New York: McGraw-Hill
- *Dictionary of Mechanical & Engineering Design,* New York: McGraw-Hill
- Fink, D. G., and Beaty, H., *Standard Handbook for Electrical Engineers,* New York: McGraw-Hill
- *Marks' Standard Handbook for Mechanical Engineering,* New York: McGraw-Hill
- Maynard, H. B. (Ed.), *Industrial Engineering Handbook,* New York: McGraw-Hill
- *McGraw-Hill Encyclopedia of Science and Technology,* New York: McGraw-Hill
- Merritt, F. S. (Ed.), *Standard Handbook for Civil Engineers,* New York: McGraw-Hill
- Stole, L. (Ed.), *Handbook of Occupational Safety and Health,* New York: John Wiley & Sons
- Tapley, B. D., *Handbook of Engineering Fundamentals,* New York: John Wiley & Sons

It is important to recognize that these listings are far from complete. Visit your local library and try to become familiar with the numerous sources of technical information that are available.

Moreover, the U.S. Patent and Trademark Office maintains a home page on the Internet from which users may obtain current information about the patent system and limited descriptions of relatively recent patents.

5.3 Protection of Intellectual Property

Consider the various types of property that one might own. Property can be legally classified as follows:

- **Real** property is tangible and is usually not movable (e.g., houses and land).[21] Ownership of real property is legally indicated by a deed.
- **Personal** property is tangible and generally movable (e.g., clothing, furniture, books). Ownership is legally indicated by a sales receipt.
- **Intellectual** property is intangible (e.g., ideas or concepts, such as inventions, works of art, music, product names, recipes). Ownership can be legally indicated via trade secrets, trademarks, copyrights, or patents.

Engineers often produce ideas that need to be protected as their intellectual property or that of their employers. Two cases that illustrate the importance of such intellectual property protection are those of the telephone (Case History 5.4 *Bell and Gray: A Search for a Better Telegraph Leads to the Telephone*) and the laser (Case History 5.5 *Laser Technology: Patent Rights Can Be Important*).

Each of the mechanisms (trade secrets, trademarks, copyrights, and patents) for protecting intellectual property offers both distinct advantages and disadvantages and also provides protection for specific types of intellectual property.

5.3.1 ▨ Trade Secrets

One type of legal protection of intellectual property provided by the government of the United States of America is known as the "trade secret." This mechanism helps an individual or company maintain the secrecy of a particular process or product. For example, the formula for Coca-Cola is a trade secret that is kept secured within a vault in Atlanta, Georgia. If anyone were to attempt to obtain this secret without permission, among the laws that would be broken is one that affords legal protection for such trade secrets. Such protection can be extremely valuable: A $125 million settlement resulted from a 1989 case in which the producer of a commercial cookie with a chewy interior sued its rivals for violating the trade secret protection of the cookie recipe.[22]

This type of protection offers both advantages and a (major) disadvantage, as follows.

---

21. One exception is a mobile home (although most mobile homes are actually quite difficult to move from one location to another).
22. Arnett (1989).

Advantages

- No time limit
- Property remains a "secret" known only to owner(s)
- Provides legal protection against others attempting to learn the secret

Disadvantage

- Belongs to owner only if he/she can keep it a secret; may be legally "reverse engineered"

The above disadvantage simply means that if the secret is somehow discovered through honest research or by accident, no law has been broken.

5.3.2 ■ Trademarks

A second type of protection for intellectual property is trademarks. According to the U.S. Patent and Trademark Office,

> A trademark relates to any word, name, symbol, or device that is used in trade with goods to indicate the source or origin of the goods and to distinguish them from the goods of others.[23]

Trademarks protect product names, logos, and other identifiable and commercially valuable property. (Brand names can be vocalized and therefore consist of words, letters, or numbers, as opposed to brand marks, which are representative symbols.[24]) For example, the names "Coke" (for Coca-Cola), "Band-aids," "Kleenex," and "Xerox" are all protected from general use as trademarks. Other companies cannot legally use such names for their products, even if the product is generically identical to the trademarked brand. (When next in the supermarket, note the generic names used for some of the above types of products.)

Successful design usually extends beyond the product itself to include such factors as the product's packaging, name, and logo—all of which can affect the degree of public acceptance enjoyed by the product (see Case History 5.6 *Logo and Packaging Design of Coca-Cola*). After all, a poorly packaged product may not be enthusiastically embraced by the public, leading to commercial failure.

As with other forms of legal protection, both advantages and disadvantages are associated with trademarks. Advantages include the opportunity to renew the trademark every five years, thereby providing a virtually unlimited time limit of protection. However, a disadvantage is that trademark

---

23. As stated in "What Is a Patent?" from the U.S. Patent and Trademark Office's Internet site (1996).
24. Sacharow (1982).

protection can be lost if the owner cannot demonstrate reasonable efforts were made to prevent the public from using the trademarked name in a generic sense. For example, the Bayer Pharmaceutical Company received trademark protection for their product known as aspirin (i.e., acetylsalicylic acid) when it was first introduced onto the market. However, as time went by, the public began to refer to all such products as "aspirin" and Bayer eventually lost their trademark protection of this name. (In Canada, the name aspirin still can only be used by Bayer.) Other former brand names that have become generic descriptors include *nylon, thermos, formica, linoleum, cellophane, shredded wheat,* and *kerosene.*[25]

Uneeda Biscuits

A product's brand name may be chosen after intensive market research, brainstorming sessions, and other efforts have been conducted in order to ensure that the most commercially viable selection is made. For example, the Uneeda biscuit slicker boy trademark was chosen to emphasize that the package used to distribute these soda crackers was revolutionary in design.[26]

Most crackers in the nineteenth century were sold from open barrels. However, once distribution methods allowed crackers to be sold beyond the local region of a bakery, new packaging was needed to maintain the product's freshness. A group of baking companies merged to form the National Biscuit Company (now Nabisco) in order to sell crackers and other products on a national scale. A new package was designed for their crackers: an inner wrapping of wax paper in a sturdy outer cardboard box that would keep the crackers fresh, clean, and free from moisture. Next, the name Uneeda was selected for their product. Finally, they searched for some striking symbol that would convey the uniqueness of their product and package to the public.

The answer was a photograph of Gordon Stille dressed for rainy weather in boots, oil hat, and slicker while holding a box of Uneeda biscuits. (Five-year-old Gordon was the nephew of Joseph J. Geisinger, a copywriter employed by the advertising firm N. W. Ayer & Son, who was working on the Uneeda trademark development project.)[27] In 1899 his image appeared on Uneeda biscuit boxes throughout the country, and the immediate success of this product revolutionized food packaging and distribution. Food products were no longer distributed mostly in bulk to local grocers; instead, brand-name packaging became the more common mode for widespread distribution and customers became brand conscious in the process.

---

25. Sacharow (1982).

26. See, for example, Morgan (1986) or Sacharow (1982).

27. The Ayer firm also created the famous *Morton Salt* umbrella girl trademark in 1911; Sacharow (1982).

The success of the Uneeda biscuit product has been called the "one biggest landmark event in the history of American trademarks."[28] Creativity in packaging and the selection of a trademark had led to revolutionary changes in both marketing and distribution.

Elmer's Glue

Sometimes, the simplest choice of all for a product's name can be very successful. For example, chemical engineer Elmer O. Pearson developed a new adhesive during the 1930s while he was employed by Borden Milk. He received recognition for this contribution when the new product was dubbed "Elmer's Glue." (Moreover, since the trademark for Borden Milk products was "Elsie the Cow," "Elmer the Bull" became a natural choice for the new glue product of the company.)

The importance of promptly registering a trademark is reflected in the following anecdote.

The AT&T Blue Bell

The famous blue bell symbol used to identify AT&T (American Telephone & Telegraph Company) was first adopted in 1889.[29] It was meant to symbolize Alexander Graham Bell, his invention of the telephone, and the company itself. (Although AT&T was not directly associated with Bell, it did acquire his firm, the American Bell Company, in 1899.)[30] For the next 64 years, the symbol underwent small design changes but remained fundamentally the same. Then, in 1953, AT&T discovered that the owners of a telephone answering service had registered a similar bell symbol as the trademark for their business. Furthermore, AT&T could not prevent the owners from licensing the use of the symbol to other answering services throughout the nation since AT&T had failed to officially register the company symbol. Eventually, AT&T paid $18,000 to the owners of the answering service to abandon their trademark, thereby allowing AT&T to register the bell system for its company. (In 1982 AT&T adopted its current globe trademark.)

28. Sacharow (1982).
29. Barach (1971).
30. Morgan (1986).

In summary, then, we see that trademarks have the following advantages and disadvantages:

Advantages

- Renewable every five years
- Unlimited time limit (simply renew)
- Provides legal protection

Disadvantages

- Belongs to owner only if it does not become used generically to identify the product "type"

5.3.3 ■ Copyrights

A third type of protection for intellectual property is copyright. Copyright protects intellectual forms of expression, such as artwork, books, and music. However, only the form of expression is protected, not the idea or concept itself. Hence, although there are many books that present the theory and applications of calculus, each author or publisher can nevertheless receive copyright protection for their particular expression of these ideas.

The advantages and disadvantages of copyright protection are as follows.

Advantages

- Renewable
- Very long lasting legal protection (For copyrights issued since 1977, this protection extends until 50 years after the death of the author.)

Disadvantages

- Time limit, although lengthy, is nevertheless finite; after the copyright protection has expired, the work enters the public domain[31]
- Only protects the specific *form* of expression, not the idea or concept itself

5.3.4 ■ Patents

The fourth type of protection for intellectual property and the one that is of most interest to engineers is that of patents. Patents are a particularly important source of technical data because of their currency (state-of-the-art technology and products) and the creativity contained within them. In addition,

---

31. For example, in 1980 a number of collected works of Sir Arthur Conan Doyle (author of the *Sherlock Holmes* stories) appeared on the market because these works had entered the public domain (Doyle died in 1930).

32. As noted in Miller and Davis (1983), this is known as the "bargain theory" of justification for the patent system. An alternative justification for the system is the "natural rights theory," which simply states that the person who originally generates intellectual property (e.g., an idea or design) should have legal ownership of it.

the U.S. patent system provides engineers with a valuable mechanism for protecting their intellectual property.

The United States patenting system is based upon the principle of *quid pro quo* (i.e., "something for something"). In exchange for publicizing his or her invention, a utility patent gives the awardee the right to exclude others from making, selling, or using the invention in the United States for a period of twenty years from the date of application. Thus, the potential of financial (and other) rewards acts as an incentive to invent.[32] (However, not everyone seeks to be rewarded for their designs; Benjamin Franklin, for example, donated all of his inventions to the public.)

Utility patents are nonrenewable; after the twenty years have passed, the invention enters the public domain. Prior to June 1995, utility patents were awarded for a period of seventeen years, dating from the date on which the patent was awarded (as opposed to the date of application). The patent system was modified so that it would be more consistent with the patent systems of many other nations and in order to eliminate the problem of so-called submerged patents.[33] A submerged patent is one that has been deliberately delayed by the inventor in order to extend the life of the protection. In the past, this could be accomplished by incorporating intentional errors in the patent application; the patent examiner would then return the application to the inventor for correction, at which time new errors would be introduced into the application.

The first recorded patent to be granted for an invention was awarded by the Republic of Florence in 1421 to the designer of a barge with hoisting gear for loading and unloading marble. Subsequent early patents were granted in Venice (1469), in Germany (1484), and in France (1543).

In 1641 Samuel Winslow was granted the first patent in colonial America (by the Massachusetts General Court) for a salt-making process. In 1783 the Continental Congress recommended the enactment (in each state) of copyright acts; then in 1789, Article I, Section 8, clause 8 of the United States Constitution empowered the federal government as follows:

> Congress shall have power . . . to promote the progress of science and useful arts, by securing for limited times to authors and inventors the exclusive right to their respective writings and discoveries.

This clause provides the Congress with the opportunity to enact statutes that will award intellectual property rights to authors and inventors; however, the statutes themselves must be passed by Congress. The first such statute, the United States Patent Act of April 10, 1790, resulted in the granting of the first U.S. patent to Samuel Hopkins for "Making Pot and Pearl Ashes" (on July 31, 1790). Beginning in 1836, the U.S. Patent Office was given the authority to examine patent applications;[34] since that time, patents have been designated

---

33. Wherry (1995).

34. The first patent act of 1790 required applications for patents to be examined by the nation's secretary of state, the secretary of war, and the attorney general.

by sequentially increasing numbers.[35] In 1870 the commissioner of the Patent Office was authorized to print copies of current patents. (Trademark protection also began in that year.) Then in 1872, the *Official Gazette of the United States Patent and Trademark Office* was first published. This weekly publication summarizes the patents and trademarks (published in two volumes) that have been issued during the preceding week. These summaries include excerpts from patent disclosures, including one claim for each newly patented design together with a representative drawing.

Abraham Lincoln's Buoying Design

One might wonder if only engineers or highly trained individuals can expect to receive patents. The answer is that anyone who is creative and develops a novel solution to a technical problem can receive a patent. For example, Abraham Lincoln, while serving as a United States congressman, was awarded patent #6,469 for "A Device for Buoying Vessels over Shoals" (1849). His invention consisted of a set of bellows attached to the ship's hull just below the waterline. Upon entering shallow water, these bellows would be filled with air, thereby buoying the vessel so that it would clear the shallow region.

Lincoln developed this idea after twice experiencing difficulties in traveling by river. Once, when Lincoln was transporting merchandise via the Mississippi River by riverboat, the boat slid into a dam; heroic efforts were required to free it. Then, years later, another ship on which Lincoln was a passenger became lodged on a sandbar in the Great Lakes.

Many creative people have received patents, even though they were not professional inventors or engineers.[36] Samuel Clemens (Mark Twain) held several patents, including one for suspenders (#121,992) and another for a game that tested players' knowledge of history (#324,535). Albert Einstein and Leo Szilard were awarded patent #1,781,541 in 1927 for a refrigeration method using butane and ammonia. Harry Houdini, the famous escape artist, received a patent #1,370,376 for the design of an escapable diving suit. Walter F. (Walt) Disney was awarded design patent #180,585 in 1957 for his mad teacup ride at Disneyland.

Zeppo Marx (one of the comic Marx brothers) held patent #3,426,747 for a clever cardiac-pulse–driven double watch system (1969); the rate of one watch, driven by the wearer's pulse, is compared to that of a second reference watch to indicate any unexpected variation in the pulse rate. Charles

---

35. Patents that were awarded prior to 1836 are identified by date and the name of the inventor(s); these are known as "name date patents."
36. See, for example, Richardson (1990) and Wherry (1995).

Darrow, the inventor of Monopoly, held the 1935 patent (#2,026,082) on this extremely popular board game. John J. Stone-Parker, together with his sister Elaine, invented a retainer in 1989 to stop ice cubes from falling out of a glass (#4,842,157); John was only four years old at the time.

Table 5.4 lists some of the important devices and processes that have been awarded U.S. patents, together with their inventors. (This list is far from complete; think about other inventions that you would add to it—particularly those devices that have been developed in the past thirty years—and the reasons for including these designs and their inventors.)

As noted in Section 5.1, the likelihood that one will be successful in engineering problem solving is greatly enhanced by familiarity with technical data and knowledge. As we will discover in Chapter 7, many patents have been granted for impractical devices that have never been manufactured. In fact, patents are awarded to about 65 percent of the applications submitted to the U. S. Patent and Trademark Office, and only about 10 percent of the patented concepts are ever manufactured.[37]

How valuable is patent protection? In recent years, Polaroid Corporation won an infringement lawsuit against Kodak in the area of instant photography.[38] The settlement by Kodak amounted to $909,500,000—nearly a $1 billion loss! Sumitomo Electric Industries in Japan suffered an even greater impact when it lost a 1987 patent-infringement lawsuit to Corning Glass Works over an optical fiber design: The company was forced to close.[39]

To summarize, then:

Advantage
- Utility patents now provide 20 years of protection during which time others are excluded from making, selling, or using the invention

Disadvantage
- Nonrenewable

In the next section, we will delve more deeply into patents since they are so critical to engineers as a source of technical knowledge and as a mechanism for protecting intellectual property.

37. Wherry (1995).
38. Wherry (1995).
39. Levy (1990).

TABLE 5.4 Selected set of patented devices and processes.

Number	Date	Inventor(s)	Invention
132	1837	T. Davenport	Electric motor
1,647	1840	S. F. B. Morse	Telegraph signs
3,237	1843	N. Rillieux	Multiple evaporation (also see #4,879)
3,633	1844	C. Goodyear	Vulcanized rubber
4,750	1846	E. Howe, Jr.	Sewing machine
4,848	1846	W. T. G. Morton, C. T. Jackson	Anesthesia
6,281	1849	W. Hunt	Dress/safety pin (also see #742,892)
17,628	1857	W. Kelly	Manufacture of steel
22,186	1858	J. L. Mason	Mason jar (screwneck bottles)
31,128	1861	E. G. Otis	Safety elevator (also see #113,555)
79,265	1868	C. Sholes, C. Glidden, S. Soule	Typewriter
88,929	1869	G. Westinghouse, Jr.	Air brake
123,002	1872	J. B. Eads	Bridge caisson
135,245	1873	L. Pasteur	Pasteurization (brewing beer and ale)
139,407	1873	E. McCoy	Lubicator (lubricating cup)
157,124	1874	J. F. Glidden	Barbed wire
174,465	1876	A. G. Bell	Telephone
200,521	1878	T. A. Edison	Phonograph
223,898	1880	T. A. Edison	Incandescent electric light
257,487	1882	W. F. Ford	Stethoscope
306,954	1884	G. Eastman	Photographic paper strip film
388,116	1888	W. S. Burroughs	Calculator
388,850	1888	G. Eastman	Roll film camera
400,665	1889	C. M. Hall	Manufacture of aluminum
493,426	1893	T. A. Edison	Motion picture projector
504,038	1893	W. L. Judson	Zipper
586,193	1897	G. Marconi	Wireless telegraphy (radio)
644,077	1900	F. Hoffman	Aspirin (acetylsalicylic acid)
686,046	1901	H. Ford	Motor carriage
766,768	1904	M. J. Owens	Glass shaping (bottle) machine
775,134	1904	K. C. Gillette	Safety razor (also #775,135)
808,897	1906	W. H. Carrier	Air conditioning
821,393	1906	O. Wright, W. Wright	Airplane with motor
942,809	1909	L. H. Baekeland	Bakelite
1,103,503	1914	R. H. Goddard	Rocket engine

continued

Number	Date	Inventor(S)	Invention
1,773,079	1930	C. Birdseye	Frozen foods
1,773,980	1930	P. T. Farnsworth	Television system (also #1,773,981)
1,948,384	1934	E. O. Lawrence	Cyclotron (acceleration of ions)
2,071,251	1937	W. H. Carothers	Nylon
2,221,776	1940	C. F. Carlson	Xerography (electron photography)
2,404,334	1946	F. Whittle	Jet engine
2,435,720	1948	E. H. Land	Polaroid camera
2,524,035	1950	J. Bardeen, W. H. Brattain	Transistor
2,708,656	1955	E. Fermi	Neutronic reactor (patent filed in 1944)
2,717,437	1955	G. de Mestral	Velcro
2,929,922	1960	A. L. Schawlow, C. H. Townes	Laser (optical maser)
3,093,346	1963	M. A. Faget *et al.*	Space capsule
3,139,957	1964	R. B. Fuller	Geodesic dome (suspension building)

5.4 Patents

5.4.1 ■ Types of Patents

Three types of patents are awarded by the U.S. Patent and Trademark Office: utility patents, plant patents, and design patents, as provided by the U.S. Patent Act (Title 35 of the U.S. Code), sections 101, 161, and 171, respectively.

Utility patents protect functional products or concepts, such as light bulbs, airplanes, electronic components, over-the-counter drugs, walkmen, robots, and so forth. As a result, this type of patent is of more concern to the engineer than design or plant patents.

Plant patents are used to protect asexually produced (e.g., produced through grafting, budding, cutting, layering, or division) plants. Certain types of fruit trees and rose bushes are protected by such patents. The first plant patent was awarded to Henry F. Rosenberg in 1931 for a climbing rose.

Design patents are used to protect the form or shape of an object. It does not protect the functional capabilities of a design, only the appearance. For example, unique patterns for silverware or shoe soles can be protected by design patents. The first design patent was awarded to George Bruce in 1842 for a novel typeface (i.e., a font style); Auguste Bartholdi received design patent #11,023 in 1879 for his design of the Statue of Liberty.

Returning to utility patents, Section 101 of the Patent Act states that:

> Whoever invents or discovers any new and useful process, machine, manufacture, or composition of matter, or any new and useful improvement thereof, may obtain a patent therefor, subject to the conditions and requirements of this title.

This portion of the statute specifies the patentability of products and processes. A *process* is a procedure or method that can be used to accomplish some task (e.g., a chemical process); whereas machines, manufactures, and compositions of matter are all tangible *products* distinguished by certain characteristics. In particular, a *machine* is defined as "a combination of heterogeneous mechanical parts adapted to receive energy and transmit it to an object"[40] (e.g., an airplane or an electric motor). A *manufacture* is anything that can be fabricated (i.e., mass produced, such as a safety pin) other than a machine or composition of matter. *Compositions of matter* are substances created from two or more materials (e.g., new chemical compounds or alloys such as nylon and steel)[41], in which the resulting compound is more than a simple mixture or recipe.[42] We also should note that new and useful improvements to such processes and products can be patented; in fact, most patents have been awarded for improvements to existing devices and concepts.

Over the years, "negative" rules of invention have developed that specify (sometimes rather loosely) what *cannot* be patented. These include illegal or immoral inventions, since there must be a utility of some merit associated with the design; simple changes in the size, weight, or form of an existing design; a mere aggregation of existing parts or components; or the substitution, omission, relocation, or multiplication of existing components or materials.[43] Moreover, one cannot patent the laws of nature, physical phenomena, or abstract ideas.[44]

One must pay a filing fee when submitting a patent application to the U.S. Patent and Trademark Office. These fees are reduced by 50 percent for applicants who are "small entities" (i.e., individuals, small-business concerns, or nonprofit organizations). As of May 1998, the filing fees for these three types of patents were as follows (with the fees for small entities given in parentheses):

Utility basic filing fee:	$790 ($395)
Design filing fee:	$330 ($165)
Plant filing fee:	$540 ($270)

In addition, one must pay the following additional fees if a patent is issued:

Utility issue fee:	$1,320 ($660)
Design issue fee:	$450 ($225)
Plant issue fee:	$670 ($335)

---

40. Ardis (1991).
41. Miller and Davis (1983).
42. Ardis (1991).
43. See, for example, Ardis (1991) and Miller and Davis (1983).
44. See, for example, Levy (1990); The use of sulfuric ether as an anesthetic was patented in 1846 by William T. G. Morton and Charles T. Jackson; however, a court eventually deemed this medical use of the substance to be unpatentable for it was a discovery and not an invention Brown (1994).

Additional fees are levied for independent claims in excess of three in number, reissues, extensions, petitions, and other actions. Since 1980, there also are maintenance fees that must be paid on any utility patent in order to prevent the patent from expiring; these fees are due 3.5 years, 7.5 years, and 11.5 years after the patent is issued. A complete list of fees is available from the U.S. Patent and Trademark Office by mail or via the Internet.

Finally, it is important to know that a complete copy of a patent can be obtained for only three dollars from the U.S. Patent and Trademark Office. Thus, for very nominal fees, one has access to a vast pool of technical knowledge.

5.4.2 ■ Criteria for Utility Patents

Patent examiners are particularly concerned with three criteria in determining if an applicant should be granted a utility patent: novelty, usefulness, and nonobviousness.

The first criterion, *novelty,* simply means that the invention must be demonstrably different from the "prior art." (All work in the field of the invention is prior art; an inventor is expected to be familiar with this prior art material.)

In the United States the person who is recognized as the first to invent (as opposed to the first to file) is granted the patent for an invention. (All inventors of a single invention must apply together for a patent.) In order to receive such recognition, the inventor/engineer must demonstrate

1. the earliest date of conception of the invention and
2. diligence in reducing it to practice without any period of abandonment.

Therefore, it is highly recommended that the inventor/engineer keep complete records during the period of development before filing the patent application. Some steps in such record keeping include:[45]

- Records should be written.
- Obtain competent witnesses of the recording:
  - Preferably two or more other persons; inventors cannot serve as witnesses to their own inventions.
  - Each witness should sign and date each page of the inventor's notes and sketches as soon as possible.
  - Witnesses should not be related to the inventor.
  - Witnesses must be available (easily located) in the future.
  - An inventor's protection that his or her witnesses will not steal the work is their signatures as witnesses.
  - Witnesses should have the technical training and knowledge to understand the invention and its use.
- Records should be kept in a bound laboratory notebook that has numbered pages.

45. See Franz and Child (1979).

- Entries should be in indelible ink and dated; any delays in conducting the work should be explained in writing.
- The advantages and uses of the invention should be identified, together with complete descriptions of the design components and specifications.
- All related papers (correspondence, sales slips, etc.) should be saved.
- If a correction is necessary, line through the error and initial and date the change, or (preferably) correct the work on a new page in the notebook and refer to the earlier incorrect entry.
- Later additions should be entered in an ink of a different color, then initialed and dated.
- Sign each page as the "inventor" and enter the date.

Furthermore, the invention must be reduced to practice or completed. Such a reduction to practice can be actual (i.e., a prototype can be built) or constructive (i.e., a complete patent application is filed that satisfies the requirements of the U.S. Patent and Trademark Office).[46]

The inventor/engineer should avoid (a) publishing a description of the invention, (b) offering a product in which the invention is incorporated, and (c) allowing anyone else to use the invention (other than on an experimental basis). In other words, disclosure before filing a patent application should be avoided. Once disclosure occurs, a clock begins to run: The inventor has one year from the date of disclosure in which to apply for patent protection.[47]

The second criterion, *usefulness*, means that a desired objective(s) must be achieved by the invention; that is, some practical utility must be associated with the invention that is specific, demonstrable, and substantial.[48] As noted earlier, immoral, fraudulent, frivolous, and antipublic policy uses are not patentable.[49] Also, theoretical designs (e.g., perpetual motion machines) that would clearly violate fundamental physical principles are not patentable; one would need to constuct a working prototype of the design and demonstrate that it actually works, thereby proving that the physical law or principle is incorrect.

The third criterion, *nonobviousness*, means that the invention must be deemed to have required more than ordinary skill to design or the mere addition/duplication of components found in a patented design. However, if a new result is achieved through the new application of an old design, then that new application may indeed be patentable. (Can you think of such an instance where an old design concept was applied to a new use?)

This third criterion often results in extensive correspondence between the inventor and the examiner as they seek agreement about whether the design is, in fact, not obvious. The evaluation of a design for inherent non-

---

46. See Case History 5.5 *Laser Technology: Patent Rights Can Be Important* for an example of a situation in which a person mistakenly believed that only *actual* reductions to practice were allowed, resulting in many years of litigation and expense.
47. Levy (1990).
48. Miller and Davis (1983).
49. Joenk (1979).

obviousness focuses upon the differences between the invention and the prior art (all earlier technical developments in the field of the invention, such as patents and other relevant technical literature) and the level of skill necessary in creating such differences.

Several factors have been considered in evaluating nonobviousness. One example is the concept of *synergism;* that is, does the combination of elements result in something more and unexpected than simply the sum of the parts? Miller and Davis (1983) note that some courts have been skeptical of the use of this concept in evaluating nonobviousness. In addition, so-called secondary considerations have been used to help evaluate nonobviousness, such as satisfying an unmet need (designs that meet longstanding unmet needs might be assumed to be nonobvious), enjoying unexpected levels of commercial success (such products may not have been obvious or they would have been developed sooner), and the level of effort exerted by others in (unsuccessfully) solving the problem.[50]

Before applying for a patent, one should consider the advantages and disadvantages of obtaining professional assistance. More than 9000 patent agents and attorneys are registered to represent inventors before the U.S. Patent and Trademark Office and an annual listing of these agents and attorneys is available from the agency.[51] A key goal of the agent or attorney is to maximize the coverage and protection of the patent by carefully writing the claims listed in the application.[52]

In contrast to utility patents, plant patent applications are examined for novelty, nonobviousness, and distinctiveness. Distinctness is determined by comparing specific characteristics (color, flavor, odor, form, habit, soil, and so forth) of the new plant to those of other plants. Finally, the three criteria used in evaluating a design patent application are novelty, nonobviousness, and ornamentality.[53]

5.4.3 ▓ Patent Disclosures

A patent disclosure is simply an application to receive patent protection for a design. A disclosure should contain the following elements:

- Title
- Abstract that summarizes the disclosure and claims
- Specification that identifies
    - Any appropriate previous patents or applications of the inventor that are related to the current application
    - The technical field to which the invention belongs

---

50. Miller and Davis (1983).

51. Entitled *Attorneys and Agents Registered to Practice Before the U.S. Patent and Trademark Office.*

52. See, for example, Nussbaumer (1979).

53. Miller and Davis (1983).

- The prior art of the invention (in order to expedite the search and examination of the application by the U.S. Patent and Trademark Office)
- The problem and the way in which it has been solved by the invention, including all advantages over the prior art
- Any figures or drawings as needed for a complete description
- The invention, in the form of a sufficiently detailed and complete description of both its form and its use(s)
- The claims to be protected by the patent
- Oath or Declaration in which the applicant states
  - Belief that he or she is the original and first inventor of the described invention
  - His or her citizenship

5.4.4 ▓ Claims

The claims contained in a patent disclosure determine the legal coverage to be provided. Each claim is written as a single sentence, beginning with the phrase "I (We) claim" or "What is claimed is." This phrase is used only once, even if multiple claims are made. All claims after the first one are written as if this phrase was prefixed to them. In addition, claims are ordered from the most general to the most specific. Remember that the claims should describe all of the real advantages of the design.

One should strive to generate as many claims as possible for an invention in order to maximize the legal protection provided by the patent. As Miller and Davis (1983) note, one must strive to satisfy two conflicting goals: The claims must be sufficiently narrow and precise, as required by patent law; however, they also should be broad enough to protect all appropriate aspects of the inventor's intellectual property. Patent attorneys are skilled in the formulation of claims, and they should be consulted to ensure that the fruit of one's work is properly protected.

One example of an extremely well-written and comprehensive set (98 in number) of claims are those contained in King Gillette's 1904 disclosure for the double-edged razor blade (see patents #775,134 and #775,135). Figure 5.9 presents a portion of the patent for Velcro (a design concept that we will review in Chapter 7). Notice how George de Mestral, the inventor, receives protection on both the *process* of manufacture and the *product* itself through the precise claims contained in his patent disclosure.

5.4.5 ▓ Conducting a Patent Search

Engineers should be familiar with the process of patent searching for (at least) two reasons:

- As noted earlier, the patent records are a rich pool of technical information. Engineers must know how to access this information.
- An engineer may want to receive patent protection for an invention.

I claim:

1. A method for producing a velvet type fabric consisting in weaving together a plurality of weft threads and a plurality of warp threads together with a plurality of auxiliary warp threads of synthetic resin material, forming loops with said auxiliary warp threads on one surface of the so woven fabric, submitting the said loops to a thermal source, thereby causing said loops to retain their shape to form raised pile threads, cutting said loops near their outer ends, thereby forming material-engaging means on at least a portion of said pile threads constituted by said cut loops.

2. A method for producing a velvet type fabric consisting in weaving together a plurality of weft threads and a plurality of warp threads together with a plurality of auxiliary warp threads of synthetic resin material, forming loops with said auxiliary warp threads on one surface of the so woven fabric, submitting the said loops to a thermal source, thereby causing said loops to retain their shape to form raised pile threads, cutting each of said loops near the respective outer end at a point between said outer end and the fabric surface, thereby forming a hook-shaped section with the free end of the respective pile thread at one side of said point at which the cut is made.

3. A velvet type fabric comprising a foundation structure including a plurality of weft threads, a plurality of warp threads, and a plurality of auxiliary warp threads of a synthetic resin material in the form of raised pile threads, the ends of at least part of said raised pile threads being in the form of material-engaging hooks.

4. A velvet type fabric comprising a foundation structure including a plurality of weft threads, a plurality of warp threads, and a plurality of auxiliary warp threads of a synthetic resin material in the form of raised pile threads, the terminal portions of at least part of said raised pile threads being in the form of material-engaging means including hook-shaped sections.

FIGURE 5.9 Excerpts from George de Mestral's U.S. patent (#2,717,437) for the invention of Velcro, entitled "Velvet Type Fabric and Method of Producing Same."

Multiple-Effect Evaporation Process

U.S. patents #3,237 (1843) and #4,879 (1846) describe the *multiple-effect evaporation process* developed by Norbert Rillieux.[54] In our discussion of freeze-dried coffee and sublimation in Section 5.1, we noted that liquids can be vaporized at lower temperatures if placed within a vacuum (or a partial vacuum) under reduced pressure. In the Rillieux method, a series of vacuum pans is used to vaporize liquid. The heat of the vapor produced by partially evaporating liquid in the first pan is used to heat liquid in the second pan and so on. Moreover, the pressure is reduced in each succeeding vacuum pan so that less heat is required to boil the liquid. By using the latent heat of the vapors in this way, the entire process becomes much more efficient. The Rillieux multiple evaporation method is now used in the manufacture of sugar, soap, glue, condensed milk, and other products, as well as in the recovery of waste products.

54. Haber (1970).

It has been called the greatest invention in the history of American chemical engineering, and it is only one example of the technical knowledge contained within the patent records.

The public file of the U.S. Patent and Trademark Office is in Arlington, Virginia. (Two identical files are maintained by the agency: one for use by the public and one for use by employees of the office, particularly patent examiners.) In addition, there are partial U.S. patent collections in many libraries throughout the nation.

The U.S. Patent Classification System contains more than 400 distinct classes. The scheme is as follows: A class is listed first, followed by a series of subclasses, each of which is further divided into even more specific categories (there are about 100,000 subdivisions within the system). Each patent is classified according to the most comprehensive claim contained within the patent. The U.S. Patent and Trademark Office offers such publications as the *Index to U.S. Patent Classification* and the *Manual of Classification* for help in searching the records. In addition, the CASSIS CD-ROM system allows one to perform a computerized database search of the patent records.

How, then, should one go about searching the patent records?

Begin with the *Index to U.S. Patent Classification* in order to find the relevant subject headings for those classes and subclasses in which to search. The *Index* is valuable in that it provides a promising list of relevant terms, synonyms, phrases, and acronyms that have been used to describe designs.[55] Next, use these terms to search the *Manual of Classification* for the general categories or classes in which a design might be found and then determine the most specific subclass that relates to your design. Use the CASSIS system to obtain a computerized listing of all patents that have been awarded in the selected subclass. Review the brief descriptions given for each of these patents in the *Official Gazette* of the United States Patent and Trademark Office. (In many libraries, the *Gazette* is available in hard copy.) Finally, review the full patent disclosures (usually archived on microfilm) for each patent that is most promising.

Additional information about patents, trademarks, and the protection of intellectual property can be found in numerous books and journal articles, some of which are listed among the references near the end of this text.

55. Ardis (1991).

- Design engineers must acquire robust (i.e., both broad and deep) technical knowledge and develop the ability to apply it whenever necessary.
- There are many sources of valuable technical information with which the engineer should be familiar: manuals, catalogs, books, magazines, journals, and patents.
- There are three distinct types of property: real, personal, and intellectual.
- Intellectual property can be protected by trade secrets, trademarks, copyrights, or patents.
- There are three types of patents: utility, design, and plant.
- Criteria used to evaluate utility patent applications include: novelty, nonobviousness, and usefulness.
- Engineers and inventors should maintain careful records in anticipation of patent applications.
- Patent protection is based upon the claims contained in the patent disclosure.

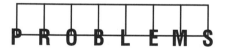
PROBLEMS

5.1 Identify an engineered product or process that depends upon a particular scientific principle or phenomenon. Describe the scientific basis underlying the product or process, and explain the use of the engineering design. (Do not use the examples of engineered products given in Section 5.1 of this chapter; include all references.)

5.2 Prepare a brief report on the photovoltaic effect and relate this effect to the development of selenium photovoltaic cells, silicon photoelectric cells, and solar cells (include all references).

5.3 Describe three engineering applications of levers, inclined planes, and/or counterweights that were not discussed in this chapter (include all references).

5.4 Prepare a brief report on the development and history of contact lenses, including the optical principles involved in their use (include all references).

5.5 Select a concert hall or theater of your choice, and prepare a brief report describing the acoustical elements in its design (include all references).

5.6 Explain the operation of the following systems and the underlying physical principles or phenomena upon which they are based (see Table 5.3):
a. Hydrometer
b. Fathometer
c. Strain gages

5.7 Describe additional engineering applications of the following physical phenomena that were not included in Table 5.3:
a. Photoemissive effect
b. Adsorption
c. Purkinje effect

5.8 Investigate and prepare a brief report on a recent court case that focused upon the possible violation of one of the following protections:
a. Trade secret protection
b. Trademark protection
c. Patent protection
Do not use one of the cases described in this chapter (include all references).

5.9 Perform a patent search and find three patents held by individuals well known for their success in fields other than technology (e.g., Mark Twain). Choose three people who are not discussed in this chapter.

5.10 Prepare a brief report on one of the important inventions listed in Table 5.4, its development, and its impact upon society.

5.11 Perform a patent search on an important invention that was patented during the past thirty years. Briefly describe the invention in terms of its function(s) and its impact upon society. Also discuss the reasons for your choice of this invention.

5.12 Many patented designs are never manufactured. Find three such designs in the patent record and suggest possible reasons for their failure to be implemented.

5.13 Perform a patent search for a design topic of your choice. Analyze and compare three designs: Which is the best and why? Be specific in your explanation and provide all documentation.

5.14 Could the collapse of "Galloping Gertie" (the Tacoma Narrows Bridge) have been prevented (see Case History 5.3)? Explain your reasoning.

5.15 If Elisha Gray has filed his patent for the telephone before Alexander Graham Bell, would Bell have been able to overturn Gray's rights to the patent (see Case History 5.4)? Explain your answer.

CASE HISTORY 5.1

A Special Mix of Experience Leads to Photocopying

Photocopying is now a common process performed by millions of people each week; however, the development of xerography (the process through which photocopies are produced) required its inventor, Chester Carlson (Fig. 5.10), to persevere for many long years.[56]

In the xerographic process, resin is fused to selected sections of a sheet of copy paper that have been sensitized by static electricity, where these sections correspond to the images on the original that are to be reproduced.

This process reflects the unique combination of Carlson's early life experiences. He worked for a printer during his boyhood (he was the sole support of his family at age 14). Upon graduation from college with a degree in physics, he was first employed by Bell Telephone Laboratories, then by a patent attorney, and finally by P. R. Mallory & Company, an electronics firm. He also continued his education through night classes, obtaining a law degree and becoming manager of Mallory's patent department.

Carlson's work experiences in printing, physics, electronics, law, and patents led to his recognition of a need: to generate high-clarity duplicates of documents quickly and easily. He built upon his experiences to formulate the solution to this need.

Paul Selenyi was a Hungarian physicist who had performed research in electrostatic imaging processes. Through library research, Carlson discovered Selenyi's work and became convinced that such a process was worthy of further investigation. Carlson's efforts bore fruit in 1937 when he produced the first copy of an image via the xerographic process. He filed his first patent application for the process in October of that year. However, more than twenty additional years of research was then necessary to reduce the concept to a practical form: the first office copier machine.

In the basic photocopying process, an electrical charge is used to transfer an image from a document to another sheet of paper. A metal drum rotates as the paper slides in contact with it. The process is as follows: The drum receives a negative electrical charge from the drum charger; the drum is a *photoconductive semiconductor;* that is, it will only conduct electricity if exposed to light. An image of the original document is then projected onto the charged area of the drum by a lens and mirror system (the optical system may include several mirrors and lamps). The negative electrical charge remains only on those areas of the drum where it has not been struck by light; any illuminated areas immediately lose their charge. Thus, the darkened areas of the image correspond to the charged areas of the drum.

---

56. Refer, for example, to the discussions given in Middendorf (1990), Macaulay (1988), and *The Way Things Work* (1967).

FIGURE 5.10 Chester Carlson with his original prototype of the photocopier. *Source:* UPI/Bettmann.

Brushes then apply positively charged toner particles to the drum's surface, where these particles only adhere to the negatively charged areas of the drum. Meanwhile, the copy paper is given a negative charge by the transfer charger before it comes in contact with the drum. The positively charged toner particles are then transferred to the negatively charged paper and fused to the paper by a heater. An erase lamp removes any remaining charge on the drum and a cleaner removes any toner residue before the process repeats itself. (A second erase lamp may be used to remove some electrical charge from the drum prior to the transfer of toner particles to the paper.) Conveyor belts are used to transport the copy paper (now containing the captured image of the original) from the machine.

The photocopying method is a very clever application of static electricity, allowing us to quickly produce copies of documents with both accuracy and clarity.

Chester Carlson persevered for twenty years in order to fulfill his dream! He did not give up, eventually becoming the founder of Xerox Corporation, and bequeathing $100,000,000 to various charities and foundations upon his death at age 62.

CASE HISTORY 5.2

Technical Knowledge and Quick Thinking Save a Life

Lawrence Kamm has designed numerous mechanical and electromechanical devices, ranging from robots and space vehicles to a heart–lung machine.

The heart–lung machine is an elegant electromechanical system that simulates the behavior of the human heart and lungs during open-heart surgery.[57] For operations of short duration, the circulation of the patient's blood is stopped through cooling or "hypothermy"; however, circulation must continue during operations of longer duration—hence, the need for the heart–lung machine.

By the 1960s heart–lung machines had the following configuration. Normally, venous blood enters the right auricle of the heart through the vena cava upon its return from passage through the body. During surgery, this blood–low in oxygen and in need of replenishment—is redirected through plastic tubing towards a glass cylinder in the heart–lung machine. In addition, blood returning to the heart through veins other than the vena cava is extracted directly from the heart itself by a pump, and then sent through a defoamer before entering the cylinder.

Rotating steel discs within the cylinder then carry thin coatings of blood into an enriched oxygen zone. The cylinder thereby acts as an artificial lung, allowing the blood to absorb oxygen before being returned to the body with the aid of a second pump (passing through both a heat regulator and a filter enroute to the body).

Finally, an auxiliary blood reservoir is attached to the cyclinder in order to replace any blood lost during the operation, and an anticoagulant prevents the blood from congealing.

This system performs the two basic functions (pumping and oxygen-enriching) of the human heart and lungs in a simple but effective manner—a truly clever engineering design.

In his book *Real-World Engineering: A Guide to Achieving Career Success* (1991), Kamm relates the following episode from his life in engineering:[58]

> The most dramatic and important debugging visualization I ever did was on the Convair heart–lung machine I had designed. It was during our second human operation. Just before connection to the patient, when the machine was circulating blood through a closed loop of tubing, great gobs of air suddenly appeared in the blood. Nothing like this had ever happened during dozens of development tests. The patient was wide open and too sick for the surgeon to close him up and try again. The machine was sterile and could not be opened for examination and modification.

57. See Kamm (1991) and *The Way Things Work* (1967).
58. Kamm (1991), pp. 14–15.

I visualized the elements and operation of the machine and saw that the effect could come from an unsymmetrical collapse of one of the pumping bladders such that a fold of the bladder would block the exit hole. I then realized that the bladders came from a new batch and therefore might have some slight differences from the ones we had been using, thus supporting the hypothesis. Further visualizing the machine, I invented the idea that the bladders could be operated with only partial collapse if the valve gear of the water engine which powered the machine were tripped by hand at half stroke. Eddie Leak, steady as a rock, and I sat on the floor for 20 minutes, reached into the mechanism with long screwdrivers, and tripped the linkages at half stroke every cycle, knowing that if we made a mistake we would kill the man on the table. He lived.

The engineering fix was easy: adding a perforated tube to limit bladder displacement. But I will be satisfied not to play Thurber's Walter Mitty again.

Kamm relates this experience as evidence that an engineer must be able to visualize mechanical structures, electrical circuits, and other systems. However, Kamm's courageous work also required a truly robust understanding of electromechanical systems in general and of the heart–lung machine in particular. This deep and broad knowledge, coupled with quick thinking and the courage to act, saved a man's life.

CASE HISTORY 5.3

The Tacoma Narrows Bridge Destroys Itself

On July 1, 1940, a sleek new cable-supported suspension bridge was opened across the Tacoma Narrows, connecting the Puget Sound Navy Yard in Bremington and the Olympic Peninsula with the mainland of Washington State (including Tacoma and Seattle). But only five months later, on November 7, the narrow two-lane bridgeway between two twin 425-foot towers would twist itself until it collapsed into Puget Sound (Fig. 5.11), becoming one of the most recorded and famous engineering failures in history as its final moments were captured on film.[59]

The main center span of the Tacoma Narrows Bridge was long (2800 feet) but unusually shallow (only 8 feet deep with a 32-foot wide roadway). Each of the two side spans extended 1,100 feet from a bridge tower to concrete

59. See Billah and Scanlon (1991), Farquharson (1940, 1949–54), Levy and Salvadori (1992), Robison (1994a, 1994b), Scanlon and Tomko (1971).

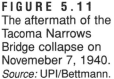

FIGURE 5.11
The aftermath of the
Tacoma Narrows
Bridge collapse on
Novemeber 7, 1940.
Source: UPI/Bettmann.

anchors onshore. The bridge's graceful and sleek appearance was achieved
through the use of stiffening girders. (Such stiffening trusses are used to offset the horizontal forces applied by a suspension system when bedrock is not
available for anchoring the structure.) Unfortunately, there are always costs
associated with every benefit. The cost for the aesthetically pleasing slender
structure was aerodynamic instability, and the bridge was more susceptible
to forces of wind than were deeper suspension bridges with open truss
designs. Even before it opened, workmen noticed the bridge deck's tendency
to oscillate vertically in mild winds, earning it the sobriquet of "Galloping
Gertie." And after it opened, travelers would enjoy the thrill of riding across
Galloping Gertie as it rose and fell in roller coaster-like motion.

Several modifications to the bridge were made in attempts to dampen its
vibratory motions, including:

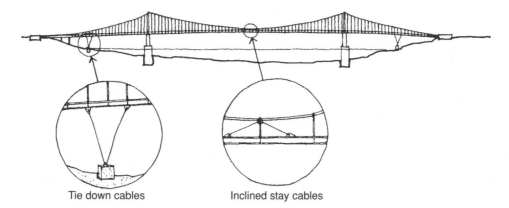

FIGURE 5.12 Tie-down and inclined stay cables meant to reduce vibrations of the Tacoma Narrows Bridge. *Source:* From *Why Buildings Fall Down* by Matthys Levy and Mario Salvadori. Copyright © 1992 by Matthys Levy and Mario Salvadori. Reprinted by permission of W.W. Norton & Co., Inc.

- Using tie-down cables to connect the bridge girders to a set of 50-ton concrete anchors lying on shore (see Fig. 5.12). Result: The cables snapped under the motions of the bridge in a windstorm.
- Running inclined stay cables between the main suspension cables and the bridge deck at midspan (Fig. 5.12). Result: The vibrations continued.
- Installing a mechanical piston-cylinder dynamic damper to absorb the vibration energy. Result: Sandblasting of the bridge (prior to painting) forced sand into the cylinder, eventually breaking the leather seals of the damper and rendering it useless.

A scale model of the bridge was constructed and tested, leading to a report in which Professor Frederick B. (Burt) Farquharson of the University of Washington and his colleagues recommended that additional modifications be made, including the installation of wind deflectors and the drilling of holes in the bridge girders (thereby allowing wind to pass through the girders rather than around it). Unfortunately, the bridge collapsed before these design changes could be implemented.

On the morning of November 7, Farquharson traveled to the site for further study. He and Kenneth Arkin (then chairman of the Washington State Toll Bridge Authority) became alarmed as they watched the bridge begin to sway and twist in the 42-mph winds. (A cable band had slipped at midspan, allowing the motions to become asymmetrical.) Arkin stopped traffic from proceeding across the bridge, leaving only one car (driven by a newspaper reporter, Leonard Coatsworth) and a logging truck (with driver Arthur Hagen and passenger Rudy Jacox) on the bridge deck. As the vibrations increased, Coatsworth, Hagen, and Jacox managed to crawl off the bridge, leaving their vehicles behind. According to eyewitnesses, Coatsworth left the bridge deck "bloody and bruised" as concrete fell all about him. Professor Farquharson then tried to retrieve Coatsworth's automobile with his daughter's dog still trapped inside but he could not as the bridge oscillations grew until the deck was rising and then falling 25 feet through the air while tilting at a 45° angle. The twisting continued until total collapse due to metal fatigue occurred at 10:30 A.M. The car, the truck, and more than 600 feet of the bridge deck fell 190 feet into the waters below. Both motion picture and still cameras recorded the entire event. (The dog was the only casualty of the disaster.)

Why did Galloping Gertie fail? Its designer, Leon Moisseiff, had developed (with his partner Fred Lienhard) the standard calculation methods for determining the load and wind forces acting on a suspension bridge. Moisseiff had earlier designed both the Manhattan Bridge in New York and the Benjamin Franklin Bridge in Philadelphia. He also served as a consulting engineer during the design and construction of other bridges, including the Golden Gate in San Francisco and New York's Bronx-Whitestone. Yet even this eminent bridge designer was "completely at a loss to explain the collapse" of Galloping Gertie.

Subsequent investigation showed that the collapse resulted from self-induced excitation, in which a flutter wake was produced in the air stream around the structure by the oscillation of the bridge itself. This is a form of resonance, in which an oscillating body is driven to ever-increasing amplitudes of motion as it is acted upon by an oscillating force with a frequency near or equal to that of the natural frequencies of oscillation of the body. However, the Tacoma Narrows Bridge collapse was *not* due to simple forced resonance in which an *external* oscillating force is applied to the body. After all, winds do not oscillate in a periodic fashion about a specific location but rather increase and decrease in a random way.[60] (Try throwing an Aerobie or a Frisbee on a gusty day and you will notice how truly random wind forces can be.)

Professor Farquharson conducted wind-tunnel testing on a model of the Tacoma Narrows Bridge. He determined that as the wind traveled across the bridge (a so-called bluff body—i.e., one that is not streamlined), its flow separated, leading to the formation of vortices. This is known as "periodic vortex shedding." These vortices in turn led to oscillations in the pressures applied by the air to the bridge, resulting in *vortex-induced vibration* of the structure. The *vertical* motions of the Tacoma Narrows Bridge were indeed due to such vibration. However, these vortex-induced vibrations quickly dissipate as self-limiting forces are created; hence, the bridge collapse was *not* due to periodic vortex shedding.

None of the amplitudes of motion were significant *except* in one mode of vibration at which the amplitude became unbounded as wind velocity increased. This single mode corresponded to the actual frequency of motion

---

60. Billah and Scanlon (1991) note that many introductory physics textbooks have mistakenly presented the Tacoma Narrows Bridge collapse as an example of forced resonance, which implies that the wind somehow managed to apply a periodic force to the bridge.

of Galloping Gertie at the time of its collapse. In this mode, an aerodynamic self-excitation developed due to a bluff-body flutter wake (distinct from an airfoil flutter wake) appearing in the air stream as its flow separated. The motion of the bridge itself created these torsional flutter vortices which then led to the final destructive twisting movement.

In addition, Galloping Gertie was inherently weak in its ability to resist twisting due to torsional forces because of its extreme slenderness (with a depth-to-length ratio of 1:350) and lack of sufficient stiffening.

One final question remains: Could the self-destructive tendency of the Tacoma Narrows Bridge to produce a flutter wake have been foreseen and therefore prevented by appropriate design modifications? Although it is impossible to foresee and avoid all possible hazards that may be associated with a design (particularly an innovative design), multiple warnings about the potentially dangerous impact of wind and vibrations on suspension bridges are contained in the historical record. For example, the Wheeling Suspension Bridge (in West Virginia) collapsed five years after it opened in 1849. The following eyewitness account[61] appeared in the Wheeling *Intelligencer* on May 18, 1854:

> About three o'clock yesterday we walked toward the suspension bridge and went upon it, as we have frequently done, enjoying the cool breeze and the undulating motion of the bridge. . . We had been off the flooring only two minutes and were on Main Street when we saw persons running toward the river bank; we followed just in time to see the whole structure heaving and dashing with tremendous force.
>
> For a few moments we watched it with breathless anxiety, lunging like a ship in a storm; at one time it rose to nearly the height of the tower, then fell, and twisted and writhed, and was dashed almost bottom upward. At last there seemed to be a determined twist along the entire span, about one half of the flooring being nearly reversed, and down went the immense structure from its dizzy height to the stream below, with an appalling crash and roar.
>
> For a mechanical solution of the unexpected fall of this stupendous structure, we must await further developments. We witnessed the terrific scene. The great body of the flooring and the suspenders, forming something like a basket swung between the towers, was swayed to and fro like the motion of a pendulum. Each vibration giving it increased momentum, the cables, which sustained the whole structure, were unable to resist a force operating on them in so many different directions, and were literally twisted and wrenched from their fastening.

Nine other suspension bridges were severely damaged or destroyed by winds between 1818 and 1889 (Farquharson, 1949, and Petroski, 1994). And in 1939 the six-lane Bronx-Whitestone Bridge in New York opened with an overall design that was quite similar to that of Galloping Gertie; it was soon discovered that winds produced substantial movements in the Bronx-Whitestone, necessitating the addition of cables and other mechanisms to dampen the vibrations.

---

61. As quoted in Petroski (1985).

- Failure to anticipate hazardous event (i.e., torsional flutter leading to metal fatigue and collapse) and design against it (Error in design due to inadequate knowledge and insufficient testing)

The Tacoma collapse led to further investigation of the effects that winds may have upon a body (and vice versa) and aerodynamics became a quintessential part of structural engineering. For example, once it was discovered that long narrow suspension bridges were susceptible to aerodynamic instability (e.g., self-excitation due to flutter), wind-tunnel testing of models was used to determine more stable design shapes.

Furthermore, the collapse of Galloping Gertie should remind engineers that they must be familiar with historical design failures if they want to avoid repeating the mistakes of the past.

CASE HISTORY 5.4

Bell and Gray: A Search for a Better Telegraph Leads to the Telephone

Two men, Alexander Graham Bell and Elisha Gray, independently and almost simultaneously invented the telephone in 1876.[62] Both inventors were trying to develop a telegraph system that would allow more than one message to be transmitted through a single wire at a time (i.e., both were trying to develop what is now known as "multiplexing"). In 1872 the Stearns duplex system allowed two messages, each one transmitted from different ends of the wire, to be sent simultaneously. Even with this new duplex system, however, transmission remained severely limited.

Gray already had patented a telegraph relay and formed the Western Electric Manufacturing Company. In 1874, he demonstrated the concept of musical telegraphy, in which musical notes were transmitted by wire. Since musical notes are distinctive, such a system could be used to transmit multiple telegraphic messages in Morse code, each message carried at the frequency of a particular note. Gray realized that human speech could be transmitted in a similar manner; however, he was persuaded by others (including his patent attorney and the journal *The Telegrapher*) that the electrical transmission of human speech would never become commercially

62. Flatow (1992), Yenne (1993), and Zorpette (1987).

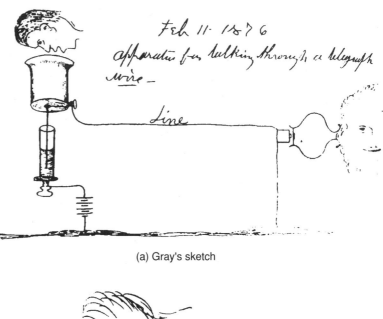

(a) Gray's sketch

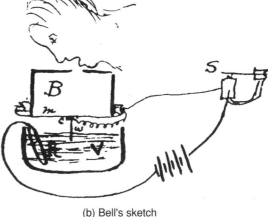

FIGURE 5.13 An example of one concept, the telephone, shared by two inventors: (a) Elisha Gray's sketch of February 11, 1876; (b) Alexander Graham Bell's sketch, created on March 9, 1876.

(b) Bell's sketch

viable. As a result, Gray concentrated on developing his system of musical telegraphy to the exclusion of the telephone concept.

However, in 1875 Gray returned to the idea of electrical speech transmission. He developed the concept of a voice transmitter that used an electrically conductive liquid, and captured this idea in a sketch on February 11, 1876 (see Fig. 5.13a). He applied for a patent on his telephone design on February 14, 1876.

Meanwhile, Alexander Graham Bell also had been working on the problem of simultaneous multiple message transmission via a single telegraph wire. Bell knew that the famous German physicist Hermann von Helmholtz had invented the electric tuning fork (a tuning fork connected to an electromagnet). With this device, one could generate an electrical current that vibrates at the same frequency as the tuning fork. Bell was convinced that a set of such tuning forks, each set to a different frequency, could be used to transmit multiple signals simultaneously through a wire (one message for each such transmission fork). The receiver would consist of a set of reeds, each of which could distinguish between multiple coded messages by tuning in to the frequency of a given transmission fork. Bell had conceived the harmonic telegraph.

Bell recognized that if the signals from a set of tuning forks could be transmitted by wire, so, too, could human speech be transmitted. Unlike Gray, Bell worked tirelessly to transform this concept of the telephone into reality. On February 14, 1876, he applied for patent protection of his concept. By March 9, 1876, Bell had designed a telephone system in which acid is used as a liquid transmitter (Fig. 5.13b). On the following day, Bell experimented with his telephone design, successfully transmitting the message "Mr. Watson. Come here. I want to see you."

Bell's application arrived at the U.S. Patent Office two hours before Gray's, and Bell was awarded the patent rights to the telephone (U.S. patent #174,465). Gray, however, was not perturbed by this unexpected turn of events since he sold his multiple telegraph system to Western Union for a substantial amount of money. Furthermore, he and others still believed that the telephone would never be more than a novelty concept. (In 1876 Western Union refused to purchase Bell's rights to the telephone for only $100,000.)

Within one year, the telephone was recognized as a major breakthrough in communication. Gray, supported by Western Union and finally realizing the commercial potential of the telephone, was persuaded to sue Bell over the rights to the design. It was determined that Bell was indeed not only the first-to-file but also the first-to-invent, meaning that he would continue to hold the patent rights to the telephone.[63] Ultimately, an out-of-court settlement was reached between the parties.

The story of the telephone illustrates several lessons relevant to this chapter:

- One must be aware of technical advances in order to be an effective design engineer; Bell's ideas for a harmonic telegraph and the telephone were based upon his knowledge of Helmholtz's electric tuning fork.
- Creative design is not necessarily unique; one should not believe that only he or she alone is developing a concept. Both Gray and Bell developed the telephone design—clearly a truly creative idea—along very similar lines.
- Patent protection can be critical to one's opportunity to prosper from a design.

63. The United States patent system awards patent rights to those who are first-to-invent, unlike other nations who use a first-to-file approach in awarding patent rights.

CASE HISTORY 5.5

Laser Technology: Patent Rights Can Be Important

The concept of light amplification by stimulated emission of radiation (LASER) began with Albert Einstein's 1917 hypothesis that radiation (e.g., light) could be used to emit energy in the form of photons from a set of electrons bathed in this radiation. Furthermore, this emitted energy would be of the same frequency and travel in the same direction as the original radiation.

In 1951 Charles H. Townes realized that this concept of stimulated emission could be used to create a beam of high-intensity energy. By 1955 Townes had constructed the first MASER in which microwave radiation was used (instead of light) to generate a concentrated beam of emitted energy. Next, in 1958, Townes and his brother-in-law Arthur L. Schawlow described the concept of a laser in a technical journal. Finally, in 1960, Theodore H. Maiman converted the theory into reality by constructing the first ruby laser.

The ruby laser consisted of a ruby rod surrounded by a spiral glass flashtube. A light from the flashtube strikes the ruby rod, exciting some atoms in the rod so that they emit photons that in turn excite other atoms. The resulting cascade of photons strikes the polished (mirrored) ends of the rod, thereby causing this light wave to be reflected back across the ruby rod. The process continues until the resulting beam of light becomes so intense that it finally is emitted from the laser.

The early ruby lasers emitted single pulses of energy, leading researchers to later develop gas lasers from which continuous output beams of light could be generated. Electricity was used to initiate the cascade of photons in these gas lasers instead of a single flash of light, leading to the continuous emission of output energy.

Today, lasers are used in surgical instruments, compact disks, holography, communication systems, computers, and in innumerable other forms to improve our quality of life.[64]

In 1957 Gordon Gould captured the concept of the laser in a notebook that he then had notarized. Unfortunately, Gould believed that he had to construct a working model of the laser before he could apply for a patent. By 1959, when he realized that this belief was in error and finally applied for a patent on his design, he found that Townes and Schawlow already had applied for patent rights to the invention. Gould then sought recognition through the courts, finally receiving limited patent rights on two aspects of the invention: laser pumping in 1977 (U.S. patent #4,053,845) and materials

---

64. Townes, together with two Soviet scientists N. G. Basov and A. M. Prokhov, was awarded the 1964 Nobel Prize in Physics for their contributions in the development of the maser and the laser.

processing with lasers in 1979 (U.S. patent #4,161,436). Although Gould's legal battles were costly (estimated to be more than $9 million) and are continuing, his royalties on these patents may amount to more than $40 million.

Patent protection can be extremely valuable to the developers of a design, both in terms of financial remuneration and appropriate recognition for one's success.[65]

CASE HISTORY 5.6

Logo and Packaging Design of Coca-Cola

The soft drink Coca-Cola is a classic example of successful trademark and packaging design.[66] John Styth Pemberton, a pharmacist in Atlanta, Georgia, created the syrup formula for Coca-Cola in 1886. Pemberton was an inveterate experimentalist, constantly seeking to find new chemical mixtures that would have commercial value. Unfortunately, none of these efforts led to success until he turned his attention towards creating a thirst-quenching drink.

Pemberton experimented with various mixtures, blending together kola nut extract (known as cola), decocainized coca leaves, sugar, and other components and then heating the blend. One day, an employee working at Pemberton's soda fountain added soda water to the syrup and served the concoction to a customer. The combination of syrup and soda water produced a truly original flavor.

Pemberton next formed a partnership with Frank Robinson (his bookkeeper) and D. Doe to market the new syrup. Robinson then made two vital contributions to the success of the product: its name and its logo. He conceived a name that would combine two of the syrup's key ingredients (coca and cola) in an alliterative and easily remembered form. Robinson then wrote this name in such an elegant and stunning style of penmanship that his handwritten version of the name "Coca-Cola" was adopted as the company's formal logo (and continues in this role more than 100 years later).

Pemberton died in 1887 and Asa G. Chandler (another Atlanta pharmacist) purchased all rights to the formula in 1888 for $2,300. Chandler created the Coca-Cola company in 1892 and the drink was registered with the U.S. Patent Office on January 31, 1893. With the development of high-speed bottling machinery beginning in 1892, sales of bottled Coca-Cola quickly grew

65. See, for example, the discussion of the Gould case in Yenne (1993).
66. Based upon Sacharow (1982) and d'Estaing (1985).

until (in 1899) Chandler sold the rights to bottle and sell the product throughout most of the nation to Benjamin Thomas and Joseph Whitehead. However, Chandler retained the manufacturing rights to the syrup. (These rights are still retained by The Coca-Cola Company. Licensed bottlers simply purchase the syrup directly from the company, then mix it with carbonated water according to company specifications, bottle the resulting soft drink and sell it.)

Other soft drinks rapidly entered the market to compete with Coca-Cola. In 1913 it was decided that customers would be better able to discern Coca-Cola among the other bottled soft drinks if it were sold in a distinctively designed bottle. The glassmaker C. S. Root was hired to create the design of the new Coca-Cola bottle. Root and his associates eventually produced the famous curved shape that was in fact a truncated version of the kola nut itself (in which the base of the nut has simply been removed).

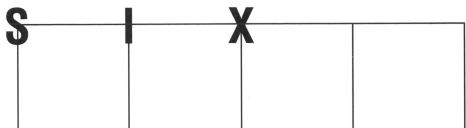

ABSTRACTION AND MODELING

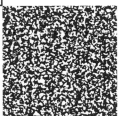

If I have done the public any service, it is due to patient thought.
Sir Isaac Newton

6.1 Abstraction

Engineers must be creative![1] Once the problem has been formulated as a set of design goals, the engineer must develop a series of alternative design solutions. During the abstraction phase of the engineering design process, one tries to generate broad, more inclusive classes or categories through which the problem situation—and various approaches to its solution—could be described.[2] The goal is to obtain a conceptual perspective or vision of the problem and its possible solutions at relatively high levels of abstraction. Such work often requires models to represent these different possible design solutions. The advantage of such an approach is that we are more likely to generate concepts that are quite varied in approach and value.

On the other hand, if we began by developing a very specific design concept as a solution to a problem, we may never move beyond the solution category or type in which this initial concept lies. Our list of alternative solutions would then be very narrow in scope and promise.

1. See, for example, Weber (1992) and Woodson (1966).
2. See, for example, Woodson (1966).

Weber (1992) has noted that the ability to think abstractly or generalize during creative problem solving distinguishes the expert from the amateur. Abstraction allows us to consider a greater range of possibilities in which the problem can be dissected into parts or subproblems, together with ways in which solutions to these subproblems can be coupled to form complete design solutions.

We wish to generate as many different alternative designs as possible in order to maximize the likelihood that we will develop the best solution to the problem. In order to accomplish this objective, we need to view the problem from a variety of perspectives. The process of abstraction enhances our ability to overcome any tendency to view the problem and its possible solution paths in a narrow way. The engineer is then more likely to generate alternative designs that are truly distinctive and not simply variations of a single solution concept or theme.

Furthermore, in terms of their relative generality, there are superior and inferior categories of solution concepts. The superior categories are more general or abstract than those that are inferior. Examples of superior categories of solution concepts include force field systems, which include gravitation and magnetism, and circular motion systems, which encompass the application of centrifugal/centripetal forces and the coriolis effect.

The first step in abstraction is to break the problem into as many functional parts, subproblems, or meaningful units as possible; this is known as *parsing*. Of course, the engineer actually begins this process of parsing during the preceding phase (problem formulation) of the design process, when he or she formulates a set of goals that should be satisfied by any viable solution to a problem. Next, one should try to classify these functional aspects of the problem into more general (i.e., superior or superordinate) categories in accordance with their distinctive characteristics. In performing this task, we might base our classification scheme upon variations of

- the general purpose that is to be achieved by a solution,
- the principles or approaches that could be used to achieve this purpose, such as a family of inventions that have been used to solve similar problems,[3]
- the contexts or operating environments in which the solution might be used, or
- the specific subtasks that must be performed (either sequentially or concurrently) in order to achieve the overall objective,

among other possibilities. We might consider, for example, which features of the individual aspects of the problem allow us to separately identify them, such as: different subtasks that are to be performed by any design solution, the materials or components that may be used in a design, the expected spatial or temporal separation of components, specific system interface issues, and so forth. If these aspects of the problem are partially or

3. See Weber (1992).

completely independent of one another, we then can form a set of subproblems that may be easier to solve individually than would be the original larger problem.

Finally, one should identify the *dimensions of variation*[4] (i.e., ways in which variation can be introduced into each of the different categories used to describe the problem or possible solution paths), together with methods to achieve such variation. In other words, identify what can be varied and suggest how such variation could be accomplished. For example, one could ask such questions as:

- Is it possible to cluster, interrelate, delete, permute, or otherwise manipulate elements of the problem (or of any proposed partial design solutions) in order to produce beneficial results?
- Could elements of the different partial solution categories be joined in some advantageous manner?

Questions such as these can lead one to both further generalize (form broader superordinate categories) and specialize (create more specific subordinate classes) with respect to the problem and possible solution paths.

Through abstraction, we view the problem and its possible solution paths from a higher level of conceptual understanding. As a result, we may become better prepared to recognize possible relationships between different aspects of the problem and thereby generate more creative design solutions.

Developing a Transportation System

As an example, one might begin by focusing upon the general purpose or objective that is to be achieved by a design solution, such as:

Design a method for transporting people from one location to another.

Rather than attempt to generate some specific design for accomplishing this task (e.g., using an airship or dirigible to transport each person), through abstraction we instead begin by focusing on several general methods of "location change," such as

- Propel (e.g., fluid motion, catapult motion, engine thrust)
- Carry [e.g., by water current (raft), air current (gliders, kites), motorized vehicles, animals]
- Attract/repel (e.g., via magnetism)
- Sink/drop (e.g., by using weights or gravity)

4. Weber (1992).

- Lift (e.g., using the bouyancy of gases)
- Slide (by reducing friction)
- Float [e.g., using air pressure (parachutes, balloons) or water pressure]
- Pull (e.g., with ropes)
- Intrinsic (i.e., locomotive ability within the system—walking, bicycling, etc.)

Next, we might consider the types of forces (e.g., gravity, magnetism, heat, wind, animal labor, and so forth) and the different classes of materials and components that could be used in these methods of transportation.

Given each of these general categories of methods, forces, materials, and other aspects of potential approaches to a solution, we next would use synthesis (the next phase of the design process, to be discussed in Chapter 7) to generate specific design concepts that are consistent with the specifications of the problem. We would try to develop such concepts by identifying possible points of intersection among these categories (e.g., lifting, sliding, and other kinds of systems that are driven by gravity, magnetism, and other types of forces, and in which belts and pulleys, sleds, balloons, and other components and materials appear).

Abstraction provides us with a perspective of the building blocks that can be used to develop a set of design solutions. Synthesis is then used to form whole solutions from these sets of building blocks or constituent parts.

6.2 Importance of Modeling in Abstraction and Design

Modeling is part of the abstraction process, since engineers use models to develop and evaluate their ideas. Models allow us to organize data, structure our thoughts, describe relationships, and analyze proposed designs. An engineer may often be confronted with problems that are unfamiliar—either the problem has never been solved before by anyone or (at the very least) it has never been solved by this particular individual. Models can help us recognize what we know and what we do not know about a problem and its solution. They can help us transform a new unfamiliar problem into a set of recognizable subproblems that may be much easier to solve. Model building and testing is a skill that should be mastered by every engineer.

In addition, proposed designs usually cannot (and should not) be built and tested to determine their practical value as engineering solutions. Often many potentially satisfactory but very different solutions can be developed for a given engineering problem; every such design cannot be built due to financial constraints. Second, prototypes often do not work properly when first constructed; they require additional refinement and revision, which necessitates more investment of both time and money. Third, some designs may be hazardous to workers or to the environment; we must minimize such hazards by eliminating all but the most promising designs prior to development. Hence, the importance of models—including computerized simulations—in both the design and analysis of engineering solutions.

6.3 Models as Purposeful Representations

A model can be a working scaled (concrete) miniature used to test an engineering solution, an abstract set of equations describing the relationships among system variables, a computerized simulation and animation of a process, a (two- or three-dimensional) graphical description of a design, or any other purposeful representation of a process, object, or system.[5]

A Competition Model

Richardson (1960) proposed the following simple mathematical model to describe the competition that may exist between two nations:[6]

$$x_1' = dx_1/dt = A\,x_2 - M\,x_1 + f_1$$

$$x_2' = dx_2/dt = B\,x_1 - N\,x_2 + f_2$$

where x_1 and x_2 denote the armament levels of each nation. A and B are the defense or reaction coefficients for each nation, indicating the degree to which each party feels threatened by the armament level of its rival. In contrast, M and N represent the expense coefficients, generally reflecting the economic costs of increased production. Finally, f_1 and f_2 are either grievance factors (indicating that the nations are competitors) or goodwill factors (for the case in which the nations are friendly to one another). This model captures some of the relationships that can act to increase or decrease the production rates x_1' and x_2' of two competitors, be they nations or companies. It allows one to predict certain types of behavior depending upon the values of the system parameters (A, B, etc.). Similar (but usually more complex) models are often used to describe the behavior of other social and economic interactions.

---

5. Our definition of a model follows that of Starfield, Smith, and Bleloch (1990). Much of the material in this chapter corresponds to the ideas expressed in this reference.
6. Richardson (1960), Saaty and Alexander (1981), and Voland (1986).

However, let us carefully evaluate the phrase "purposeful representation" as a definition of a model. First of all, what is the purpose of a model? A model is used to obtain greater insight and understanding about that which is being represented. A model is an abstraction of a problem and its proposed design solution(s). Thus, we create a model whenever we design a system or process that is too complex, too large, or insufficiently understood to implement without further evaluation. A model is meant to elucidate relationships and interdependencies among system components and variables that may not be recognized without the use of the model. Different interdependencies and combinations of components can be proposed and quickly evaluated via modeling (particularly with the aid of computerized models). As we test and refine the model, its precision and value as a representation of the real-life process or system is enhanced.

Furthermore, our understanding of the problem itself and the goals that should be achieved by a viable design solution increases as we develop the model. (See Case History 6.1 *Florence Nightingale: Hospital Reform via Modeling, Design, and Analysis* for an example of successful systems modeling.)

6.4 Model Formats and Types

6.4.1 ▥ Formats

Models can be abstract or concrete. Mathematical/symbolic, graphical, and computer-based (e.g., simulation, finite element, CAD) representations are examples of abstract models of a system or process.[7] An abstract model should describe a concept in sufficient detail to allow its evaluation and, if necessary, refinement.

If the evaluation determines that the concept is valid, the abstract model may be used as the basis for a concrete representation of the design (such as a clay mock-up of an automobile or a prototype of a new wheelchair configuration) that can then be tested and modified. Alternatively, the abstract model [e.g., a computer-aided design (CAD) description of a circuit board] may be used as the basis for manufacturing the design itself without generation of a concrete model.

Some physical (concrete) models can be quite crude, composed of clay, cardboard, rubber bands, glue, and other materials that are easily available. Although crude, such models can be very valuable. They provide us with the opportunity to test a design concept quickly, perhaps thereby discovering unexpected behavior or unanticipated difficulties before more money, time, and effort is invested in developing a full-scale prototype of the design.

---

7. Models can be iconic, analogic, or symbolic. Each of these types of models is an abstraction of reality.

Modeling the Statue of Liberty

The sculptor Auguste Bartholdi used several plaster and clay models of various sizes to create the Statue of Liberty.[8] He began by first producing many drawings of his initial ideas for the statue's form. He then fashioned numerous small clay figures with which he progressively refined his conception, ending with a four-foot tall clay model that captured his vision.

Next, he created a series of plaster models, each increasing in size over its predecessor and allowing Bartholdi to continue to modify and refine the figure. The third plaster model was 38 high, one quarter of the 151-ft statue to be constructed. This 38-ft model was the last one that could be built as a single piece; the next full-size plaster model would need to be constructed in sections by measuring the relative locations of thousands of reference points on the quarter-size model and then scaling these measurements by a factor of four. Each reference point, measured with respect to a three-dimensional frame surrounding the quarter-size model, was scaled and transfered to a full-size frame that would encompass the full-size plaster sections. In this way, the workers could create a grid pattern of the points that defined each section of the statue. The full-size plaster model was completed in this way.

Each section of the full-size model was then covered with a wooden framework to form a mold around the model. Each section of this wooden mold was then laid on its back so that copper sheets, each only 3/32 in. in thickness, could be pressed and hammered against the mold. To avoid making the copper brittle because of work hardening, the workers periodically applied a blowtorch to the metal to soften it; this process is known as annealing. The work side (i.e., the surface on which hammering and pressing were performed) of these sheets formed the interior surface of the statue; This process of forming a statue by sculpting it from the inside is known as repousse ("push back" in French). Finally, the 350 copper plates were riveted together to form the finished statue, supported by a freestanding wrought-iron skeletal framework designed by Gustave Eiffel (who some years later designed the Eiffel Tower). This skeleton supported the entire 100 tons of copper sheeting so that no single sheet needed to support another.

Clearly, this massive effort could not have been successful without the use of models. Today, most engineering projects still rely upon the modeling skills of the designers.

6.4.2 ▨ Types

Models are sometimes divided into three types: iconic, symbolic, and analogic.[9] *Iconic* models are visually equivalent but incomplete two-dimensional (2D) or three-dimensional (3D) representations, such as maps and world globes; 3D physical models of proposed bridges, highways, and buildings; and 3D models generated via CAD. These models actually resemble the process or system under consideration.

8. Boring (1984) and Shapiro (1985).
9. Our discussion is based in part on Woodson (1966).

In contrast, *analogic* models are functionally equivalent but incomplete representations; they behave like the physical process or system being modeled even though they may or may not physically resemble the reality. Miniature airplanes dynamically tested in wind tunnels, and computerized simulations of manufacturing processes are examples of analog models. These models have been designed to behave in the same way as the real processes or systems under certain conditions, and they therefore have functional modeling capability.[10] (Of course, many analog models are also iconic in that they do resemble the proposed design.)

Symbolic models are higher-level abstractions of reality, such as the equations $F = M * a$ or *Area of circle* $= \pi * r^2$. These models include the most important aspects of the process or system under consideration in symbolic form, neglecting all (presumably) irrelevant details. The symbols can then be manipulated and quantified as necessary to increase one's understanding of the physical reality.

The engineer may need to develop detailed, sophisticated models of a system. Mathematical models (i.e., sets of equations) can be used to describe and predict the behavior of systems such as (a) a set of masses, springs, and dampers, (b) the orbital behavior of a planet and its moons, (c) two interacting populations, one of which is the prey and the other of which is the predator, (d) the reaction between chemical compounds, and (e) a traffic-control flow process. Such mathematical/symbolic models are based upon physical laws (e.g., conservation of energy and conservation of momentum), intuition (e.g., an increase in the predator population will have a negative effect upon the rate of increase in the prey's population), and empirical observations (e.g., Hooke's law for ideal springs). These models may then be solved in accordance with a given set of assumptions and conditions, allowing one to predict the expected behavior of the system under investigation.

6.5 Finite Element, Process Simulation, and Solid Models

Finite element models, solid models, and computer simulations of systems are some of the most popular forms in which engineering designs are represented for further analysis and refinement.

In the finite element method (FEM),[11] a system is described as a collection of interdependent discrete parts. Figure 6.1 shows a finite element model of the *Columbia* space shuttle in which the entire structure has been decomposed into small sections or elements. The extent of decomposition depends upon the level of detail that one seeks in the analysis: More elements of smaller size will generally provide a more precise description of the system

10. Many analog models are useful as measurement devices.
11. See, for example, Bickford (1991).

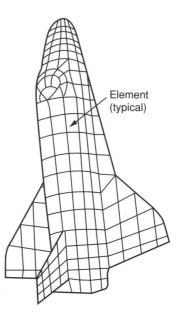

Element
(typical)

FIGURE 6.1 Finite element configuration of the space shuttle *Columbia. Source:* Bickford, 1991.

and its dynamic behavior under a given set of specified conditions. However, modeling and analysis become more tedious and time consuming as the number of elements increases.

Each region within a finite element model is usually represented by a set of dynamic equations that describe relationships and constraints among the system variables (the reaction of the elements to external forces acting on them, conservation of energy, conservation of matter or mass, and so forth). As the entire system is converted into a set of discrete elements, these equations may change from one form (e.g., ordinary differential equations) to another (e.g., algebraic equations) that will be easier to analyze and modify as needed.

Finite element models are used to describe fluid flow, heat transfer, dynamic mechanical responses, and other phenomena in systems that would be difficult (if not impossible) to analyze in any other way. Once the system under investigation is broken into discrete elements, variables of interest within each element are identified and the mathematical relationships between them are developed (e.g., conservation equations describing the flow of mass and the flow of energy from one element to another in a thermal fluid system). One may use the mathematical models for the continuous system as a starting point and convert these continuous equations into their corresponding discrete forms.

A computer model of these finite element equations is then developed, including any necessary boundary conditions or constraints acting on the system and its component elements. Boundary conditions are simply mathematical statements about the condition of the system at certain points in time and/or space. For example: the amount of mass flowing through an opening in a pipe equals 15 kg/hr at the beginning of the process; the final velocity of a vehicle is equal to 50 mph at the end of its journey; and the rate of heat loss from one element to its neighbor equals 0.02 kilojoules per hour. These data points describing the variable values at certain times or places are then inserted into the governing system equations so that any unknown constants in the equations can be determined. The final computer model is then tested and used to investigate different possible configurations of the system of interest to the engineering team.

A finite element model is only one type of computer simulation. Other forms used to mimic the behavior of real-life systems include process simulations and solid models. Process simulations begin by breaking the entire process into a set of subordinate activities (such as specific assembly operations in a manufacturing plant or the sequential use of buttons on a product's control panel by an operator) and then creating a computerized model of these activities and their interdependencies. In contrast to FEM, one does not always need to develop a set of descriptive equations to perform a process simulation. Commercial simulation software often allows one to simply define and insert (through the use of software menu options) specific objects in the computer model that will represent important system and process components (e.g., workers, workstation sites and operations, materials flowing through the system from one site to another, and the distribution of finished products). Modern computer process simulations can be very visual and intuitive by dynamically displaying a system's behavior through the use of moving or changing screen images. In addition, alphanumeric information describing the system's behavior often is calculated, collected, displayed, and updated on the computer for use by the designer. (See Case History 6.2 *Simulation and Interface Modeling at Xerox* for an illustration of how important computer simulations can be to a product's success.)

Solid modeling software allows engineers to create detailed 3D descriptions of a design (Fig. 6.2). Some CAD software allows the engineer to perform finite element analysis upon the solid model and create concrete prototypes of the design by controling and directing fabrication equipment.

Finite element models, process simulations, and solid models provide engineers with the opportunity to quickly and inexpensively modify a design and examine its expected performance under a variety of conditions. As computer modeling software becomes more sophisticated, engineers will be able to develop and investigate even more complex designs with ease and precision. However, to model a design properly, one must recognize the assumptions and limitations inherent in any approximation of a system or process.

6.6 System and Process Models

System models[12] can vary in type. Some are *deterministic,* meaning that they will always perform according to an expected pattern and produce the

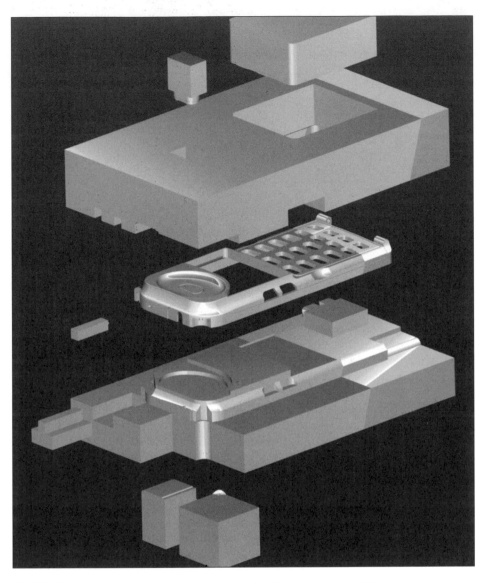

FIGURE 6.2 Rendered image from AutoCad®. *Source:* Lee, 1999.

expected answer. For example, if we model a falling object (in a gravitational field) with equations based upon the laws of physics (and given a set of conditions, such as the initial velocity), we can predict with certainty its subsequent locations, velocities, and accelerations as time progresses.

12. We use the word *system* very broadly to denote both single concrete entities (such as a spring, or a diode, or a person) and assemblies of such objects (e.g., an electrical circuit, an automobile, or a social system).

Other models are *stochastic,* that is, they allow us to predict the behavior of a system or process, but only with a certain degree of uncertainty in the results. Stochastic models are based upon statistics and probability; they are used to determine the average behavior of a system, based upon various sample sets of empirical data.

Traffic-Pattern Models

A stochastic model of the traffic-flow pattern on a highway can be developed from a set of observations, wherein the traffic patterns are studied at specific times on particular days. This model can then be used to predict—with a certain likelihood or probability—whether traffic might be slow moving on a given Thursday morning. Of course, the model could not be used to predict that an accident will occur on that day, resulting in a traffic jam.

Similarly, quality assurance engineers use stochastic models to determine the likelihood that a batch of manufactured units (e.g., computer chips) is free of defects, based upon a sample set of units taken from the batch and analyzed for defective behavior.[13]

Process models can also vary in type: Some are *prescriptive* whereas others are *descriptive.* Figure 1.1 presents a prescriptive model of the engineering design process; that is, it proposes a procedure that we should follow in order to perform design. Prescriptive models can be very helpful in that they provide general guidelines and rules about how a process should be performed in order to achieve a desired objective. However, their value is limited in that they provide only general guidelines; the specific process under consideration may need to differ in important ways from the general process prescribed by the model. In particular, one often takes shortcuts based upon his or her experience with a process rather than exhaustively performing a recommended procedure. For example, a mechanic may not follow some step-by-step procedure for diagnosing an automobile that refuses to start if he or she recognizes certain important symptoms (e.g., the horn and headlights will not operate) that suggest a specific diagnosis (a dead battery).

13. If a sample set of 50 units is free of defects, can one then be certain that the entire batch of 1000 units is free of defects? No, but one can determine the likelihood that the batch contains no defective units. Of course, the size of the sample and the size of the batch have a significant effect on this determination; that is, how large should the sample set be in order to ascribe a likelihood of 99 percent to the conclusion that the batch is free of defects?

Other process models are descriptive, meaning that they describe the actual procedure that is followed to achieve a desired goal. For example, there are descriptive models of the design process that illustrate the methods used by experienced engineers in particular disciplines to generate innovative solutions. These descriptive models have indicated that, with experience, engineers often focus on certain primary goals as they develop alternative designs (where they decide which goals are of primary importance in their design effort).

Both descriptive and prescriptive models of processes are important. Descriptive models provide us with empirical data about how a process is actually performed, sometimes successfully and sometimes unsuccessfully. These data then allow us to improve the prescriptive models that provide guidelines for achieving success.

6.7 Approximations

All models are approximations. One must choose those aspects of a system, object, or process to include in its model; all other aspects are then deemed to be negligible and are excluded from the model. This is not done through guesswork, but rather through a careful evaluation of the problem and the level of detail (i.e., the resolution) needed in the model for it to be a valid representation of reality. The engineer may need to make general estimations or projections of quantities as he or she develops an appropriate model. Such estimation requires that one be informed; if you do not have all pertinent information, then acquire it—do not guess!

In order to determine what is significant about a design and what will therefore be included in the model, an engineer should first consider the specific purpose for the modeling effort: Why does the system need to be investigated and what specific contributions are to be provided by the model?

The engineer must also determine the minimum level of detail needed in the model; that is, the resolution needed to properly describe the system under consideration. A model should never be more complex than is absolutely necessary in order to achieve its purpose. It should never include more than the essential details needed to properly describe the process or system being modeled. This rule—that only essential information should be considered when solving a problem—is known as Occam's razor.[14] All unnecessary detail should be removed from the model so that one may view the problem with clarity. One may devote more time and effort than is needed in analyzing the model if irrelevant details are included and, more important, more data will need to be collected to provide this greater

---

14. In the fourteenth century William of Occam (Ockham) proposed the heuristic or general rule of thumb *Non sunt multiplicanda entia praeter necessitatem* or "Things should not be multiplied without good reason" (Starfield et al, 1990). In other words, do not clutter the problem with unnecessary and insignificant detail, but *do* consider all necessary information.

resolution. On the other hand, a model must be sufficiently detailed and comprehensive so that the effect of all significant characteristics on the behavior of the system will be recognized when the model is analyzed.

6.8 Developing a Model

Our next task, then, is to determine how one goes about developing a model of a process or system.

Begin with the purpose for developing the model: What is to be gained from the model? Be explicit: Identify the specific result that the model is expected to provide. For example, will your model (a) predict some system variable as a function of time, (b) provide a description of an object with sufficient detail for it to be manufactured, or (c) prescribe a process that should be followed in order to achieve a particular goal?

Next, consider both the types of models and the formats that could (or should) be used to represent the system or process (e.g., deterministic or stochastic; prescriptive or descriptive; abstract or concrete; iconic, analogic, or symbolic), together with the reasons for selecting model types and formats. For example, if you are dealing with a situation in which uncertainty is present and in which statistical data could be useful, a stochastic model would be preferable to a deterministic one. If the design is to be an innovative manufactured product, an abstract model followed by a concrete mock-up of the system might be preferable.

As you develop a model, continue to ask two fundamental questions:

- Is the model now useful relative to the purpose for which it was developed? If the answer is no, then identify the model's shortcomings— why is it not useful?—and modify it accordingly.
- Does the model accurately describe the system or process under consideration? Again, if the answer is no, identify those characteristics in the process or system that are either not included in the model or are incorrectly imbedded within the model. Remember that all models are approximations, reflecting simplifying assumptions of the modelmaker, and that some inaccuracy will be present. Eliminate those sources of inaccuracy that lead to error in the model's ability to properly describe the system or process, but also remember Occam's razor.

Unfortunately, it can be difficult for the design engineer to recognize flaws in his or her model of a concept. One always should be critical of one's own work and be willing to correct it as needed. In particular, focus on the assumptions that you have made in developing your model. As new information becomes available, include it in your model where appropriate.

As described in Chapter 3, a heuristic is a rule of thumb that offers a promising approach to a solution because it has led to success in the past. However, it does not guarantee success even though it has been verified through many problem-solving efforts and trials. Nevertheless, heuristics should be applied when their use appears to be promising. Some helpful guidelines or heuristics for developing a model have already been mentioned in this chapter; others include:

- Restate the problem and explain it to another person (not someone on the design team); this may lead to a deeper understanding or another perspective of the task(s) to be performed.
- Identify boundaries or constraints on your model: Are these realistic and consistent with the problem to be solved?
- Your model reflects various simplifying assumptions: Are these assumptions reasonable? If not, adjust your model (i.e., discard any unreasonable assumptions and make the model more complex).
- Try to develop some additional simplifying assumptions and modify your model accordingly.
- Are all known facts and data properly imbedded within your model? If not, why?
- Do you need additional information? If so, obtain it.
- Are there similar problems/models available? If so, how does your model differ from these other models? Why?
- Use another (different) format to model the process or system. Compare the two formats: Which is the more accurate representation and why?

SUMMARY

- In abstraction, one tries to generate as many different solution categories as possible. During the next phase of the design process (synthesis), one will then develop detailed design concepts in each of these categories.
- A model is a purposeful representation of a process, object, or system.
- Models help us to organize our thoughts, data, and knowledge.
- All models are approximations due to simplifying assumptions. An engineer must be cognizant of all assumptions that were made in developing the model.
- The engineer must determine the resolution (the minimum level of detail) needed in the model for it to properly describe the system under consideration.
- A model should never include more than the essential details needed to properly describe a process or system (Occam's razor).
- Models can be abstract or concrete. Furthermore, they may be iconic, analogic, or symbolic.
- System models can be deterministic or stochastic, whereas process models can be prescriptive or descriptive.

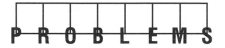

PROBLEMS

6.1 Consider the following mathematical model of the predator–prey system (Voland, 1986, p. 219):

$$x_1' = a\,x_1 - b\,x_2;$$
$$x_2' = c\,x_1 - d\,x_2;$$

where x_1 represents the population of the prey (e.g, rabbits) and x_2 denotes the predator population (e.g., foxes). x_j' indicates the rate of change (dx_j/dt) in x_j, where x_j refers to both x_1 and x_2. The coefficients $a, b, c,$ and d are constants with values that depend upon a number of factors. This model assures an unlimited food supply (e.g., grass) for prey and that predators only feed on prey. Evaluate these equations intuitively: Do they seem to be reasonable representations of the two predator–prey subsystems? Are there other assumptions in this model? Explain your reasoning.

6.2 Develop an iconic model, an analog model, *and* a symbolic model for a process or system of your choice. Discuss the relative value (strengths and weaknesses) of these formats for modeling the process or system. Do any of your models overlap in type; that is, does the iconic model also contain any functional (analogic) representations of the reality? Explain.

6.3 Florence Nightingale (see Case History 6.1) demonstrated the value of statistical models of social phenomena and professional practices. Prepare a brief report on another example of mathematical modeling that led to (a) a deeper understanding of a system or process and (b) the development of an engineering design that satisfied human needs associated with that system or process.

6.4 Florence Nightingale did not subscribe to the theory of germs during the early years of her work in hospital design and practice (Case History 6.1). Nevertheless, her many contributions were invaluable in controlling the spread of contagious diseases throughout a hospital because they fell within an acceptable "tolerance limit" surrounding their theoretical foundations. Such tolerance limits define the *allowable margin of error* such that any design solution, although based upon incorrect assumptions or faulty reasoning, will be successful if it lies within these limits.[15] The tolerance limits of early hospital design were based largely upon empirical evidence that Nightingale and others had observed about contagious disease; although she did not subscribe to the theory of germs, her work was consistent with this theory for she had concluded that fresh air and ventilation were absolutely necessary in a hospital environment.

Similarly, we all sometimes must work with an incomplete and/or incorrect understanding of a problem as we develop a process or system. The challenge is to then develop engineered designs that will lie within the tolerance limits of our knowledge and be valuable even if some of our initial assumptions are incorrect.

Prepare a brief report on a successful engineering solution that was (at least partially) based upon incorrect assumptions. Explain why this solution was successful even though its theoretical foundations were faulty.

CASE HISTORY 6.1

Florence Nightingale: Hospital Reform via Modeling, Design, and Analysis

Today, industrial engineers focus upon the interactions between people and the environments in which they live and work. These engineers apply their expertise in ergonomics (discussed in Chapter 4), work design, operations research, facilities layout, and other areas in order to design both large-scale systems (e.g., hospitals, manufacturing plants, and offices) and small-scale personal environments (e.g., computer keyboards, chairs, sinks, and wheelchairs) that will optimize the safety, comfort, and productivity of the user. Florence Nightingale—known for her pioneering work in the field of nursing—also was one of the first de facto industrial engineers.[16]

Nursing was established as a recognized profession by Nightingale (1820–1910) after her heroic work during the Crimean War. She popularized the use of statistical data and graphics for both modeling and analyzing social patterns. In addition, she evaluated and designed hospital wards and procedures in order to improve the level of care provided to patients. Thus, Nightingale acted as an industrial engineer in both her statistical analysis and design work, long before industrial engineering was established as a profession.

15. The danger of working within so-called tolerance limits is that such boundaries are ill-defined. As we will discover in Chapter 9, many engineering failures are the result of invalid assumptions and faulty reasoning!
16. See Nightingale (1863), Cohen (1984), Darr and Rakich (1989), d'Estaing (1985), Huxley (1975), Smalley and Freeman (1966), Thompson and Goldin (1975), Wadsworth et al. (1986), and Woodham-Smith (1951).

THE CRIMEAN WAR: SANITARY REFORMS SAVE LIVES

British troops serving in the Crimea were in critical need of medical care in 1854 when Nightingale volunteered to lead a group of nurses to the war zone. Upon her arrival, she discovered truly appalling conditions: filthy hospital barracks infested with rats and fleas, overcrowding, laundry caked with organic matter and washed only in cold water, insufficient medical supplies, and other intolerable elements. One result of these conditions was that seven of every eight deaths among British soldiers were directly due to disease and not to injuries sustained in battle.

In response, Nightingale invested her own funds (together with funds generated through public donations) to provide adequate hospital supplies for the troops. She insisted that other reforms be instituted throughout the barracks (e.g., boiling water must be used for washing laundry, and additional kitchens must be installed to meet the nutritional needs of patients). These basic sanitary reforms led to an amazing reduction in the mortality rate among the hospitalized troops. The rate was 42.7 percent in February 1855; Nightingale's reforms were begun in March, and by the spring the rate had plummeted to 2.2 percent!

Nightingale's innovations and dedication were publicized in the *Times* and in Longfellow's famous 1857 poem, making the "Lady with a Lamp" an international heroine. Upon her return to England following the war, Nightingale became determined to improve the unsatisfactory conditions that existed in hospitals (both military and civilian) throughout England.

Her knowledge of hospital practices was not based solely upon her experiences during the Crimean War. As she explained to the Royal Sanitary Commission of 1857:

> . . . for thirteen years I have visited all the hospitals in London, Dublin, and Edinburgh, many county hospitals, some of the Naval and Military hospitals in England; all the hospitals in Paris and studied with the "Soeurs de Charité"; the Institution of Protestant Deaconesses at Kaiserwerth on the Rhine, where I was twice in training as a nurse, the hospitals at Berlin and many others in Germany, at Lyons, Rome, Alexandria, Constantinople, Brussels; also the War Hospitals of the French and the Sardinians.[17]

She was to bring this uniquely rich and varied knowledge to her remaining life's work: revolutionizing hospital and nursing practices throughout the world.

GRAPHICAL MODELING LEADS TO HEALTH REFORMS

Nightingale admired the work of Lambert-Adolphe-Jacques Quetelet (a Belgian astronomer/statistician widely recognized as the founder of social statistics), and so she carefully prepared graphical summaries of her Crimean War data in order to demonstrate the effectiveness of sanitary

---

17. As quoted in Woodham-Smith (1951 p. 225).

reforms. She invented the polar-area or radial chart, in which the size of the area shown on the chart corresponds to the size of the statistic that it represents (Fig. 6.3). With these charts, the mortality rates due to "zymotic" disease (which at the time referred to all epidemic, endemic, and contagious diseases) were easily seen to far outweigh deaths due to wounds and other causes more directly related to war. (In January 1855, the rate of mortalities due to contagious diseases was so high that, if left unchecked, it would consume the entire British Army stationed in the Crimea within one year!)

Nightingale also used bar charts (based upon the work of William Farr in graphical statistics) to emphasize the unusually high rates of mortality among military personnel serving on the home front. These charts clearly indicated the need for reform since the rate of deaths among the military due to disease far exceeded the rate among the corresponding civilian population. (In fact, the military's mortality rate was twenty times greater than the civilian rate.)

These early graphical models of statistical data were striking in their simplicity and clarity. Equally significant, Nightingale was tenacious in her efforts to achieve reform. Eventually, the British government instituted reforms in several areas, including the establishment of a new sanitary code for the army, revised procedures for collecting medical data for statistical analysis, and design modifications in military barracks and hospitals. Nightingale was the primary activist behind each of these improvements.

INNOVATIVE HOSPITAL DESIGN: FLORENCE NIGHTINGALE AS ENGINEER

Nightingale also established the first training school for nurses in 1860. Requirements for graduation included (a) a full year's instruction in medical practices and related procedures, (b) satisfactory performances on both written and oral examinations, and (c) practical experience in hospital wards. Such standards were revolutionary in an occupation that had been viewed with disdain by most physicians and society at large when not performed by religious nursing orders. The benefits of such training soon led to the establishment of other nursing schools. In the United States, Nightingale's recommendations were adopted during the Civil War by Dorothea Dix, who led the American revolution in nursing practices. Three nursing schools (at Massachusetts General Hospital, Bellevue Hospital, and New Haven Hospital) opened in 1873; by 1910, there were 1,129 such schools throughout the United States.

Nightingale wrote several books on nursing practices and hospital design. Hospitals had been constructed with materials that retained moisture easily and became dirty, attracting fungi. Wards inevitably were overcrowded, with insufficient food and medical care for patients. Indeed, people were loathe to enter hospitals as patients since hospital mortality rates were so high. In response, Nightingale stated the following principle in her 1859 book *Notes on Hospitals*:

FIGURE 6.3 Nightingale's chart of mortality rate at Scutari, the primary British hospital during the Crimean War. *Source:* From Nightingale, 1858, reprinted in *Florence Nightingale on Hospital Reform,* edited by C.E. Rosenberg, 1989, with permission from Garland Publishing, Inc.

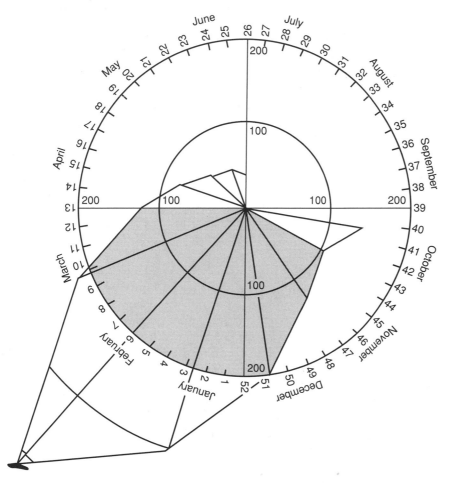

It may seem a strange principle to enunciate as the very first requirement in a Hospital that it should do the sick no harm. It is quite necessary nevertheless to lay down such a principle, because the actual mortality in hospitals, especially those of large crowded cities, is very much higher than any calculation founded on the mortality of the same class of patient treated out of hospital would lead us to expect.

Nightingale enumerated many defects (as she perceived them) in hospital design and construction; these included

- poor site locations that could inhibit a patient's recovery due to impurities in the air, inaccessibility for friends and medical staff, and difficulty in transporting patients to the site (Hospital were often near the sewer outlets, far from the center of town.);
- poor ventilation due to the height and width of hospital wards, location of beds along "dead walls" (i.e., walls without windows or vents), more than two rows of beds between windows, and other factors;
- inadequate drainage systems, water closets, sinks, and so forth;
- the use of absorbent materials (such as plaster) for walls, floors, and ceilings, leading to high levels of moisture, dirt, and fungi growth;
- inadequate kitchens, laundry facilities, and nursing accommodations.

The corridor designs of the eighteenth and early-to-mid nineteenth centuries had paired or double (back-to-back) wards separated by privy facilities. The Royal Victoria Hospital at Netley, under construction when Nightingale returned from the Crimea, is a striking example of the poor corridor layout that blocked nurses' views of patients and allowed foul air to travel between wards.

Nightingale specified various ways in which hospitals could be designed and constructed better to ensure that patients would be provided with proper ventilation, improved water-supply systems, adequate nutrition, enhanced medical care, and so forth. She corresponded extensively with engineers, architects, and others in construction work and she embraced the pavilion system for hospital architectural design in which patients are housed in separate pavilions or detached housing blocks with well-ventilated "open" wards. (Although Nightingale did not subscribe to the then-developing theory that germs were the underlying cause of many illnesses, her focus upon the need for proper sanitation and ventilation was nevertheless consistent with many future health practices based upon the germ theory.)

Nightingale recommended that each ward in the pavilion be oblong with ventilation provided by windows along both long walls and by doors at the narrow ends. She also provided specifications for the number of floors per pavilion, number of wards per floor, the size of wards, and the number of beds per window—all in order to provide patients with adequate ventilation, sufficient light for reading, and other necessities. The design of each ward also included an adjacent head nurse's room with a window that overlooked the entire ward (thereby allowing proper supervision of patients' needs) and—at the opposite end of the ward—a bathroom, lavatory, and water closets. She also recognized the need for efficiency in performing tasks by noting:

> Every unneeded closet, scullery, sink, lobby, and staircase represents a place which must be cleaned, which must take hands and time to clean, and a hiding or skulking place for patients or servants disposed to do wrong. And of such no hospital will ever be free. Every five minutes wasted upon cleaning what had

better not have been there to be cleaned, is something taken from and lost by the sick.[18]

She provided guidelines for building materials that would resist absorption of moisture and allow for thorough cleaning (e.g., parian cement for walls and ceilings; oak, pine, or tile for floors). She also specified layouts for laundries, bathrooms, lavatories, water-supply systems, and kitchens so that no unnecessary duplication of services or facilities would occur. And she recommended that open-air gardens and sheltered exercise grounds be included in hospital layouts.

Furthermore, Nightingale designed layouts for improved ventilation systems in which foul-air shafts (located on alternating wall spaces between windows) would be used to remove air and Sherringham's ventilating inlets (located near the ceiling and opposite each shaft) would be used to provide fresh air to each ward. In addition, she provided guidelines for furniture selection, dishes (which were to be made of glass or earthenware), and other details of ward design.

Numerous hospital plans were submitted to Nightingale for her review. These included the plans for Coventry Hospital, Birkenhead Hospital, the Edinburgh Infirmary, the Chorlton Infirmary, the Swansea Infirmary, the Staffordshire Infirmary, and others. Between 1859 and 1864 Nightingale supervised the construction of Herbert Hospital, the first facility built entirely according to her specifications.

Nightingale was an extremely influential figure in medical practices throughout her life. And through her books and other works, she continues to influence the medical profession. The basic ward layout of the pavilion design has undergone little modification until recent times, attesting to the farsighted contributions made by Nightingale.

When compared with today's sophisticated hospital layouts, Florence Nightingale's recommendations were somewhat rudimentary. Nevertheless, her work was extremely important in that she

- identified numerous shortcomings in hospital design and practice and developed a set of corresponding improvements,
- demonstrated the importance of satisfying the patient's primary needs via thoughtful design,
- focused upon the interactions between a system and its users (an early instance of ergonomics or consideration of human factors in engineering design), and
- emphasized the need to eliminate unnecessary tasks in order to ensure that work is performed in an efficient and appropriate manner.

Although Florence Nightingale was not an engineer by training, her work emphasized the need for proper engineering design principles, mathematical/graphical modeling, and analysis. More important, she teaches us that we as engineers must anticipate the effects that a design may have upon our clients and then work to ensure that these effects are beneficial.

18. Nightingale (1863).

CASE HISTORY 6.2

Simulation and Interface Modeling at Xerox

Modeling and simulation became extremely important to the Xerox Corporation during the 1980s.[19] In 1959 Xerox introduced the first plain-paper copier (the Xerox 914). By 1976 the company had captured 82 percent of the worldwide copier market with sales of $6 billion. However, its patents on the photocopying process had expired in 1969, leading many other companies (such as Canon, IBM, Kodak, Minolta, Ricoh, and Sharp) to enter the market. By 1982, although its sales had increased to $8 billion, Xerox retained only 41 percent of the market.

In response, the company reorganized and refocused its attention on the operability of its machines. The company had developed sophisticated machines for high-production applications—machines that could reduce, enlarge, collate, staple, and perform many other functions. Traditionally these machines were operated by workers whose sole responsibility was to produce copies as needed. The features and functional capabilities of the high-production machines were transferred gradually to the low-production machines used by casual operators (e.g., secretaries, clerks, office workers). Unfortunately, this upgrading in the capabilities of the low-production copiers did not include a redesign of the machines' interfaces and operating instructions. In other words, the fact that distinctly different user populations—with differing sets of responsibilities and expertise—operate the low-production and high-production machines was not taken into account. The low-production copiers became increasingly more difficult for the casual operators to use, maintain, and service.

To correct this situation, Xerox design teams interviewed customers, salespeople, and service representatives to determine specific areas in which improvements were needed. These teams also observed casual operators working with copiers. Based upon the information and knowledge so obtained, Xerox designers developed a conceptual model of the proper operation, maintenance, and repair of copiers—a model that could be described clearly in illustrations and text and that could be supported through the physical design of the machines by providing operators with color-coded directions, iconic interfaces, and control panels.

To further develop the new user-friendly copiers, the designers contacted the Xerox Palo Alto Research Center (PARC). PARC had developed the Xerox Star (which eventually became the Xerox Viewpoint) computer interface system in which icons, windows, and pull-down menus could be

---

19. See Wasserman (1989).

manipulated with an electronic mouse. (This iconic interface design was later successfully commercialized in Apple's Macintosh computers.) In response to the needs of the Xerox designers, PARC developed the Trillium simulation software that allowed alternative copier designs to be developed and evaluated in hours rather than months. With Trillium, the engineers were able to develop effective iconic information displays and graphic control panels for the redesigned copiers.

By 1986 Xerox had rebounded in the market with its new copier designs and interfaces, capturing 55 percent of the market.

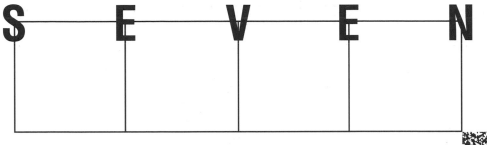

SYNTHESIS

Genius is one percent inspiration and ninety-nine percent perspiration. Thomas Alva Edison

O B J E C T I V E S

Upon completion of this chapter and all other assigned work, the reader should be able to

- Discuss the role that synthesis plays in design as entities are combined in appropriate ways to create an engineered product or system.
- Identify the barriers or conceptual blocks to creative thinking.
- Explain why one must be prepared to transform accidents into design solutions.
- Apply such strategies as brainstorming, bionics, synectics, adaptation, and inversion to stimulate creativity in engineering problem solving.
- Use morphological charts to perform synthesis by joining preliminary ideas or subsolutions to form total design concepts.

7.1 Synthesis in Engineering Design

Abstraction (see Chapter 6) provides us with a perspective of the building blocks that can be used for design solutions. Synthesis is the formation of a whole from a set of such building blocks or constituent parts.

Deductive and inductive logic, specialized analysis, precedent, group concensus, and established habits are all ways of thinking that have a place in engineering design. However, creative thinking is distinguished in engineering by its ability to synthesize, or combine, ideas and things into new and meaningful (i.e., practical) forms.

This new and meaningful synthesis may be experienced as a flash of insight, but it is generally true that it follows careful, and perhaps laborious, preparation. A key element in this preparation are the design specifications (discussed in Chapter 4), which are established by analyses of human needs, physical constraints, and other conditions. After the effort is made by an engineer to acquire a considerable amount of available knowledge and understanding pertaining to a particular problem, a time of mental relaxation is usually needed. When conscious effort is relaxed and the mind is refreshed, a synthesis may suddenly occur in the designer's mind "with definite awareness of assured suitability for a particular new development."[1] Such moments have been frequently documented.

Somewhat differently from those who work in music, sculpting, or other artistic endeavors, the engineer must be creative within the design specifications or boundaries of the problem to be solved. As a result, a sort of *constrained creativity* is needed in which imagination is not only balanced with practical considerations, but is actually driven by them.

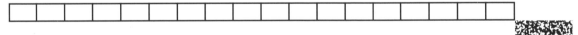

Creative Yet Impractical Patented Designs

Many ideas for new products have been failures because the inventor failed to recognize one or more elements of impracticality in the design.

In Chapter 5 we noted that the records of the United States patent system are a rich source of technical knowledge. However, one must be skeptical when reviewing patents; many designs that have been awarded patent protection either do not operate as described or are clearly impractical. Because of their impracticality, none of these ideas has ever served anyone's needs. They remind us that engineers must be both creative and practical![2]

Figure 7.1 presents some of the design failures that have been patented over the years. These include:

- A 1923 design for a bottle to transport messages by sea (a bell dangling from a support shaped like a question mark is included to attract the attention of passersby).
- A wooden roller coated with emory, which was designed for shaving without soap or water. The driving wheel from a sewing machine is connected by belt to

1. Edel (1967).
2. Examples of patented but impractical designs are described by Brown and Jeffcott (1932), Feinberg (1994), Jones (1973), Lasson (1986), Richardson (1990), and Woog (1991).

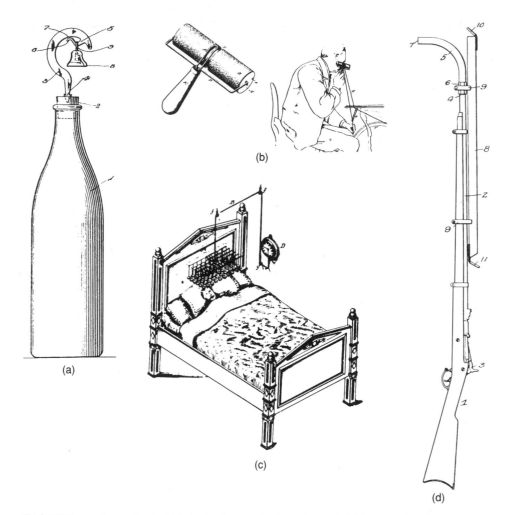

FIGURE 7.1 Patented ideas for impractical products: (a) Message bottle (patent #1,469,110); (b) Roller system for shaving (patent #646,065); (c) Alarm systems (patent #256,265); and (d) Rifle for shooting around corners (patent #1,187,218).

the roller and the user then spins the roller at high speed by footpedaling this wheel (1900).

■ An 1882 patent (#256,265) for a wake/alarm system. The user would be awakened when a set of suspended corks were released above his or her head.

■ A 1916 patent for a curved extension piece and periscope that were to be attached to a rifle. The curvature of the barrel extension then was expected to allow an infantryman to aim safely and shoot the weapon around the corner of a foxhole or other protective structure.

Other unusual or apparently impractical designs include: an 1895 design (U.S. patent #536,360) that allows one railroad train to pass over another during a head-on collision; an 1879 device (U.S. patent #221,855), consisting of a parachute connected to a person's head and oversized elastic shoes, to be used when escaping from a fire; an 1897 design for a cow costume in which two hunters could hide while hunting for game (U.S. patent #586,145); and a 1991 "collapsible riding companion," essentially a mannequin's head and torso that could be attached to the seat of an automobile in order to deter criminals (U.S. patent #5,035,072).

In contrast to the above cases, the development of 3M's Scotch tape is an example of creative thinking driven by truly practical considerations (see Case History 7.1 *Scotch Tape*).

The design engineer must be able to view the problem from different perspectives and overcome various barriers to creativity that may limit both the number and the variety of solutions that are proposed. Yet—simultaneously—he or she must always remain focused upon elements of practicality if the proposed designs are to be feasible (see Case History 7.2 *Control via Frequency Hopping—Designed by a Movie Star*). How, then, can this balance between creativity and practicality be accomplished?

Numerous methods exist for stimulating one's inherent creative abilities while focusing upon the practical aspects of an engineering design problem. We will review some of the most successful methods in this chapter. No single method may work for everyone; however, one or more of these methods may indeed work for you. Experiment with each method and try to determine those that seem to enhance your creative ability in problem solving.

7.2 Barriers to Synthesis

Each of us must overcome barriers to synthesis and creative thought. Most engineers are analytical by nature; when a new idea or concept is proposed, we immediately begin to evaluate its strengths and weaknesses. Such preliminary analysis may preempt our opportunity to build upon an initially impractical idea, molding it into a revised form that is indeed feasible. As we will discover in this chapter, many innovations in engineering began with a silly concept that was later refined and distilled into the final (practical) solution through synthesis.

Before we consider various techniques for stimulating creativity, we should begin with a discussion of some of the more common barriers to creative thinking (sometimes called conceptual blocks[3]). These include

---

3. See, for example, Adams (1974) and Fogler and LeBlanc (1994).

knowledge blocks, perceptual blocks, emotional blocks, cultural blocks, and expressive blocks.

7.2.1 ▨ Knowledge Blocks

As we noted in section 5.2, engineers need to be knowledgeable about scientific principles and phenomena if they expect to generate creative yet practical design solutions to technical problems. Engineering is the application of science to technical problem solving. Most engineered solutions are based upon one or more scientific principles that were applied by the designers in a creative manner (recall Tables 5.2 and 5.3). Without a reasonably deep scientific knowledge base from which to draw ideas, we will have difficulty in solving problems in new and wonderful ways; in other words, we will suffer from knowledge blocks (see Case History 7.3 *Radar in the Battle of Britain* and Case History 7.4 *V-2 Rocket*).

7.2.2 ▨ Perceptual Blocks

Sometimes one is unable to discern important aspects of the problem that is to be solved. Such inability may be due to

a. *Stereotyping elements* in a proposed solution or in the problem itself, thereby limiting our ability to recognize other interpretations of these elements.

   For example, one might view a pencil only as a writing instrument; however, a pencil can also be used for

   ▪ Firewood,
   ▪ A shock absorber or bumper (assuming that there is an eraser attached to it),
   ▪ A roller—for example, one or more pencils could be placed beneath a heavy object (such as a refrigerator), which could then be rolled to a new location,
   ▪ A support rod for a structure,
   ▪ A weapon, or
   ▪ A source of graphite

b. *Delimiting the problem,* in which one imagines that additional constraints exist beyond the actual design specifications, thereby unnecessarily restricting the range of possible solutions.

   For example, imagine that we are asked to design a product that will prevent blockages in the downspouts of home rainwater gutter systems. These blockages may be due to leaves, pine needles, balls, dead animals, or other types of debris. We first acquire detailed information about standard downspouts and gutter systems, from which we then develop a set of design goals and specifications.

   We may also assume that any successful design must be retrofitted to existing gutter systems. However, the best solution may be one in which existing downspouts are replaced with some more effective method for directing collected rainwater away from a house. The imaginary constraint of retrofitting (partly due to the original phrasing of the problem statement in which we were asked to prevent "blockages in the downspouts") will prevent us from even considering an approach in which the gutter system is eliminated in its entirety.

c. *Information overload* Very often, engineers must wade through vast amounts of data in order to develop a sufficiently detailed understanding of a problem and the context or environment in which the solution will be used. Unfortunately, the very quantity of available information can sometimes prevent one from developing an accurate understanding of the goals that must be achieved by a solution, thereby limiting one's ability to develop feasible designs. It may be too difficult to distinguish between what is significant and what is insignificant among the data. One may also generate an extensive set of design specifications based upon the data, not recognizing that many of these constraints are unnecessary; for example, they may simply represent some limitations of specific solutions contained within the database, rather than global limitations that must be placed on all solutions.

For example, imagine that we are given detailed and extensive information about the existing water reservoir system for a town.[4] Data is in the form of surveyors' reports, extensive topographic maps of the region, numerous computer-generated tables describing characteristics of primary and secondary water sources, records of power supply and use, engineering reports, catalogs and handbooks of water pumps, etc. We are then informed that a localized source of pollution is infecting the current water-supply system. We are assigned to redesign the system so that it will avoid the pollution source. We are informed that, because of cost constraints, our design must use as much of the existing system network of pipes and channels as possible.

Unfortunately, the sheer volume of data in this situation may prevent us from even considering certain approaches (such as eliminating the source of pollution rather than redesigning the water-supply system).

A perceptual block prevents us from clearly viewing the problem and the various paths that could lead to viable solutions.

7.2.3 ▨ Emotional Blocks

Our creativity may be inhibited by an emotional block. Such blocks tend to limit our willingness to try new and uncertain approaches in solving a problem. Emotional blocks include the following:

- Fear of failure and the need for approval
- A need to follow only prescribed paths and methodologies

4. Adapted from a problem presented at the 1982 New England DesignFest by Thomas R. Long of West Virginia University.

- A tendency to accept the status quo in both problem formulation and the types of solutions that are generated
- Impatience, leading to a quickly developed solution that may not solve the problem in an effective or optimal manner

Creativity, like prospecting for oil or minerals, requires some degree of risktaking and perseverance as one examines different solution paths. Many of these paths will lead to dead ends, but one must be willing to prospect in more than one area if he or she expects to be successful.

7.2.4 ■ Cultural Blocks and Expectations

Different companies can have very different cultural predilections that are reflected in their work environments and in the products or services that they provide. For example, the culture within one company may discourage employees from considering new ways of performing routine tasks because it does not want to "rock the boat," believing that the status quo in daily operations should be maintained in order to avoid risks. In contrast, another company may provide incentives for creative work by its employees in the form of financial bonuses, extra vacation time, or promotions.

In an inhibiting environment, designs may be generated that are narrow in scope and reflect the particular atmosphere in which the engineers work. Although we all should apply our life experiences in solving problems (as Chester Carlson did in developing the xerographic process; see Case History 5.1), we also need to overcome the limitations that specific work environments might place upon our creativity.

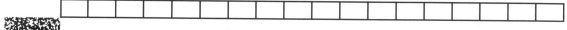

Inflatable Space Station

A conceptual design was submitted to NASA for an inflatable space station.[5] As proposed, the station, which would have resembled the inner tube of an automobile tire, would be inflated in space after first being folded and transported in the nose cap of a rocket to the desired location.

This proposal was developed by a well-known tire company, reflecting (at least in part) the traditions and expertise of that industry.

5. See "Goodyear Proposes Expandable Structures as Space Stations," *Missiles and Rockets* **8** (May 29, 1961, p. 24), and Ferguson (1992).

Of course, sometimes the best solution to a problem will be developed by a particular industry because of its inherent cultural perspective and historical expertise. Moreover, the need to overcome cultural predilections must be balanced with the practical necessity to work successfully with others in developing the best possible solution to a problem.

We also need to recognize that there can be design expectations or preconceptions held by our clients and customers. Sometimes a creative engineer may not recognize that a truly innovative design is simply too unusual to be readily accepted by the public or sponsoring agencies.

The Rejected Golden Gate Bridge Design

Joseph P. Strauss served as chief engineer during the construction of the Golden Gate Bridge in San Francisco. He had been a very successful builder of bascule drawbridges in which counterweights are used to lift the draw spans via a rolling action. Between 1924 and 1932 he sought support for his proposed design of the Golden Gate Bridge in which two bascule bridges would be connected via a massive suspension span.[6] However, support did not materialize for this creative but quite unusual concept; it apparently was simply too different from traditional bridge designs. Finally, in 1932 he altered his design to that of a conventional suspension bridge (but with a center span of 4200 feet), which was then completed in 1937.

7.2.5 ▨ Expressive Blocks

Expressive blocks inhibit one's ability to communicate effectively with others *and* with oneself. We need to model the problem and its possible solutions in the most appropriate form (see Chapter 6). Furthermore, we need to be cautious about the terminology used to define a problem or to describe solutions. Consider the following assignment:

> Design a self-propelled vehicle (to be made only of paper, paper clips, pencils, rubber bands, and glue) that will climb 30 feet along a taut vertical rope. The vehicle should be able to carry a payload of (at least) one ounce.

The difficulty is in the use of the word *climb.* Our creativity may be limited by this word as we design various climbing mechanisms (that would necessarily require much energy to perform the so-called climbing task) rather

---

6. See the discussion in Ferguson (1992); also see Van der Zee (1986) and Cassady (1979).

than focusing upon the design of a simpler type of self-propelled vehicle that will simply *traverse* the rope for the required distance of 30 feet (e.g., a simple slingshot type of design).

7.3 Design by Accident

Sometimes an engineering innovation will occur by accident. For example, someone might observe a new and unexpected result while conducting an experiment or a new idea may spring to mind while one is pondering a given problem. However, one must be prepared to take advantage of such accidents or the opportunity may be lost forever. (In addition, such "accidents" may actually be the result of previous work that has led the engineer to deeper understanding of the problem and the ways in which it can be solved.)

Slinky

One incident of design by accident occurred in 1943 when marine engineer Richard James unintentionally dislodged a very flexible steel spring from its shelf.[7] (He had been searching for a means to protect radios and other shipboard instruments from damage.) He watched as the spring flipped over again and again on a pile of books. Later at home, James and his family enjoyed playing with the spring. After World War II, he founded James Industries to market the "Slinky Spring" as a toy.

However, the subsequently phenomenal success of the Slinky required one more clever development: its use as the centerpiece in a wide range of Slinky pull-toys, including Slinky trains, toy soldiers, seals, worms, dogs, and caterpillars. This extension of the Slinky concept was due to Helen Malsed. After giving her son Rick a Slinky Spring for his birthday, he exclaimed "Boy, would that ever go if it had wheels!" His father then added wheels to the spring and Helen attached a string to form the world's first Slinky pull-toy. Helen's imagination led to the creation of other pull-toys and she contacted James Industries to properly license her designs. When Arthur Godfrey enthusiastically played with a Slinky pull-toy train during his televised Christmas show in 1954, sales surged upward as the Slinky pull-toy became the most popular pull-toy of all time.

A simple accident, together with creativity, insight and thought, had led to commercial success for both James Industries and Helen Malsed—and to great joy for millions of children.

---

7. See Woog (1991).

Teflon

Another accident led to the discovery of Teflon; however, as noted by Roberts, the innate curiosity and intelligence of Roy J. Plunkett were needed to transform this accident into a discovery.[8]

In 1938 Plunkett was working with his assistant Jack Rebok on the development of a nontoxic refrigerant in his role as a research chemist at the Du Pont Company. One day, Plunkett and Rebok opened a tank of gaseous tetrafluoroethylene. The weight of the tank indicated that it was (or, at least, should have been) full; nevertheless, no gas was released. After testing the valve to ensure that it was working properly, Plunkett sawed the tank apart in order to investigate the mystery of the missing gas. He discovered a white waxlike powder had formed on the inside of the tank. Somehow, in this one tank, the gas molecules had combined to form a polymer.

Tests indicated that this new inert polymer—dubbed "polytetrafluoroethylene"—was extremely resistant to acids, bases, and heat. With further investigation, Du Pont chemists discovered a method (albeit expensive) for manufacturing polytetrafluoroethylene.

By this time (World War II), work had begun elsewhere on the atomic bomb. The production of U-235 for such bombs required the use of the gas uranium hexafluoride. Unfortunately, no gasket could long resist the corrosive effects of this gas. A new gasket material, highly resistant to corrosion, was needed. General Leslie R. Groves (head of the United States Army unit involved in the development of the atomic bomb) discovered that Du Pont had a new corrosion-resistant polymer and they might be used to contain uranium hexafluoride. Gaskets were then made from the polymer (now known worldwide as Teflon) and they did indeed contain the corrosive gas.

Of course, Teflon has since been used to provide kitchen pots, pans, and utensils with nonstick surfaces. However, Teflon also has been used for more critical applications. For example, Teflon is used to protect space vehicles and other systems from extreme heat. Artificial bones, heart valves, tendons, dentures, sutures and other medical devices composed of polytetrafluoroethylene are not rejected by the human body because of Teflon's natural inertness. Many lives have been saved because Roy Plunkett did not simply discard that "empty" tank of tetrafluoroethylene in 1938, but instead took the next step and investigated the situation.

Charles Goodyear's discovery of the vulcanization process for curing rubber is a famous example of exhaustive experimentation coupled with a fortuitous accident. It is also a lesson about the need for each of us to sometimes collaborate with others who can provide expertise that we may lack (see Case History 7.5 *Goodyear and the Rubber Vulcanization Process*).

8. See Roberts (1989).

It is important to recognize that most accidents are simply mishaps or true disasters that do not lead to valuable innovations. Engineers certainly cannot depend upon accidental occurrences to provide solutions to technical problems. However, like Roy Plunkett, we must be prepared to take advantage of an accident in order to transform it into a discovery. Such preparation requires an extensive knowledge of science, mathematics, and engineering fundamentals, together with curiosity, analytical skills, and imagination. The next section focuses upon ways in which we can train ourselves to become more imaginative problem solvers.

7.4 Creativity Stimulation Techniques

Many different strategies for overcoming barriers to creativity have been proposed during the last sixty years. Many of these methods have been quite successful in producing innovative solutions to a broad range of problems. We will not attempt to present an all-inclusive list here, but instead focus upon those techniques that have been particularly successful for various individuals.

Remember that no single technique will be effective for everyone. Experiment with different strategies and find one that works for you.

7.4.1 Brainstorming

The most successful and well-known strategy for generating creative solutions to a problem is brainstorming, which was developed by A. F. Osborn during the late 1930s.

In this method, a group of people are selected to work on a problem for a limited amount of time. The group is often composed of people with a variety of backgrounds who bring different types of expertise to the problem-solving effort. For example, a group might consist of engineers from various disciplines together with managers, salespeople, maintenance and repair people, and distributors. The challenge is to generate as many ideas as possible in the given amount of time. Quantity—not quality—is sought in the solutions that are proposed; it is assumed that if 100 preliminary ideas can be generated, at least some of those concepts will be promising. If there are five practical yet innovative concepts for solving a challenging problem among a set of 100 ideas generated during a two-hour period, the effort can be deemed successful and well worth the small investment of time and energy. (It is important to recognize that brainstorming may not result in well-formulated, final design concepts but rather promising ideas that usually need further development and refinement.)

Brainstorming can also be effective when used by only one or two people.

In order to generate as many ideas as possible within a small amount of time, it is critical that each idea not be evaluated at the moment of its conception. Criticism should be deferred until after the brainstorming session has ended. Not only will such criticism require time to perform, but it may also inhibit people from making suggestions that they feel are too silly or impractical to propose. As we noted in Section 7.2, many times an apparently silly concept is in fact the innovative solution that was so long sought. Consider the following problem.

Design a method for transporting people from one location to another.

Some methods spring immediately to mind, but others may not. A brainstorming session might result in the following ideas:

- Use a conveyor belt
- Use a giant slingshot
- Use balloons
- Use a rocket
- Use kites
- Carry
- Float/paddle
- Use a hoist
- Use skis (water or snow)
- Use rollerblades
- Use land skis with rollers
- Use gliders
- Use horses
- Slide downhill
- Use trains
- Use a ladder

In summary, when brainstorming one should:

- Seek quantity of concepts, not quality.
- Defer judgment or analysis of concepts.

7.4.2 ▨ Bionics

A particularly successful strategy for solving technical problems is bionics. In this method, one searches for an existing solution within nature that can be adapted to solve the problem under consideration. The natural world is filled with many wonderful examples of tasks that are performed in effective and elegant ways.[9]

Table 7.1 presents numerous examples of products and solutions to technical problems that are analogous to natural phenomena.[10]

---

9. Some people believe that the development of radar and sonar during World War II was based upon bats' use of echoes for navigation, but in fact these two innovations are not examples of bionics in action but rather the result of science directly applied to engineering problem solving (See Case History 7.3).

10. See Alger and Hays (1964), Middendorf (1990), Voland (1987), and Woodson (1966).

TABLE 7.1 Developing analogies between natural phenomena and technical concepts via bionics.

Natural Phenomena	Technical Concept/Product
Feather oil of ducks	Anti-wetting agent
Bamboo, bones	Tubular structures
Muscles attached to bones	Levers, fulcrums
Trap-door spiders	Doors, hinges
Venus's-flytrap	Trigger mechanisms
Evaporation through skin surface	Cooling
Squid siphon	Jet propulsion
Heart	Squeeze pump
Birds' legs, claws	Retractable landing gear, automatic clasps
Beetle's eye	Aircraft altitude and ground-speed indicator
Massasauga rattlesnake	Heat sensor
Human brain	Artificial neural networks
Beehives	Storage containers

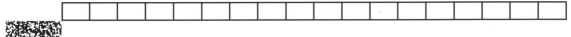

Velcro

A truly classic example of bionics leading to a successful design is the development of Velcro by Georges de Mestral.[11] (You may recall that the claims from de Mestral's patent disclosure were given in Chapter 5.) In 1948 de Mestral became intrigued with the cockleburs that clung to his dog and to his own clothes. (He and his dog had been hunting in the Swiss Alps.) With the aid of a microscope, he discovered that each burr contained hundreds of tiny hooks with which it could cling to other objects. This method of fastening thereby ensured that the plant's seeds, which are contained within the burr itself, will be dispersed over a very broad area.

de Mestral then strove to develop a manmade version of this natural and very effective means of fastening two objects together. By 1956, he had created two different types of woven nylon strips that were based upon his observations of the cocklebur.

11. See Travers (1994) and Roberts (1989).

Thousands of tiny hooks on one strip intertwined with thousands of equally minuscule loops on the other strip whenever the strips were pressed together, resulting in the two strips becoming firmly fastened to one another. Through the use of bionics, de Mestral had created a new and very effective temporary fastener: Velcro.

Velcro-type strips are now manufactured by several firms since the 1957 patent has long since expired. However, the name "Velcro" remains a registered trademark; it was derived from the French *velours,* velvet, and *crochet,* hook.

Vortex Noise and Owls' Wings

Engineers at the Tullahoma Space Institute at the University of Tennessee were seeking ways to reduce vortex noise generated by an airplane at it travels through the air. This was a particularly vexing problem in noise control since vortex noise was created in and transmitted through the atmosphere, making the conventional sound absorbers used in enclosed environments to dampen unwanted noise impractical.

The engineering team decided to focus upon natural solutions to this problem by investigating birds' flight. They discovered that the owl must fly silently through the air if it is to capture rodents; otherwise, the rodent will be forewarned of the owl's approach and scurry to safety.[12]

The researchers then conducted a series of tests to determine how owls fly without generating any significant vortex noise. Their investigation led to the discovery that the leading edge of an owl's wing does not remain smooth but instead becomes serrated during gliding flight, resulting in less vortex noise and a silent approach.[12]

A solution to the problem of airplane vortex noise had been found through bionics.

Improved Chain Saw Cut

Another example of bionics in action is the development of an improved chain saw design by Joe Cox in 1935.[13] Charlie Wolf patented the first truly practical chain saw design in 1920. Unlike earlier patented designs, Wolf's saw was portable, easy to use, and allowed one to reverse the chain after it became dull (i.e., the chain operated in

12. Lumsdaine and Lumsdaine (1990).
13. See Woog (1991).

either direction). Unfortunately, the chain saw initially was seen by loggers as a threat to their jobs and it was not successful on a large scale.

Fifteen years later, Cox (a mechanic, handyman, and logger) sought to develop an improved chain system. The crosscut approach used by power saws at that time generated considerable waste and quickly dulled the chains. One day while chopping wood, Cox noticed that the larvae of the timber beetle cut through wood quickly and with little waste by using the two cutters on their heads in a left-to-right, back-and-forth motion. Borrowing this concept from nature, Cox then designed and developed a chain with two cutters on a single head. In 1947 Cox successfully introduced his chain design to the timber industry. His basic design continues to be the prototype of most chain saw systems currently in use.

7.4.3 ▨ Checklisting

Another strategy for creativity developed by A. F. Osborn is checklisting,[14] where, basically, one uses words and questions to trigger creative thought.

Often these triggers focus upon possible changes in an existing product, concept, or system that may have been generated through brainstorming or bionics. Such changes may focus upon *quantity* (increase, reduce), *order* (reverse, stratify), *time* (quicken, synchronize), *state* or *condition* (harden, straighten), or the *relative motion or position of components* (attract, repel, lower). Table 7.2 presents a set of checklisting triggers that can be used to stimulate creativity.

One particularly useful checklisting question to ask about an existing product is, What's wrong with it?, as shown in Case History 7.6 *So What's Wrong with Our Toaster?*

7.4.4 ▨ Synectics

Synectics was developed by W. J. J. Gordon as another strategy for creativity. The method has nine phases (see, for example, Edel, 1967, and *Gordon,* 1961) which include the use of analogies and metaphors to trigger ideas and two other unique activities: making the familiar strange and making the strange familiar.

These last two activities represent Gordon's recognition that we are sometimes confronted by a problem that is either so familiar that it becomes very difficult to view it in any type of innovative form or so unusual that it is difficult to relate it to our past experiences.

Imagine that we are asked to develop a new product for fastening sheets of paper together. This problem is one that is "familiar" in that we immediately might think of paper clips and staples. Unfortunately, we then have difficulty in thinking of other possible solutions.

14. See, for example, Osborn (1957) and Edel (1967).

TABLE 7.2 Checklisting triggers.

Examples of trigger questions

What is wrong with it?	What does it fail to do?
What is similar to it?	In what way is it inefficient?
Why is it necessary?	In what way is it costly?
What can be eliminated?	Who will use or operate it?
What materials could be used?	Are there any other (possible) applications?
How can its assembly be improved?	What is it not?
Can any components be eliminated?	Can it be misused?
Is it unsafe?	

Examples of trigger words

Accelerate	Cool	Harden	Lower	Soften
Add	Curve	Heat	Renew	Solidify
Alter	Deepen	Increase	Repel	Stratify
Animate	Destroy	Influence	Reverse	Strengthen
Attract	Direct	Interchange	Rotate	Stretch
Cheapen	Elongate	Interweave	Roughen	Substitute
Combine	Encircle	Lessen	Shorten	Subtract
Condense	Energize	Lift	Slow	Thin
Converge	Fractionate	Lighten	Smooth	Widen

Trigger-word categories

Change *quantity*	Increase, reduce, lengthen, shorten, deepen, combine, fasten, assemble, fractionate
	Examples: hearing aids, pocket flashlights
Change *order*	Reverse, stratify, arrange
	Examples: rear-drive vehicles, reversible belts, plywood
Change *time*	Quicken, slow, endure, renew, synchronize
	Examples: rechargable batteries, self-winding watches
Change *state*	Harden, soften, straighten, curve, roughen, smooth, heat, cool, solidify, liquefy, vaporize, pulverize, lubricate, moisten, dry
	Examples: water softeners, powdered eggs, food concentrates
Change *relative motion/position*	Attract, repel, lower, rotate, oscillate
	Examples: photocopiers, Velcro

We decide to "make the familiar strange." Instead of sheets of paper, we think of two giant flat asteroids floating in space that need to be somehow locked together. These asteroids and the corresponding sheets of paper may or may not need to be aligned with one another. Remember that the original problem did not specify whether such alignment needed to be achieved. One method might be to drive a harpoon through both asteroids and then tie them together. We then ask ourselves, Is there an equivalent technique for fastening paper sheets? Another approach would be to coat one asteroid with an adhesive and then force the two asteroids together. A third method might use magnetism (i.e., we could attach magnets to the asteroids). Many other ideas might be forthcoming, now that we have reformulated the original (familiar) problem into a more unusual structure.

7.4.5 ■ Analogies, Adaptation, and Duplicated Design

Bionics and synectics both rely upon the use of analogies for generating new ideas. The method of analogies, sometimes known as the method of adaptation, recommends that we seek different types of analogies to the problem under consideration that can then be used as the basis for design solutions. Gutenberg, for example, used the analogy of a wine press in his development of the printing press.[15]

At least four types of analogies have been used to develop innovative engineering solutions. These include:[16]

- The *direct* analogy: The current problem is directly related to a similar problem that has been solved.

   Imagine that we need to develop a new wear-resistant shoe. After considering the use of certain types of longlasting materials and other ways to increase durability, we decide to use the method of analogies by drawing a direct analogy between our shoe problem and automobile tires (shoes for cars). We discover that certain tire tread patterns have been designed to increase durability by maintaining the effects of friction. We adapt these patterns for use in our shoe design.

- The *fantasy* analogy: When confronted by a problem that cannot be solved easily, imagine that a solution already exists.

   A manufacturer of electronic equipment was developing a new product.[17] During the design phase, it was discovered that a special type of solder would probably be required to securely connect the electronic components to the system's printed circuit board (i.e., the standard types of solder could not be used).

   After considerable effort in trying to solve this subproblem, one engineer suggested that a special solder known as stratium be used in the system. The

15. Middendorf (1990).
16. One discussion of direct, personal, and symbolic analogies is in Woodson (1967).
17. An imaginary example adapted from a case related by P. H. Hill, professor emeritus of Tufts University.

other now frustrated engineers agreed with this recommendation and the design group moved forward to work on other aspects of the system. Upon completion of the new design, it was determined that a traditional type of solder could indeed be used. The engineer who had suggested the use of stratium was then asked by his colleagues, "What is stratium? We never heard of it." He replied that there was no such thing as stratium. He had simply suggested it in order to force the engineering team to move onward to other aspects of the design. By imagining that a solder such as stratium existed, the team did not continue to work on a difficult subproblem that was delaying their progress on the rest of the design. Moreover, it is often the case that a particular subproblem does not need to be solved in the final design configuration, just as the development of a new type of solder was unnecessary in this case.

- The *symbolic* analogy: When confronted by a problem that cannot be easily solved, use a poetic or literary analogy to generate ideas (e.g., similes or metaphors).

  Imagine that we are trying to develop a new adhesive, and consider the phrase "clings like a barnacle." This phrase may lead us to focus upon a barnacle's method of fastening itself to a rock or to the hull of a ship in a very effective way (i.e., we are led to a bionics approach.)

  The adhesive used by barnacles hardens underwater and may be thixotropic; that is, whenever a force is applied to dislodge the barnacle, the adhesive bond hardens; once the force is no longer applied, the bond softens.[18] A commercial adhesive that behaves in a similarly dynamic way could be very useful.

- The *personal* analogy: One imagines being part of the system in order to view the problem from a different perspective.

  If we are trying to design a more durable computer diskette, we might try to imagine ourselves as the diskette itself in various situations in which it might be harmed (e.g., in a book bag, tossed about on the ground, lost beneath a desk, underwater). This might then lead us to an understanding of the hazards that must be overcome for the diskette to be protected from harm (e.g., damage due to sudden impacts; damage due to dust, moisture, and grime).

Analogies can be very effective in leading us to new ideas and applications based upon existing concepts (see Case History 7.7 *Running Shoes with Waffle Soles*).

Sometimes one can adapt an existing design to generate new solutions to an unrelated problem or mold an old and rejected concept into a truly feasible form.

---

18. Kivenson (1977).

The Toy That Imitated Full-Scale Construction

Alfred Carlton Gilbert designed a succession of products that were built upon the Erector Set toy that he and his wife Mary developed during the early years of this century.[19]

Gilbert had been a champion athlete (he held the world records for the pole vault, chin-ups, and running long jump). In 1911, two years after establishing the Mysto Manufacturing Company (later the A. C. Gilbert Company) with John Petrie, Gilbert became fascinated with the construction of power lines by the Hartford Railroad. The erection of steel girders during this construction led Alfred and Mary to create a set of small cardboard girders from which many different structures could be built. With the aid of a machinist, a steel set soon followed, together with nuts and bolts for assembly. A commercial version of the Erector Set was first offered to the public in 1913. The product was a smashing success: 30,000,000 sets were sold during the first twenty years of production as children (and many adults) discovered the myriad configurations and structures that could be constructed from this wonderful toy.

By 1916 Gilbert had developed a small electric motor for use with his Erector Set. He used this motor as the basis for many other products, including the nation's first inexpensive small fan (the Polar Cub Fan). The Gilbert motor was later used in the development of the first practical heart–lung machine.

In addition to the Erector Set and its small motor, Gilbert developed the first hand-held vacuum cleaner, the first electric coffee maker, a food processor, and numerous educational toys.

In addition to designs based upon analogies and adaptations, sometimes virtually identical designs are developed independently by different people. The virtually simultaneous and independent development of the telephone by Alexander Graham Bell and Elisha Gray (see Case History 5.4) is one example of such duplicated discovery and design. Another example of duplicated design is described in Case History 7.8 *Pneumatic Tires.*

7.4.6 ▩ Obtain a Fresh Perspective

This is a very simple idea but one that can be very effective: Simply describe the problem that you are struggling to solve to someone who is not involved in the design effort. This other person may be able to provide new insights and a new perspective. Furthermore, sometimes the task of explaining a problem helps us to more clearly understand the objectives that we seek to achieve with a solution.

---

19. See Woog (1991).

A New Twist on the Elevator Concept

Some years ago in southern California, a hotel's management was confronted with a serious problem: Bottlenecks were developing in the lobby because the hotel's elevators could not accommodate the large number of guests that were now using the facility. Somehow, the capacity of the elevators needed to be increased. Unfortunately, additional elevator shafts would necessitate a reduction in the number of hotel rooms, with a corresponding loss of income.

The managers of the hotel were discussing this problem when they were overheard by the doorman. The doorman remarked to them that if elevators could be added to the outside of the building, no loss of income or upheaval would be necessary; furthermore, he suggested that the elevators include glass walls in order to provide travellers with a pleasant view.

The doorman provided the managers with a different frame of reference—his own, outside the building! A new outside elevator was installed, beginning the trend toward external elevators (for both economic and aesthetic reasons) now seen in many modern hotels.[20]

7.4.7 ▧ Inversion

The inversion strategy states that one should concentrate on ways to make a product or system less effective and then invert these ideas to form ways in which the product can be improved.

For example, it may be very difficult to think of 25 ways in which energy can be saved in a building; however, if we try to think of 25 ways through which energy can be wasted, we will probably find that some of these wasteful ideas can be inverted and used as the basis for energy-saving improvements.

Moreover, there is another type of inversion that one can use to generate more effective solutions to a problem: Simply invert the problem statement itself. One illustration of this approach is described in Case History 7.9 *Jokes for Trash.*

7.4.8 ▧ Idea Diagrams

Idea diagrams allow us to organize and correlate ideas as they are generated. In this approach, one seeks to divide solutions into categories and subcategories that become increasingly more well-defined.

---

20. Weldon (1983) and Fogler and LeBlanc (1995); as Fogler and LeBlanc note in their telling of this story, knowledge of engineering design was needed to transform this idea into reality.

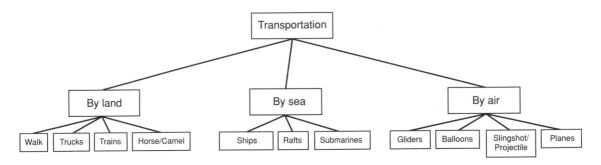

FIGURE 7.2 A partial idea diagram.

Let's return to the problem of transporting people (e.g., military troops) from one location to another. The partial idea diagram given in Figure 7.2 shows that we first can divide solution paths according to three possible transportation modes: by land, by sea, or by air. Next, these three modes can each be subdivided into partial solution concepts (truck, ship, plane, etc.). Additional levels in the tree diagram would correspond to more well-defined solution concepts.

7.5 Morphological Charts

Assume that the engineering design team has used abstraction, together with various creativity techniques, to develop a set of partial solutions to a problem. Each of these subsolutions corresponds to a particular idea for achieving one of the goals or functions desired in the final design. The next task—synthesis—is to join these individual subsolutions together in order to form alternative total designs. A morphological chart is one tool that can be used for this purpose.[21]

Morphological charts allow one to systematically analyze the different shapes or forms that might be given to engineering designs. The rows of the chart correspond to the different functions or goals that should be achieved by any viable solution to the problem under consideration, and the columns correspond to the different ways or means through which each of these functions could be achieved. As a result, the chart can provide the designers

21. See, for example, Cross (1989).

with an overview of the creative concepts that they have generated, together with the opportunity to form new and possibly wonderful combinations of these concepts.

For example, a simple morphological chart could be used in the design of a table lamp. Possible design solutions then could be formed by selecting one idea from each row and joining these ideas together to form a cohesive whole, as shown in Figure 7.3 (only a portion of the chart is shown in this figure). Each of the resulting combinations by definition must correspond to a total design solution, since it is composed of ideas for achieving all of the desired goals.

The challenge in using a morphological chart to perform synthesis is to recognize and pursue only those combinations that are most promising and to discard any combinations that are impractical. With experience, the engineer can develop the ability to recognize infeasible or even impossible combinations of ideas.

Once the most promising combinations of preliminary ideas have been identified, the engineer is ready to perform the next stage of the design process—analysis—in which these alternative total solutions or designs will be compared and evaluated.

FIGURE 7.3 Example of a simple morphological chart used to combine ideas for achieving desired goals in a table lamp. Circled options together would form one design solution. (Only a small portion of the chart is shown.)

Desired functions or goals	Partial concepts or means to achieve each goal		
■ ■ ■	■ ■ ■	■ ■ ■	■ ■ ■
Stability	Large base	Weighted Base	Tie-down straps
Adjustable Height	Flexible neck	Sectional designs	Adjustable legs
Lightweight	Use lightweight materials	Use less materials	■ ■ ■
■ ■ ■	■ ■ ■	■ ■ ■	■ ■ ■

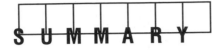

SUMMARY

- Synthesis is the formation of a whole from a set of building blocks or constituent parts. In engineering design, entities are combined in appropriate ways to create a product or system that solves a problem. Such problem solving performed in new and different ways clearly requires creative thought.
- There are numerous barriers to creativity, including knowledge blocks, perceptual blocks, emotional blocks and cultural blocks.
- One must be prepared to transform accidents into design solutions.
- Many techniques and strategies for stimulating creativity and overcoming blocks to imagination have been developed. These techniques include brainstorming, bionics, checklisting, synectics, the method of analogies, explaining the difficulty, inversion, adaptation, and idea diagrams.
- Engineering design requires that we be creative yet practical. One or more of the above creativity stimulation techniques can be used when we are confronted by a block to our creativity.
- A morphological chart can be used to perform synthesis by joining preliminary ideas or subsolutions together to form total design concepts.

PROBLEMS

7.1 The Leidenfrost phenomenon is as follows (Hix and Alley, 1965):

> If a liquid is placed on a surface that is then rapidly heated well beyond the boiling point temperature of the liquid, it will not wet the surface because a layer of vapor forms between the surface and the liquid. For example, water placed on a 300° F surface will evaporate slowly as it dances about on a layer of steam. However, if the surface were heated only to 230° F, the water will evaporate quickly. The difference between the temperature of the liquid and that of the surface must reach a minimum threshold value. For water, this threshold value is 45° F. (p. 130)

> Propose two possible applications of the Leidenfrost phenomenon in technical problem solving. In each case, carefully describe both the problem to be solved and your proposed solution based upon this phenomenon. Sketch your ideas as appropriate.

7.2 Bowls of ancient crystallized honey have been successfully reconstituted with water (Kivenson, 1977). Propose a potential application of this concept in technical problem solving; in essence, use bionics to adapt this phenomenon to solve a specific problem.

7.3 Working in a three-person group, use brainstorming to generate 25 ways in 30 minutes to solve the following problem:

> Dogs are often chained in backyards for their protection and the protection of others. Unfortunately, the chain can become entangled in bushes, trees, dog houses, and other obstructions, thereby restricting the dog's motion and resulting in property damage and perhaps even harming the dog itself. Develop a product that will eliminate this problem.

7.4 Use inversion to develop 10 ways in which a product can be improved. Submit your original list of ways in which the product can be made less effective, together with your final list of recommended improvements.

7.5 A natural phenomenon that may have practical applications is the bottle-nosed dolphin's ability to minimize the frictional resistance of water by rippling the surface of its skin, thereby allowing it to travel at high speeds (Kivenson, 1977). Use this concept as the basis for a design solution to a problem of your choice (i.e., use bionics).

7.6 Transparent Scotch tape was dependent upon the earlier development of another product, waterproof cellophane. Describe two engineering designs (other than those discussed in this text) that were similarly dependent upon the development of other products. Identify both the designs and the products upon which they were based. (Include references to support your answers.)

7.7 What if the water supply to your home or apartment was unexpectedly shut off for several weeks? List the effects—both positive and negative—that such an occurrence might have upon your daily routine. Next, design ways to overcome any and all difficulties caused by this event. Describe these solutions in words and/or with sketches.

7.8 What if automobiles and other vehicles were banned from city streets? Identify three different problems that would result from such a ban. Use one of the creativity techniques presented in this chapter to generate three design solution concepts for each of these problems (for a total of nine concepts). Describe each concept in words and/or with a sketch.

7.9 What if you needed to communicate with someone who is living on a deserted island? You know the location of the island but you are unable to travel to it. Design three different methods of communication. Describe these methods in words and/or with sketches.

7.10 What if your toothpaste suddenly solidified in the tube? Design a method for removing one ounce of "paste" whenever necessary without contaminating or otherwise harming the remaining supply in the tube. Describe the method in words and/or with a sketch.

7.11 What if you are driving an automobile through a duststorm? The dust becomes so thick that you cannot see through it, yet you must continue onward to your destination. Design a method that will allow you to continue your journey. Describe the method in words and/or with a sketch and specify all assumptions that you made while developing this method.

CASE HISTORY 7.1

Scotch Tape

In 1905 the Minnesota Mining and Manufacturing Company (3M) entered the field of sandpaper production. Since sandpaper consists of an abrasive material fastened to paper stock, the company gradually developed expertise in adhesives or glues. One of the early contributions of the firm was the development of waterproof sandpaper in response to the needs of automobile workers. Sandpaper was used to finish the paint applied to car bodies; however, until the 1920s, only "dry sanding" could be used for this purpose since the sandpaper would immediately disintegrate if moistened with water. Many autoworkers developed lead poisoning from the dust generated by dry sanding. In response to this need, researchers at 3M developed waterproof sandpaper, thereby allowing wet sanding operations, which generated considerably less dust.

In 1925 Richard Drew (a laboratory technician who completed correspondence school courses in engineering) was assigned the task of testing this new waterproof paper at various automobile shops in the St. Paul, Minnesota, region. At these sites, Drew noticed that in order to create a two-tone paint design on an automobile, workers first would apply paint of one color, then cover or mask this paint with newspaper or butcher paper before applying the second color. If everything went well, this use of paper produced a sharp clean edge between the two paints. Unfortunately, if glue was used to attach the paper to the auto body, scraping was sometimes required to remove the glue—resulting in some paint being scraped off as well.

Drew recognized that a weak adhesive tape would eliminate this problem since it could be removed from the car without damage to its finish. After two years of experimentation, he eventually discovered that crepe paper could be used as a successful base surface for his new "masking tape." The new product was an immediate success. However, company legend states that autoworkers became frustrated with the early version of this tape in which adhesive had been applied only to the edges. Apparently, some decision maker within the company believed that the tape would be easier to use in autobody painting if no adhesive was applied to its middle. Auto painters discovered that the adhesive applied only to the edges was insufficient to support the heavy paper, which would tear loose because of its weight. Supposedly, the use of insufficient adhesive on the tape was interpreted by at least one painter as evidence of the company's stinginess, leading to the popular appellation "Scotch" tape for the product.[22]

Drew realized that his tape was imperfect. For example, since the crepe paper was not waterproof, it was useless for applications in moist environments. Drew then had another brainstorm: Why not use cellophane instead of crepe paper as the base for the tape?

Cellophane was developed by Jacques Edwin Brandenberger (a dye chemist who had worked in the textiles industry). In 1900 he began experiments to develop an easily cleaned protective coating for cloths. Both liquid viscose and thin solid viscose were found to be unsatisfactory for this application; however, he decided that cellulose film itself could be a valuable product. By 1912 he had patented both the process and the machinery for producing thin cellulose film. A French rayon manufacturer, the Comptoir de Textiles Artificiels, through a newly created subsidiary known as La Cellophane, then provided the financial backing for the commercial development of his work. The Du Pont Chemical Company worked with Brandenberger and Comptoir to obtain the U.S. patent rights to the process and its manufacture, resulting in the formation of the DuPont Cellophane Company in 1923. Then in 1926 two Du Pont researchers, William Hale Charch and Karl Edwin Prindle, discovered how to successfully apply a coating to cellophane film that would make it waterproof. Shortly thereafter, Drew thought of his idea to use this waterproof film as the base for his Scotch tape.

Unfortunately, one could not simply transfer the adhesive used on the Scotch crepe-paper tape to cellophane. When this was attempted, Drew noted that

> it lacked the proper balance of adhesiveness, cohesiveness, elasticy, and stretchiness. Furthermore, it had to perform in temperatures of 0 to 110 degrees F (and) in humidity of 2 to 95 percent.[23]

Notice how, in his statement, Drew focused upon the very *practical* and specific goals that his new transparent tape needed to achieve. Within one year, he had developed the first version of a new transparent pressure-sensitive cellophane tape that was indeed capable of satisfactorily achieving each of the above goals. Scotch tape was here to stay—thanks to creative thinking that was driven by practical considerations!

22. See, for example, Petroski (1992) and d'Estaing (1985) for discussions about the development of Scotch tape and Jewkes et al. (1969) for a brief history of cellophane's development and manufacture.
23. Petroski (1992, p. 82).

CASE HISTORY 7.2

Control via Frequency Hopping—Designed by a Movie Star

Frequency hopping is the antijamming technique used in today's satellite and defense communication systems. The concept was devised by Hedy Lamarr, a major Hollywood motion picture star of the 1940s and 1950s.[24]

Lamarr was Austrian by birth and in 1940 she wanted to contribute to the war effort against the Nazis. She conceived of a way in which radio signals could be transmitted without fear of being jammed or deciphered by the enemy: Simply transmit the signals on a series of predetermined frequencies, each frequency being used for only a fraction of the entire message. By such hopping from frequency to frequency, the broadcast message could be received only by someone who knew the synchronized pattern of hops.

Lamarr then collaborated with George Antheil, a composer of film scores. Antheil designed a mechanism in which slotted paper rolls were used to synchronize hops among 88 frequencies. His design reflected his familiarity with player pianos in which slotted paper rolls are used to automatically control the movement of the 88 keys. Lamarr[25] and Antheil were awarded a patent for their design in 1942 (Fig. 7.4).

Although never used during the war, Lamarr's concept reappeared in a similar electronic frequency-hopping communication system developed independently by Sylvania (Lamarr's original patent had long since expired).

Lamarr demonstrated remarkable creativity with her concept of frequency hopping—an antijamming communication technique that is at once both simple and elegant.

CASE HISTORY 7.3

Radar During the Battle of Britain

Engineers must be knowledgeable in the latest developments and technological breakthroughs in their fields and allied disciplines as well. Otherwise, they may "reinvent the wheel" because they are unaware that it

24. See, for example, MacDonald (1992) and Stanley (1993).
25. In the patent, Lamarr used her real name of Hedy Kiesler Markey.

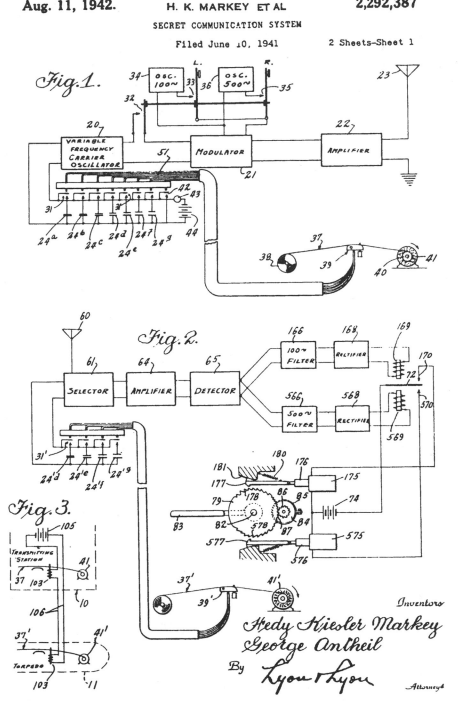

FIGURE 7.4 Illustration from the patent by Hedy Kiesler Markey (Hedy Lamarr) and George Antheil for a "secret communication device" that used frequency hopping. *Source:* U.S. patent #2,292,387; drawing by H. K. Markey.

has already been invented, or, worse yet, never apply new knowledge to problems that need to be solved.

Technical information is available through a multitude of sources: professional journals, popular magazines, textbooks, monographs, computer bulletin boards, reference works, and so forth. International developments should also be followed; never assume that only those breakthroughs that are occurring in your own country or company are of interest. Such cultural arrogance can quickly lead to professional disaster. Important advances in engineering are occurring throughout the world and we must all learn from one another.

The use of radar during the Battle of Britain in 1940 provides one example of why engineers and others must be knowledgeable in the latest international developments in the use of technology.[26]

RADAR DETECTION

In radar detection, shortwave radio waves are transmitted in pulses. If these pulses strike an object they are reflected back to a receiver that can then determine the location of the object.

The similarity in the reflection of both light and electromagnetic waves was first noted by Heinrich Hertz in 1887. By 1904 a device designed to prevent collisions between objects by using radio echos had been patented by a German engineer named Hulsmeyer. Marconi proposed in 1922 that ships traveling in fog might be able to detect other ships with equipment that could sense reflected electric waves. By 1935 the French had installed a detector on the ocean liner *Normandie*.

Others also designed and used primitive radar equipment. Reflected radio waves were used to measure the height of the Kennelly-Heaviside layer in Britain by Sir Edward Appleton and M. F. Barnett in 1924 and in the United States by Gregory Breit and Merle A. Tuve (using pulses) in 1925. [The Kennelly-Heaviside layer is a highly ionized region of the earth's upper atmosphere.]

The British developed their radar system for detecting aircraft between the years 1935 and 1940. In February 1935, Sir Robert Watson-Watt proposed a general scheme through which radio waves could be used for such detection; by the summer of that year, he had developed—with government sponsorship and six assistants—the first practical British aircraft detection radar system. Watson-Watt and his research team had succeeded in building a transmitter of sufficiently high power, modulating this transmitter with short pulses, and creating the necessary receivers.

THE BATTLE OF BRITAIN

By 1939 a radar system that included several 245-foot towers had been constructed along the eastern and southern coasts of England. The system could detect aircraft travelling 100 miles away at an altitude of 10,000 feet.

26. See d'Estaing (1985) and Jewkes et al. (1969).

Meanwhile, radar equipment for military purposes had also been developed in Germany, the United States, and Holland. In fact, by 1939 the Nazis were using radar to detect aircraft; however, German radar did not continue to develop as quickly as that of Great Britain and the United States once World War II began because the Nazis believed that the war would end quickly and that a hurried development of military radar systems would be unnecessary.

However, Germany could have obtained detailed information about the British radar system directly from professional journals. Prior to the Nazi invasion of France, Maurice Deloraine, the European representative of the ITT, visited the British radar facility. Afterward he asked his staff in Paris to collect all relevant nonclassified material on the subject of radar. To his dismay, Deloraine discovered that the British secrets had been described in various scientific publications.

Fortunately for the Allies, the Nazis failed to read and correlate this technical information in order to understand both the concept of radar and its successful application by the British. Germany believed that the 245-foot British towers were simply part of a radio-based aircraft guidance facility rather than a sophisticated detection system. As a result, more than 2,300 Nazi aircraft were destroyed during the Battle of Britain compared to only 100 British fighters—a monumental imbalance that certainly was a major factor in Great Britain's victory.

CASE HISTORY 7.4

V-2 Rocket

In contrast with Case History 7.3, a revealing comment made by Wernher von Braun following World War II clearly indicated that not all Germans chose to ignore important sources of technical knowledge.

Von Braun and other German scientists had very quickly developed the V-2 rocket during the war; similar U. S. efforts had failed to progress so fast. Von Braun explained the Germans' rapid progress when he simply noted, "Why, we studied the patents of your great Dr. Goddard,[27] of course."[28]

In other words, American scientists did not fully appreciate the vast technical knowledge that was contained in their own patent system.

27. Robert Hutchings Goddard (1882–1945), an American physicist and one of the principal figures in rocket development.
28. See Kamm (1991, p. 11).

CASE HISTORY 7.5

Goodyear and the Rubber Vulcanization Process

Charles Goodyear's discovery of the vulcanization process for curing rubber is a famous example of exhaustive experimentation coupled with a fortuitous accident.[29] It is also a lesson about the need for each of us to sometimes collaborate with others who can provide expertise that we may lack.

Vulcanization, or curing of natural rubber, is achieved by first mixing the rubber with sulphur (and other ingredients such as zinc oxide and litharge) and then heating the compound. There are a number of variations of this process, all resulting in a permanently elastic solid substance that is generally unaffected by moderate changes in temperature.

Goodyear was the son of a hardware manufacturer. In 1826 America's first hardware store was established by Charles and his wife Clarissa. Their objective was to create an effective retail distributor for the products manufactured by his father.

By 1834 Charles had left the hardware business to begin his career as an inventor. He purchased a rubber life preserver and designed a new valve for the device. Although he failed to sell his valve design to the manufacturer, Roxbury India Rubber Company, he did discover that such manufacturers were in desperate need of an improved rubberlike material. Natural rubber's temperature-dependent shortcomings (i.e., becoming inelastic in cold temperatures and turning into a soft tacky mass in warm weather) had soured the American public on the substance.

Goodyear embarked on a series of trial-and-error experiments, mixing rubber with virtually every substance that he could find, including castor oil, ink, soup, and cream cheese! After five years of such work, he discovered that the mixed vapors of nitric and sulfuric acids did, in fact, produce the desired result: cured rubber. Unfortunately, this discovery only led to the commercial failures of Goodyear's business ventures. First, in 1837 his company failed due to an economic panic by the public. A second business opportunity with Nathaniel Hayward withered in the warm summer of 1839 when 150 mailbags made from his acid-cured rubber for the United States Post Office melted because the rubber had not been fully treated by the acid vapors.

Nevertheless, Goodyear persevered in his search for an effective curing process. He was both resolute and tenacious in this effort. Hayward had mixed rubber with sulphur, leading Goodyear then to focus on such rubber–sulphur mixtures until one day in 1839 he accidentally spilled a mixture of rubber, sulphur, and white lead on a hot stove. Rather than simply discard the partially burned compound, Goodyear examined it carefully, discovering that the unburned portion had retained its elasticity! In other

29. See Travers (1994) and Singer et al (1958).

words, it had survived the heat without melting into a gooey mass. He then exposed the fragment to cold winter temperatures; rather than becoming brittle, it retained its elasticity. Goodyear had discovered the vulcanization process for curing rubber.

Next, in 1841 he developed a fabrication method using a heated cast iron trough, in which continuous sheets of cured rubber could be produced. Financial support came from New York rubber manufacturer William Rider. Vulcanization was now an industrial process.

Unfortunately for Goodyear, he never achieved commercial success in the rubber industry. He granted manufacturing licenses for others to use the vulcanization process, but the income from such licenses was far below fair market value. In addition, others infringed upon his patents, leading to legal battles in which Goodyear was represented by Daniel Webster. Although Goodyear recovered his patent rights through these proceedings, his legal costs were more than the income he eventually received from his discovery.

Another factor in Goodyear's difficulties was the initial reluctance of the American public to accept vulcanized rubber as a superior form of the material. Goodyear tried to introduce his process in England, where rubber enjoyed a positive reputation. He asked Stephen Moulton (who was about to visit England) to show samples of cured rubber to appropriate British manu-facturers and others who might be interested in using the method for a fee. Moulton could not reveal the vulcanization process and one major manufac-turer (MacIntosh & Company) suggested that Goodyear first apply for a British patent so that they could then evaluate his new process better.

It was now 1842. In England the market for lightweight clothing was growing, and the heavier double-texture articles in which uncured rubber was used for insulation were becoming unpopular. Clearly, a successful vul-canization process would be most welcome, relieving rubber manufacturers from the threat of economic disaster.

About this time, one of Goodyear's cured rubber samples that had been taken to England by Moulton reached Thomas Hancock. Hancock immedi-ately noted the sulfuric character of the compound and began to perform a type of reverse engineering on the vulcanization process. Hancock noted that

> I made no analysis on these little bits nor did I procure either directly or indi-rectly, any analysis of them . . . and I considered the small specimens given me simply as proof that it was practical . . . I knew nothing more of the small spec-imens than I or any other person might know by sight and smell.

Although Hancock did indeed begin to investigate the curing process on his own and did not "steal" Goodyear's work, he did have two distinct advantages over Goodyear: (a) Hancock was a genuine scientist and (b) he had worked in the rubber industry for decades. Eventually, Hancock did ascertain the secret of vulcanization through his experimentation and secured the British patent rights for the process before Goodyear could obtain the funds necessary to apply for a patent. Goodyear had lost the sub-stantial British market for his process.

Other methods of vulcanization were quickly found. For example, the cold-cure method discovered by Alexander Parkes in 1846 consisted of placing rubber strips in a solution of sulphur chloride in naphtha or carbon disulphide. It became popular in the manufacture of thin rubber products such as single-texture clothing, balloons, and gloves.

The so-called vapour curing process used sulphur chloride vapor rather than a liquid bath; it was developed by W. Abbott in 1878. Finally, Stephen Moulton developed another high-temperature curing process in which sulphur is replaced by lead hyposulphite.

Meanwhile, because of his various business failures, Goodyear often found himself in debtors' prison. When he died in 1860, he was $200,000 in debt.

By the 1870s, in order to meet increasing market demand, efforts to cultivate natural rubber on plantations had begun. Of course, demand grew to enormous levels when rubber became the principal component of pneumatic automobile tires near the beginning of the twentieth century. Eventually, rubber plantations could be found in Java, Ceylon, Malaya, the East Indies, and many other areas throughout the world. An entire industry had become well established.

Notice that Goodyear's struggles contain several important lessons for all of us:

- Perseverance can be critical in trial-and-error experimentation.
- We need to take advantage of opportunities when they occur (recall that Goodyear did not discard the burned rubber–sulphur–lead compound but instead carefully examined it).
- We need to recognize our personal limitations and collaborate with others who can aid us. Goodyear would have benefited greatly from collaborations with appropriately gifted partners who could provide the scientific, business, and financial expertise that he lacked. (Although Goodyear did collaborate with others on occasion, for one reason or another these efforts did not lead to success. In such circumstances, one needs to nurture new professional relationships that may be necessary and fruitful.)

CASE HISTORY 7.6

So What's Wrong with Our Toaster?

During the 1960s a kitchen appliance manufacturer discovered that its toaster products were losing their market share. In order to reverse this development, management assigned the following task to the company's engineers: Develop a better toaster that will revolutionize the marketplace and expand our market share.[30] (Of course, one should note that this problem statement was given in terms of a particular solution—design a toaster—as opposed to a proper focus upon function; recall the *Lawn Mowers and the Yo-Yo* anecdote in Chapter 3.)

The engineering team worked on this problem for some time, gradually developing variations on the basic toaster design (such as toasters with lights, sound cues, unusual shapes, and so forth). However, none of these concepts were so innovative and desirable as to revolutionize the marketplace.

Finally, one of the engineers suggested that everyone in the group answer the question, What's wrong with our current toaster design? One by one, each person noted various shortcomings of the existing toaster: It was not very durable, it was cumbersome to store when not in use, and so on. Finally, one senior engineer (who apparently truly disliked toasters) began to list his criticisms, some of which were

- Toasters were difficult to clean.
- They trap bread.
- They often burned the bread or failed to toast the bread sufficiently.

After several minutes, he had gained momentum in his enthusiastic critique and found additional criticisms:

- He could not toast English muffins.
- He could not toast bagels.
- He could not toast a tuna sandwich.
- He could not toast underbaked pie.

Finally, another engineer in the room had to stop this seemingly endless tirade against the toaster by shouting, What do you want—an oven?! A new product appeared on the market shortly thereafter: the toaster oven!

CASE HISTORY 7.7

Running Shoes with Waffle Soles

Bill Bowerman served as coach for NCAA championship teams in track and cross-country at the University of Oregon. In 1964 he formed a partnership with one of his former student-athletes, Phil Knight, to sell custom-designed

---

30. Based upon an undocumented but reliable source.

shoes. (Two years earlier, Knight had formed Blue Ribbon Sports to sell track shoes; Blue Ribbon later changed its name to Nike.)

In 1971 Bowerman was testing various sole patterns for running shoes for greater cushioning of the athlete's foot while firmly gripping the ground. While eating a waffle one morning, he realized that the waffle's form could be the very sole pattern for which he had been searching.[31] He poured urethane into the waffle iron, and then cut the resulting waffle pattern into the shape of a sole. Further testing verified the value of this sole pattern; indeed, the University of Oregon's cross-country team won that year's NCAA championship with the waffle soles. The following year, the waffle pattern was introduced to the public.

CASE HISTORY 7.8

Pneumatic Tires

An old and rejected concept may sometimes be the answer to a modern problem. For example, the concept of the pneumatic or air-cushioned tire was "rediscovered" independently more than forty years after its initial development and commercial failure.[32]

The pneumatic tire was developed into a practical and commercially viable form by the veterinary surgeon John Boyd Dunlop in 1888. Many years earlier, in 1845, Robert W. Thomson had actually constructed inflatable "aerial wheels" by saturating layers of canvas with rubber that was then vulcanized and encased in leather to form a primitive inner tube. This tube was then encased in leather, riveted to a carriage wheel, and inflated with air. (Solid steel tires mounted on wooden wheels were used on vehicles at that time. The steel tires tended to skid and failed to damp vibrations, leading to very uncomfortable rides.) Unfortunately, Thomson's pneumatic tires—although technically an improvement over steel tires in terms of both comfort and control—never became commercially successful.

By 1846 Thomas Hancock was manufacturing tires made of solid vulcanized rubber. Many bicycles were mounted with these solid rubber tires, which reduced skidding, during the 1870s and 1880s. However, cyclists were still jarred by each jolt of the road.

Dunlop reduced the vibrations of his son's tricycle by rediscovering Thomson's concept of an air-cushion tire. Dunlop first enclosed an inflatable

31. See Woog (1991).
32. See Travers (1994), d'Estaing (1985), and Singer et al (1958).

rubber inner tube in an external canvas casing. He then added thick rubber strips to both protect the canvas and serve as the tread. Finally, he wrapped the entire canvas jacket with tape and fastened the tire to the wheel rim with rubber adhesive. Two years later, Charles K. Welch used wire embedded in the casing to attach a tire to the wheel and William E. Bartlett introduced the "Clincher" design in which beaded edges on the casing are used to fasten the tire to the rim.

Dunlop also added a one-way valve that allowed the user to adjust the inflation of the inner tube. Dunlop developed his air-cushion tire design with no knowledge of Thomson's earlier work on the same idea; he patented it in 1888 and created the first tire company.

Cyclists discovered that Dunlop's pneumatic tires reduced not only road vibrations but also the amount of effort needed to pedal a bicycle. The popularity of Dunlop tires continued to increase and, in 1894, they were selected for use on the first mass-produced motorcycles (the Hildebrand and the Wolfmuller).

Tire technology progressed rapidly following Dunlop's work. In 1891 the removable pneumatic tire (the first tire that could be repaired easily by the user after a blowout) was developed by Edouard and Andre Michelin. Four years later, the Michelin brothers mounted such tires on a racing car called the *Lightning,* which then competed in the Paris–Bordeaux–Paris race—the first automobile so equipped.

Other companies soon joined Dunlop and Michelin, including Benjamin F. Goodrich in 1896, the Goodyear Tire Company (named in honor of Charles Goodyear, inventor of the vulcanization process for rubber) in 1898, and Harvey Firestone in 1903.

The Michelin company was responsible for a number of innovations, such as the use of so-called dual tires (side-by-side sets of tires) on trucks and buses to support heavy loads beginning in 1908 and the first steel-ply tire (the *Metallic* in which the rubber was reinforced by steel wire) in 1937. Michelin was the first to market the radial-ply belted tire in 1948, the same year in which the tubeless tire was introduced. (The belted radial design had been patented in 1914 by Christian H. Gray and Thomas Sloper, but it was never marketed.)

Although technically an advancement, Thomson's air-cushioned tire had failed to generate public support during the 1840s. Yet forty years later, Dunlop's reconceptualization and refinement of the design was a resounding success. One factor in this reversal of public acceptance was the continued development of mechanical road vehicles (e.g., motorcycles, bicycles, automobiles) during the latter part of the nineteenth century; people were simply ready to embrace a new tire design that would match the other advancements then appearing in new road vehicles. Additional factors probably included the increased prominence of vulcanized rubber goods throughout society, Dunlop's focus on the specific needs of cyclists, and the improvements made in pneumatic tires by Dunlop, the Michelins, and others.

CASE HISTORY 7.9

Jokes for Trash

Richard Von Oech has related the story of a city in Holland where much of the populace was failing to deposit their trash in the cans provided by the city. Streets were becoming filled with litter. The sanitation department had to find a solution to this increasing problem. The fine for littering was doubled, but to no effect; the littering continued unabated. Next, patrols were increased to enforce the antilittering law—but again to little effect.

Finally, a startling idea was suggested: What if new trash cans were designed that would pay people whenever trash was deposited? After some discussion, the city officials realized that the government could not afford to implement this plan. However, a modified version of this plan was put into action. New trash cans were constructed in which tape recorders were inserted. Whenever litter was deposited in the can, its recorder would be activated so that it could relate a joke to the law-abiding citizen. Once implemented, this solution resulted in much less litter as people enthusiastically deposited their litter in cans in order to hear new jokes.[33]

Notice two very important lessons appear in this episode:

a. In order to generate a solution, a "what if" scenario was needed. Such scenarios often lead to new lines of thought or new ways of looking at the problem.

b. The solution was developed once the problem was "inverted" from one of intimidation (of lawbreakers) to one of reward (for responsible behavior).

Inversion of a problem statement can lead to creative design solutions.

---

33. See Von Oech (1983) and Lumsdaine and Lumsdaine (1990).

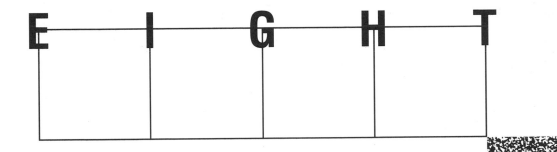

E I G H T

ETHICS AND PRODUCT LIABILITY

The greatest of faults, I should say, is to be conscious of none.
Thomas Carlyle

To do injustice is more disgraceful than to suffer it.
Plato

O B J E C T I V E S

Upon completion of this chapter and all other assigned work, the reader should be able to

- Explain the practical value of professional codes of ethics.
- Describe early forms of guidelines for engineering practice.
- Discuss the legal obligations and requirements that must be satisfied by professional engineers.
- Describe various cases in which ethical violations occurred.
- Apply the NSPE's Code of Ethics to hypothetical situations in which one is confronted by an ethical choice or dilemma.
- Explain the need for design engineers to be familiar with product liability law.
- Distinguish between manufacturing, design, and warning defects.
- Define the three major categories under which product liability lawsuits are filed: negligence; strict liability and implied warranty; and express warranty and misrepresentation.
- Describe how privity protected manufacturers from lawsuits.
- Discuss the need to design against foreseeable uses and misuses of a product.
- Explain why changes in a product during its useful lifetime must be anticipated in the design process.

8.1 Ethical and Legal Considerations in Design

Engineering designs sometimes fail because of purely unethical behavior by those who are responsible for their development and implementation. Such behavior, if deliberate, reflects the flawed value system of the person who is acting irresponsibly. However, unethical actions often are unintentional, which is one reason why engineers should become familiar with the codes of ethics that have been developed by professional engineering societies.

These codes provide us with guidelines for dealing with dilemmas in which the ethical action may not be obvious. It is absolutely necessary to become familiar with these codes and guidelines because engineers are indeed liable—both legally and morally—for the consequences of their work.

One may be confronted by an ethical dilemma without recognizing it as such or without any idea as to which course of action should be followed. In such cases, an engineer who desires to follow the ethical path might fail to do so. Codes of ethics, together with various case studies, can be used to acquaint oneself with some important aspects of ethical decision making in engineering. As a result, one is more likely to make informed and ethical choices.

As further evidence of the need to focus upon ethical considerations in engineering design, consider the following statement by the Accreditation Board for Engineering and Technology (ABET):[1]

> An understanding of the ethical, social, economic, and safety considerations in engineering practice is essential for a successful engineering career. Course work may be provided for this purpose, but as a minimum it should be the responsibility of the engineering faculty to infuse professional concepts into all engineering course work.[2,3]

In other words, the study of ethics must be included within any college engineering program that seeks to be accredited. This emphasis upon ethics in educational curricula simply reflects its importance in daily engineering practice. The study of engineering codes of ethics should not be treated in a casual or cavalier manner. Ethical behavior can be the difference between success and failure, triumph and tragedy, life and death.

8.1.1 A Personal Guideline

In the last 150 years, engineering and technology have transformed society in innumerable ways, most of which are positive but some of which are not.

1. ABET is the organization responsible for accrediting engineering programs in universities throughout the United States.
2. "Guide on Professionalism and Ethics in Engineering Curricula," Accreditation Board for Engineering and Technology, July 1989.
3. Ertas and Jones (1993).

For example, a century ago, most people did not travel far from their birth-place. The development of high-speed, reliable transportation (in the form of automobiles, trains, planes, and other systems) has increased employment opportunities and career choices for most people in this country far beyond the limited choices that were available to the average person in 1900. Unfortunately, it also can be argued that the opportunity to commute to more distant locations has contributed to both a breakdown of traditional family life (as more time is spent away from the home) and an increase in stress among the populace. Such negative effects of motorized vehicles were largely unexpected; however, the negative consequences of our engineering achievements should be anticipated *whenever possible* and then either minimized or eliminated.

One simple personal guideline may be helpful as we review engineering codes of ethics and various case histories:

> When making any professional decision, consider the possible consequences of that decision in terms of the impact that it may have on other people. Then imagine that this impact will not be felt by strangers, but by your own loved ones. Would you still make the same decision?

We sometimes overlook the possible negative consequences of our work on others because we fail to personalize these consequences (i.e., people who may be affected by our work are seen as faceless and distant strangers), leading to less motivation to protect these people from possible harm. We are more likely to make correct decisions if we imagine that a dear friend or relative will suffer the negative consequences of our efforts.

Everyone seeks happiness and safety. Among our responsibilities as engineers is to protect others from disappointment and harm whenever they use the technical fruits of our labor.

8.1.2 ■ Early Codes for Engineering Practice

Codes of ethics for engineers have existed for many years in one form or another. Hammurabi, king of Babylon nearly 4000 years ago, instituted a very direct code for builders, as follows:

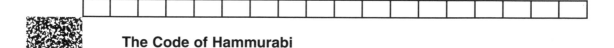

The Code of Hammurabi

If a builder has built a house for a man and has not made the work sound, and the house which he has built has fallen down and so caused the death of the householder, that builder shall be put to death. If it causes the death of the householder's son, they shall put that builder's son to death. ... If it destroys property, he shall replace anything it has destroyed; and because he has not made sound the house which he has built and it has fallen down, he shall rebuild the house which has fallen down from his own property. If a builder has built a house for a man and does not make his work perfect and the wall bulges, that builder shall put that wall into sound condition at his own cost.[4]

One can imagine how carefully builders performed their work in Babylon.

Steamboat Explosions

During the early part of the nineteenth century, steam engines were redesigned by people such as James Watt, Oliver Evans, and Richard Tevethick to become both portable and powerful. This work allowed such engines to be installed in mines, factories, trains, and (steam) boats.

Unfortunately, an unexpected negative development soon appeared: Steamboat owners were competing with one another for lucrative trade agreements by racing along the riverways. Safety valves were disabled to increase steam pressure, boilers were forced to operate beyond their capacities, and explosions would often occur as a result of these actions. Between 1816 and 1848, 2563 people were killed (and 2097 injured) in 233 such boiler explosions.

The United States Congress enacted an inspection law in 1838 (the year in which 151 people were killed by a single explosion aboard the steamboat *Moselle*). Henceforth, ships, boilers, and engines would be inspected for safety. Unfortunately, some inspectors accepted bribes and others were not trained in how to conduct a safety inspection. Explosions continued to occur.

Alfred Guthrie then entered the fray. He sought to identify the causes of these explosions by inspecting—at his own expense—approximately 200 steamboats. (This was a problem of explanation, as defined in Chapter 2—that is, *why* did the explosions occur?). He presented his conclusions in a report that was then used by the U.S. Senate as the technical foundation for a new law that established an effective regulatory agency in 1852. Guthrie became the first supervisor of this agency and boiler explosions became very rare events.

8.1.3 ▓ Legal Constraints

Hammurabi's code for builders reflected the impact that poor engineering work could have upon society. The steamboat code is an example of the many engineering and manufacturing standards that have resulted from much effort by many people.

---

4. See Harper (1904) and Martin and Schinzinger (1989).

Today, there continue to be very broad legal constraints and practical guidelines within which engineers must work; these include:

- **Professional registration**   One must satisfy certain minimum legal requirements to be registered as a professional engineer. These requirements relate to one's education, training, practical experience, and ability to pass a professional engineering examination. (Registration requirements vary from state to state.)
- **Specific laws, regulations, and ordinances**   There are various federal, state, and local laws, regulations and ordinances such as building codes that engineers must satisfy in their work.
- **Contract law**   Engineers often enter into contractual agreements when performing their work; such agreements are governed by contract law. Rights and responsibilities of all parties are specified in a contract. Any breach of these contractual responsibilities may lead to a civil lawsuit. Of course, one should recognize the legal constraints and responsibilities of any contract before agreeing to it.
- **Tort law**   In addition, tort (noncontract) law governs the work and liability of engineers in situations that do not involve contractual agreements. Our legal system recognizes the need to protect the public from poor engineering work even in the absence of a contractual agreement. Engineers should be familiar with the law of torts that governs their work (including such legal principles as negligence and strict liability discussed later in this chapter).

Engineers must be aware of these legal constraints, including any changes that occur in laws governing their work. (See Case History 8.1 *A Tidal Wave of Molasses Strikes Boston* for a discussion of the disaster in 1919 that led to state certification of engineers in the construction industry and a more uniform, dynamic level of professionalism throughout engineering.)

8.2 Modern Engineering Codes of Ethics: Self-Regulation

Although engineers are certainly required to obey the laws of the land, the engineering profession itself is also self-regulating, meaning that various professional societies seek to ensure that all members of the profession behave in an ethical, competent, and proper manner.

Unethical actions may lead to loss of one's professional reputation, the need to defend oneself in civil and/or criminal lawsuits, and other severe consequences. Professional engineering societies provide guidance for ethical behavior in the form of specific codes of conduct. These professional standards need to be thoughtfully applied in situations wherein a decision will be based upon various ethical considerations. Codes are neither fail-safe nor comprehensive, in that not all possible ethical dilemmas may be treated within a given code. However, codes do

- encourage engineers to behave according to the accepted standards of the profession;
- provide guidelines about these standards and their application;
- assure lawmakers that engineering societies can be trusted to regulate the actions of their own members; and
- encourage professional societies to support those members who do act in an ethical manner but then suffer negative consequences (e.g., loss of employment) because of this action.

The Code of Ethics of the National Society of Professional Engineers (NSPE) is representative of the major elements found in most engineering codes of conduct and is included as an appendix to this text.[5]

8.3 Common Violations

Common types of violations of a professional code of ethics include the following.[6]

8.3.1 ▨ Failing to Protect the Public

One may fail to protect the safety, health, welfare, and property of the public by *not notifying employers or clients of such dangers* (II.1.a).

Asbestos and Johns–Manville Corporation

The Johns–Manville Corporation (now simply known as the Manville Corporation) once was the largest producer of asbestos. Asbestos fibers can result in an incurable cancer known as asbestosis. One study has indicated that 38 percent of asbestos insulation workers die of cancer (11 percent specifically from asbestosis). (In his youth, the film actor Steve McQueen held a summer job in which he worked with asbestos. He succumbed to asbestosis at age 50.)

Johns–Manville was aware of this hazard as early as the 1930s but failed to inform its own workers and the general public for another thirty years. Eventually, the

---

5. Many other engineering societies (e.g., ASCE, ASChE, ASME, IEEE, and IIE) have developed their own specific codes of ethics, but there is significant overlap among these various codes.

6. The particular paragraph(s) from the NSPE Code of Ethics corresponding to each type of violation is given in parentheses. This format is adapted from that used by Ertas and Jones (1993).

company was confronted with civil lawsuits resulting in damages of $2.5 billion to be paid to victims and their families over a 25-year period.[7] (See Case History 8.2 *Industrial Murder* for another example of this type of violation.)

One may also fail to protect the safety, health, welfare, and property of the public by *approving documents or work that are in violation of professional standards* (II.1.b; II.2.b).

The *Challenger* Disaster: Part II

An earlier case history focused upon the role that O-ring selection played in the *Challenger* tragedy. However, various engineers gave warnings of the danger to the management of Morton–Thiokol (the booster rocket manufacturer) and to NASA representatives. Management was informed that the probability of O-ring failure was expected to increase in cold weather because of the corresponding decrease in the pliability of the rings and putty seal. Allan J. MacDonald served as M–T's director of the solid-rocket booster project and the company's representative at Cape Kennedy. He arranged a teleconference between NASA and M–T engineers to discuss his concerns that a cold weather launch could be hazardous. One engineer recommended that no launch be attempted in temperatures lower than 53 degrees Fahrenheit.

Morton–Thiokol was in the process of negotiating a renewal of its contract with NASA for supplying booster rockets. One M–T senior executive told the vice president of engineering "to take off your engineering hat and put on your management hat." Company executives overruled the concerns of their own engineers by voting that the seals could not be shown to be unsafe. (Note the subtlety of this finding.)

Morton–Thiokol's vice president for booster rockets signed the recommendation to launch when MacDonald refused to sign it. The *Challenger* crew was never informed of the engineers' concerns for their safety.

8.3.2 ▨ Unethical Disclosure of Facts and Information

Another violation of professional ethics involves the *revelation of facts, data, or information obtained in a professional capacity without the prior consent of the client or employer* (II.1.c). Physicians and lawyers are expected to maintain the confidentiality of their dealings with clients. So, too, is an engineer

---

7. See Brodeur (1985) and Martin and Schinzinger (1989).

expected to protect privileged information such as trade secrets. See Case History 8.3 *Industrial Espionage in Silicon Valley* for an example of this type of violation.

Maintaining confidentiality can be difficult sometimes, even though one intends to do so. For example, a research engineer may move from one employer to another as his or her career advances, often working in the same area of research at both companies. This engineer may then apply knowledge that was acquired while working for a previous employer to tasks being performed at his or her current firm. Yet the former employer cannot expect the engineer to "forget" all technical knowledge that was acquired at that location. Unfortunately, it is not always a simple task to differentiate between what is to remain confidential and what is not. The following two cases illustrate this difficulty.

Secret of the Space Suit

Donald Wohlgemuth was a chemical engineer who had served as manager of the space suit division at B. F. Goodrich.[8] Eventually, he accepted a new position as manager of engineering at International Latex Corporation (ILC). Among the programs that he would direct in his new position was one in which space suits for the *Apollo* astronauts were to be developed. Goodrich sought a restraining order to prevent Wohlgemuth from working for ILC or any other company that was competing with Goodrich in the development of space suit technology.

The Ohio Court of Appeals did issue an injunction that prohibited Wohlgemuth from disclosing any of Goodrich's trade secrets. However, the court refused to issue a restraining order since this would violate Wohlgemuth's right to achieve career advancement within the industry.

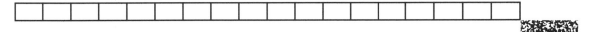

Titanium Oxide—Keep It a Secret!

In contrast to the B. F. Goodrich case described above, consider the following situation.[9] A chemical engineer was hired by American Potash and Chemical Corporation (APCC) after responding to an APCC advertisement in which industrial experience with titanium

8. See Baram (1968) and Martin and Schinzinger (1989).
9. See Carter (1969) and Martin and Schinzinger (1989).

oxide was noted as a desirable attribute sought in applicants. The engineer had previously worked for E. I. Du Pont de Nemours and Company where he had been in charge of efforts to produce titanium oxide.

Du Pont sought an injunction that would prohibit the engineer from participating in any titanium oxide projects at APCC. The court issued the injunction, noting that it was indeed inevitable that Du Pont trade secrets eventually would be revealed by the engineer.

These cases suggest that both the engineer and his or her employers need to identify *carefully* what is to be considered privileged information and therefore subject to protection of confidentiality.

8.3.3 ■ Failure to Include All Pertinent Information in Professional Reports

Failing to be truthful in professional reports, statements, and testimony or failing to include all pertinent and relevant information in such communications is also unacceptable behavior (II.3.a; III.3.a). Unfortunately, one may sometimes mistakenly believe that he or she is helping an employer by falsifying a report. This is never true!

The Ford Motor Company Case: Falsifying Tests Is NOT Helpful!

The interests of both the employer and the public often coincide, as seen in the following case.[10] Automobile manufacturers are required by the 1970 Clean Air Act to submit to the EPA the results of 50,000-mile emission tests on any new engine. Only one tune-up may be performed on each engine during these tests. In 1972 test results were submitted by the Ford Motor Company on its new 1973 model engines. Unfortunately, unbeknown to top-level management, four employees had ordered or directly performed illegal maintenance on the engines during testing. This maintenance included repeated replacement of spark plugs and points, cleaning of carburetors, and resetting of ignition timing.

A computer specialist discovered this illegal maintenance while examining computerized records of the testing. He wrote a memorandum outlining his findings to Lee Iacocca (then president of Ford Motor Company). Iacocca immediately informed the EPA and withdrew Ford's application for certification of its engines.

---

10. Martin and Schinzinger (1989).

Production schedules were delayed while new tests were conducted on an emergency basis. Ford also received a significant amount of negative publicity and was fined $7,000,000 for both civil and criminal violations.

The overzealous employees who were responsible for these illegal acts of engine maintenance failed to meet their obligations both to their employer and to the general public. Although they may have perceived themselves as loyal employees acting in the best interests of the company, in fact they were disloyal.

8.3.4 ▨ Other Common Violations

Other ethical violations sometimes include:

- Performing work for which one is not qualified (II.2.a).
- Expressing a professional opinion that is not founded upon both adequate knowledge of facts and technical competence in the field (II.3.b).
- Issuing a statement or other communication without identifying all interested parties (who may have influenced the content of the communication by paying for it or otherwise affecting the impartiality of the engineer) (II.3.c).
- Failing to act as a faithful agent or trustee of one's employer or client (II.4).

Please note that in the above list of violations we have focused upon only a small portion of the guidelines contained within the NSPE Code of Ethics. The reader should carefully review the entire code.

8.4 Resolving Conflicts among the Guidelines

One might wonder about situations in which two or more of the guidelines in the Code of Ethics conflict; for example, should one act as a faithful agent of his or her employer (paragraph II.4 of the code) if one knows that the actions of the employer will endanger the health and safety of the public (paragraph II.1.a of the code)? One general answer is that paragraph II.1.a is *most* important in resolving such conflicts because it explicitly states that "Engineers shall at all times recognize that their primary obligation is to protect the safety, health, property, and welfare of the public."[11] In other words, as engineers our foremost responsibility is to act in the best interests of the public. All other considerations are secondary to this goal. See Case History 8.4 *The BART Case* for an illustration of this critical message.

---

11. Of course, conflicts unrelated to the health and safety of the public may occur, in which case one may decide to seek appropriate advice and counsel (e.g., from a professional society) in order to resolve the conflict.

8.5 Product Liability: The Legal Costs of Failure

Much of the responsibility for ensuring the safety of a product in its manufacture, its use, and its disposal belongs to the engineer who designed it.[12] Engineers must consider the legal, ethical and moral implications of a design throughout its development if failures are to be avoided.

Today we live in a much safer world than did our parents or grandparents. In 1970, 10,000 television sets caught fire and 70,000 children were injured by toys. Overall, 110,000 people were permanently disabled and 30,000 people were killed that year in incidents involving various products throughout the United States. The astounding aspect of these numbers is that they excluded food, drugs, motor vehicles, firearms, cosmetics, cigarettes, and radiological hazards. In other words, the products involved were those that would normally be considered safe to use by most of us.

Standards for product design and manufacturing, food preparation and storage, waste disposal, drinking water, workplace conditions, pollution controls, highway safety, the control of toxic substances and their use, and other aspects of modern life have been developed and refined in recent years. Many of these standards now have the force of federal law. In 1906 the United States Congress passed the Pure Food and Drug Act in order to ensure that food and drugs were prepared and stored properly. (Six years earlier, lead arsenate was the pesticide of choice for apples, potatoes, and many other foods, and infant mortality in the United States was more than 13 percent.)[13] In 1947 agricultural and other widespread uses of pesticides were restricted by the Federal Insecticide, Fungicide, and Rodenticide Act.

Other legislation followed, including the Flammable Fabrics Act (1953), the Refrigerator Safety Act (1956), the National Traffic and Motor Vehicle Safety Act (1966), the Federal Coal Mine Health and Safety Act (1969), the Poison Prevention Act (1970), the Occupational Safety and Health Act (1970), the Federal Railroad Safety Act (1970), the Consumer Product Safety Act (1972), and so forth. Such legislation provides legal guidelines for socially acceptable products and behavior; in particular, product liability laws specify some of the legal boundaries within which the engineer and the manufacturer must work.

Since our knowledge and awareness of product hazards is ever increasing, these legal boundaries undergo continuing change with which the engineer must be familiar—that is, product liability laws are dynamic. For example, paint containing lead was first banned in 1978, as were patching compounds

---

12. This section is based largely upon Keeton, Owen, Montgomery, and Green (1989), Middendorf (1990), Lowrance (1976), and Weinstein, Twerski, Piehler, and Donaher (1978).
13. Lowrance (1976).

containing asbestos; before that time, such products were officially deemed to be safe. Similarly, lawn darts were banned in 1988, after it was recognized that more than 600 injuries occurred each year because of these 12-inch, four- to eight-ounce shafts striking people.[14]

As with most things in life, certain costs and benefits are associated with product liability laws. The costs include the greater financial burdens and other constraints that are placed upon companies and individuals who develop new products and the threat of lawsuits in which unreasonable damages may be sought by an injured party. However, the benefits include longer and happier lives for many people who otherwise would suffer injury or death because of a poorly engineered product.

Product liability can be defined as an action in which an injured party (plaintiff) seeks to recover damages for personal injury or loss of property from the seller or manufacturer (defendant) of a product where it is alleged that the injury or loss occurred because of a defect in the product. In addition, lawsuits can be filed in which a commercial business suffers a financial loss due to the inadequate performance of a product (e.g., a retailer's reputation may be harmed as a result of a defective product line).

8.5.1 ■ Sources of Product Defects

Three primary sources of defects often appear in product liability lawsuits: manufacturing defects, design defects, and warning defects.

Manufacturing defects are those that fail the manufacturer's own standards. These are usually one-of-a-kind defects affecting a single unit among the mass population of manufactured units (i.e., one heating coil is damaged in one toaster from among a population of 10,000 toasters). A manufacturing defect is a serious but localized flaw.

Design defects are the result of inferior design and/or inadequate standards. All units are defective (e.g., an entire model line of automobiles are recalled to the factory for repair of a particular design defect). Design defects are global, affecting the entire product population.

Finally, *warning defects* refer to the absence of adequate warnings that allow the user to guard against danger.

A 1977 survey of product liability lawsuits found that 45 percent were based upon manufacturing defects, 37 percent focused upon design defects, and 18 percent involved warning defects.

Figure 8.1 presents a typical risk-versus-cost curve for a product. Notice that there are both primary (*P*) and secondary (*S*) costs associated with a product. Primary costs include direct manufacturing and development expenses, whereas secondary costs denote such expenses as those associated with litigation, manufacturing down time, and loss of customer goodwill

14. Hunter (1992).

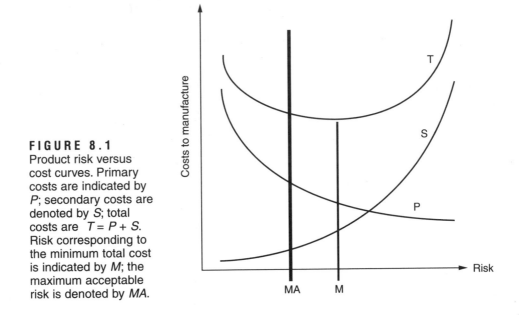

FIGURE 8.1
Product risk versus cost curves. Primary costs are indicated by *P*; secondary costs are denoted by *S*; total costs are $T = P + S$. Risk corresponding to the minimum total cost is indicated by *M*; the maximum acceptable risk is denoted by *MA*.

(Martin and Schinzinger, 1989). Primary costs increase as one tries to improve safety, whereas secondary costs increase with less safe products. The minimum total cost *M* may correspond to a higher level of risk than that which is minimally acceptable (at *MA*), in which case one must develop the design with *MA* as the target point (with a correspondingly higher cost than that associated with the point *M*).

8.5.2 ■ Limited Versus Absolute Liability

Product liability is *not* absolute liability. It is necessary for the plaintiff to demonstrate that the manufacturer violated a legal responsibility to the user. This is accomplished by examining either the conduct of the defendant, the quality of the product, or both.

Furthermore, the defendant must demonstrate that the defect or malfunction caused the injury or loss of property—that is, there would have been no injury or loss if the defect had been absent.

However, there is a legal evidentiary doctrine known as *res ipsa loquitur*, which means "the thing speaks for itself" in Latin. Under this doctrine, the existence of a defect can be inferred when it is the most reasonable conclusion that can be made even in the absence of evidence that such a defect does exist.

The Patent Danger Rule

In *Campo v. Scofield* (1950), the court stated that manufacturers should not be responsible for dangers that are obvious to a user.[15] The plaintiff claimed that the manufacturer of an onion topping machine was negligent in failing to include safety guards in the design. The plaintiff had been feeding onions into the machine when his hands became caught in the steel rollers, resulting in severe injuries. The court concluded that

> the manufacturer of a machine or any other article, dangerous because of the way in which it functions, and patently so, owes to those who use it a duty merely to make it free from latent defects and concealed dangers. Accordingly, if a remote user sues a manufacturer of an article for injures suffered, he must allege and prove the existence of a latent defect or a danger not known to the plaintiff or other users.

Patent Danger Versus Reasonable Design

The patent danger rule eventually was rejected by the New York Court of Appeals in *Micallef v. Miehle Company* (1976).[16] Mr. Micallef was an operator of a photo-offset printing press. It was standard practice that operators would remove any foreign objects (called "hickies" by printers) that inadvertently come in contact with the plate by inserting a piece of plastic against the rapidly rotating plate. This is known as "chasing the hickie" and eliminates the need to shut off the machine, thereby saving the three hours that would be required to restart the machine.

Unfortunately, one day the plastic became caught in the machine and drew Micallef's hand into the press. Although the danger was obvious to Micallef and other operators, the New York court recognized that it also would be reasonable to expect the manufacturer to include safety guards in order to prevent such accidents. As a result, it concluded that the patent danger rule (i.e., liability for nondisclosure of concealed dangers) would be only one of several criteria used to determine if a design was defective.

15. See, for example, Weinstein, Twerski, Piehler, and Donaher (1978), Kantowitz and Sorkin (1983) and Sanders and McCormick (1987).

16. Weinstein, Twerski, Piehler, and Donaher (1978), Kantowitz and Sorkin (1983), and Sanders and McCormick (1987).

The patent danger rule is still used as the sole criterion for safe design by some courts, whereas others will consider the reasonableness test (i.e., a product is defective if it presents an *unreasonable* danger to the user even if that danger is obvious).

8.6 ■ Principles of Product Liability Law

Product liability lawsuits are filed under one of the following legal principles: negligence, strict liability and implied warranty, and express warranty and misrepresentation. Under negligence, the conduct of the defendant is examined. Under strict liability and implied warranty, the quality of the product is evaluated. Under express warranty and misrepresentation, both written and oral promises concerning the relative safety of the product and its use are examined.

8.6.1 ■ Negligence

Under the principle of negligence, the conduct of the defendant is examined in order to determine if such conduct resulted in an unreasonable exposure of the plaintiff to risk. It is not necessary to establish that the defendant intended to do harm to the plaintiff; it is only necessary to convince a jury and/or judge that negligence occurred.

The judge and jury act as "reasonable people" who measure the defendant's conduct against an expected norm. In establishing this norm, three factors are considered:

- **Probability that harm will occur**   The larger this probability, the more responsibility is placed upon the manufacturer to ensure the public's safety.

  As an example, assume that identical injuries resulted from the use of a kitchen knife and a paper clip. Both products cracked because of flaws in the metal. The knife manufacturer would be expected to invest more effort (in terms of time, personnel, and money) than the paper clip manufacturer to ensure that such an accident did not occur because an accident involving a knife would be more likely to cause injury than one involving a paper clip.

- **Gravity of the harm**   If the injury due to one product is expected to be more severe than that due to another product, then the manufacturer of the first product has greater responsibility to protect the public from such harm.

  Assume that two lawsuits are brought to court, one involving a knife and one involving a lawnmower. Although both products have sharp-edged blades, the lawnmower would be expected to cause more severe injuries in the event of an accident. Hence, the lawnmower manufacturer would be expected to invest greater effort in ensuring that its product is safe.

- **Burden of precaution**   This factor is used to limit the expectations that can be placed upon a manufacturer of a naturally hazardous product. For example, a manufacturer cannot be expected to produce a 100 percent safe lawnmower; to do so (if it were possible) might require a financial investment that would make the final product economically infeasible.

One might ask whether meeting industry government and standards would protect a manufacturer from the charge of negligence. The answer is no; such standards are floor standards (not ceiling standards) that establish the minimum threshold that must be met by the product. It is necessary to meet such standards, but not sufficient. The entire set of industry standards may be found to be inadequate, thereby "raising the floor" that must be met. (See Case History 8.5 *A Helicopter Crash* for an instance in which floor standards were found to be inadequate.)[17]

8.6.2 ▨ Strict Liability and Implied Warranty

Strict liabilty is a tort[18] concept in which a manufacturer or seller is held responsible for any injury that results from a defect in the product. Conduct of the manufacturer or seller is not an issue; behaving in a reasonable and responsible manner is not sufficient.

An *implied warranty* is a related concept under contract law that falls into two distinct classes: merchantability and fitness. *Merchantability* means that the sale of the product implies that it is safe. The plaintiff usually attempts to demonstrate the presence of a manufacturing defect that led to the injury or loss (i.e., a one-of-a-kind type of flaw). *Fitness* refers to the situation in which a representative of the manufacturer has made a recommendation regarding the use or application of the product in response to an inquiry by the user. If it can be demonstrated that the recommendation was incorrect and led to the injury, the manufacturer can be held liable.

8.6.3 ▨ Express Warranty and Misrepresentation

A third category of product liability law is known as express warranty, which focuses upon any promise (written or oral), any description, or any sample/model that affirmed the worthiness of a product and influenced

---

17. A more detailed discussion of product standards and their development is included in Chapter 9.

18. Tort law focuses upon any wrongs or injuries for which one may seek redress through legal means (Middendorf, 1990) in contrast to contract law in which the court examines the various contractual obligations placed upon the parties involved.

the sale. A defect does not need to be established if the plaintiff can demonstrate that (a) the defendant expressly represented or warranted that the product would perform in a specific manner; (b) the product failed to meet the conditions of the warranty; and that (c) the injury resulted from this failure.

Warning labels and instruction manuals must be prepared carefully to guide the consumer in the proper use and maintenance of products. Warnings must be intelligible (i.e., they should be understandable to those who may not read English or who have a limited education). Furthermore, they must be both adequate and complete in warning the user of all dangers, their consequences, and any measures that should be taken if an accident occurs. Finally, they should be placed in a visible location and remain readable for the life of the product.

8.6.4 ▨ Privity

The concept of *privity* refers to a direct contractual relationship between two parties.[19] Privity was long used to shield manufacturers from lawsuits by requiring that the product have been purchased by the buyer directly from the defendant. If the manufacturer sold its products through a distributor, only the distributor (who had sold the product directly to the buyer) could be sued in the event of a mishap. And, of course, the distributor could claim that he had sold the product in good faith and therefore was not responsible for the injury or loss. Privity was essentially a wall of protection surrounding the manufacturer.

MacPherson v. Buick Motor Company

The privity wall of protection began to develop cracks in 1916 with the landmark case of *MacPherson v. Buick Motor Company.* A wheel dislodged from the new automobile purchased by Mr. MacPherson, leading to an injury. This occurred in the early days of the automotive industry when many small car manufacturers were producing relatively unsafe vehicles. The case came before Judge Benjamin Cardozo (later a U.S. Supreme Court justice), who declared that

> if the nature of a thing is such that it is reasonably certain to place life and limb in peril when negligently made, it is a thing of danger; and if to the element of danger

---

19. Middendorf (1990).

there is added knowledge that the thing will be used by persons other than the purchaser, then the manufacurer of the thing of danger is under a duty to make it carefully.

This decision struck down privity as a defense in cases of negligence. Manufacturers were now held liable for their conduct even if they did not sell the product directly to the buyer.

Automotive (and other) manufacturers began to perform extensive testing of their products, thereby providing evidence that they were indeed striving to develop safe vehicles. Such testing did result in safer automobiles.

Although, as a result of the above case, the wall of privity protection surrounding manufacturers did develop cracks, it did remain. Negligence is difficult to prove in court (particularly if the plaintiff was also negligent in some way). Then in 1960 another famous case, *Henningsen v. Bloomfield Motors, Inc.,* removed privity protection from cases involving implied warranty.

Henningsen v. Bloomfield Motors, Inc.

Five years before the court finding, Ms. Henningsen was driving her new 1955 Plymouth (10 days old with 488 miles on the odometer) at 20 miles per hour when she heard a loud snap beneath the hood and the steering wheel spun in her hand. The car crashed into a wall, resulting in extensive damage to the vehicle.

Negligence could not be proved because of the extensive damage to the car. Instead, a lawsuit was filed under breach of implied warranty. The court concluded that

where the commodities sold are such that if defectively manufactured they will be dangerous to life or limb, then society's interests can only be protected by eliminating the requirement of privity between the maker and dealers and the reasonably expected ultimate consumer.

In 1962 a similar case, *Greenman v. Yuba Power Products, Inc.,* struck down privity protection under strict liability. (Recall that strict liability is tort law, whereas implied warranty is contract law.)

Greenman v. Yuba Power Products, Inc.

Mr. Greenman was operating a combination lathe, saw, and drill-press machine when a piece of wood struck his forehead, resulting in a serious injury. He alleged that the product had been improperly designed since it lacked adequate set screws to prevent such an accident. The case was eventually appealed to the California State Supreme Court, which concluded that

> a manufacturer is strictly liable in tort when an article he placed on the market, knowing that it is to be used without inspection for defects, proves to have a defect that causes injury to a human being. [Furthermore,] …the purpose of such liability is to insure that the costs of injuries resulting from defective products are borne by the manufacturers that put such products on the market rather than by the injured persons who are powerless to protect themselves.

With this case, the wall of privity protection about manufacturers had crumbled. Every product, by reason of its existence and sale, was now expected to be safe to use. Most product liability lawsuits are now filed under the tort category of strict liability.

8.7 Designing Against Foreseeable Uses and Misuses

One objective of the engineer should be to design against any *misuse* of a product; in other words, try to anticipate foreseeable misuses of the product by the user and make it impossible for such misuses to occur. Of course, one cannot expect to eliminate all possible misuses of a product, but the attempt must be made.

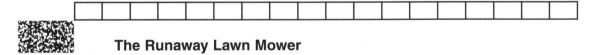

The Runaway Lawn Mower

In one case, a self-propelled lawn mower was neglected by its operator. The motor's vibrations caused the control lever to move into the engaged position. The mower injured the plaintiff. The manufacturer should have designed against such a shift in the position of the control lever.

In fact, manufacturers also are responsible for injuries or losses that result for foreseeable *uses* of a product. Consider the following anecdote.

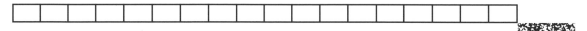

The Dangerous Door

A shipment of doors were laid in 42-inch-high stacks that were then placed in cardboard packing held in place by two steel bands. Unfortunately, the doors were not solid: Each door had a large opening for a glass window that was to be installed at a later date. The wooden ends were exposed, thereby giving the impression that the doors were entirely solid. The cardboard containers were simply marked "fine doors."

The accident occurred when a stevedore walked across the doors while carrying a 100-lb sack of flour. He fell through the nonsolid (open) portion of the doors, sustaining injuries. The court concluded that the manufacturer was indeed liable for failing to anticipate such a foreseeable use of the door shipment.[20]

8.8 Anticipating the Effects of Change in a Product

Design engineers also must anticipate the changes that may occur in a product during its useful lifetime. All products age with resultant changes in their characteristics. These changes may result in injury or loss if one fails to anticipate such consequences. Consider the following case.

The Altered Hammer

A farmer was using a hammer when a metal chip struck him in the eye.[21] There were no metallurgical flaws in the hammer; however, the metal of the hammer had hardened as it was used. (This phenomenon is known as work hardening.) This made the

---

20. See Middendorf (1990).
21. Martin and Schinzinger (1989).

hammer more likely to chip upon striking a harder object. The hammer, work-hardened to a Rockwell hardness of C-52, was used by the farmer to drive a clevis pin with a Rockwell value of C-57. Chipping was not unexpected under these circumstances.

The court concluded that the manufacturer was indeed liable since (a) chipped hammers had been returned to the factory, (b) the work hardening phenomenon was well known to metallurgists, and (c) the use of the hammer to drive a hard metal object should have been anticipated.

Engineers also need to anticipate the consequences of a design modification, as illustrated by the redesign of the Ford Pinto.

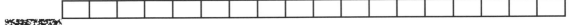

Redesign of the Ford Pinto

During crash-testing of the Pinto, it was found that the windshield shattered in front-end collisions. In order to eliminate this problem, the drive train was moved towards the rear of the vehicle. Unfortunately, the differential was then very near the gas tank; in many rear-end collisions, the tank collapsed and exploded. A redesign meant to eliminate hazards led to exactly the opposite result.[22]

Designers also must consider the disposal of a product after its useful life has ended. Both people and the environment must be protected from the disposal of a hazardous product.

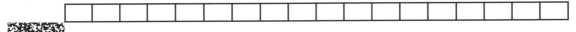

Unsafe Refrigerators

Until 1956 more than a dozen children perished each year because they would climb into a discarded refrigerator. The door latch then would shut, trapping the child inside. In that year, Public Law 930 was passed and required that all new refrigerator doors should need no more than 15 pounds of force to be opened from the *inside*. Of course, today, magnetic gaskets are used to hold doors shut and require little force to open.

22. See, for example, Martin and Schinzinger (1989), Camps (1981), Strobel (1980), and Cullen, Maakestad, and Cavender (1987).

SUMMARY

- Codes of ethics provide guidelines for dealing with various types of professional dilemmas.
- Early guidelines for acceptable engineering practice include Hammurabi's ancient code for builders and the early United States steamship code.
- Specific legal obligations and requirements must be satisfied by professional engineers.
- The NSPE's Code of Ethics can be helpful if one is confronted by an ethical choice or dilemma.
- Design engineers must be familiar with product liability law in order to understand the legal boundaries for socially acceptable products.
- Three types of defects can occur in a product: manufacturing, warning, and design. The first is localized to a single unit, whereas the second and third affect the entire product population.
- There are three major categories under which product liability lawsuits are filed: negligence, strict liability and implied warranty, and express warranty and misrepresentation.
- Privity no longer serves to protect manufacturers from lawsuits.
- One must design against foreseeable uses and misuses of a product.
- Possible changes in a product during its useful lifetime must be anticipated in the design process.

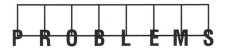

PROBLEMS

8.1 Consider the applications of technology and engineering to transform society in various ways since 1900.
   a. Describe three beneficial changes in modern society that are the result of technology/engineering.
   b. Describe three negative changes in modern society that are the result of technology/engineering.
8.2 Identify the specific section(s) in the NSPE Code of Ethics that relate to the following actions:
   a. Whistle-blowing when confronted by unethical behavior within your firm
   b. Disclosing a potential conflict of interest
   c. Accepting a gift from a client or contractor
   d. Preparing a misleading proposal
   e. Failing to reveal a hazard in a design

f. Working on a project for which you are not qualified by education or experience

g. Approving a report of work that was not directly supervised by you or under your direct control

h. Criticizing another engineer's work

8.3 The *Challenger* disaster is an example of the tragedies that can result from lapses in both engineering practice and professional ethics. Prepare a brief report outlining the technical and ethical issues that were involved in this disaster. (Use library resources to obtain relevant information for preparing your report.)

8.4 Review the case entitled *Titanium Oxide—Keep It a Secret!* Do you agree with the court's decision to issue an injunction against the engineer? Explain your reasoning and offer counterarguments that might be made by the court.

8.5 Compare and contrast the Ford Motor Company case involving engine emission tests with the BART case. In what ways are these two cases similar? In what ways do they differ?

8.6 Research a product liability case of your choice and write a detailed report describing all important aspects of the case.

8.7 Discuss the relative merits of the patent danger rule and the reasonableness test for determining if a product is defective.

CASE HISTORY 8.1

A Tidal Wave of Molasses Strikes Boston

At 12:40 P.M. on January 15, 1919, a huge tidal wave of molasses spread through the streets of Boston's North End immediately engulfing people, wagons, horses, automobiles, and anything else lying in its path.[23] More than 12,000 tons of this brown, gooey substance formed a wall that was 15 feet high and 160 feet wide as it roared through the streets at 35 miles per hour. By the time the wave had subsided, 21 people were dead and more than 150 others were injured.

The molasses had been stored in a massive cylindrical steel tank that was 50 feet high with a diameter of 90 feet. The tank had been constructed from seven sheets of steel, each approximately seven feet high and formed into circular rings. The vertical seams of the top six sheets were each sealed with six

23. See Brown (1919), Frye (1955), and Weingardt (1994).

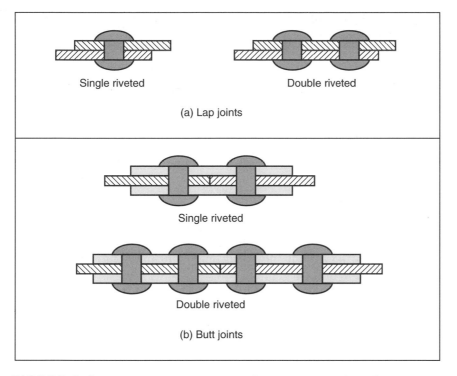

FIGURE 8.2 Riveted lap and butt joints. *Source:* Adapted from Giesecke et al., 1993.

rows of rivets along a set of lap joints. (As the name implies, "lap joints" simply refer to configurations in which the connecting pieces overlap with one another prior to riveting. Alternatively, the two pieces may abut one another in a so-called butt joint configuration, in which case they are covered by splice plates before being riveted together. Figure 8.2 shows examples of both lap and butt joints.) In contrast, the riveted vertical seam of the bottom sheet was composed of butt joints and splice plates. Each steel ring was then fastened to its neighbors along horizontal single-riveted lap joints.

The tank eventually burst, unleashing 2.3 million gallons of molasses, because it had been *underdesigned*. The walls at the bottom of the tank required the greatest strength because they would experience the highest pressures applied by the tank's liquid contents. Unfortunately, a manhole with a 21-inch diameter had been cut in the bottom steel ring directly below the vertical lap joint of the second ring. The manhole and lap joints were the weakest parts of the two bottom rings, and the tank failed at these locations as indicated by two major fractures found among the wreckage.

One fracture ran through the manhole opening in the bottom ring to the vertical lap joint of the second ring; the second fracture ran diagonally across the bottom sheet up towards the lap joint of its neighbor. The single-riveted horizontal lap joints connecting the second sheet to its neighbors also separated, and some of the rivets along the vertical joints were sheared. In addition, some of the steel plates between the rivets were torn. These fractures then led to the entire tank being torn asunder.

The bottom ring needed to be properly reinforced in order to overcome its weakening by the addition of the manhole opening; this was not done (first error). The investigation also found that the thicknesses of all seven steel sheets were actually 5–10 percent less than described in the original building permit documentation (second error). Although the manufacturer (Hammond Iron Works) had never constructed a tank of this enormity, the design calculations were never verified by an experienced engineer (third error).

SUMMARY OF FAILURES AND TYPES OF ERRORS

- Failure to properly reinforce structure (Error in design due to conceptual misunderstanding, invalid assumptions, and faulty reasoning)
- Failure in executing design specifications (Error in implementation: materials selection and inspection)
- Failure to verify calculations (Error in implementation)

The Boston molasses disaster gradually led to many positive and far-reaching changes in construction engineering. These included:

- Perhaps most important, all professional engineers would need to be certified by the state or states in which he or she worked.
- All calculations would henceforth be included with the design plans submitted for building permits.
- Building permits would not be issued unless plans were sealed by a registered professional engineer.

These changes created a more uniform and *dynamic* level of professionalism throughout the construction industry—a level that rises with the addition of new knowledge and improved technical capabilities.

CASE HISTORY 8.2

Industrial Murder

Silver can be recovered from used photographic and x-ray plates by soaking these plates in a cyanide solution. Workers must wear protective clothing (e.g., rubber gloves, aprons, and boots) and use respirators in order to prevent contact with deadly cyanide gas.

Film Recovery Systems was one company that was active in this recovery work. Unfortunately, employees were not provided with respirators but only inadequate cloth gloves and paper face masks.[24] Workers frequently became physically ill.

Finally, an autopsy revealed that one employee had died of cyanide poisoning, leading the authorities to file murder charges against certain company executives. The Illinois statute under which these charges were filed states "a person who kills an individual without lawful justification commits murder if, in performing the acts which cause the death ... he knows that such acts create a strong probability of death or great bodily harm to that individual or another" (Frank, 1987).

The company president, the plant manager, and the plant foreperson were shown to be familiar with the hazards associated with cyanide and its use at their facility. In 1985, each was convicted of industrial murder, fined $10,000, and given a 25-year prison sentence.

CASE HISTORY 8.3

Industrial Espionage in Silicon Valley

Silicon Valley is a nickname given to the Santa Clara Valley in Northern California, reflecting the intense concentration of computer chip manufacturers throughout the region.

Development of computer chips can require huge investments of money, time, and effort. In addition, chip technology can rapidly become obsolete. Companies sometimes save millions of dollars in development costs by using a process known as reverse engineering. In this process, a competitor's chip is analyzed and a similar or better chip is then developed. Reverse engineering is a legal way to acquire trade secrets (as discussed in Chapter 5).

Unfortunately, people sometimes seek to learn trade secrets in illegal ways.[25] For example, Peter Gopal was a semiconductor expert who started a consulting company in 1973. He established relationships with various people throughout the computer chip industry who were willing to sell their companies' trade secrets. One, an employee of National Semiconductor

---

24. See Frank (1987) and Martin and Schinzinger (1989).
25. See Halamka (1984), Hiltzig (1982), Samuelson (1982), and Martin and Schinzinger (1989).

Corporation, provided Gopal with confidential documents outlining details about the circuitry of the company's products. This information was then sold to Intel Corporation. Gopal also managed to steal information from Intel that was then sold to National Semiconductor. He did this by simply bypassing the elaborate security system that was in place at Intel, instead concentrating on one of Intel's subcontractors at whose facilities were stored materials for manufacturing from which detailed information about Intel technology could be extracted. He bought this material from a supervisor at the subcontractor's site, thereby obtaining Intel secrets.

The police conducted an extensive undercover operation with the cooperation of both Intel and National Semiconductor that eventually resulted in Gopal's arrest.

CASE HISTORY 8.4

The BART Case

One of the most famous instances of whistle-blowing is that of the BART case. Whistle-blowing here is defined as the act of informing appropriate persons other than those specified by the chain of command of significant ethical or moral lapses within an organization in the hope that this act will result in the correction of a dangerous, unjust, illegal, or otherwise intolerable situation.

BART is an acronym for San Francisco's local railway known as the Bay Area Rapid Transit System. The system includes numerous high-tech features. For example, trains were to be controlled automatically and redundancy systems were installed in place of the more traditional fail-safe mechanisms. A redundancy system is one in which critical components are intentionally duplicated, thereby allowing the system to switch to a back-up unit if any component fails to operate. In contrast, a fail-safe design simply shuts the system down in the event of a critical component failure.

During the design and development of the BART system, three engineers (Max Blankenzee, Robert Bruger, and Holger Hjortsvang) warned management of some inherent dangers within the system. These problems included inadequacies in

- the design and testing of the automatic train control mechanism,
- the preparation of each human operator in the use of the control mechanism, and
- the monitoring of the contractors who were building the BART system.

These engineers then communicated their concerns in writing to supervisors through three levels in the company's chain of command. Management gave no indication that any further investigation of the perceived hazards would be performed.

Without going into the details of the case, we will simply note that the engineers eventually contacted some members of the BART board of directors, thereby violating the expected chain of command sequence for relaying information to higher levels of management.[26] One of the directors released copies of the engineers' memos and the report of a private engineering consultant who had also been contacted by the three engineers to the press.

The three engineers were dismissed from the company for insubordination, incompetency, and other perceived violations. However, their concerns about potential hazards in the design, testing, and operation of the automatic control system were found to be justified. Design modifications were then made in this control system, but only after several accidents had occurred.

The engineers sued BART for breach of contract, deprivation of their rights under the First and Fourteenth Amendments, and harming their opportunities for further employment. The Institute of Electrical and Electronics Engineers (IEEE) filed an *amicus curiae* (friend of the court) brief on behalf of the engineers, noting that every engineer is expected to promote the public welfare. An out-of-court settlement was reached.

Eventually, each of the three engineers received IEEE's Award for Outstanding Service in the Public Interest for "courageously adhering to the letter and spirit of the IEEE code of ethics."

CASE HISTORY 8.5

A Heliocopter Crash

On July 9, 1962, Cloyd Berkebile perished when the helicopter that he was piloting crashed into a hillside.[27] His wife sued the manufacturer (Brantly Helicopter Corporation) of the two-seat copter under strict liability law.

---

26. See Anderson, Perrucci, Schendel, and Trachtman (1980) and Martin and Schinzinger (1989).
27. See, for example, Weinstein, Twerski, Piehler, and Donaher (1978) and Kantowitz and Sorkin (1983).

An autorotation system allows an aircraft to glide slowly to the ground whenever the power fails during a climb. Brantly had modified their original design of the copter's autorotation system before the product was distributed, and the company felt confidant that their design was safe and dependable, and one that "beginners and professionals alike agree ... is easy to operate," as their advertisements claimed.

Purchasing the helicopter in January of 1962, Berkebile was flying solo when the seven-foot outboard motor section of one of the three main rotor blades suddenly separated, leading to the fatal crash. The Brantly copter required that the pilot activate the autorotation sytem within one second of power failure. The plaintiff (Mrs. Berkebile) claimed that this was insufficient time for the average pilot to respond. The defendant responded that its system satisfied the standards set by the Federal Aviation Administration and that Mr. Berkebile had taken off with a nearly empty fuel tank.

The final court decision held that FAA regulations set a lower limit—but not a ceiling—for such autorotation systems and that under strict liability law Brantly was indeed liable for the accident by failing to produce a reasonably safe product.

Full compliance with government standards does not ensure that one has satisfied his or her duty to the user under the strict liability concept.

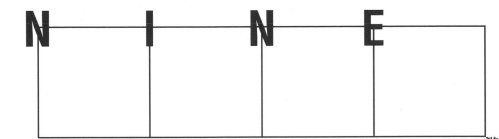

NINE

HAZARDS ANALYSIS
AND FAILURE ANALYSIS

A wise man learns from the mistakes of others, nobody lives long enough to make them all himself.   Author unknown

O B J E C T I V E S

Upon completion of this chapter and all other assigned work, the reader should be able to

■ Explain the three levels of failure: physical features, errors in process, and errors in perspective or attitudes.
■ Distinguish among such product hazards as entrapment, impact, ejection, and entanglement.
■ Apply hazard and operability study (HAZOP), hazards analysis (HAZAN), fault tree analysis (FTA), failure modes and effects analysis (FMEA), and Ishikawa diagrams to engineering design analysis.
■ Describe how exposure to a hazard can be reduced through the use of guards, interlocks, and sensors, or via safety factors, quality assurance, or intentional redundancy.

9.1   Failures in Engineering

We begin our discussion of failure analysis in engineering design by considering the following brief case histories.[1]

---

1. See Bignell, Peters, and Pym (1977).

The Hixon Level Crossing Incident

Most of us are familiar with automatic half-barriers at railroad crossings with roadways. Trains are given priority at such intersections because they normally cannot stop as quickly as road traffic. Automatic half-barriers provided a cost-effective means for controlling the flow of road traffic until a train had safely passed the crossing. Since these barriers were automatic, no attendants would be needed at the crossings. Moreover, two sets of half-barriers were used—rather than full barriers that would extend across the entire highway—in order to block oncoming traffic from entering the intersection while simultaneously allowing any vehicles already on the crossing and traveling in the opposite direction to continue safely off the tracks. In the system used at England's Hixon Level Crossing a set of flashing warning lights and bells would be triggered for eight seconds by any approaching train, after which the half-barriers would be lowered across the roadway during the next eight seconds. The barriers remained in the lowered position for eight more seconds before the approaching train arrived at the crossing. Thus, any vehicle on the crossing had a minimum of 24 seconds to exit safely before the train would reach the intersection.

On January 6, 1968, a road transporter (a 32-wheeled trailer and two six-wheeled tractors totaling 148 feet in length) carrying a 120-ton transformer began to move through the Hixon Crossing. It was escorted by several police vehicles. Traveling at only two miles per hour, this carrier needed *one full minute* to cross the intersection safely. A telephone system had been installed near the intersection to allow communication between road traffic (in this case, the trailer crew and police escort) and the railway operators. Unfortunately, members of the trailer crew and escort did not realize that the telephone had been provided as an additional safety precaution and so they did not use it before proceeding through the crossing.

A train approached the intersection just as the carrier was lumbering across the tracks. The train, unable to stop in time once the crossing came into its view, collided with the carrier. Eleven people perished and 45 others were injured. The 24-second interval of the half-barrier system, coupled with human error, had proven insufficient to avert a disastrous collision.

The Disaster at Aberfan

For generations, waste from the coal mine near Aberfan in South Wales had been stacked in "tips" along the side of Merthyr Mountain. Mine refuse (e.g., discarded rock, ash, and dirt) was transported by railway trams to a location where a crane could tip the material in each tram car onto the mountainside. Tram tracks were gradually extended over the waste itself as it collected along the mountainside. This method of tipping allowed more material to be discarded adjacent to the mountain than might be collected into a single pile on level ground.

Between 1914 and 1958 a series of six tips had been created and abandoned along the mountainside. Each of these tips eventually became so unstable that it had to be abandoned. Then, beginning in 1958, Tip 7 began to be formed. Unlike its predecessors, this new tip included residue from a mining technique known as the froth-flotation process (a technique with which one can extract even small amounts of coal). This ashy residue, dubbed "tailings," was moistened with water before being tipped so that it would lay flat after tipping. Moreover, rainwater collected into streams, across which Tip 7 was formed. This combination of wet tailings and surface water contributed to the final deadly landslide of Tip 7 on October 21, 1968, when 140,000 tons of waste rushed down the mountainside into Aberfan, killing 144 people (including 116 children).

Could this landslide have been foreseen and avoided? In 1927 the hazard of water collecting in a tip and the subsequent need for drainage in order to avoid slipping were discussed in a professional engineering society presentation. In 1939, 180,000 tons of residue slid down a hill, resulting in a second document in which the danger of sliding tips was noted. In addition, both Tips 4 and 5 had reached hazardous states before being abandoned, with much of Tip 4 cascading down the mountainside in 1944. Finally, Tip 7 itself partially slid down the mountainside in 1963, leading to concerns among the residents and local officials about the danger that it posed to the village. Nevertheless, none of this prior knowledge was used to prevent the final deadly landslide in 1968.

The Progressive Collapse at Ronan Point

Ronan Point was a 22-story high-rise apartment building in East London. Its slab and joint design was based upon the use of large reinforced concrete floor and wall panels that are fabricated offsite and then transported to the construction locale. Moreover, in the Ronan Point design, the walls of each apartment acted as load-bearing supports for the floors and walls of the rooms above—unlike other high-rise designs in which a steel framework is used to support each floor independently of other floors.

Disastrous events began to unfold in an 18th-floor apartment (Flat 90) on April 15, 1968. A brass nut, used to connect the hose running between a kitchen stove and the gas supply line, contained a manufacturing defect: It was too thin at one point, eventually cracking during the night of April 15th and allowing gas to fill the apartment. The tenant awoke that morning and struck a match at 5:45 A.M. to light the stove, igniting the gas. The explosion blew apart the walls of Flat 90, thereby eliminating the structural support for the apartments directly above this unit. The living rooms in these upper apartments fell, and the resulting debris led to a progressive collapse of the floors below. In essence, twenty-two floors of apartments partially collapsed in a domino effect, resulting in four deaths and seventeen injuries. (Fortunately, most residents were not in their living rooms, where the damage was focused, when the explosion occurred; otherwise, the death toll could have been much higher.)

Each of the preceding disaster stories illustrates the need to anticipate hazards and to take actions against them. In the Hixon Level Crossing accident, the danger of a very slow moving and massive vehicle could not be overcome even with timed delays, barriers, and a communication (telephone) system in place. In the case of the Aberfan landslide, multiple warnings of the danger posed by huge unstable collections of waste along the mountainside were ignored. Finally, the design of the Ronan Point high-rise failed to provide sufficient overall structural support of surrounding units in the event of a localized failure in which some walls of a single apartment were destroyed. These incidents provide us with evidence that, when trying to minimize hazards and avoid failures, engineers should strive to

- expect the unexpected,
- respond to warning signals, and
- focus on both the local and global aspects of a design.

There is a limit to what one can do to prevent accidents; however, ignoring the lessons of the past will ensure that we and our clients will experience failure.

Engineering designs fail whenever they do not perform their intended function(s) or fall short of fulfilling the user's reasonable expectations. Such failures may be temporary or permanent, and their consequences may range from the trivial to the deadly serious. But all share one common trait: They are undesirable and should be avoided if at all possible.

We will use the terms *failure analysis* and *hazards analysis* to describe two distinct types of activities: The former will be used to denote a diagnostic evaluation of the factors that led to an actual design failure, whereas the latter will be reserved for determining the potentially dangerous aspects of a design that could lead to disaster. Hence, failure analysis will include all post-failure forensic or diagnostic activities, and hazards analysis will refer to all pre-failure preventive actions. Both types of analysis are critical if one is to produce safer engineered products. Skills in failure analysis will allow the engineer to reconstruct the events that led to a specific engineering failure. Moreover, these diagnostic skills will help one to become more adept at preventive hazards analysis—that is, recognizing and avoiding conditions that could lead to disaster in future designs.

As Witherell (1994) notes, we should not expect to prevent all engineering failures; every human-made design will eventually cease to function properly. Instead, we should strive to *avoid* unexpected failures, particularly those that can result in the loss of human life. Such avoidance requires that (a) we learn from the recorded history of engineering failures in order to avoid repeating the mistakes of the past and that (b) we remain alert to any new hazards that might be imbedded in our current designs. Through such activities as proper inspection, maintenance, replacement, and redesign, hazards often can be eliminated from an engineering solution before failure can occur.

A failure in engineering can be costly beyond the financial burden directly associated with the failure itself (e.g., replacement costs or damages resulting from the failure). For example, an engineering firm deemed to be responsible for a major engineering disaster may lose its credibility in the marketplace, resulting in lost income and missed opportunities for future contracts. Nonengineering employees may lose their jobs, and members of the engineering staff may lose both their jobs and their professional reputations. However, the greatest costs lie in the loss of human life, the emotional burden that must then be carried by those who were responsible for this loss, and the damage done to the public image of the engineering profession itself. For all of these reasons, failures must be avoided!

Finally, it is important to recognize that many engineering failures are reflections of human error. They are not due to failures in materials, manufacturing, or machinery, but instead result from errors made by one or more people involved in the development, analysis, and/or implementation of a design. Consequently, these failures can be avoided if we perform our work with care, knowledge, precision, and thought.

9.2 Three Levels of Failure

Professional codes of ethics and product liability statutes provide us with both moral and legal incentives for ensuring that our engineering designs will be safe and effective (Chapter 8). Given these incentives, we next must ask how one actually goes about avoiding failure in engineering design.

The first step is to be aware of the potential causes of failure. Recall that in Chapter 2 we identified three levels at which sources or causative elements for engineering failures occur: physical flaws in the design itself, errors in the process used to develop and maintain the design solution, and errors in the attitudes or perspectives held by those responsible for the design and its operation. The tangible or physical flaws are those elements in the design that directly led to its failure. These first-level flaws are the most obvious reasons for the failure. Such physical flaws often result from errors in the process through which an engineering solution is designed, developed, and fabricated. These process errors comprise the second level of causative elements and may in turn reflect underlying flaws of perspective—that is, flaws in the engineer's professional behavior and attitudes (the third and most basic level). Catastrophic and unexpected failures can be avoided only if we are aware of each of these different types of error.

9.2.1 ■ Physical Flaws

Physical flaws refer to any tangible characteristics of a design that led to its inability to perform a desired function. After failure has occurred, this physical source for the breakdown usually is determined first.

We will focus on only a few of the physical flaws (overloading, fatigue, corrosion, and electrical hazards) that can lead to failure in order to demonstrate how failure analysis can be used to develop an understanding of hazardous conditions that should be avoided in engineering design.

Overloading and Fatigue   Structural designs are developed to withstand certain levels of both static and impact loads. A weight, when supported by a structure, applies a static load. Buildings and bridges are designed to support their own weights (the "dead load") together with the weights that may be applied by people, vehicles, and other objects (the "live load"). In addition, these structures must be able to resist impact loading—that is, a sudden force that is applied momentarily to the structure and which may be quite large in comparison to the static loads. The structure should be able to dissipate the energy of the impact loading without cracking and permanently (plastically) deforming.

Structural failure can also occur due to cyclic loading and unloading. Unfortunately, a structure can experience such fatigue failure even if the cyclically applied loads lie below those levels that are considered safe for static and impact loading. In these cases, the structure undergoes microscopic plastic deformation as the cyclic loading continues (i.e., it does not return to its original form) until total failure occurs, usually at a geometrically weak point in the structure, such as the location of a hole or some other discontinuity in its shape. Stresses can concentrate at these locations and infinitesimal cracks may then appear and grow under continued cyclic loading.

BOAC Comet Jetliner Explosions

In 1954 metal fatigue resulted in the explosions of two BOAC Comet jetliners while in flight. Extensive testing of the fuselage design finally revealed that cyclic pressurization and depressurization of the cabin led to deadly cracks forming around a window until the structure eventually broke apart.

Other incidents in which metal fatigue was a factor include the 1979 collapse of the Kemper Arena roof in Kansas City, Missouri, which involved fatigued bolts, and the 1989 crash of a Boeing 747 in Japan.[2]

Designs should be tested and inspected periodically to ensure that such fatigue failure will not occur. (See Case History 9.1 *The* Alexander Kielland *Floatel* for a detailed example of fatigue failure.)

Brittle fractures also lead to failure. Glass, for example, is a very strong but very brittle material; it does not exhibit plastic deformation and shatters rather easily unless it is reinforced with other material. A portion of the King's Bridge in Melbourne, Australia, failed in 1962 when the girder framework underwent brittle fracturing.[3]

Corrosion   The 1992 crash of a Boeing 747-200 into an apartment complex in the Netherlands was due to corroded engine mounting bolts, leading to a loss of both engines (see Case History 9.2 *Two Boeing 747 Crashes*).[4] Materials corrode as their surfaces gradually deteriorate. If the level of corrosion rises beyond some acceptable point, failure has occurred.[5]

What is acceptable will depend upon the function to be achieved. Even a slightly corroded paint and metal finish of an automobile may be unacceptable for aesthetic reasons.

Corrosion can be quite general on a surface. For example, atmospheric corrosion refers to the deleterious effect that air and impurities can have on a material's surface. Galvanic corrosion occurs between two dissimilar metals with electrical potential differences; the metal with less electrical resistance (i.e., the one acting as the anode) undergoes corrosion as current flows from it to the other (cathode) metal.

In contrast, crevice corrosion is localized and occurs at the sites of metallurgical defects or other breaks in the surface of a metal where liquid can become lodged.

Stress corrosion can lead to cracking and failure. In the case of the Silver Bridge disaster (Case History 9.3 *The United States' Worst Bridge Disaster*), an eyebar underwent stress corrosion, eventually resulting in the bridge's total collapse.

Electrical Hazards   Electrical malfunctions can lead to fires, shocks, electrocution, loss of system control, and other threats to safety.

2. See, for example, Schlager (1994) and Levy and Salvadori (1992).
3. Schlager (1994).
4. See, for example, Ertas and Jones (1993) and Walton (1991).
5. Schlager (1994).

The *Apollo 1* Disaster

An electrical malfunction led to the deaths of three American astronauts in the *Apollo 1* disaster in 1967.[6] On January 27, 1967, Virgil I. "Gus" Grissom, Edward H. White, and Roger B. Chaffee perished when the interior of the *Apollo 1* command module became engulfed in fire. The three astronauts were conducting a routine ground test of the module prior to a manned space flight scheduled for the following month. It is believed that poor design and installation of the almost twelve miles of electrical wiring within the capsule led to an arc that started the fire. The interior of the capsule was pressurized with pure oxygen in order to closely simulate the conditions of space flight, wherein external pressure would be far less than that within the module. This level of oxygen allowed the fire to spread rapidly. The astronauts had no fire-extinguishing equipment on board, and they could not open the escape hatch before succumbing to the fire. (The six-bolt hatch required 90 seconds to open. Furthermore, since the hatch opened inward, even if it had been unbolted the astronauts could not have opened it because of the high internal pressure within the capsule.) Emergency crews were not on full alert since an unfueled module test was under way.

Subsequently, fifteen hundred modifications were made in the design of the *Apollo* command module, including electrical wiring covered with fire-proof coatings, the elimination of nearly all combustible materials, the development of a more temperature-resistant material for the astronauts' space suits, the use of nitrogen and oxygen at atmospheric pressure within the capsule during ground testing, and a fast-release escape hatch that swung outward.

An unexpected electrical malfunction, combined with a set of unfortunate circumstances, had caused the first loss of life in NASA's space program. But the tragedy led to a much safer capsule design and the eventual landing of Apollo astronauts on the moon.

9.2.2 ■ Errors in Process

Physical flaws in a design often are the result of errors that have occurred in a process, be it design, analysis, manufacturing, maintenance, or some other set of expected actions. Two particular types of procedural errors are *errors of omission* (in which one fails to do what is necessary) and *errors of commission* (in which one performs an action that should not be executed).

Errors can include incorrect calculations, invalid assumptions, faulty reasoning, miscommunication, and failure to follow established procedures.

6. Schlager (1994).

Engineers need to design the processes of operation for a system just as carefully as they design the system itself, anticipating the types of human errors that may lead to disaster.

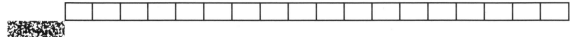

Chernobyl

The 1986 Chernobyl nuclear power plant disaster in the Ukraine was the result of design limitations coupled with process errors during an experiment.[7] Four RBMK reactors were operating at the Chernobyl facility. The former Soviet Union selected the RBMK reactor design because it could perform two tasks: the generation of electricity for domestic needs and the production of plutonium for use in a nuclear weapons arsenal.

Unfortunately, the RBMK design did not include a containment shell that could have prevented radioactive substances from being released into the atmosphere in case of an accident; however, the design allowed the rate of nuclear fission to increase with the loss of cooling water. (The water-moderated reactors used in the United States reduce the fission rate if any loss in cooling water occurs and they include containment shells.)

On April 26, 1986, an experiment was conducted at the Chernobyl facility. Plant operators were testing to see if the kinetic energy of the spinning turbine blades would be sufficient during a power outage to support the cooling pumps in the plant until the emergency generators could take over this task after about one minute had passed. In order to perform this experiment, the operators decided to disable the coolant system. The test also was performed with the reactor operating below 20 percent of its capacity, a level at which any rise in temperature leads to an increase in power output, further elevating the temperature.

As the reactor lost power, the control rods were removed from the nuclear core, leading to an immediate increase in the fission rate. The operators then tried to reinsert the control rods into the channels in order to gain control of the situation. However, the heat had deformed the channels so that the rods could not be reinserted properly. The increasing heat led to steam dissipation from the reactors, further reducing water levels in the core and increasing the power output by a hundredfold in less than a second.

Errors in process led to the deaths of 31 people (299 others were injured), the evacuation of 335,000 people, and extensive loss of crops and farm animals.

Workers and others must be informed when a change is made in a system. Kletz[8] describes a plant accident that resulted from a simple failure in

7. Kletz (1991) and Schlager (1994).
8. See Kletz (1991) and *Petroleum Review* (1974).

communication. Sulfuric acid and caustic soda (alkali) were normally stored in similar containers but in different locations in the plant. One day, it was decided to place one container of each substance in each of these locations for greater convenience. A worker, uninformed of this change, inadvertently mixed the acid with some alkali, leading to a violent chemical reaction.

This case emphasizes the fact that communication is critical! If any change occurs in a work environment or in a specific operating procedure—or if there is any indication of a potential but unexpected hazard—this information must be transmitted immediately to all those who need it. An appropriate communication system (even if it is only something as simple as a bulletin board) should be designed and used for such transmission.

We include among errors in process any failure to anticipate the conditions in which a design will operate. For example, in 1965 three water-cooling towers at Pontefract, England, were unable to withstand the impact of winds, leading to their collapse; in 1973 the meteoroid shield and a solar array were lost from Skylab (NASA's orbiting space laboratory) because of the aerodynamic forces on the vehicle during its launch.[9] These cases illustrate the need for the engineer (as much as humanly possible) to anticipate the actual conditions under which a design will operate and prepare the design to function properly under those conditions.

Other disasters that were the result of flawed processes of operation are discussed in Case History 9.4 *Disasters on the Railroads* and Case History 9.5 *Carelessness Sinks the Ferry*.

9.2.3 ▨ Errors in Perspective or Attitude

Errors in process or procedure often result from a flawed perspective or value system held by the participants in the process, leading to unacceptable and unprofessional behavior. Such errors in perspective comprise the third and most basic level at which sources of failure can be found. In Chapter 2 we reviewed several engineering failures that were the direct result of overconfidence, indifference, arrogance, selfishness, or other forms of focusing upon oneself rather than upon others. Moreover, in Chapter 8 we discussed numerous cases in which an error in judgment or an error in moral priorities led to disaster and product liability lawsuits. Unfortunately, many engineering failures have occurred because of such flawed personal values (see, for example, Case History 9.6 *The* Titanic *Disaster*).

9.2.4 ▨ Sources of Failure: Some Examples

Numerous specific sources of failure[10] can be identified and placed in each of the following three general levels of causative elements. (Please recognize

---

9. Schlager (1994).

10. See, for example, Walton (1991) and Witherell (1994).

that the sources of many failures will actually overlap two or more of these general levels.)

Level One: Physical Flaws

- Metal fatigue
- Corrosion
- Toxicity
- Exposed moving parts
- Excessive noise or vibrations
- Electrical hazards
- Inadequate structural integrity, leading to collapse

Level Two: Errors in Process

- The problem is misunderstood, leading to an incorrect and hazardous design solution
- The design and its implementation are based upon invalid assumptions
- Errors in calculation
- Incomplete or improper data collection upon which design decisions have been based
- Incorrect or faulty reasoning used to develop an engineering solution
- Miscommunication of essential information, constraints, and/or expectations
- Information overload
- Errors in manufacturing
- Error(s) in assembly of the final design
- Improper operation or misuse of a product by the user, misuse that might have been foreseen and prevented by the design engineer
- Failure to anticipate unexpected operating conditions or other developments
- Improper storage
- Errors in packaging
- Carelessness
- Inadequate training of personnel
- Errors in judgment

Level Three: Errors in Perspective or Attitude

- Unethical or unprofessional behavior
- Inappropriate priorities, objectives, and values
- Isolation from others who will be affected by one's work
- Lack of motivation
- Indifference and callousness to others' difficulties or needs
- Overconfidence
- Impulsive behavior or decision making

Any convergence or simultaneous occurrence of two or more of the above causation elements in a single design will often lead to failure. Case History 9.7 *DC-10 Cargo Door Design* describes a situation in which numerous errors converged for disastrous effect.

9.3 Dealing with Hazards

How, then, do we avoid failure? By anticipating and controlling hazards that may be embedded in our designs before they can harm someone.

A *hazard* is any condition with the potential to cause an accident, whereas exposure to such a hazard is known as the corresponding *danger*. One also tries to estimate the *damage* that may result if an accident occurs (i.e., the severity of the resulting injury), together with the likelihood that such damage will occur (the *risk* associated with the hazard). The designer seeks to develop safety mechanisms that will control any hazards that may exist, thereby minimizing the danger, and/or decreasing both the level of damage and the level of potential risk.

See Case History 9.8 *Another Air Disaster* for an instance in which the failure to anticipate a hazardous event (a complete loss of an aircraft's hydraulic control system due to an exploding rear engine) and design against it (perhaps by including an appropriate back-up system) led to disaster.

9.3.1 ▓ Hazards

Some hazards are inherent within a design; for example, the blade of a lawn mower is hazardous by its very nature. Other hazards are contingent upon some set of conditions, such as improper maintenance, unsafe design, or inadequate operating instructions.

Several distinct types of hazards can be associated with engineered designs. For example, machines may have one or more of the following hazards:[11]

- **Entrapment hazards** in which part or all of a person's body may be pinched or crushed as parts move together. Gears, sprockets, rollers, flywheels, belts and pulleys, and other moving components pose such hazards if exposed.
- **Contact (tactile) hazards** in which contact with a hot surface, a sharp edge, or an electrically charged element can cause injury.
- **Impact hazards** in which a person strikes a portion of an object (such as sharp edges and bolts) or a part of the device strikes the person.

11. See Hunter (1992) and Lindbeck (1995).

- **Ejection hazards** in which bits of material (e.g., pieces of metal or particles of wood) from a workpiece or loose components from a machine strike a person.
- **Entanglement hazards** in which a person's clothing or hair can become entangled in a device.
- **Noise and vibration hazards,** which can cause loss of hearing, a loss of tactile sense, or fatigue. Moreover, an unexpected sound may cause a person to respond in a reckless or dangerous manner.

Engineered systems should be examined for all potential hazards and modified as needed in order to ensure the safety of the user. For example, during the kinematic design of machines one should focus upon the various interactions among moving parts in order to eliminate or minimize entrapment and entanglement hazards.

Many other types of hazards may be associated with a design. For example, one may need to eliminate the danger of stored energy in a design because this energy, if unexpectedly released, could injure an operator. Such energy—stored in mechanical springs, electrical capacitors, pressurized gas containers, and other components[12]—should be (a) released harmlessly once the device is deactivated so that the system will be in a so-called zero energy state or (b) contained safely by adding appropriate protective elements (such as locks or guards) to the design.

In addition, environmental and biological hazards associated with a design, its manufacture, operation, repair, and disposal should be identified and eliminated. One type of environmental hazard that has received some attention in recent years is that of oil spills. One quart of oil can pollute as much as 750,000 gallons of water, leading to ecological disaster. In 1989 the Exxon *Valdez* tanker spilled more than ten million gallons of crude oil into Alaska's Prince William Sound, only one of 26 spills of similar magnitude that occurred throughout the world between 1978 and 1990. Other instances of environmental and biological disaster include contamination due to the disposal of radioactive waste near the Columbia River in the state of Washington; mercury poisoning of the residents of Minamata, Japan, as the result of industrial pollution; and ecological contamination from the widespread use of the insecticide DDT.[13]

Finally, recall our discussion in Chapter 8 of the need to foresee all possible uses, misuses, and abuses of a design. Hazards may be lurking within any of these activities.

How, then, does one perform an effective hazards analysis?

9.3.2 ■ A General Methodology for Dealing with Hazards

The following five-step methodology for dealing effectively with hazards has been adapted from the work of Hunter.[14]

---

12. Hunter (1992).
13. Schlager (1994).
14. Hunter (1992).

Step 1: Review existing standards. One should first determine if standards and requirements exist for the product or system under development. As noted in Chapter 8, any industrial or governmental standards represent the minimal requirements (i.e., floor standards) that must be satisfied by a design.

There are many types of standards, including product performance standards, packaging standards, and personal exposure standards (e.g., standards relating to one's exposure to noise or radiation).[15] In the United States, more than 300 organizations have developed specific standards for those products and processes with which they are associated; these organizations include:[16]

- **American National Standards Institute (ANSI)** ANSI publishes broadly accepted standards for products and processes that have been reviewed and/or developed by appropriate groups such as the American Society of Mechanical Engineers (ASME) and the Underwriters Laboratories (UL). Engineers should become familiar with the current ANSI standards that are relevant to their work.
- **American Society for Testing and Materials (ASTM)** ASTM focuses upon the definition of material properties and the appropriate methods for testing these properties. Standards for steel, textiles, fuels, and many other materials can be found in the *Annual Book of ASTM Standards.*
- **Society of Automotive Engineers (SAE)** SAE produces numerous standards for self-propelled vehicles, such as the ergonomics publication SAE J833 *Human Physical Dimensions.*
- **Underwriters Laboratories, Inc. (UL)** Standards for the testing and certification of devices are developed by UL in accordance with specific performance criteria. For more information, contact Underwriters Laboratories, Publications Stock, 333 Pfingsten Road, Northbrook, IL 60062.
- **The United States Government** Many U.S. governmental organizations and agencies issue standards, requirements, and regulations for products and processes. Some important examples include the Department of Defense (DOD), the General Services Administration (GSA); the Occupational Safety and Health Administration (OSHA); the Consumer Product Safety Commission (CPSC); the National Highway Transportation Safety Board (NHTSB); the National Aeronautics and Space Administration (NASA); and the National Institute of Standards and Technology (NIST).

There also are many important codes developed by other organizations, such as Building Officials and Code Administrators International, Inc. (which publishes the BOCA National Building Code) and the National Fire Protection Association (which developed the National Fire Codes, including

15. Lowrance (1976).
16. Hunter (1992).

NFPA 70 National Electrical Code and NFPA 101 Code for Safety to Life from Fire or the Life Safety Code).

Remember that all standards are dynamic and should be expected to change with time. Therefore, it is important for engineers to be familiar with the *current* sets of standards for the design, manufacture, and operation of the particular product or process under consideration. Numerous publications can be consulted in order to ensure that the minimal requirements set by existing standards will be satisfied by a design. The *Index and Directory of U.S. Industry Standards,* (published by Information Handling Services, Englewood, Colo., and available from Global Engineering Documents, 2805 McGraw Avenue, Irvine, CA) identifies thousands of these publications in which current standards for products and processes are described.

Step 2: Identify known hazards.   Given a set of standards for a product type, one should be able to identify the recognized hazards usually associated with the design (e.g., lawn mowers include several inherent hazards commonly recognized by designers). The engineer should then incorporate appropriate elements in the design to either eliminate or minimize the threat posed by these hazards. (Such elements of protection will be discussed shortly.) Many of the existing standards for the product will correspond to these protective aspects of the design. Nevertheless, the engineer should strive to go beyond the existing standards—if it is feasible to do so—in order to eliminate the threat of all known hazards from the product.

Step 3: Identify unknown hazards.   Next, one must seek the unknown or hidden hazards in the design. These hazards include those which have not yet been recognized in the form of industrial or governmental standards that must be followed in order to eliminate them. The engineer must follow a systematic approach to identify these undiscovered booby traps lurking within the design and in its use or misuse by the operator.

Obviously, this step is critical if one is to produce an acceptably safe design. No design is absolutely hazard-free, *but every design should be generally recognized as safe* (known as *GRAS*) when finally implemented.[17] Several techniques can be used to identify the unknown hazards, including hazard and operability studies (HAZOP), hazards analysis (HAZAN), fault tree analysis (FTA), and failure modes and effects analysis (FMEA), each of which will be reviewed in the next section.

Step 4: Determine characteristics of hazards.   One next attempts to determine the frequency, relative severity, and type (e.g., entanglement, contact, impact) of each hazard. By doing so, the designer can focus initially upon those hazards that can result in the most damage (e.g., such dire consequences as disabling injuries or death) and/or that have the greatest risk

17. Hunter (1992).

associated with them (i.e., those that are most likely to occur). Later efforts in striving to develop an acceptably safe design should be directed towards those other hazards that are less dangerous and less likely to occur.

Step 5: Eliminate or minimize hazards. Once the hazards have been recognized and ranked in terms of their relative severity and frequency of occurrence, the designer should focus upon ways through which the most dangerous hazards can be eliminated. If it is not possible to eliminate a hazard, one must then strive to minimize the threat that it presents to the user. Such actions might include:

- Controlling certain hazards through the use of shields, interlock systems, and other safety features.
- Introducing a safety factor into the design—that is, intentionally over-designing a solution to increase the likelihood that it will not fail while in use.
- Developing quality assurance programs in which testing is performed in an effort to identify defective units before they enter the marketplace.

9.4 Hazards Analysis

9.4.1 Hazard and Operability Studies

One fundamental approach to identifying hidden hazards in a design is known as HAZOP (hazard and operability study).[18] In HAZOP, a team of investigators usually is formed, often including design engineers, process engineers, representatives of management, and a HAZOP specialist who can guide the team in its work. The team begins by considering any deviations from the desired performance of a system that may occur, the consequences of such deviations, and the possible causes for such deviations.

HAZOP is a qualitative yet systematic approach that focuses upon the answers to the question "What if (some deviation from desired performance occurs)?" Deviations can be represented in terms of such qualitative descriptors as "more," "less," "none," "part of," "other than," and so forth. These are known as guide words in HAZOP, and they can be used to trigger one's thinking along a particular direction (somewhat similar to the use of trigger words and questions in the creativity technique known as checklisting, which was discussed in Chapter 7). For example, one might describe possible deviations as "there is more waste from the process than is acceptable," "there is less deceleration than necessary," "there is no power output," "part of the

18. See Kletz (1992).

composition of a chemical product is incorrect," or "there is behavior exhibited by the system other than what is expected." Although these descriptions are not quantified, they are sufficient for all members of the HAZOP team to understand the type of deviation that is under consideration.

The team may then try to anticipate the consequences of each deviation; for example, they may ask such questions as:

- Will someone be harmed? Who? In what way? How severely?
- Will the product's performance be reduced? In what way? How severely? What will be the effect of such a reduction in performance?
- Will costs increase? By how much?
- Could there be a cascading effect in which this one deviation leads to another, and then to another, and so on? What, specifically, is this cascading effect?

Once the deviations from acceptable performance and their possible consequences have been delineated, the team then seeks to identify the possible causes for these deviations. They work their way through the design, focusing upon specific scenarios (e.g., leaks, blockages, fractures, inadequate assemblies, environmental conditions, poor maintenance) that could lead to each possible deviation.

Finally, the team develops a set of specific actions that could be performed to eliminate or minimize these deviations. Such actions might include back-up systems, locks, guards, relief valves, increased inspection and maintenance procedures, enhanced operator training seminars and documentation, more precise and comprehensive communication systems, more informative displays, more effective control devices, and/or a system redesign.

In summary, HAZOP is a method for identifying hazards within a design, and it is a particularly fruitful method for identifying flaws within the design for a process.

9.4.2 ▓ Hazards Analysis

Formal hazards analysis (HAZAN) seeks to identify the most effective way in which to reduce the threat of hazards within a design by estimating the frequency and severity of each threat and developing an appropriate response to these threats. Although there are some similarities between HAZOP and HAZAN (e.g., both focus upon hazards, and both try to anticipate the consequences of such hazards), nevertheless there are clear distinctions between the two methods. For example, HAZOP is qualitative in nature, in contrast to HAZAN, which is quantitative. Furthermore, HAZOP usually is performed by a team, whereas HAZAN is performed most often by one or two people. Most important, HAZOP principally is used to identify hazards within a design, whereas HAZAN is a method for assessing the importance of these hazards.

Kletz[19] summarizes the stages of HAZAN in the form of three brief questions:

- How often? (i.e., how frequently will the hazard occur?)
- How big? (i.e., one focuses upon the possible consequences of the hazard)
- So what? (i.e., should action be taken to eliminate or reduce the hazard?)

HAZAN is based upon probabilistic analysis in estimating the frequency with which some threat to safety may occur, together with the severity of its consequences. Through such analysis, engineers can focus their initial efforts toward reducing those hazards with the highest probabilities of occurrence and/or the most severe consequences. However, one should resist the temptation to simply accept other hazards with very low probabilities of occurrence. Some engineers accept small hazards in their designs because they believe that such risks are both acceptable and inevitable. Many times, however, such risks can be reduced substantially or eliminated with very little effort. Case History 9.9 *Deadly Fire in a London Subway* describes a hazard that was ignored because its occurrence was thought to be very unlikely until 31 people lost their lives in a tragic incident.

9.4.3 ■ Fault Tree Analysis

In fault tree analysis (FTA),[20] first developed at Bell Laboratories and later used extensively by NASA, one begins by postulating an undesirable event and then seeks the underlying conditions that could lead to this event. For example, the undesirable event might be a blowout of an automobile tire; all possible sequences of events that could lead to such a blowout are then considered in a systematic manner by creating a fault tree diagram.

Fault tree analysis is similar to HAZOP since one begins, in both techniques, by identifying an undesirable event (deviation), after which the underlying reasons for this event are sought. Furthermore, FTA often is used as part of HAZAN since it is a technique through which a system can be analyzed methodically and in which probabilities are used to identify those events that may occur most frequently.

Figure 9.1 presents some of the basic logic symbols used in fault tree diagrams. In the fault tree diagram shown in Figure 9.2, some of these symbols are used to trace the conditions that could lead to an airplane crash. Rectangular blocks denote events or "faults," the causes of which are then sought by working downwards through the system structure. Preliminary conditions that could lead to these faults are noted, together with their underlying causes. This process continues until one reaches the basic or most fundamental faults that could trigger an entire sequence of events.

19. Kletz (1992, p. 56).
20. Kletz (1992) and Kolb and Ross (1991).

Combination event: Event (often a fault) due to a combination of more fundamental or basic component faults

 Basic fault: Component fault (usually with assignable probability of occurrence)

 Undetermined fault: Fault of undetermined causes (causes remain undetermined because of a lack of data, limited time, or lack of necessity)

 Normal event: Event that is expected to occur

 Reference key: Continuation key to another part of fault tree

 AND-gate: All inputs events must occur for output event to occur

FIGURE 9.1 Examples of fault tree symbology. *Source:* Based upon Kolb and Ross, 1991.

OR-gate: Any one (or more) input event(s) is sufficient to trigger output event

Various gates (e.g., AND- and OR-gates) can be used to trace these sequences of events in diagrammatic form, allowing one to specify logical combinations of conditions that could lead to a particular event. All input or prerequisite conditions to an AND-gate must be satisfied before the gate is triggered, causing its output event to occur. In contrast, the output event of an OR-gate will be triggered if any of the gate's input conditions has been satisfied. By tracing the logical sequencing of events and their preliminary conditions in this way, the engineer can identify potential trouble spots.

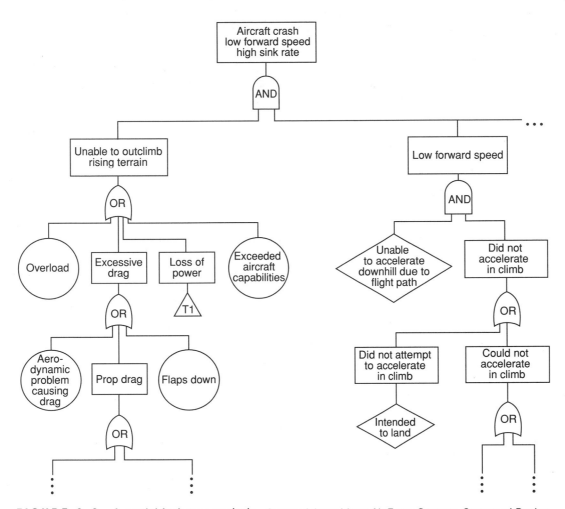

FIGURE 9.2 A partial fault tree analysis. *Source:* Adapted from Air Force Systems Command Design Handbook.

Moreover, probabilities of occurrence for basic faults often can be obtained from test data or from historical records of similar designs, allowing the engineer to determine the likelihood of certain events occurring within a specific time period (due to the existence of one or more basic faults within the system). The designer then can focus upon those faults that are most severe in their consequences and/or most frequent in their occurrences. Appropriate modification of the design should be made in order to eliminate or reduce the likelihood of undesirable events.

9.4.4 ■ Failure Modes and Effects Analysis

In contrast to FTA, when using failure modes and effects analysis (FMEA)[21] to troubleshoot a design, one begins by focusing upon each basic component— one at a time—and tries to determine every way in which that component might fail. (All components of a design should be included in the analysis, including such elements as warning labels, operation manuals, packaging, and advertising.)[22] One then tracks the possible consequences of such failures and develops appropriate corrective actions (e.g., a redesign of the product, the creation of warning labels, improved packaging). In other words, one traces a sequence of events from the bottom upward (i.e., from single components through combinations of these parts to a failure of the system itself), exactly the opposite of the top-down approach used in the FTA method.

Charts of the system's components usually are generated in FMEA. Kolb and Ross (1991) suggest that a format be used through which all components or parts can be listed, together with the following data:

- **Part number** Any code commonly used to identify the component should be noted.
- **Part name** The descriptive name of the component should be given.
- **Failure mode(s)** All ways in which the part can fail to perform its intended function should be identified. More than one failure mode may exist for a given part. Moreover, some failures may be gradual and/or partial, whereas others occur immediately and completely.
- **Failure cause(s)** The underlying reasons leading to a particular failure mode should be recorded.
- **Identification method** The ways in which one could recognize that a particular failure mode has occurred should be described.
- **Back-up protection** Any protective measures that have been designed to prevent the failure should be identified.
- **Effect/hazard** The (usually undesirable) consequences of a failure mode should be described.
- **Severity-frequency-imperilment (SFI)** The severity of a failure mode's consequences upon people, property, or in other ways (e.g., loss of future income) should be noted. Often this is achieved through the use of some severity scale, such as:

  Level 1: Minor repairs necessary, no injuries or property loss
  Level 2: Major repairs necessary, no injuries or property loss
  Level 3: Some property loss, no injuries
  Level 4: Minor injuries and/or large property loss
  Level 5: Major injury
  Level 6: Death or multiple major injuries
  Level 7: Multiple deaths

21. Kletz (1992) and Kolb and Ross (1991).
22. As correctly noted by Kolb and Ross (1991).

TABLE 9.1 Portion of an FMEA sheet.

Part Number	Part Name	Failure Mode	Failure Cause	Identification Method	Back-up Protection	Effect/ Hazard	SFI	Notes
949	Filter	Fails to control flow of coolant	Jams	System overheats	None	Fire; Engine stops; Damage	—	Need warning signal & auto. shut-off; stop jamming!
872	Ratchet	Slips	Loose	Load slips	None	Supported load slips or released	—	Need back-up lock/support
—	—	—	—	—	—	—	—	—
—	—	—	—	—	—	—	—	—

Source: Adapted from Kolb and Ross, 1991.

The preceding scale is given only as an illustrative example. More formal severity scales exist for use under specific conditions or in achieving particular purposes. For example, the National Electronic Injury Surveillance System (NEISS) of the Consumer Product Safety Commission uses a specific severity scale to estimate the relative values of loss associated with various physical injuries (e.g., sprains, punctures, concussions, and amputations).[23]

The expected frequency of occurrence should be estimated for each failure mode, based upon historical or test data.

One also should note the ability of those affected by a failure to respond with appropriate actions that would minimize the negative impact of the failure; this ability corresponds inversely to the imperilment of those affected. For example, a low level of imperilment would correspond to a situation in which a person has more than sufficient time to react to a hazard in order to avoid injury. A higher level of imperilment would correspond to a situation in which only an expert operator would have sufficient time to avoid injury, and the highest level of imperilment would correspond to a situation in which no time is allowed to avoid injury.

■ **Notes** Any suggested actions that could be taken to minimize or eliminate a hazard, or prevent a failure mode, should be noted, together with any other pertinent comments or remarks about the component.

By using the preceding (or some similar) format, an inventory of all parts, failure modes, and hazards can be captured on an FMEA chart. As an illustration of this format, we have applied it to single modes of failure for a filter and a ratchet as shown in Table 9.1. (A more comprehensive analysis would list

23. Kolb and Ross (1991).

multiple modes of failure, together with their possible causes, for these components.) For simplicity, we have intentionally neglected to include values for severity, frequency of occurrence, and imperilment (SFI).

Fault tree analysis is a top-down approach, whereas FMEA is directed from the bottom upwards. A combination of the two methods sometimes can be most beneficial in identifying the hazards associated with a design and the underlying reasons for these hazards.

Using any of the methods discussed here, hazards analysis is a critical activity in engineering design. Case History 9.10 *The MGM Grand Hotel Fire* describes a situation in which unanticipated system failures led to the deaths of 85 people—a tragedy that might have been prevented with appropriate hazards analysis.

9.4.5 ■ Ishikawa (Fishbone) Cause–Effect Diagnostic Diagrams

A simple graphical tool that can be used to diagnose a design for inherent weaknesses or hazards is the Ishikawa cause–effect diagram, also known as the fishbone diagram because of its appearance. Figure 9.3 presents an example of an Ishikawa diagram for an automobile design undergoing development. (The question marks in the diagram indicate that more information needs to be added.) The system (in the example, the automobile itself) is represented by the primary spine of the diagram. Each secondary rib denotes a specific component or subsystem in the design (e.g., fuel system, tires, exhaust). Finally, one lists as many potential flaws, weaknesses, or hazards as possible along each of these secondary ribs. For example, tires can be affected negatively by valve leaks, low pressure levels, and punctures.

The value of an Ishikawa cause–effect diagram in hazards analysis is that one can obtain a broad perspective of a system, its components, and its inherent trouble spots quickly and easily.[24] A fault tree could then be generated in order to (a) focus upon possible causation factors for each flaw or weakness, (b) identify interdependencies of the system components and possible behavior patterns, and (c) incorporate probabilities of occurrence for each event.

9.5 Hazard Avoidance

Once the known and unknown hazards in a design have been identified, work can begin on eliminating or reducing these hazards.[25]

It always is preferable to eliminate the hazard from the design if possible. For example, the ground terminal on three-prong electrical plugs is

---

24. An Ishikawa diagram also can be used to generate specific design goals for a system. One first identifies the weaknesses or flaws in the current design by use of the diagram, then the weaknesses are inverted to form a set of corresponding specific goals for an improved design.

25. See, for example, Hunter (1992).

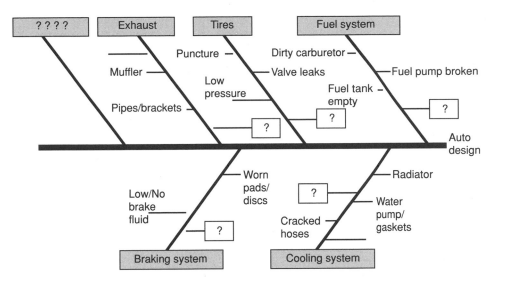

FIGURE 9.3 An example of an Ishikawa diagram.

round whereas the two conducting terminals are flat, thereby ensuring that the plug will be inserted correctly into an outlet. Similarly, 3.5-inch computer diskettes are designed so that they can be inserted easily into disk drives in only one way (the correct way), thereby eliminating damage to the disk or drive. Unfortunately, sometimes the elimination of hazards is not possible, in which case one must reduce the operator's (and other people's) exposure to the hazard by various means.

For example, one might use shields, interlock systems, and other safety features to protect or distance the operator from the hazard, thereby reducing her or his exposure to the danger. Shields or guards prevent accidental contact between the operator and the hazardous parts of a device (such as the self-adjusting guards on hand-held power tools). Interlocks prevent the operation of a device until the user is safely out of the danger zone. Sensors can be used in a device to detect the presence of a person, causing the machine to shut off.

Moreover, a safety factor can be incorporated into the design by intentionally overdesigning a solution to increase the likelihood that it will not fail while in use. A safety factor is the ratio between the presumed strength of a design (i.e., its ability to perform its given function) and the load expected to be applied to it when in operation. Hence, a design is theoretically safe when its safety factor is greater than one since its capability should then exceed any loads that it will need to support. Unfortunately, one never can be certain of the actual loads that will be applied to a design once it is in operation; the unexpected may occur. In addition, the presumed strength of a design may be in error due to invalid assumptions, incorrect calculations, a manufacturing defect, or some other error. Hence, it is important to provide a sufficient margin of safety in order to ensure that no one will be harmed and that nothing will be damaged if unexpectedly large loads are applied or if a design is unintentionally weak. For example, high-speed elevator cables are designed with a safety factor equal to 12, whereas structural steel designs call for safety factors of 1.65 for tensile stresses.[26]

One also can develop quality assurance programs in which testing is performed in an effort to identify and remove defective units before they enter the marketplace (see Chapter 11). Moreover, redundancy (back-up safety or other critical systems) can be included in a design, in case the primary system fails to function. (See Case History 9.11 *Deadly Boeing 747 Blowout* for a situation in which redundancy might have saved nine lives.)

Finally, if a hazard cannot be eliminated or further reduced, one should prepare an appropriate set of warnings and instructions for the operator so that he or she can take precautions to avoid the danger. Warnings must be intelligible to the user, easily visible, and complete, noting all hazards, the possible consequences of disobeying the warning, and the preventive measures that one should take in order to avoid harm. Such factors as size, color, shape, placement, and durability of the warning must be considered in order to ensure that the user is aware of the message and obeys it.

Warning labels usually begin with a signal word that indicates the level of possible damage (the severity of the injury) and the risk (the likelihood that injury will occur). The word *caution* is used for situations in which relatively minor injuries could result, *warning* for situations in which severe injury or death could occur, *danger* to alert users of situations in which severe injury or death will occur if the warning is not heeded.[27]

Next, the nature of the hazard is indicated, followed by the damage that could result if one fails to heed the warning. Finally, instructions are provided for avoiding the hazard. For example, Wolgalter, Desaulniers and Godfrey (1985) provide the following example of a minimal warning:

Signal word	→	DANGER
Nature of hazard	→	High voltage wires
Possible damage	→	Can kill
Instructions	→	Stay away

Warning labels should be brief. Otherwise, the reader can succumb to warning overload in which the message contains so much detailed information about every possible hazard that it goes unheeded.

---

26. Kaminetzky (1991), Shigley and Mitchell (1983), and Walton (1991).
27. See Fowler (1980) and Sanders and McCormick (1987).

However, warnings also should be comprehensive and identify the specific hazards associated with a given situation or product. Consider the following warning:[28]

> Caution: Flammable mixture. Do not use near fire or flame. Caution! Warning! Extremely flammable! Toxic! Contains naphtha, acetone, and methyl ethyl ketone. Although this adhesive is no more hazardous than dry-cleaning fluids or gasoline, precautions must be taken. Use with adequate ventilation. Keep away from heat, sparks, and open flame. Avoid prolonged contact with skin and breathing of vapors. Keep container tightly closed when not in use.

This warning might seem sufficiently comprehensive. However, in the case of *Florentino v. A. E. Daley Manufacturing Company,* the court found that the above warning was inadequate because it failed to note that one should not use the product near closed and concealed pilot lights. Clearly, it can be difficult to find the proper balance between desired brevity and necessary specificity when designing a warning label, but it must be done.

It is best to eliminate a hazard if at all possible. Otherwise, one should strive to reduce the exposure of the user to the hazard by (a) using appropriate guards, sensors, interlocks, and other mechanisms to distance the user from the hazard, (b) increasing the safety factor for the design, (c) incorporating redundancy into the design, or (d) using quality assurance efforts to minimize the number of defective units that enter the marketplace. The use of warnings is the least effective approach to hazard reduction, and it should be used only as a last resort or in combination with other approaches.[29]

SUMMARY

- Commercial products should be evaluated for such hazards as entrapment, impact, ejection, and entanglement.
- Hazards analysis (HAZAN), fault tree analysis (FTA), and failure modes and effects analysis (FMEA) can enhance the safety of engineering designs.
- It is best to anticipate and eliminate hazards. Further, engineers reduce the exposure of the user to hazard by using mechanisms to distance the user from the hazard, increasing the safety factor, incorporating redundancy, and using quality assurance efforts to minimize the number of defective units that are manufactured and sold.

28. See Kreifeldt and Alpert (1985) and Sanders and McCormick (1987).
29. Hunter (1992).

- The use of warnings is the least effective approach to hazard reduction; it should be used only as a last resort or in combination with other approaches.
- Safety factors are used to ensure that engineering designs will be able to withstand expected loads when in operation.

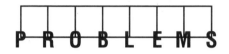

PROBLEMS

9.1 a. Evaluate a product of your choice and identify possible hazards in its use and misuse. Sketch the product and label all such design hazards.
   b. Redesign the product in part (a) so that the identified hazards are eliminated. Sketch your redesigned version of the product, labeling all improvements in safety.

9.2 Using one of the techniques described in this chapter, perform a hazards analysis upon one of the following products and redesign it so that hazards in its use and misuse are eliminated. Sketch your redesigned product, labeling all improvements in safety.
   - Lawn tractor
   - Videocasette recorder
   - Refrigerator
   - Table lamp
   - Doghouse
   - Bicycle

9.3 Investigate a design failure of your choice and identify the cause(s) for this failure. Select a case that is not discussed in this text.

9.4 Research and prepare a brief report on a product which endangered users because of one of the following types of hazards:
   - Entrapment
   - Ejection
   - Entanglement
   - Noise and/or vibration
   - Environmental or biological hazard
   Discuss the specific hazard, its underlying cause, and its elimination or minimization through a redesign of the product (if such redesign occurred).

9.5 Identify some of the most important safety standards that must be met in the design of one of the following products or systems. List each of these standards and explain its importance in the development of an effective design. (Hint: Consult the specified ANSI reference guide that corresponds to this product.)
   - Portable metal ladders (ANSI Standard A14.1)
   - Walk-behind power lawn mowers (ANSI Standard B71.3)

- Woodworking machines (ANSI Standard O1.1)
- Laundry machinery (ANSI Standard Z8.1)
- Bakery equipment (ANSI Standard Z50.1)

9.6 Perform a preliminary hazard and operability study (HAZOP) on a process that you can carefully observe (e.g., the tasks performed at a local hospital; the operations of a grocery or retail store; the processing of books that have been returned to a library). Report your findings in tabular form and suggest ways in which to improve the observed process.

9.7 Perform a fault tree analysis (FTA) upon a product or system of your choice. Consider two or more possible deviations from the system's desired behavior and identify the underlying causes that could lead to these deviations. Present your analysis in the form of a fault tree diagram, indicating the sequence of events that could lead to the deviations that were considered.

9.8 Perform a failure modes and effects analysis (FMEA) on a system of your choice. Determine the possible failure modes for each component in the system and present your results in the form of a FMEA chart. Include as many details in your analysis and the resultant chart as possible.

9.9 Prepare an Ishikawa cause–effect diagram for one of the following systems. Consider all components and the ways in which each component might fail to perform its function(s).
- Desk chair
- Tennis racket
- Dishwashing machine
- Sewing machine
- Mechanical pencil
- Door lock
- Automobile theft-prevention system
- Backpack or book bag
- Photocopier
- Bicycle

9.10 Evaluate an existing product and identify the safety features of its design. Prioritize and list these elements from most important to least important. Submit this list, together with a brief explanation of your rationale for this particular order of priority.

9.11 Identify an existing design in which redundancy is used to increase its safety and/or reliability. Describe the redundant (primary and secondary) systems used in the design.

9.12 Evaluate a warning label on a product of your choice. Discuss the adequacy or inadequacy of this label for protecting the user from harm. Design an improved warning label for this product and compare your version with the current label; identify relative weaknesses and strengths in both versions.

CASE HISTORY 9.1

The *Alexander Kielland* Floatel

The *Alexander Kielland* originally was designed to serve as a five-legged semi-submersible offshore drilling rig, to be towed on five pontoons from one ocean location to the next as it searched for oil. At each site, the rig would be submerged partially and stabilized by filling its pontoons with water. If oil was found, the drilling hole would be capped until a permanent rig could be constructed, and the semi-submersible would move on to drill new holes elsewhere.

However, technology in offshore oil drilling operations progressed so quickly that the *Alexander Kielland* was deemed obsolete even before its construction could be completed in 1976. Although never to perform its original function as a semi-submersible, it was retrofitted and given a new role as a "floatel," a floating hotel for oil workers who would commute by helicopter between their drilling rigs and the floatel. The reconfigured *Alexander Kielland* could house up to 366 guests comfortably; it included a motion picture theater, saunas, and other amenities.

At approximately 6:20 P.M. on the evening of March 27, 1980, a pre-existing crack in one of its leg struts led to the capsizing of the entire floatel. Winds were between 40 and 65 miles per hour with waves up to 25 feet high. Two hundred and twelve men were aboard the *Alexander Kielland* when disaster struck that night; only 89 survived.[30]

Struts are used to interconnect the legs of a structure, thereby increasing stability. During construction of the *Alexander Kielland,* a plate had been welded onto one of its leg struts to allow attachment of a hydrophone to the rig. (A drilling rig uses a hydrophone to navigate above a submerged well cap. The cap transmits an electronic signal to the hydrophone, allowing the rig to position itself properly with respect to the cap.) This welding resulted in a three-inch crack in the strut, a crack that must have existed before the rig's final coat of paint was applied since paint was found inside the crack following the disaster.

Once the *Alexander Kielland* was deployed at sea as a floatel, this pre-existing and undetected crack inexorably grew with each wave that struck the leg. After millions of such wave cycles, the crack had grown so large that the strut itself finally failed—that is, *fatigue failure* had occurred due to the cyclic loads being applied to the structure.

However, even if one of its legs failed, the *Alexander Kielland* should not have capsized so quickly. It had been designed to list slowly in such an instance, providing those on board with the opportunity to reach safety before it finally capsized. Its retrofitting into a floatel may have made its deck top-heavy, leading to the unexpectedly swift overturning.

30. See Petroski (1985) and Schlager (1994).

- Poor welding and inspection leading to fatigue failure (Error in implementation: manufacturing, inspection)
- Potentially hazardous retrofitting (Error in design due to invalid assumptions and faulty reasoning)

Frequent inspection for crack growth in a metal structure must be performed if fatigue failure is to be avoided, particularly if extensive cyclic loading is expected. Following the *Alexander Kielland* incident, more stringent regulations for the inspection of offshore drilling rigs were instituted. In addition, rigs were designed to more effectively withstand fatigue loading by the ocean's waves.

CASE HISTORY 9.2

Two Boeing 747 Crashes

In 1992 the crash of a El Al Israel Airlines Boeing 747-200F (freighter) jet in Amsterdam underscored the difficulty in accurately assessing the relative desirability of alternative design goals.[31]

An engine that slowly tears itself loose from an aircraft's wing can be more hazardous than one that quickly separates from the plane (e.g., a gradual separation may lead to more severe wing damage and/or the ignition of fuel stored within the wing). Because of such concerns, Boeing aircraft engines are designed to separate entirely and rapidly from the plane if they begin to break away for any reason (e.g., a direct impact between the engines and the ground in a forced belly landing). This is accomplished through the use of *fuse pins,* which are designed to withstand typical in-flight forces but fracture completely if an unexpectedly high load suddenly is applied to them.

Unfortunately, an engine may tear loose if one or more of its fuse pins become cracked or weakened. Hence, an alternative design is found in planes manufactured by Airbus: Fuse pins are not used on these aircraft and engines are expected to remain attached to the planes under all circumstances.

The dilemma, then, is to decide which of these two approaches is inherently safer: (a) use fuse pins to design against additional damage to a plane because of a slow separation of an engine or (b) avoid fuse pins because they may become cracked, thereby causing the loss of an engine.

31. See Acohido (1994a).

Fuse pins are basically hollow steel cylinders that are used to secure an engine strut to a wing. The danger of cracks appearing in these pins because of corrosion had been recognized as early as 1978. Machining of the old-style pins (those manufactured before 1981) created tiny imperfections in the metal; moisture due to condensation, liquid cleaners, and other sources then could collect in these tiny crevices, leading to corrosion and gradual crack growth. As a result, these pins required relatively frequent inspection and if necessary, replacement.

New-style fuse pins then were designed in which inserts were used in order to eliminate machining of the internal surfaces. Unfortunately, these new-style pins also became subject to corrosion when the inserts were attached, removing bits of the anti-corrosive primer coating that had been applied to the pins. This problem first became apparent in the mid-1980s when an inspector noticed that an engine was sagging on a 747 due to cracking of a new-style pin. In May 1991, the Federal Aviation Administration ordered airlines to inspect all new-style pins on their 747s for damage. In December of that year in Taiwan, a China Airlines 747-200 F with recently installed new-style pins lost its right inboard engine. This engine then struck and dislodged the right outboard engine, sending the plane into its final fatal arc. All five people on board were killed.

Finally, after another near-accident due to a crack in a new-style pin, Boeing prepared an advisory that was to be issued on October 8, 1992. In this message to all airlines operating 747s, Boeing would recommend that all new-style pins again be inspected for cracks and corrosion.

Tragically, the Amsterdam crash occurred days before this advisory was to be issued. On October 4, 1992, a right inboard engine broke free on the El Al Boeing 747-200F and struck the right outboard engine, tearing it loose. As in the crash of the China Airlines jet only 17 months earlier, the loss of both engines sent the plane into an arc until it finally plunged through a 10-story apartment building. All six crew members, together with more than 50 people on the ground, perished.

SUMMARY OF FAILURES AND TYPES OF ERRORS

- Failure to inspect and correct system components in a timely manner (Error in implementation: maintenance and repair)
- Failure to anticipate hazardous event (i.e., one engine striking another so that both are lost) (Error in design due to inadequate knowledge, invalid assumptions, faulty reasoning, and/or incomplete analysis)

Subsequent inspection of 747s found numerous fuse pins with cracks and/or significant damage due to corrosion. Replacement of these damaged components may have prevented additional tragedies.

In addition, the possibility that one engine may dislodge another is now clearly recognized. And once a problem is recognized, one can begin to develop its solution.

Finally, we are left to ponder whether the advantages of fuse pins that are subject to corrosion outweigh their danger while work continues on the development of a truly corrosion-proof pin.

CASE HISTORY 9.3

The United States' Worst Bridge Disaster

The total collapse of the Point Pleasant Bridge (linking Point Pleasant, West Virginia, with Gallipolis, Ohio) on December 15, 1967, remains the worst bridge failure in U.S. history.[32] The collapse occurred during rush hour while the bridge was jammed with traffic. An estimated 75 vehicles plummeted 80 feet into the Ohio River and its shoreline below the bridge, resulting in 46 deaths and scores of injuries.

The impressive structure was nicknamed the "Silver Bridge" in recognition of its rust-resistant aluminum paint. The 1,460-foot-long suspension bridge with a central span of 700 feet opened in 1926 using a chainlink support structure rather than the more traditional wire-strand cable system. Each link in the chain was formed from a pair of steel eyebars, each of which was between 25 and 30 feet in length. Twelve-inch diameter pins were inserted through the holes in the enlarged ends of the eyebars, thereby fastening the links sequentially to one another to form the bridge chains. Finally, bolted cap plates were used to hold the pins in place.

In addition to possible aesthetic advantages over wire-strand cable systems, the Silver Bridge chain design offered presumed financial and structural enticements. Indeed, it did prove to be less costly than comparable cable systems. It also was used as part of the top chord of the bridge's stiffening trusses.[33]

---

32. See Dicker (1971), Levy and Salvadori (1992), Petroski (1985), Robison (1994b), and Scheffey (1971).

33. Stiffening trusses tend to offset the horizontal forces applied by a suspension system when bedrock is not available for anchoring the structure. (Concrete troughs on reinforced concrete piles were used as the anchors of the Point Pleasant Bridge.) The applied live load is distributed more evenly throughout the overhead cable or chain via such trusses, thereby minimizing the downward deflection of the bridge deck at its point of contact with the load.

Unfortunately, the design's presumed advantages ultimately were outweighed by its significant shortcomings. A crack on a single eyebar along the northwest portion of the chain system was discovered following the collapse. Once this crack formed fully, the eyebar failed, causing a chain reaction as the cap at the joint was pried loose by the twisting motion of the remaining eyebars under unbalanced loads. Other links then broke loose along the length of the now twisting chain, leading to the total collapse of the bridge.

Why did the eyebar crack? Although friction between the critical eyebar and a pin apparently resulted in a high concentration of stresses at the point of failure, this applied stress was still well below the limit that theoretically could be supported by the steel eyebars. However, following the collapse it was discovered that the bridge's heat-treated carbon steel eyebars were quite susceptible to stress corrosion. This type of gradual corrosion weakened the eyebar until it failed in brittle fracture.

The cracked eyebar presumably was composed of ductile (elastic) steel, yet it fractured in a brittle manner, suddenly and without warning. Stress corrosion can lead to such fracturing if there are microscopic holes in the material. These holes can grow as stresses are applied to the structure if the material lacks "toughness," that is, if it is not sufficiently resistant to stress corrosion. (Ductility and toughness decrease with temperature; at −30°F, steel can be shattered like glass.) Metallurgical examination of the eyebar revealed that below its ductile surface lay more brittle material. When formed into an eyebar, the steel had undergone forging and hammering that increased its hardness. Heat treating was then used to restore its ductility. Unfortunately, the subsequent cooling of the eyebar was performed too rapidly and its inner region remained brittle. Tiny holes in the brittle layer gradually grew under the stress of numerous winter/summer temperature variations. This cyclic temperature-induced loading would have been insignificant if the inner material had been soft; but because the steel was relatively brittle, it became susceptible to stress corrosion, leading to its final and fatal cracking.

Although its unusual design made it difficult to inspect the bridge, this gradual cracking due to stress corrosion could only have been detected in 1967 by completely disassembling the entire eyebar joint, according to the final report on the disaster by the Federal Highway Administration (FHWA).

Furthermore, the large vibrations of the bridge (due to the inability of the steel to stretch sufficiently in order to dampen such motion) may have contributed to metal fatigue and the final disastrous failure.

Experts disagreed for some years following the collapse of the Silver Bridge about the source of the initial eyebar crack. However, it now seems reasonable to conclude that stress corrosion was the primary cause of this crack with other factors (e.g., the inelasticity of the high-strength steel and a high concentration of forces at a single point) contributing to the failure.

Finally and perhaps most important, one could argue that the bridge design itself—and not the crack—was the true cause of the collapse. As noted by Petroski (1985) and Levy and Salvadori (1992), there were no alternate paths along which the loads could be redistributed once the first link failed. This lack of intentional redundancy in the bridge's design meant that once one link had failed, total collapse became inevitable. Thus the failure was not due to the eyebars' lack of ductility that allowed stress corrosion to develop; instead, it resulted from the absence of alternate load paths in the bridge design itself.

SUMMARY OF FAILURES AND TYPES OF ERRORS

- Failure to include alternate load paths in a structural design (Error in design due to invalid assumptions and faulty reasoning)
- Failure in fabrication processes to prevent flaws in material, leading to stress corrosion and brittle fracture (Error in implementation: manufacturing)

Specific responses to the Point Pleasant disaster included the dismantling and replacement of the Saint Marys Bridge in West Virginia (the twin of the Silver Bridge). Numerous other bridges either were closed or their use was restricted.

More global consequences included the first nationwide inspection of all bridges (more than 700,000 in all). Another was a renewed focus upon the properties of materials that may be used in structural designs. In addition, bridge inspection procedures were reevaluated and upgraded.

Most important, the Silver Bridge collapse reminds us of the need to include alternate load paths in structural designs if we hope to avoid disaster.

CASE HISTORY 9.4

Disasters on the Railroads

In 1915 a railroad signalman (whom we will call signalman A) rode on a slow northbound train to his station at Quintinshill, England (near the border with Scotland), where he was about to begin his shift. He did not know that England's most tragic railway accident was about to occur because he would forget about this train upon which he was riding.[34]

An express train was traveling north toward Glasgow. Normally, the slower northbound passenger train would be moved to a side loop line in

---

34. See Hamilton (1969) and Kletz (1991).

order to allow the express to pass; however, freight trains were already located on both side lines. Instead, the slower train was backed onto the southbound line as signalman A debarked from it.

Signalman B then went off duty, having been replaced by A. According to standard operating procedures, B should have placed a "reminder collar" on the signal lever to remind A of the presence of the stopped train on the southbound line; however, B failed to tag the lever with the reminder collar and he did not mention the presence of the train on the up line to A.

Signalman B may have thought that there was simply no need to remind A of the location of this train; after all, A had just left this train moments earlier and surely knew of its presence. Unfortunately, A forgot about this waiting train and allowed a southbound train carrying military troops to continue forward until it collided into the waiting passenger train. The accident then grew to disastrous proportions when the northbound express train crashed into the tangled wreckage of the first collision. Fire spread through the wooden coaches of the troop train. The final death toll was 226 people, mostly soldiers.

Since England was involved in World War I at the time, the loss of so many military personnel was particularly devastating. Both signalmen were imprisoned for their actions.

SUMMARY OF FAILURES AND TYPES OF ERRORS

- Inadequate efforts to ensure that procedural operations are obeyed (Errors in design and implementation: invalid assumptions and faulty reasoning)

The failure of signalman B to collar the signal lever was in fact a failure to follow standard operating procedures (SOPs). Many such procedures are designed to overcome occasional memory lapses that can lead to system failure. Since anyone can have a faulty memory at times (or can be easily distracted), these procedures must be obeyed.

Nevertheless, people do sometimes fail to follow procedures. In recognition of this fact, modern engineered systems often include automatic failure-avoidance devices that are designed to prevent accidents such as the Quintinshill train wreck. For example, many railroad lines are now track-circuited, meaning that the presence of a train on a line completes an electrical circuit that then prevents the clear signal to be given to any other train.

Unfortunately, accidents can occur even in the presence of such failure-avoidance systems. In 1979 a freight train was blocking the path of an oncoming passenger train. Once again the signalman on duty forgot about the presence of the waiting freight train. Track-circuiting prevented the clear signal from being given to the passenger train. Unfortunately, the signalman thought that the electronic track-circuiting system was faulty and used a green hand signal from his signal box to direct the passenger train onward until it crashed into the freight train. As Kletz (1991) notes, this accident was due to another failure in following standard operating procedures: The signalman should have checked the electronic display in his signal box, which clearly indicated the presence of the waiting freight train. Furthermore, he should have given the green hand signal while standing beside the track itself (not from inside his signal box), which then might have allowed him to see the freight train.

Simply stated, automatic failure-avoidance devices should be incorporated into engineering designs whenever possible. However, most automatic systems are designed to be manually overridden in the event of a system failure. Because of this, operators must be strongly encouraged to follow those procedures that are meant to prevent accidents due purely to human error. In many instances, such encouragement should be *more* than simply vocal recommendations: Operators should be required to obey procedures through the system's inherent design. For example, in the event of a failure in an automatic track-circuiting system, signal operators should be required to manually throw a switch at trackside before giving a green hand signal to an oncoming train. Such a requirement would force signal operators to stand beside the tracks in such situations. Although accidents due to operator errors might still occur, their frequency likely would be reduced.

CASE HISTORY 9.5

Carelessness Sinks the Ferry

In 1987 the automobile ferry *Herald of Free Enterprise* left Zeebrugge, Belgium, and headed across the English Channel towards Dover with its inner and outer bow doors open. One hundred and eighty-six passengers and crew members lost their lives when it sank beneath the waves.[35]

The assistant bosun failed to close the doors because he slept through the public address announcement that the ferry was about to set sail. No adequate monitoring system was used to ensure that the doors had been closed. (There even may have been an unfounded sense of security that the doors could remain open without danger for ships had successfully—although inadvertently—sailed with open doors in the past.) The captain sailed unless informed of a reason not to proceed. The official report stated that "from top to bottom the body corporate was infected with the disease of sloppiness."[36]

---

35. See Kletz (1991).
36. MV Herald of Free Enterprise: Report of Court No. 8074: Formal Investigation, Department of Transportation, HMSO (1987), as quoted in Kletz (1991).

- Inadequate hazard prevention due to failures in design, monitoring, inspection, and system management (Errors in design and implementation: invalid assumptions and faulty reasoning)

Anyone can be lax at times about performing a routine task. When such tasks are critical to the safe and proper operation of an engineered system, an appropriate monitoring process (automatic and/or manual) should be designed and used.

CASE HISTORY 9.6

The *Titanic* Disaster

In 1912 the *Titanic* was acclaimed as the first unsinkable ocean liner.[37] The worst possible scenario was that a collision (be it with another ship or with an iceberg) might occur at a point in the hull where two of the *Titanic*'s sixteen watertight compartments met, flooding these two compartments. However, the ship had been designed to float even if four of its vertically aligned compartments were to become flooded; hence, the confidence of all involved that this was indeed a truly unsinkable ship.

Unfortunately, the *Titanic* struck an iceberg in April 1912 on its maiden voyage. If this had been a direct collision at the point of the bow, it is likely that no more than two compartments would have flooded. Instead, the collision was indirect as the ship's crew, who had spotted the iceberg too late to avoid it completely, desperately turned the ship away from the berg. In the indirect collision, the iceberg acted as a giant razor, ripping open five of the ship's compartments and dooming the great ship to the ocean floor.

Yet this unexpected calamity need not have led to the loss of 1,522 lives. This tragic loss of life was due not to the sinking of the ship but to the inadequate number of lifeboats carried on board. Although the *Titanic* was lavishly designed and constructed for the comfort of its passengers (particularly those in first class), an abundance of lifeboats was not among its many luxuries. British regulations required the *Titanic* to provide only 825 lifeboat seats—far too few for the 2,227 passengers and crew members on board that night. (Indeed, the tragedy could have been much greater: The actual capacity of the *Titanic* was 3,547!)

37. See Davie (1986), Lord (1976), Martin and Schinzinger (1989), and Wade (1980).

The shortage of lifeboats was the result of two factors: First, the then-current safety regulations were not written for any ship as vast as the *Titanic*; the shipline's compliance with these regulations meant that only the minimal number of lifeboats had to be carried by the giant ocean liner. Second, and perhaps more important, the number of lifeboats was kept to a minimum because the ship itself was thought to be unsinkable; after all, who aboard the *Titanic* would ever need a lifeboat?

SUMMARY OF FAILURES AND TYPES OF ERRORS

- Inadequate precautionary measures (Errors in design and implementation: faulty assumptions, arrogance, and overconfidence)

The designers of the *Titanic* failed to learn a lesson from a past tragedy. Some years before, most of the passengers and crew of the steamship *Arctic* perished because there were simply too few lifeboats on board. One might argue that it was reasonable to assume that lifeboats would never be necessary for all of the *Titanic*'s passengers and crew. However, such an argument fails to recognize that overconfidence in engineering design has often led to disaster. We must learn from mistakes and not simply rationalize that they could never happen to us.

The sinking of the *Titanic* and its huge loss of life rocked the civilized world in much the same way that the *Challenger* space shuttle disaster stunned the public nearly 74 years later. Both vessels were thought to be safe, but such thoughts were in fact based upon faulty assumptions and a misunderstanding of the limits of technology. People were shocked to discover that technology can indeed fail and that there remain inherent risks in all the works of man. Overconfidence can quickly lead to disaster.

CASE HISTORY 9.7

DC-10 Cargo Door Design

Dan Applegate served as the senior engineer during the development of the DC-10 fuselage by Convair (a subcontractor for McDonnell–Douglas). After carefully analyzing the fuselage design, Applegate recognized that the following sequence of events could occur:[38]

38. See Eddy, Potter, and Page (1976), Godson (1975), Newhouse (1982), Perrow (1984), Martin and Schinzinger (1989), and Serling (1994).

1. Due to various design flaws, the plane's cargo doors could open abruptly during flight.
2. The cargo hold below the passenger cabin would then depressurize.
3. Next, the floor of the passenger cabin would collapse.
4. The collapse of the cabin floor would then break the jet's hydraulic control lines running below this floor.
5. The plane then could no longer be controlled by the pilot.

Applegate sent a memorandum to Convair's vice-president, detailing his concerns and recommending that the cargo doors and cabin floor be redesigned to eliminate these hazards. Management did not dispute the technical aspects of Applegate's analysis; however, they chose not to inform McDonnell–Douglas of these hazards because of the financial liabilities that Convair (and McDonnell–Douglas) would incur if all planes were grounded while design modifications were made.

The 1972 Near Disaster   McDonnell–Douglas became aware of the danger during a 1970 pressurization test in which a rear cargo door was blown out, causing the cabin floor to buckle. A safety device was added that was meant to prevent pressurization of the cabin if the door was not properly sealed. Nevertheless, in 1972 a door did open on an American Airlines flight departing from Detroit, leading to near disaster. The sudden decompression caused a 17-foot hole to form in the fuselage, the cabin floor to partially collapse, and some of the hydraulic control lines to break. Control over both the rudders and stabilizers immediately was lost, only control of the ailerons and (to a limited extent) the elevators remained. Captain Bryce McCormick, the heroic pilot of the now disabled jumbo jetliner, managed to land the plane safely by first shutting down the center engine and then increasing the thrust provided from the other two engines.

Following the 1972 near disaster, the National Transportation Safety Board (NTSB) recommended that the Federal Aviation Administration (FAA) insist that the door be redesigned. McDonnell–Douglas installed a metal support plate to prevent further accidents (a less costly solution than totally redesigning the cargo door). In addition, a one-inch diameter peephole was added to the door, which would allow baggage handlers to see whether a lockpin was seated properly once the door had been closed, thereby (presumably) ensuring that the door itself was secured. Finally, a warning light was expected to alert the pilot if a door had not been sealed properly.

The 1974 Turkish Airlines Crash   Even the simple addition of a metal plate takes time to accomplish (e.g., 90 days for installation on all DC-10s of United Airlines, 268 days for American Airlines planes, and 287 days for Continental Airlines planes). Tragically and most important, the support plate was never installed (due to both assembly and inspection errors) on all DC-10s, including the Turkish Airlines jumbo jetliner that crashed near Paris on March 3, 1974, killing 346 people. A baggage handler had improperly closed the aft cargo door prior to take-off from Orly Airport in Paris. The plane's departure had been rushed so that it would be allowed to land in

London before an air traffic controllers' job action could result in cancellation of the flight. The baggage handler was unfamiliar with the DC-10 door design and he could not read either of the two languages in which were printed instructions for properly closing the door (a communication error); ironically, he could speak in two other languages. He forced the door closed, thinking that the lock had properly seated, which it had not, and failed to check the lockpin position by peering through the peephole. (Apparently he did not realize the function of the peephole due to failures in communication and training.) In addition, the warning light system malfunctioned by sending a false signal to the cockpit instrument panel, indicating that the door had been secured properly and locked (an implementation error).

The final failure was that of the safety system meant to prevent pressurization of the cabin if a door remained unsealed (an implementation error causing a malfunction). As the giant plane lifted off, the cargo door abruptly opened, leading to the chain of events that had been detailed long before by Applegate. All control systems—both primary and backup—were disabled and the jetliner plunged into the Ermemonville Forest at nearly 500 mph.

It was the first crash of a DC-10, but it was not totally unexpected. Numerous design, implementation, and management actions had converged for disaster.

SUMMARY OF FAILURES AND TYPES OF ERRORS

- Failure to inform customers/public (Errors of intent and implementation)
- Failure to anticipate consequences of design (Error in design: faulty reasoning)
- Failure to retrofit and install modified design elements (Error in implementation: assembly and inspection)
- Failure to inform and instruct operators (baggage handlers) (Error in design and implementation: communication and training)
- Warning system failure (Error in implementation resulting in malfunction)
- Pressurization safety system failure (Error in implementation resulting in malfunction)

A backup system has been added to all DC-10s that now allows pilots to continue to control their planes even if all hydraulic systems fail.

Ethical behavior is essential if one is to avoid harming others directly or indirectly. Engineering codes of ethics provide us with guidelines for such behavior.

Engineering systems do not always behave according to the designer's expectations; they may contain hidden or unrecognized hazards. As a result, it is incumbent upon the engineer to evaluate a design for all such hazards, using the most appropriate tools and methods available (computer simulation and analysis, mathematical modeling, laboratory experiments, field tests, etc.). Once a hazard has been identified, it must be eliminated (or at least minimized) without delay.

CASE HISTORY 9.8

Another Air Disaster

In 1989 United Airlines Flight 232 departed from Denver on its way to Philadelphia (with a stopover in Chicago). When its number 2 tail engine exploded during the flight, all hydraulic control lines were severed. There were no manual backup control mechanisms since no human being could be expected to control such a large craft without the aid of hydraulic power.

The pilots then used the only mechanism left to them for controlling the plane: throttle adjustment of the remaining two wing engines. By alternately reducing and increasing the relative thrust of each engine, the pilot managed to steer the jetliner towards the nearest airport in Sioux City, Iowa. Subsequent investigation found that the cockpit crew had performed a truly amazing feat in directing their craft to the airport with virtually no control capabilities. Unfortunately, the pilot could not prevent the plane from rolling during its emergency landing in Sioux City as he made a final attempt to use the engine throttles to reduce the plane's high speed. One hundred and eleven people were killed as the plane rolled over and its fuel ignited. However, the valiant efforts of the crew had saved the other 185 lives that were on board that day.

The rear engine exploded because of an undetected crack in a fan disc. The disintegrating metallic bits of the disc severed the hydraulic control lines. Following the accident, metallurgists concluded that the crack first appeared in the disc during its final machining some 18 years earlier. During its years of operation, numerous inspections were performed on the fan disc. In such inspections, a fluid (actually a fluorescent penetrant or FPI) is applied to the metal so that it will collect in any existing surface cracks; these cracks, now filled with fluorescent fluid, can then be detected under ultraviolet light. Despite such precautions, the lethal crack in Flight 232's fan disc somehow escaped detection until it was too late.[39]

Finally, the triple hydraulic failure of Flight 232 was thought to be impossible by the McDonnell–Douglas Corporation, the plane's manufacturer.

SUMMARY OF FAILURES AND TYPES OF ERRORS

- Failure to anticipate hazardous event (i.e., a complete loss of the hydraulic control system due to an exploding rear engine) and design against it (perhaps by including an appropriate backup system); (Error in design: invalid assumptions and faulty reasoning)
- Failure in machining processes to prevent fracture of disc; failure to detect preexisting crack in fan disc (Error in implementation: both manufacturing and inspection)

39. See Mark (1994b).

An industrywide systems review task force (with Airbus, Boeing, General Electric, Lockheed, McDonnell–Douglas, Pratt–Whitney, and Rolls–Royce among its membership) was formed following the crash of Flight 232 with the goal of developing improved aircraft system designs meant to prevent accidents (i.e., hazard avoidance via engineering design). In addition, new and more effective inspection methods are being sought.

CASE HISTORY 9.9

Deadly Fire in a London Subway

The London Underground subway system had numerous insignificant fires over the years. Since no one perished in any of these fires, there was an inherent belief by those who operated the system that such fires posed no real danger. In 1987 the truth became apparent when a passenger dropped a lit match onto an escalator at the King's Cross underground station.[40] Normally, a metal cleat would have stopped the match from falling between the escalator's treads and skirting board; however, this cleat was missing, thereby allowing the match to ignite grease and dust along the running track. The escalator's wooden treads, skirting boards, and balustrades next caught on fire. Flames then engulfed the ticket hall located above the escalator. Eventually, this fire claimed the lives of 31 people.

No automatic fire-extinguishing systems were in operation and the manually operated water valves were unlabeled. Since the possibility of a truly serious fire seemed so remote, no extensive efforts were made to prevent such an incident. As Kletz (1991) has suggested, the risk could have been reduced if all-metal escalators, nonflammable grease, automatic sprinkling systems, and other safety measures (including timely replacement of missing cleats) had been used.

SUMMARY OF FAILURES AND TYPES OF ERRORS

- Inadequate maintenance of safety equipment (Error in implementation: maintenance)
- Use of combustible materials (Error in implementation: materials selection)

40. See Kletz (1991).

- Inadequate hazard prevention systems (Error in design due to invalid assumptions and faulty reasoning

Most of us become resigned to accepting small risks in our lives because we believe that such risks are both acceptable and inevitable. Many times, however, such risks can be reduced substantially or eliminated with a little effort.

CASE HISTORY 9.10

The MGM Grand Hotel Fire

The second worst fire in terms of deaths (85) and injuries (more than 600) in the history of the United States occurred on November 21, 1980, in Las Vegas, Nevada, at the MGM Grand Hotel.[41] Several interdependent systems at the hotel had failed to contain the fire and protect people from the deadly smoke. The systems were the following:

- **Fire-supression system**   Water sprinklers were never installed at several locations (including the deli) throughout the hotel. These sites were expected to be populated on a 24-hour basis. It therefore was assumed that since someone would always be present at these locations, any fire would be noticed quickly and extinguished without the need for sprinklers. Unfortunately, once the hotel was in operation, this assumption proved to be unfounded: sometime around 7:00 A.M. on November 21, the deadly fire started in the closed and deserted delicatessen. By the time it was discovered, the fire had grown out of control.
- **HVAC system (heating, ventilation, and air conditioning)**   HVAC systems often include fire dampers or hinged louvres that are designed to close automatically and block the flow of air through certain passages in the event of a fire. Such dampers normally are held open by combustible links that will melt in sufficient heat. However, some of the dampers in the MGM Grand Hotel HVAC system were held open permanently by bolts or metal wire links. A HVAC design feature meant to limit the spread of fire had been rendered useless. Once the fire began, air and smoke flowed freely through these openings to other locations in the hotel.
- **Fire-zone system**   Large structures often are divided into distinct fire zones, which restrict fire and smoke from spreading beyond a given

41. See Ford (1994).

location. These zones are constructed with fire- and smoke-resistant materials. Elevator shafts, air shafts, and other pathways between zones are designed to protect the integrity of the zone boundaries. Unfortunately, combustible materials were used to enclose some of the fire zones at the MGM Grand Hotel; in addition, the elevator and air shafts failed to prevent smoke from spreading across zone boundaries.

- **Fire alarm system** The MGM Grand Hotel used a two-layer fire alarm system. A primary (localized) alarm was expected to detect a fire and notify the security office, thereby providing security personnel the opportunity to verify the danger *before* informing the hotel's occupants (via the public address system and other alarms) about the situation. Unfortunately, the alarm system itself was overcome by the fire so that guests and other occupants in the hotel's tower were not alerted immediately to the danger.

- **Egress system** The egress system at the hotel was particularly flawed in its design. Occupants could exit from the tower only via stairwells with self-locking doors. Once people entered a stairwell and the doors locked behind them, they could not exit before reaching the street level. Many people perished as these stairwells became smoke-laden death traps.

In summary, then, a combination of system failures occurred during this tragedy.

SUMMARY OF FAILURES AND TYPES OF ERRORS

- Self-locking doors for stairwell exits (Error in design: invalid assumptions and faulty reasoning)
- Inadequate water-sprinkler system (Error in design: invalid assumptions and faulty reasoning)
- Inadequate alarm system (Error in design: invalid assumptions and faulty reasoning)
- Fire dampers permanently opened (Error in implementation: maintenance, possibly due to conceptual misunderstanding)
- Use of combustible materials (Error in implementation: materials selection)

The MGM Grand Hotel did *not* violate any existing fire codes and regulations. These codes (and many others throughout the nation) simply did not require an existing structure to be retrofitted with new fire prevention devices that had been developed since the building's original construction. After the Las Vegas disaster, many codes were changed to include such retroactive modifications to existing buildings (e.g., installation of improved smoke-exhaust systems, alarm systems, and smoke detectors).

In addition, smoke became recognized as equally or more dangerous as the fire itself. As Ford (1994) notes, there was a shift in priorities "from saving buildings to saving lives" following the MGM Grand Hotel fire, as reflected in the development and installation of improved smoke-detection and smoke-control systems.

CASE HISTORY 9.11

Deadly Boeing 747 Blowout

Prior to the introduction of the Boeing 747, plug-type cargo doors that opened inward were used on jetliners. As the fuselage was pressurized, these plug doors would be pressed firmly against the door frame, thereby ensuring an effective and airtight seal. Of course, a cost was associated with this benefit of presumed safety: Potential cargo space was lost to the relatively large inside clearances required by the plug doors.

The 747 design uses two gull-wing–type cargo doors located front and aft on the starboard (right) underside of the fuselage. These doors open outward and upward, thereby eliminating the need for the extensive internal clearances and guiderail tracks required by plug doors. Once the door is shut, electric motors rotate a set of C-shaped latches to their closed positions about pins on the door sill. Locks then are engaged to prevent the latches from backwinding and allowing the door to be opened inadvertently. Finally, electrical power to the motors is blocked by a cut-off switch so that the latches are locked securely.

The unexpected flaw in the gull-door latch system was its dependence upon the single cut-off switch. If this switch failed after take-off and pressurization of the fuselage, the entire system then became susceptible to any stray electrical signal that might actuate the motors and begin to unseal the door by backwinding the latches. Once the seal was broken, the immense internal pressure inside the fuselage would force open the cargo door, severely disabling the plane.

The above scenario did indeed occur on February 24, 1989, during United Airlines Flight 811 from Honolulu, Hawaii, to Auckland, New Zealand.[42] While traveling more than 500 mph at 22,000 feet above sea level, the 747-100's forward cargo door suddenly opened and the resultant depressurization caused a 13-foot by 15-foot starboard portion of the fuselage to blow away, together with nine passengers (Fig. 9.4). Although the two starboard engines then failed, Captain David Cronin managed to land safely in Honolulu before more lives were lost.

The danger associated with the 747's latch system—and a way to at least minimize this danger—were already known by the time of Flight 811's tragic loss. In 1987 eight locks on the forward cargo door of a Pan Am 747-100 passenger jet were found to be damaged. It then was discovered through tests by Boeing that the latch motors' cut-off switch could jam and become disabled, thereby allowing the latches to backwind if a stray electrical signal actuated the latch motors. Furthermore, the aluminum door locks were unable to prevent the latches (when motorized) from fully rotating and opening the door.

42. See Acohido (1994b).

FIGURE 9.4 The 13-foot by 15-foot gash in the starboard side of Flight 811's fuselage caused the sudden opening of a cargo door and resulted in the loss of nine passengers. *Source:* AP/Wide World Photos.

Many damaged locks were then discovered during an inspection of Pan Am's entire 747 fleet (damage apparently resulting from backwinding of the latches). Boeing recommended that 747 operators install steel braces to reinforce the aluminum locks on their planes and the FAA required that all airlines perform this installation within the next two years.

Pan Am immediately installed the steel braces. Unfortunately, given the FAA's relatively leisurely two-year time period for installation, United Airlines failed to add the steel braces to their 747s before the Honolulu tragedy occurred.

Until the recovery of Flight 811's lost cargo door by the United States Navy on September 27, 1990, the NTSB thought that the cause of the accident was twofold: (a) the door's aluminum locks had been damaged in fourteen separate instances of malfunction during December 1988, and (b) the ground crew failed to latch the door properly prior to take-off. However, recovery of the door itself revealed that the latches had backwound against their locks. Finally, in 1992, the NTSB reversed its earlier ruling, now finding that a faulty switch or wiring had led to actuation of the door latches.

SUMMARY OF FAILURES AND TYPES OF ERRORS

- Failure to provide sufficient redundancy in safety system—that is, absence of any backup to a single critical safety component (the cut-off switch); failure to anticipate hazardous event (Error in design: invalid assumptions, faulty reasoning and incomplete analysis/testing)

- Failure to act in a timely manner once a hazard became known (Error in implementation: corrective retrofitting, possibly due to conceptual misunderstanding or invalid assumptions)

After Flight 811's loss of nine passengers, the FAA required that steel braces be installed on all 747 lock systems within thirty days. In 1991 an electrical signal caused the latches on the cargo door of a United Airlines 747-200 to rotate and the door to open while the plane was being prepared for takeoff. United then discovered 21 electrical short circuits that inadvertently could activate the latch motors. United ground crews were ordered to cut off all electrical power to the latch system by opening two circuit breakers after the door was closed. Thus, a manual override was used as a precautionary backup to the cut-off switch, in effect replacing the switch.

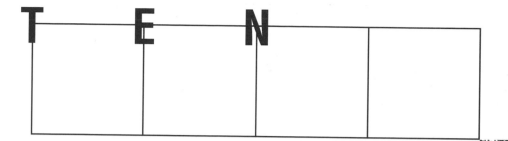

DESIGN ANALYSIS

You have erred perhaps in attempting to put color and life into each of your statements instead of confining yourself to the task of placing upon record that severe reasoning from cause to effect which is really the only notable feature about the thing. Sir Arthur Conan Doyle, *The Adventure of the Copper Beeches*

Upon completion of this chapter and all other assigned work, the reader should be able to

- Apply design evaluation via such tools as decision matrices and the Kepner–Tregoe analysis method to
  - prioritize design goals
  - rank each design alternative in terms of its ability to achieve the desired goals
  - determine the so-called best solution to the problem
  - identify the relative strengths and weaknesses (including potential hazards) of the so-called best design
- Maintain the highest level of critical objectivity when developing a decision matrix in order to avoid misleading or incorrect results due to subjectivity.
- Recognize the importance of the time value of money and identify many of the factors that can affect the final cost of a design.

10.1 Evaluating Alternative Designs

Once a set of alternative design solutions to a problem has been generated, one is confronted with the task of determining which alternative is preferable, why is it preferable, and what is wrong with it.

Remember: There is no such thing as a perfect design solution to an engineering problem. The design that is ultimately adopted as the "best" solution to the problem represents a set of compromises that the engineers have made in order to solve the problem in the most effective yet economically feasible manner. Each design alternative will have particular strengths and weaknesses that will need to be compared and evaluated in order to determine the best solution.

Several different techniques have been developed to aid the design engineer in his or her quest to determine this best solution. Most of these techniques are similar in that they require the engineer to:

- prioritize (or weight) the design goals against which each alternative will be evaluated
- formulate a scheme by which ratings can be assigned to each design concept
- combine the prioritized weightings of the goals with the ratings given to the designs to generate a combined score for each of the alternative solutions
- compare the total scores of all the design alternatives in order to identify the best overall solution

- carefully re-evaluate this so-called best solution in order to determine if any remaining weaknesses in the design can be either minimized or eliminated and to anticipate any hazards that may be associated with its implementation—that is, refine the final design selection as appropriate

Most of these techniques use a decision table or matrix of one form or another in order to compare the design alternatives in a quasi-objective and quantitative manner.

All of these techniques are, in fact, subjective and easily influenced by the engineers' biases and intuitive judgments. The engineer may believe that one design is clearly superior to the others; however, he or she needs to verify this belief in a thoughtful and methodical manner. The rationale for using a decision table to evaluate design alternatives is to provide us with a tool that requires us to pause and reflect upon the relative importance of each design goal and the ability of each design to achieve these goals. Although subjectivity will remain present in the evaluation, its influence upon the final decision will be less significant than if one relied solely upon intuition.

Let us now focus on the development of such a table.

10.2 Rank-Ordering the Design Goals

It is critical that one first rank-order the design goals since any subsequent evaluation and comparison of alternative solutions will be very dependent upon the priorities given to each of these goals. One method, discussed by Cross (1989), of rank-ordering design goals is shown in Table 10.1, which lists some of the goals to be achieved by a new computer system:[1]

Performance   One of the general design goals common to most engineering solutions. (see Chapter 4)

Minimum maintenance   Another general design goal.

Aesthetics   We would like the system to be pleasing in appearance.

Minimum cost   Another general design goal.

Availability of parts   Parts should be readily available.

Ease of use   It should require little or no special training or skills.

Versatility   The system should be useful for performing a wide range of computer-based tasks (e.g., numerical calculations, word processing, generation of graphical images).

Portability   The system should be easily transportable.

1. Whenever possible, each goal should be stated in a form that allows one to quantitatively assess the ability of each design concept to achieve that goal (Cross, 1989); for example, state the quantitative design specification that corresponds to a given goal so that one can then more precisely measure the degree to which a design succeeds in satisfying that spec. Of course, some goals will necessarily remain qualitative in type and therefore will need to be evaluated using less precise measures of assessment.

TABLE 10.1 Rank-ordering the list of design goals for a computer. A row-goal is given a value of 1 if it is more important than a column-goal; otherwise, it is given a value of 0.

Goals	Performance	Minimum maintenance	Aesthetics	Cost	Availability of parts	Ease of use	Versatility	Portability	**Total**
Performance	—	1	1	1/2	1	1	0	0	4.5
Minimum maintenance	0	—	1	0	0	0	0	0	1
Aesthetics	0	0	—	0	0	0	0	0	0
Cost	1/2	1	1	—	1	1	0	0	4.5
Availability of parts	0	1	1	0	—	0	0	0	2
Ease of use	0	1	1	0	1	—	0	0	3
Versatility	1	1	1	1	1	1	—	1	7
Portability	1	1	1	1	1	1	0	—	6

Source: Adapted from Cross, 1992.

In Table 10.1 each goal is systematically compared to the others, one at a time, using the following schema:

- The goal (along the row associated with it in the rank-ordering table) is awarded a score of 1 if it is perceived to be more important than another goal (associated with a particular column).
- The goal is awarded a score of 0 if it is deemed to be less important than the column goal.
- If two goals are deemed to be of equal significance, they are each given a score of 1/2.

TABLE 10.2 Final rank-order of design goals for a computer.

	1. Versatility
	2. Portability
(Tie)	3. Performance cost
	5. Ease of use
	6. Availability of parts
	7. Minimum maintenance
	8. Aesthetics

TABLE 10.3 Establish range for weighting factors of design goals, including general categories of significance.

71–100	Critical
31–70	Important
1–30	Optional

This method of carefully comparing simple pairings of goals provides the engineer with the opportunity to (a) narrow his or her focus and (b) obtain a better estimate of the relative significance of each goal than might be generated if all goals were considered simultaneously.[2]

The scores in the given row for each goal are then summed and the goals are listed in final rank-order according to their total scores (see Table 10.2). One now has a relative rank-order of the goals. The next step is to assign absolute weights of significance to the goals.

10.3 Assigning Weighting Factors to Design Goals

There are many ways to assign absolute weights of significance (the so-called weighting factors) to each design goal. One approach is simply to place each goal somewhere along an absolute scale after discussing the relative importance of the goals with all members of the design team (and other people—such as customers, management, and marketing—who may be able to provide insight about the relative significance of each goal).

In order to accomplish this task, one should

- Decide upon the range of values (e.g., 1–100) that will be used for the weighting factors.
- Separate the range into general categories of significance, as shown in Table 10.3. In this table, we have defined the range 71–100 for weighting factors judged critical for success, that is, reserved for those goals that must be achieved if a design is to represent a valid solution to a problem. Such critical goals usually relate to technical requirements for the design. (For example, any viable design for a stair-climbing wheelchair must necessarily satisfy the technical goal "climbs stairs.")

---

2. Notice that the scores (e.g., "0-1/2-0-1-1-0") entered in the particular row for a certain goal are opposite those in the corresponding column for that goal (i.e., "1-1/2-1-0-0-1").

Optional goals (in the range 1–30) are those that are desirable but not necessary for success. These goals often relate to the aesthetic aspects of a design.

Finally, all goals judged to be neither critical nor optional are placed in the midrange (31–70) of important goals, that is, goals that should be achieved.

■ The engineering design team should then evaluate each goal, beginning with the most important goal(s). The team should try to assign a weighting factor to this goal that will accurately reflect each team member's assessment (e.g., a weighting factor that is the average of the assigned weights given to the goal by all members of the design team). The team should then consider the least important goal(s), assigning it an appropriate weighting factor. The team should continue to alternate between the most important and least important goals among those that have not been assigned a weighting factor.

We recommend such alternating assignments at the extremes of the weighting range because it is often easier for people to reach agreement concerning the weights of significance for the most important and least important goals than to assign weighting factors to goals considered sequentially in their order of importance. In addition, design goals may be numerous (e.g., twenty or more in number). After the critical and optional goals at the extremes have been assigned appropriate weighting factors, there may remain a large number of midrange (important) goals that need to be given weights. In such cases, one can simply assign an average weight to these numerous midrange goals. Any effort to order these midrange goals in a numerical fashion will be subjective and imprecise; an average weighting will simply reflect that all such goals should be achieved by any viable design solution.

For example, in the case of the computer system design described in Tables 10.1 and 10.2, the team might readily agree that the most important goal, versatility, should be given a weight of 100, and that the least important goal of aesthetics should be assigned a weighting factor of 15 (i.e., midway within the optional goal subrange). After some discussion, the team then decides that "portability" is almost as critical to the design's success as "versatility" and assigns it a weighting factor of 95. More discussion results in no other goals being labeled as optional; the least important goal among those remaining to be assigned weights is "minimum maintenance," given a weighting factor of 40. Next, "performance" and "cost" are assigned the same weight (80) since it may not be possible to accurately assess the relative significance of these two critical goals (i.e., perhaps one goal deserves a weighting of 82 and the other should be assigned a score of 79; however, it is impractical to try to assign weighting factors with such accuracy due to the inherent imprecision and subjectivity of the process).

This process continues until all goals have been assigned weighting factors (see Table 10.4).

TABLE 10.4 Assign relative weighting factors to design goals.

	100	Versatility
	95	Portability
	90	
Critical	85	
	80	Performance; cost
	75	
	70	
	65	
	60	Ease of use
	55	
Important	50	
	45	Availability of parts
	40	Minimum maintenance
	35	
	30	
	25	
	20	
Optional	15	Aesthetics
	10	
	5	
	0	

10.4 Rating the Alternative Designs

Next, one needs to rate the relative success or failure of each design alternative in its ability to achieve each goal. Rating or utility factors are assigned to each design as follows:

- If possible, use measurable parameters to estimate the success or failure of each design. For example, a design's success in "minimizing cost" can be evaluated quantitatively in terms of its cost relative to the other designs. Similarly, "availability of parts" might be transformed into a more quantitative performance parameter such as "average number of suppliers" for particularly critical components of a computer system.

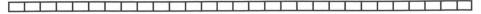

TABLE 10.5 An example range (1–10) for design-rating factors.

10 = Excellent
8 = Good
6 = Satisfactory
4 = Mediocre
2 = Unacceptable
0 = Failure

- For those goals that cannot be converted into a truly quantifiable form, an arbitrary rating scale (such as that shown in Table 10.5) can be used.

Rating factors are then assigned to each alternative solution, reflecting the solution's ability to achieve each of the design goals.

10.5 Decision Matrix

The weighting factors for the design goals, together with the rating factors for each alternative concept, are then inserted within a decision matrix,[3] as shown in Table 10.6. Each weighting factor WF is multiplied by the corresponding rating factor RF for each design, producing a so-called decision factor DF:

$$DF = (WF) \times (RF) \tag{10.1}$$

These decision factors are then summed for each design and stored in the right-most column of the matrix. In Table 10.6, we see that the XKZ-Model 10 design alternative received a total score of 3640, outscoring the other three designs.

If two or more designs have total decision factor scores that are within 10 percent of one another, the results should be interpreted as a tie. (After all, there is much subjectivity associated with this evaluation method.) These equally ranked designs should then be compared on a goal-by-goal basis, focusing upon the individual ratings achieved by each design on the most important goals. If one design is clearly superior to the others in terms of its ability to achieve a critical goal but not as effective as the other designs in achieving less important goals, thereby resulting in the overall tie score, then it probably should be selected as the best solution.

3. See, for example, Voland (1987) and Hill et al. (1979).

TABLE 10.6 Decision matrix.

Design Alternatives	Versatility	Portability	Performance	Cost	Ease of use	Availability of parts	Minimum maintenance	Aesthetics	Total
	Weighting factors								
	100	95	80	80	60	45	40	15	Total
XKZ-Model 10	7/700	6/570	8/640	7/560	7/420	9/405	6/240	9/135	3640
Porta-Tec	4/400	3/285	6/480	6/480	7/420	8/360	9/360	4/60	2845
Logic-Pack	6/600	7/665	9/720	8/640	4/240	6/270	4/160	5/75	3370
Compu-Save	**9**/900	6/570	7/560	3/240	6/360	6/270	2/**80**	8/120	3100

Goals

Rating Factor   Decision factor = weighting factor × rating factor.

10.6 Kepner–Tregoe Decision and Potential Problem Analyses

Kepner–Tregoe decision and potential problem analyses[4] modifies the above approach in the following way:

- First a statement that summarizes the decision to be made is developed.
- Next, the design goals are divided into "MUSTS," which are mandatory, and "WANTS," which are desirable. (MUSTS essentially correspond to the critical goals described in section 10.3.)
- Each alternative is then evaluated in terms of its ability to achieve the MUST goals. If the goal is satisfied, the alternative is assigned a "GO," meaning that it can now be evaluated in terms of its ability to achieve the WANTS. Otherwise, the design is assigned a "NO GO" and is no longer considered a viable alternative.
- Each WANT goal is next assigned a weighting factor (e.g., on a scale of 0-10) indicating its relative importance. These assignments can be made using the technique described in section 10.3.
- Each alternative solution is then evaluated in terms of its ability to achieve the WANT goals, assigning rating factors to the alternatives.
- The products of the weighting factors and rating factors are calculated to form decision factors that are then summed. The total decision factors are then compared to ascertain the best solution to the problem.

4. Kepner and Tregoe (1981); Fogler and LeBlanc (1992).

- Finally, any risks or hazards associated with the winning alternative(s) are considered. All possible negative consequences should be identified, together with an estimate of their probabilities for occurrence and their relative severities. The threat associated with each potential risk is then equal to the product of its probability and its severity. The total threat associated with each alternative is then obtained by summing the individual threats. If the threat associated with the best solution is deemed to be too high, one may choose to develop the second-best alternative instead and/or redesign the best solution to reduce its inherent risks. This final step is the most distinctive, and most significant, element of Kepner–Tregoe decision analysis; it should be performed in every engineering design effort, no matter what other analytical steps may have been used to compare and contrast alternative designs in order to identify the best solution to a problem.

  Kepner and Tregoe (1981) have noted that the simple calculations of threats as the products of probabilities and severities may be inappropriate, leading one to give insufficient attention to a risk of disastrous proportions (i.e., a high degree of severity) if such a risk has a very low probability of occurrence. Clearly, one needs to carefully consider any deadly risk associated with a design—even if that risk is quite unlikely to occur—and not simply conclude that since the probability of disaster is so low one can ignore the possibility entirely. Instead, the design engineer must strive to eliminate or minimize such a risk.

Table 10.7 presents an example of the Kepner–Tregoe method in which four proposed designs for allowing passengers on a ship or airplane to escape unharmed are compared. The engineering team performing this analysis decides that there are two MUSTS that need to be satisfied by any design if it is to be acceptable; these MUST goals are:

1. All of the life support needs of the passengers must be satisfied.
2. The design must be economically feasible.

Kepner–Tregoe decision and potential problem analysis results in a NO GO being assigned to the design entitled "Escape Pods." The remaining three alternatives are then evaluated in terms of their abilities to satisfy the desired WANTS of the engineers. As seen in the table, the "Converta-Craft" solution achieves the highest total score.

Finally, we recall that Kepner–Tregoe decision and potential problem analysis requires that one consider the risks of each alternative. Table 10.8 summarizes the risks associated with each design, together with their probabilities of occurrence and their relative severities. The total threat associated with the Converta-Craft design is seen to be so high that the engineers' second alternative, "Lifebelts," may be a better overall choice.

Too often engineers end their analysis of alternative designs with a decision matrix or its equivalent, simply identifying the presumably best solution without focusing on the various risks that may be embedded in it. The final step of Kepner–Tregoe decision and potential problem analysis, in

TABLE 10.7 Kepner–Tregoe decision matrix.

Musts		Alternatives						
Provides life support needs Economically feasible		Floatation device GO GO		Lifebelts GO GO		Escape pods GO NO GO	Converta-Craft GO GO	
Wants	**Weights**	*Rating*	*D=R×W**	*Rating*	*D=R×W*		*Rating*	*D=R×W*
Compact	5	9	45	2	10		5	25
Environmentally safe	9	2	18	8	72		8	72
Reusable	7	3	21	9	63		9	63
Other	3	3	9	2	6		5	15
Total			93		151			175

* Decision factor = rating factor × weighting factor.

TABLE 10.8 Kepner–Tregoe evaluation matrix of adverse consequences.

Adverse Consequences	Probability P	Severity S (Scale: 1–10)	Threat $(= P \times S)$
Flotation device			
▪ May fail due to leakage	0.05	9	0.45
▪ May not be activated properly	0.10	5	0.50
▪ Passengers may crowd onto a single unit, leading to failure	0.10	9	0.90
			Total 1.85
Lifebelts			
▪ May be misused by passengers	0.10	7	0.70
▪ May be lost or misplaced	0.05	9	0.45
			Total 1.15
Converta-Craft			
▪ Unreliable under certain conditions (components may jam, engines fail, etc.)	0.10	8	0.80
▪ There may be insufficient time to activate device	0.15	9	1.35
			Total 2.30

which one compares and contrasts the specific risks associated with each of the proposed solutions, should always be performed during engineering design analysis (even if methods other than Kepner–Tregoe decision analysis are used to identify the so-called best design solution). This small amount of additional effort may prevent truly disastrous consequences once the design is implemented.

10.7 Value and Subjectivity of a Decision Matrix

The value of a decision matrix is that it forces us to view the various design alternatives in a careful and thoughtful manner. It also requires us to prioritize the goals to be achieved by a solution, revisiting our original formulation of the problem and perhaps revising this formulation. Finally, it allows us to consider the relative strengths and weaknesses (including potential hazards) of the so-called best design. Overall, it is a very valuable evaluation tool for engineering design.

However, if the engineer fails to maintain the highest level of critical objectivity when developing a decision matrix, the results can be misleading or even skewed to verify his or her initial intuitive choice of the best design, thereby defeating the purpose of the entire evaluation effort.

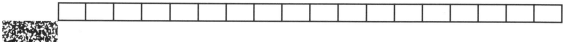

Satellite Defense Systems

Two teams of engineers are evaluating a set of alternative defense system designs for a satellite.[5] These designs include:

- The use of decoy satelites
- A satellite in which critical subsystems are duplicated in order to increase the probability that the system will continue to operate even if damaged, that is, intentional redundancy of critical elements (see Chapter 9),
- A system with multiple defensive mechanisms to protect itself from damage
- A satellite with a hardened external surface or skin
- A system controlled by mobile remote ground stations

5. For purposes of illustration, this example focuses upon imaginary satellite defense systems. For information about actual systems, refer to Gorney, Blake, Koons, Schultz, Vampola, and Walterscheid (1991).

TABLE 10.9 Decision matrix for alternative defense systems.

(a)

Design Alternatives	Goals					
	Survivability	Minimum cost	Reliability	Operating life	Minimum maintenance	
	Weighting factors					
	100	80	80	60	40	Total
A. Decoys	9	4	7	5	6	2320
B. Redundant subsystems	5	7	9	8	6	**2540**
C. Multiple shields	6	5	6	7	4	2140
D. Hardened surfaces	5	8	6	5	8	2240
E. Mobile remote stations	7	2	8	8	4	2140

(b)

Design Alternatives	Goals					
	Survivability	Minimum cost	Reliability	Operating life	Minimum maintenance	
	Weighting factors					
	100	40	80	20	40	Total
A. Decoys	9	4	7	5	6	**1960**
B. Redundant subsystems	5	7	9	8	6	1900
C. Multiple shields	6	5	6	7	4	1680
D. Hardened surfaces	5	8	6	5	8	1720
E. Mobile remote stations	7	2	8	8	4	1760

These system designs are evaluated and compared with respect to five goals: survivability of the system, minimum cost, reliability, operating life (i.e., expected lifetime for each system), and minimum maintenance.

One engineering team performs the evaluation that is summarized in Table 10.9a, whereas the second team's results are presented in Table 10.9b. These two evaluations are based upon different sets of weighting factors for some of the design goals, reflecting the differing judgments of the engineers about the relative importance of the design goals. The absolute values of the two analyses differ, with one analysis resulting in the selection of the system with intentional redundancy and the other analysis leading to the choice of the decoy system.

This example illustrates two important aspects of the decision matrix when used to compare alternative designs. First of all, any use of decision matrices is inherently subjective and dependent upon the opinions of the evaluators. Second, in both Table 10.9a and 10.9b, we note that the total decision factors of the decoy and redundancy designs lie within 10 percent of one another, indicating that these designs are essentially tied with each other in both evaluations. The basic equivalence of the two designs (in terms of their overall success) is reflected in the relative sensitivity of their total decision factors to any modifications in the goals' weighting factors.

10.8 Economic Analysis

Many dimensions in an engineering design must be analyzed before one can be satisfied that the design is satisfactory as a solution to a problem; these dimensions include the design's ability to perform under the expected operating conditions, the relative hazards associated with the design, and the economic viability of the design. Although a detailed discussion of economic analysis is beyond the scope of this book, a brief review of cost estimation, return on investment, and the time value of money will be given.

10.8.1 ■ Cost Estimation

Costs associated with the design and development of a new device, system, or process must be carefully estimated. Such costs may include the construction of new manufacturing facilities or the conversion of existing facilities, environmental impact statements, depreciation of capital equipment (i.e., the money that will be necessary to replace worn or discarded equipment), taxes (which can be affected by the method one uses to calculate depreciation costs), materials, direct labor (i.e., labor devoted to a particular product), packaging, distribution and transportation expenses, marketing, repair, and scrap losses. Overhead costs (costs not directly related to the development, manufacture, and marketing of a product such as supplies, maintenance, indirect labor, security, legal representation, and rent) must also be taken into account. Finally, one needs to be able to estimate the profit that could be generated by a product, based upon its expected selling price and sales volume.

Consider the situation summarized in Tables 10.10 and 10.11[6] illustrating the types of expenses that might be considered in a cost estimate for an

---

6. Adapted from data in Peters and Timmerhaus (1991, pp. 210–211), and an example provided by Professor Ronald Willey of the Department of Chemical Engineering at Northeastern University, Boston, Massachusetts.

TABLE 10.10 Fixed cost investment (FCI)

I. Direct Costs (DC)	Approximate Percentage	Example
A. Purchased equipment (PE)	15–40% of FCI	$50,000
Installation of PE	25–50% of PE	12,500
Instrumentation and controls	6–30% of PE	15,000
Piping (installed)	10–80% of PE	14,000
Electrical wiring	10–40% of PE	17,000
B. Buildings (related to project)	10–70% of PE	35,000
C. Service facilities and yard equipment (e.g., remove soil, reseeding)	40–100% of PE	30,000
D. Land	4–8% of PE	3,500
Subtotal		177,000

II. Indirect Costs (IC)		
A. Engineering and Supervision	5–30% of DC	$20,000
B. Construction Expenses and Contractors Fee	6–30% of DC	32,000
C. Contingency (set aside)	5–15% of FCI	20,000
III. **FCI = DC + IC**		**$249,000**

IV. Working Capital Investment (WCI)	10–20% of TCI	$40,000
(Necessary, since cash flows may lag expenses for labor, materials, etc.)		

V. Total Capital Investment (TCI) TCI = FCI + WCI		$289,000

initiative in the chemical processing industry. This table is divided into two parts: fixed cost investment (Table 10.10) and total operating costs (Table 10.11). Included in this example are some typical ranges of certain expenses, given in terms of their percentage values of other quantities such as:

DC = Direct costs
FCI = Fixed cost investment
M&R = Maintenance & repairs
PE = Purchased equipment
TOC = Total operating costs

DS = Direct supervision
IC = Indirect costs
OL = Operating labor
TCI = Total capital investment
WCI = Working capital investment

TABLE 10.11 Total operating costs (TOC).

Item	Approximate Percentage	Example
A. Direct Operating Costs		
1. Raw materials	10–50% of TOC	$140,000
2. Operating labor (OL) (8 hrs/day, $30/hr, 250 days/yr)	10–20% of TOC	60,000
3. Direct supervision (DS)	10–25% of OL	15,000
4. Utilities (Electricity, water, …)	10–20% of TOC	65,000
5. Maintenance & repairs (M&R)	2–10% of FCI	18,000
6. Operating supplies	10–20% of M&R	2,500
7. Laboratory supplies	10–20% of OL	9,000
8. Patents and royalties	0–6% of TOC	0
B. Fixed Charges		
1. Depreciation	10% of FCI	24,900
2. Local taxes	1–4% of FCI	6,000
3. Insurance	0.4–1% of FCI	2,000
4. Rent	8–12% of land, bldgs.	4,000
C. Plant overhead (managers, etc.)	50–70% M&R+OL+DS	55,000
D. General Expenses		
1. Administrative costs (central offices, etc.)	15% of M&R+OL+DS	13,950
2. Selling	10–20% of TOC	60,000
3. Research and development	5% of TOC	25,267
4. Financing	0–10% of TCI	30,000
E. TOC = SUM (Costs/year)		$530,617

10.8.2 ▇ Return on Investment

One very basic estimate of the economic viability of a design is the return on investment (ROI) that it can be expected to generate. ROI is simply the percentage of the initial investment that will be generated as a profit if a design is implemented. Companies have a broad range of investment opportunities for their funds; the expected ROI for a design must be competitive with the other available investment alternatives if the design is to be funded.

For example, consider a situation in which the ROI for a design must equal 15 percent or more for the project to be approved for funding. If the initial investment for a proposed design is estimated to be $300,000 and the expected profit on this investment is $36,000, then the ROI is simply the ratio of the expected profit to the initial investment:

ROI = Profit/Investment
      = 36,000/300,000
      = 0.12

or 12 percent. Since this ROI is below the minimum threshold of 15 percent, the project will not be approved for funding and further development. The ROI could rise above the threshold level if design modifications can be made to reduce the investment costs and/or increase the expected profit. The danger in such modifications is that certain critical design goals (e.g., public acceptance, safety, performance) may be adversely affected in order to satisfy the economic requirements. It is the responsibility of the engineer to ensure that the product will remain functionally viable and marketable in its modified form.

The calculation of ROIs is important in determining which design alternatives are most promising from a financial perspective. Of course, the data used in estimating the initial investment and expected profit figures should be as reliable as possible.

10.8.3 ▨ Time Value of Money

When performing a cost analysis of a design, one should begin by recognizing that money has a time value associated with it. A specific amount of money P, invested today at an annual interest rate i, will be more valuable after n years have passed. In fact, the future worth of P is given by the simple formula:

$$F = P(1 + ni) \tag{10.2}$$

However, if the interest is *compounded* over the n years (i.e., if interest is allowed to accrue on both the principal P and the interest already earned on P), the formula becomes

$$F = P[(1+i)^n] \tag{10.3}$$

"$[(1+i)^n]$" is known as the single-payment compound interest factor; it allows us to convert a present worth P to an equivalent future worth F under the condition of compound interest.

For example, consider a proposed five-year project that will require an initial investment of $300,000. (For simplicity, we will assume that these are the only costs associated with the project.) At the end of the five-year period, the (future) worth of the initial $300,000 investment, with an expected compound annual interest rate of 10 percent, is given by

$F = \$300,000 \times [1.10]^5$
$\quad = \$483,153$

Thus, when estimating the funds needed to support a project, one must consider the time period during which the investment will be made and the impact that this time will have on the value of the investment.

Many other formulae are available for converting between future worth F, present worth P, and annual payments A. Table 10.12 presents some of these conversion formulae. As an example, the annual payment A needed to recover an investment P and interest i in n years is given by the formula:

$$A = P\{[i(1+i)^n]/[(1+i)^n -1]\}$$

If P is equal to \$300,000, i equals 0.10, and n equals 5, then

$$A = (\$3 \times 10^5)\{[(0.10)(1.10)^5]/[(1.1)^5 -1]\}$$
$$= (\$3 \times 10^5)\{0.161051\}/[0.61051]$$
$$= \$79,139 \text{ (approximately)}$$

One should note that there is a self-consistency between these formulae, in that certain conversions can be obtained from others. For example, the uniform-series present worth factor F_{AP} (which allows one to convert an annual payment A to an equivalent present worth or principal amount P)

TABLE 10.12 Conversion formulae for economic analysis

Formulae

1. $F = P[(1+i)^n]$
2. $P = F[(1+i)^{-n}]$
3. $A = P\{[i(1+i)^n]/[(1+i)^n -1]\}$
4. $P = A\{[(1+i)^n -1]/[i(1+i)^n]\}$
5. $A = F\{i/[(1+i)^n -1]\}$
6. $F = A\{[(1+i)^n -1]/i\}$

Conversions

1. Compound future worth F of present worth P
2. Present worth of some future worth F
3. Annual payment A needed to recover an investment P and interest i in n years
4. Present worth P of n annual payments A at interest i
5. Annual payment A needed to generate future worth F
6. Future worth F of n annual payments A

Conversion Factors

- $F_{PF} = (1+i)^n$
- $F_{FP} = (1+i)^{-n}$
- $F_{PA} = i(1+i)^n/[(1+i)^n -1]$
- $F_{AP} = [(1+i)^n -1]/\{i(1+i)^n\}$
- $F_{FA} = i/[(1+i)^n -1]$
- $F_{AF} = [(1+i)^n -1]/i$

Names of Conversion Factors

- Single-payment compound interest factor
- Present worth factor
- Capital recovery factor
- Uniform-series present worth factor
- Sinking fund factor
- Equal payment series future worth factor

can be obtained by multiplying the equal payment series future worth factor F_{AF} by the present worth factor F_{FP}; that is,

$$F_{AP} = F_{AF} * F_{FP}$$
$$= \{[(1+i)^n -1]/i\} * (1+i)^{-n}$$
$$= [(1+i)^n -1]/[i(1+i)^n] \qquad (10.4)$$

The difference between the present worth and future worth of money must be taken into account when one evaluates the costs for successfully developing and marketing a design. One needs to compare costs and potential profits on a equal basis (e.g., comparing only the present worths of two designs or only the future worths of these designs, as opposed to some mix of these two worths).

There are three popular approaches to estimating costs, depending upon the depth of information that one has available about each of the above factors. In *methods engineering,* one determines the total cost per part for a given sequence of manufacturing (and other) operations. One seeks to compare different methods for producing and marketing a product in order to find that method of operation which will optimize profit, efficiency, use of available facilities, and aspects of the production process. *Costs by analogy* is an alternative method in which one uses historical cost data for past products of similar type and adds an allowance for cost escalation and for any differences in size that may exist between the past and present products. *Statistical cost analysis* uses historical data to formulate functional relationships between system costs and product parameters (such as power, weight, and accuracy); these resultant mathematical formulae can be used to calculate costs for designs with specific parameter values. Whereas methods engineering focuses upon the specific components of a design, statistical cost analysis seeks to develop a broad economic model of the entire design. All three methods require that one have appropriate data available upon which to base one's estimates.

Finally, the *rule of thumb* method is an alternative to such detailed economic analyses. It can be used during the early stages of the design process when one does not have complete data relating to the various factors that affect a design's final cost. In this method, one calculates the retail costs for each off-the-shelf (standard) component in a design (as if you were buying these components from a retail outlet without any reduction in price due to large sales volumes), together with a rough estimate of the costs for raw materials (for those components that will need to be custom-made). One then multiplies the sum of these off-the-shelf and raw materials costs by a factor of 4, thereby obtaining a rough estimate of the retail price that will be needed in order to recoup one's investment in the product and its manufacture, together with a small profit. Such work will provide the engineer with an estimate of the economic viability of a design (particularly if competition exists in the form of similar products in the marketplace). If the expected retail price is simply too high to interest a customer in purchasing the product, then the design will need to be modified or discarded.

- Design evaluation via such tools as decision matrices and the Kepner–Tregoe analysis method allow engineers to
  - prioritize design goals
  - rank each design alternative in terms of its ability to achieve the desired goals
  - determine the so-called best solution to the problem
  - identify the relative strengths and weaknesses (including potential hazards) of the so-called best design
- Kepner–Tregoe potential problem analysis, in which one compares and contrasts the specific risks associated with each of the proposed solutions, should always be performed.
- The engineer must remember to maintain the highest level of critical objectivity when developing a decision matrix in order to avoid misleading or incorrect results due to subjectivity.
- Costs associated with the design and development of any new device, system, or process must be carefully estimated.

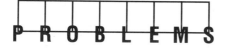

PROBLEMS

10.1 There are many types of respirators. Develop a decision matrix from which one can select the best alternative for an assumed application by considering such factors as comfort, likelihood that the user will wear the respirator, environment of use, types of fumes and dust to be avoided, and so forth. State all assumptions that you make about the respirator(s) and the application(s) for which they are being considered.

   Write a statement summarizing your results. Submit this statement together with the decision matrix.

10.2 Collect information describing three or more of the current commercial designs for one of the following types of products. Select appropriate goals to be achieved by these designs and create a decision matrix to evaluate and compare these alternative solutions. Provide a brief explanation of the values that you assign to the weighting and rating factors in your matrix.

   - Air conditioners
   - Backpacks
   - Bicycles
   - Camcorders
   - Automobiles
   - Batteries
   - Blenders
   - Cameras

- Can openers
- Copiers
- Exercise equipment
- Garage-door openers
- Microwave ovens
- Roof shingles
- Telephones
- Toys
- Videocassette recorders
- Computers
- Dishwashers
- Fax machines
- Lawn mowers
- Refrigerators
- Snow throwers
- Televisions
- Vacuum cleaners
- Wristwatches

10.3 Apply Kepner–Tregoe potential problem analysis to the situation in Problem 10.1.

10.4 Research a problem for which (at least) three solutions have been developed. (Consider the patent records, marketing journals and magazines, professional literature, retail outlets, etc., to obtain data about each solution.)

Develop a decision matrix for this problem, evaluating each of the design alternatives. Write a statement summarizing your results. Submit this statement together with the decision matrix.

10.5 Apply Kepner–Tregoe potential problem analysis to the situation described in Problem 10.4.

10.6 Select one of the product types listed in Problem 10.2. Obtain information about three or more current commercial designs for this product. Perform a Kepner–Tregoe potential problem analysis upon these three alternative designs; compare and contrast the specific risks associated with each of these three solutions.

10.7 Select two of the product types listed in Problem 10.2. Identify the most important goals that should be achieved by these solutions. Which goals are important to both types of product? What is their relative importance to each product? Which goals are important to only one type of product? Explain your reasoning.

10.8 Develop a fixed cost estimate for an existing large-scale system of your choice. If you cannot obtain actual dollars values for certain costs, use a "best estimate" (i.e., best guess) and complete your analysis. Explain your estimates when necessary and note all sources of actual data.

10.9 Given that the ROI for a project equals 0.15 and the expected profit is $67,000, calculate the initial amount of money that must have been invested in the project.

10.10 For Problem 10.9, what must the expected profit be for the ROI to equal 0.20? Show all calculations.

10.11 Calculate the present worth P for an amount that has a future worth F (in three years) equal to $75,000 with an interest rate of 0.08. Show all calculations.

10.12 Calculate the annual payment A needed to recover an investment equal to $250,000 with an interest rate of 0.06 for five years. Show all calculations.

10.13 Given that four annual payments of $40,000 will be needed to generate a future worth F at an annual interest rate equal to 0.065, calculate the value of F.

10.14 Calculate the present worth P of three annual payments of $10,000 at an interest rate of 7 percent.

10.15 Calculate the future worth F of three annual payments of $10,000 at an interest rate of 7 percent.

CASE HISTORY 10.1

Paper Towels

A manufacturer of paper towels is attempting to decide which of five designs to produce. A marketing survey indicates that the attributes of most importance to potential customers are absorbency (the ability to absorb water and oil), toughness (the ability of the towel to withstand frictional wear against a rough surface), the use of recycled materials (paper, ink), wet strength (the ability to support a load when moist), ease of tearing (which corresponds to the number and degree of perforations between sheets), and cost.[7]

The engineers at the firm rank-order these goals based upon customer expectations, as shown in Table 10.13. Relative weighting factors are then assigned to these goals (Table 10.14) and a decision matrix is generated in which five towel designs are compared (Table 10.15). Based upon the rating factors assigned to each design, the choice of the best alternative appears to be a double-ply design composed of recycled paper.

An alternative evaluation is shown in Tables 10.16, 10.17, and 10.18, indicating the subjectivity inherent within this evaluation method. Note the differences in the rank-ordering of the design goals (particularly cost per sheet), the weighting factors assigned to these goals, and the resultant choice of a best alternative (a thick, single-ply design made from recycled paper). Clearly, engineers must be cautious in ranking goals and assigning relative values for the weighting and rating factors in a decision matrix if the resultant choice of a best design is to be valid.

7. This evaluation is for purposes of illustration only and does not necessarily reflect the critical elements that should be used to compare paper towels; however, *Consumer Reports* did consider such factors as wet strength, absorbency, toughness, and cost in its February 1998 comparison of commercial paper towel products.

TABLE 10.13 Rank-ordering the goals for paper towels.

Design Goals	Wet strength	Absorbency	Toughness	Cost/sheet	Recycled materials	Perforations	Total
Wet strength	—	1	0	1	0	0	2
Absorbency	0	—	1/2	1	1	1	3.5
Toughness	1	1/2	—	1/2	0	0	2
Cost/sheet	0	0	1/2	—	1	0	1.5
Recycled materials	1	0	1	0	—	1/2	2.5
Perforations	1	0	1	0	1/2	—	2.5

TABLE 10.14 Assigning relative weighting factors to goals.

100 = Absorbency
80 = Recycled materials; perforations [Tie]
60 = Wet strength; toughness [Tie]
40 = Cost/sheet

TABLE 10.15 Decision matrix for paper towel designs.

	Goals						
	Wet strength	Absorbency	Toughness	Cost/ sheet	Recycled materials	Perforations	
	Weighting factors						
Design Alternatives	60	100	60	40	80	80	Total
A. Double ply, recycled	7/420	8/800	8/480	5/200	9/720	3/240	**2860**
B. Double ply	7/420	9/900	8/480	4/160	2/160	3/240	2360
C. Single ply, thick	5/300	6/600	6/360	6/240	3/240	7/560	2300
D. Single ply, thick, recycled	4/240	5/500	6/360	7/280	9/720	7/560	2660
E. Woven, thick	8/480	7/700	7/420	2/80	2/160	5/400	2240

TABLE 10.16 Alternative rank-ordering of the goals for paper towels.

Design Goals	Wet strength	Absorbency	Toughness	Cost/sheet	Recycled materials	Perforations	Total
Wet strength	—	1	0	0	0	0	1
Absorbency	0	—	1/2	0	1	1	2.5
Toughness	1	1/2	—	0	0	0	1.5
Cost/sheet	1	1	1	—	1	1	**5**
Recycled materials	1	0	1	0	—	1/2	2.5
Perforations	1	0	1	0	1/2	—	2.5

TABLE 10.17 Assigning relative weighting factors to goals.

100 = Cost/sheet
80 = Recycled materials; perforations; absorbency [Tie]
40 = Toughness
20 = Wet strength

TABLE 10.18 Alternative decision matrix for paper towels.

	Goals						
	Wet strength	Absorbency	Toughness	Cost/sheet	Recycled materials	Perforations	
	Weighting factors						
Design Alternatives	20	80	40	100	80	80	**Total**
A. Double ply, recycled	7	8	8	5	9	3	2560
B. Double ply	7	9	8	4	2	3	1980
C. Single ply, thick	5	6	6	6	3	7	2220
D. Single ply, thick, recycled	4	5	6	7	9	7	**2700**
E. Woven, thick	8	7	7	2	2	5	1760

CASE HISTORY 10.2

Reclining Chairs

Reclining chairs[8] may differ in their materials (e.g., frames made of hardwood or plywood, solid foam or shredded foam) and styles (such as the full chaise design in which both the seat and leg rest are covered by a single pad, and the hidden chaise design in which the seat, the leg rest, and the gap between them are covered by separate pads), among other characteristics. Tables 10.19 through 10.21 illustrate the development of a decision matrix to compare a set of imaginary alternative recliner designs in terms of their relative abilities to achieve six selected goals: comfort, minimum cost, frame strength, durability, ease of use, and warmth.

TABLE 10.19 Rank-ordering the goals for reclining chairs.

Design Goals	Frame strength	Durability	Comfort	Warmth	Ease of adjustment	Minimum cost	Total
Frame strength	—	1/2	0	1	1/2	0	2
Durability	1/2	—	0	1	1/2	0	2
Comfort	1	1	—	1	1	1	5
Warmth	0	0	0	—	0	0	0
Ease of adjustment	1/2	1/2	0	1	—	0	2
Minimum cost	1	1	0	1	1	—	4

TABLE 10.20 Assigning relative weighting factors to goals.

100 = Comfort
85 = Minimum cost
60 = Frame strength, Durability, Ease [Tie]
30 = Warmth

8. This evaluation is for purposes of illustration only and does not necessarily reflect the critical elements that should be used to compare recliner designs; see *Consumer Reports* (February 1998) for an evaluation of recent recliner models.

TABLE 10.21 Decision matrix for reclining chair designs.

Design Alternatives	Goals						Total
	Comfort	Minimum cost	Frame strength	Durability	Ease of adjustment	Warmth	
	Weighting factors						
	100	85	60	60	60	30	
A. Hardwood frame, solid foam, full chaise	8	2	8	9	5	8	2530
B. Hardwood frame, solid foam, hidden chaise	7	4	8	8	7	7	2630
C. Hardwood frame, shredded foam, full chaise	8	3	8	6	5	8	2435
D. Plywood frame, solid foam, hidden chaise	7	8	7	7	9	7	**2970**
E. Plywood frame, shredded foam, full chaise	8	7	7	5	5	8	2655

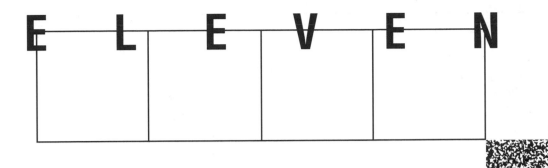

E L E V E N

IMPLEMENTATION

What is the worth of anything
But for the happiness 'twill bring?   R. O. Cambridge, *Learning*

O   B   J   E   C   T   I   V   E   S

Upon completion of this chapter and all other assigned work, the reader should be able to

- Explain why poor implementation of a design concept may lead to failure or disaster.
- Describe why engineers should determine the processes, materials, and equipment for fabrication, assembly, and other manufacturing operations.
- Apply the basic principles and guidelines that one should follow to design for X, where X may represent manufacturability, assembly, reliability, quality, packaging, maintainability, disassembly, and recyclability.
- Recognize that materials chosen for an engineering design must match the product's performance, environmental, processing, and financial requirements.
- Distinguish between the general characteristics of metals, polymers (plastics), ceramics, and composites.
- Describe some of the fabrication processes commonly associated with metals and polymers.
- Select materials for a design by considering the various relevant properties of material candidates, together with the fabrication processes that may be required.
- Describe some of the obstacles that may confront an engineer during the construction of large-scale, one-of-a-kind systems.

11.1 Transforming a Design Concept into Reality

We now turn our attention to the final phase of the design process, *implementation*, in which a design becomes reality by converting raw materials into the desired product or system.

Consider some of the many personal items that you have used today: toothpaste, a ballpoint pen, a wristwatch, clothing, paper, lightbulbs, a microchip in your calculator, and so forth. Think of the large-scale systems that we all share and rely on daily: roadways, buildings, buses, wastewater treatment facilities, electrical power networks, and so forth. The world around us is filled with objects and systems that have been fabricated by other people—often in very ingenious ways—to satisfy society's needs.

The manufacture of mass-produced objects (such as microchips or integrated circuits) or popular commodities (such as paper) can be surprisingly complex. Paper made from the cellulose in wood requires a series of specific operations:[1] Trees are first cut into wood chips, after which the lignin or natural glue is removed from the chips through a chemical bath. The resulting wood pulp is then washed, bleached, beaten, squeezed, dried, and rolled into paper sheets of the desired consistency.[2]

Similarly, a series of tasks was necessary to produce the tiny microchip in your calculator, including[3] creation of a schematic design for the chip's electrical circuit; projection of this design drawing onto a glass surface (known as a "mask"); transferring this circuit image from the glass onto a semiconducting silicon wafer; etching the wafer and "doping" it with impurities to create specific regions of high conductivity; coating the wafer with metal; etching the metal coating; testing the wafer circuits; and cutting the wafer into individual microchips. Moreover, each of these tasks must be performed with great precision, and in an absolutely clean environment free of particles that might contaminate the chip. (See Case History 11.2 *Microchip* for details of this manufacturing process.)

Some designs (such as airplanes) are mass-produced in very limited quantities because of substantial costs, limited demand, size, or other reasons. For example, a single Boeing 747 jet is composed of more than four million parts, and about 10,000 workers are needed to build one of these planes. Nevertheless, the 747 was indeed a mass-produced design, with an average rate of seven planes per month being achieved some years ago.[4]

The fabrication of one-of-a-kind, large-scale systems also requires completion of a series of tasks, often performed simultaneously, as illustrated by the following case.

1. See, for example, *How Things Are Made* (1981).
2. *How Things Are Made* (1981).
3. Cone (1991) and Jeffrey (1996).
4. *How Things Are Made* (1981).

Empire State Building

Construction of the 102-story Empire State Building began in 1929 by clearing the site of existing buildings. A simple yet elegant design was developed for what would become (for many years) the world's tallest building. Next, a foundation was laid to help support the huge weight of the building, and the steel framework was created one floor at a time. As the framework for each story was completed, a precast concrete floor was laid. The building's facade or outer skin of brickwork and masonry was added, then the interior walls, electrical systems, plumbing, and other detail work were completed. Once building had begun, the project's 3500 workers managed to achieve the incredible construction rate of six floors per week. (One example of the speed with which project tasks were completed: Some steel girders were being connected to form the framework of the building only four days after the steel itself had been formed in a Pittsburgh foundry.) The entire 1250-ft. structure was completed in a mere 17 months.[5]

One might ask, Given the speed of construction of the Empire State Building, were there any significant errors committed by the designers or workers in their haste? The question might be answered by two anecdotes, both involving aircraft. The building itself is so well designed and constructed that it survived a tragic direct hit at the 79th floor by a B-25 bomber traveling at 250 mph through fog and low hanging clouds on July 28, 1945. The crash created an 18-ft. × 20-ft. hole in the side of the building and killed the plane's pilot, two passengers, and ten other people. One of the plane's engines traveled through the building and onto a neighboring structure, while the second engine fell into an elevator shaft. Nevertheless, the concrete flooring absorbed much of the impact, thereby reducing the potential damage to the structure itself.[6] Given this incident and the nearly 70-year history of the Empire State Building, it would appear that the structure's basic design and construction is indeed sound.

However, there was one unforeseen design "error" in the building: A mast was added to the top of the structure so that airships (i.e., dirigibles or blimps) could be tied to the building to load and unload their passengers. However, the first—and only—attempt to do so ended in the airship being overturned by the powerful winds swirling up around the building. The idea was abandoned and the mast eventually became part of a huge 222-ft. television antenna.[7]

The engineer must consider the practical challenges associated with the fabrication of a design during the earlier phases of the design process. He or she must avoid developing an abstract or theoretical design concept which cannot be transformed into reality because of such unforeseen obstacles as excessive manufacturing costs, the lack of suitable materials, or the establishment of unreasonable physical demands upon those who must fabricate or construct the envisioned solution to a problem.

5. Lewis (1980).
6. Levy and Salvadori (1991) and Lewis (1980).
7. Lewis (1980)

Recall, too, that the design process is iterative, meaning that if one discovers new or important information during any of the process's later phases, earlier phases may need to be redone. If at the end of the design process it is determined that an unforeseen constraint is associated with the practical implementation of a design, the engineer must add this constraint to the list of design specifications developed during phase two (*problem formulation*) of the process and modify the design accordingly.

This final stage of the design process in which a dream is transformed into reality can be extremely difficult, both conceptually and in practical terms. Successful implementation means that one has overcome these difficulties.[8]

11.2 Concurrent Engineering and Design for X

Throughout this text, we have described the design process as a sequential series of activities that one should perform in order to create an engineering solution. In reality, design engineers cannot wait until all decisions have been made about a product's critical characteristics (such as geometric shape, component parts, and operation) before turning their attention to the design's fabrication, assembly, distribution, maintenance, repair, disassembly, and recycling (or disposal). Decisions to incorporate certain features or functions in a design often will predetermine the manufacturing processes, materials, and assembly operations that must be used to fabricate the product. Other elements in a design may limit one's options to recycle or dispose of the product after its operating life has ended.

Engineers should determine the processes, materials, and equipment that will be used for manufacturing, assembly, and other operations as they develop the design concept itself. Such simultaneous development of all aspects of a design—from the initial concept to its manufacture, maintenance, and disposal—is known as *concurrent engineering*. Many individuals, each representing one or more areas of product development, such as conceptual design, marketing, manufacturing, assembly, packaging, and recycling, work together as a team throughout the entire design process to ensure that the final product will exhibit the highest levels of performance and quality, achieved at the lowest possible cost and in the shortest production time.

There is an ever increasing need to produce engineering designs quickly and effectively. Concurrent engineering has reduced the time required for producing a new design by establishing more effective communication links among all those involved in the effort and by allowing critical engineering

---

8. See Case History 11.3 *Panama Canal* for an illustration of engineers working diligently to overcome major obstacles in their path.

issues to be resolved much earlier in the design process, thereby reducing the need for corrective actions to be taken after substantial amounts of time, effort, and money have been invested. It has been estimated[9] that concurrent engineering has resulted in manufacturing costs being reduced by 30 percent to 40 percent, and scrap/rework efforts being reduced in some instances by 75 percent.

Certain guidelines and strategies have been developed to aid engineers as they develop a design concept and the processes through which it will be transformed into reality. These strategies are sometimes called the "design for X" or "DFX" methods, where X may represent such considerations as manufacturability, assembly, reliability, quality, maintainability, disassembly, and recyclability.

Most important, as one uses these strategies to develop the appropriate modes of production for a design, care must be taken to avoid creating oppressive or dehumanizing working conditions for the laborers.

11.2.1 ■ Objectives of DFX

DFX seeks to achieve the following objectives, all of which can be critical to a design's ultimate success or failure:

- Superior (or at least satisfactory) performance of all design functions— that is, satisfy the customers' needs
- Minimum cost
- Minimum maintenance and repair
- High levels of quality and reliability
- Fastest time to market
- Environmentally safe products and processes
- Designs that can be updated and upgraded as necessary

Any design that falls short of these objectives may fail in the marketplace because it does not satisfy the customers' expectations, it is too costly to produce, or it cannot compete successfully against other similar products.

11.2.2 ■ Design for Manufacturability and Design for Assembly

Classical guidelines for successful manufacturing operations include the following recommendations:[10]

- **Employ a division of labor.** Rather than expect one person (a craftsperson) to perform all of the many different tasks required for the fabrication of a product, design the process so that each worker performs

9. Walker and Boothroyd (1996).
10. See, for example, *How Things Are Made* (1981).

only a limited number or subset of these tasks. The process should be more efficient and the quality of the finished product should be enhanced as each person becomes increasingly more skilled in performing his or her specialized work.

■ **Use interchangeable (standardized) parts.**   Components manufactured in large quantity are designed to be identical to one another in order to simplify assembly and replacement operations. Eli Whitney is usually credited with first achieving the goal of interchangeable or standardized parts when he manufactured a group of muskets for the United States government, muskets that could be assembled from a set of (nearly) identical components. (The parts were not actually interchangeable among the 10,000 muskets produced by Whitney.)

■ **Use assembly line operations.**[11]   Once truly interchangeable parts could be produced and labor was divided among many workers, the creation of the moving assembly line seems to have been inevitable. Henry Ford (originally a machinist and steam engine mechanic) instituted such a process for the manufacture of his Model T automobiles during the early 1900s. As the unfinished auto traveled by conveyor belt passed each worker's location, a part (e.g., engines, radiators, fuel tanks) was added to the car. Each worker performed the same tasks on each car. This well-designed sequencing of operations reduced costs significantly in the production of each automobile, allowing Ford to dramatically reduce the car's retail price (from $850 to $290) over a 17-year period. He became the dominant auto manufacturer of his day and thereby confirmed the value of the assembly line method.

■ **Use machines whenever appropriate.**   Machines often are used to perform many of the tedious and/or unsafe tasks involved in the manufacture of a product. These machines may be quite complex (e.g., robots that are capable of completing many different assembly operations in a predetermined sequential pattern during the fabrication of a computer circuit board) or very simple (e.g., a flow gauge that continuously monitors the rate at which fluid is traveling through a pipe).

Modern manufacturing operations often incorporate design for manufacturability (DFM) and design for assembly (DFA) principles, such as:[12]

■ **Use modular design and subassemblies.**   In order to achieve greater efficiency and flexibility in manufacturing operations, modern engineers often design products that are constructed from modular components or subassemblies. Subassemblies may be used in more than one product, thereby allowing volume discounts in costs and labor. Furthermore, modular designs allow one to replace (if necessary) a

---

11. Claypool (1984).

12. See, for example, Anderson (1990), Dieter (1991), *How Things Are Made* (1981), and Machlis et al. (1995).

single failed subassembly rather than an entire product unit. Modular systems also can be more easily redesigned to respond to changing customer needs, allowing one to update and upgrade the product.

In order to achieve modularity in design, many companies now group similar parts to form a single cluster or family type. These similar components are identified and coded according to certain shared attributes (e.g., geometric shape; locations of holes, slots, and other features; size; material). This practice, known as *group technology* (GT), thereby creates a set of standard part families that are specified by the company itself to support its own internal manufacturing operations. Since they are expected to be used in many of the company's present and future products, these selected group parts can then be either fabricated or ordered in large quantities, resulting in a substantial reduction of costs while increasing the availability of these components for use in the firm's manufacturing operations.

Group technology reduces the need to continuously reconfigure (i.e., break down and set up) manufacturing workstations in order to process dissimilar parts. If a workstation is dedicated to the fabrication and assembly of parts from a single family, the station's machinery, tooling, jigs, and fixtures can remain largely unchanged since these parts will share common attributes with one another.

These dedicated workstations sometimes are clustered together with a corresponding set of material handling processes to form *automated manufacturing cells.* These cells or islands of automation in which a set of closely cooperating machines are used to fabricate and assemble products can increase production efficiency by reducing material handling operations and workstation set-up times.[13] Although largely automatic and computer-controlled in their mostly repetitive processing operations, these cells nevertheless are ultimately supervised and maintained by human operators who are able to respond to changing or unexpected conditions.

GT and automated manufacturing cells have led to the strategy of *flexible manufacturing* through which a company is able to respond better to changing market demands by varying production rates and product designs without suffering major disruptions in its manufacturing operations. The rate of production is adjusted to satisfy the current market demand for a particular design, without the need for large production runs (meant to reduce workstation and other set-up costs) and their resultant (often substantial) inventories of unsold units that can be costly to maintain.

- **Use rapid prototyping.** The prototype of a design is an initial working model. Computer modeling and simulation systems now allow many design prototypes to be created quickly and easily. For example,

13. Williams (1991).

in stereolithography a plastic prototype of an object is created from its 3D computer model by sequentially constructing thin cross-sections of the model from liquid polymer; layer after layer of the object is formed by tracing its cross-sectional outlines in the polymer with a laser until the operation is complete.[14]

Moreover, entire factory processes can be simulated with the use of commercial software so that the engineering team can attempt to optimize the flow of materials and components through the system by identifying and eliminating potential bottlenecks and other problems.[15]

- **Minimize the number of parts.** Minimize the total number of parts required in a design, thereby reducing the financial expenditures, labor, and time that must be devoted to fabrication, assembly, transportation, cleaning, inspection, replacement, and disposal.

- **Minimize part variations.** For example, one may be able to use a single type of fastener throughout a product rather than a range of differently sized screws, bolts, and nuts (unless such variation is absolutely necessary).

- **Design parts to be multifunctional.** For example, a single screw, when properly located, can be used to both fasten and align components, thereby eliminating the need for additional parts.

- **Avoid separate fasteners.** For example, screws can be replaced by molded snap fits unless the screws will serve as more than simple fasteners. Separate washers and nuts might be eliminated by using self-tapping screws, threaded holes, or welded nuts.

- **Minimize the number of assembly operations.** Design parts so that both separate motions and errors are minimized during assembly. For example, gravity-feed chutes or other fixtures might be used to guide parts together during assembly. In addition, interlocking parts can be designed with holes, chamfers, tapers, and other features to ensure proper positioning when assembly is about to occur, without requiring the operator to carefully position these parts with respect to one another.

- **Maximize tolerances for easy assembly.** Since all fabrication processes are inaccurate, determine the maximum acceptable variation (i.e., the tolerance) in the size or shape of each component, then design these parts to ensure that they will fit together if they lie within these tolerance ranges. Conversely, if tolerances are too large, the parts may be fitted together so loosely that the design fails to function or the components separate during the product's operating life.

- **Provide access for easy assembly.** The assembler must be provided with unobstructed access to the product undergoing assembly, with sufficient space for the required tools and for operations that must be performed.

14. Dieter (1991).
15. Walker and Boothroyd (1996).

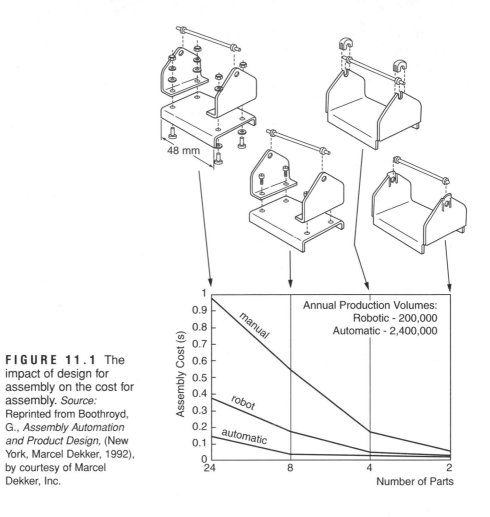

FIGURE 11.1 The impact of design for assembly on the cost for assembly. *Source:* Reprinted from Boothroyd, G., *Assembly Automation and Product Design,* (New York, Marcel Dekker, 1992), by courtesy of Marcel Dekker, Inc.

Figure 11.1 illustrates the impact that DFA can have on the assembly costs for a product as the component parts are reduced in number. Each component and its functionality in the overall design are examined to determine if it can be combined with other components or eliminated entirely. The result often is a simplified and cost-effective design configuration without a reduction in overall product performance and functionality.

11.2.3 ▨ Design for Reliability

Reliability[16] has been defined as "the probability that a device will satisfactorily perform a specified function for a specified period of time under given operating conditions."[17] Many products exhibit reliability behavior that can

16. Anderson (1990), Kolarik (1995), and Kuo and Zhang (1995).
17. Smith (1993).

be described by the so-called bathtub curve shown in Figure 11.2, divided into three phases, which are not necessarily drawn to scale. As can be seen, there is a high number of units that fail early in their expected operating lifetime because of manufacturing defects, damage while in distribution or storage, or errors in installation. Many of these potential early failures can be identified before they enter the marketplace by subjecting all units to a series of stress tests (extremes of temperature, humidity, voltage, mechanical load, vibration, and so forth). During the third phase of the product's expected operating lifetime, most of the long-term surviving units will fail due to wear and deterioration.

During the product's expected useful operating lifetime (phase two in the diagram), some units can be expected to fail at a constant (and we hope small) rate λ. These failures often are the result of taxing the product beyond its performance capabilities; that is, the stress or load to which the unit is subjected is greater than its strength or capacity to perform at that level. The product should be designed so that its capabilities will be able to satisfy the expected demands of operation without the risk of failure. To do so, the engineering team must be able to anticipate most of the operating demands that the product will experience and be aware of the performance limitations of the design through data collection, experimental investigation, and analysis.

As can be seen in Figure 11.3, a stress-strength diagram, the potential stresses or loads that may be applied to a unit often have a range of possible values, reflecting the variations that may occur in operating environments, users' expectations, and so forth. (Different customers may subject a product to differing loads.) Moreover, different units may exhibit differing levels of stength or performance capacity, resulting in a range of possible values. If these ranges in strength and stress overlap, some units with strengths lying

FIGURE 11.2 Typical bathtub curve behavior.

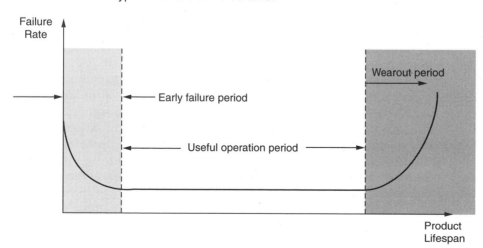

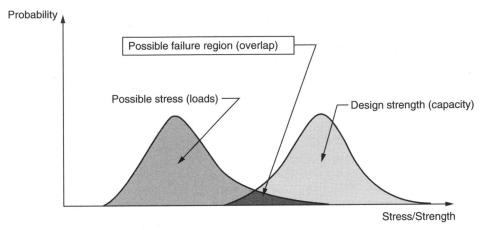

FIGURE 11.3 Overlap of stress (load)-strength (capacity) probability distributions.

in this overlap range may be unable to survive the loads that are applied to them. Clearly, the solution to this dilemma would be to eliminate overlap between the strength and stress ranges by overdesigning the product—that is, increasing its capacity to perform the expected tasks to a level well above any possible loads that may be applied to it. Unfortunately, extensive overdesigning is usually not possible because of the increased costs that would be associated with it. Instead, one tries to minimize the overlap between the strength and stress ranges, thereby reducing the likelihood of product failure without unduly increasing cost.

Engineers often can increase the reliability of a system through component selections, intentional design redundancies, preventive maintenance, and corrective maintenance. For component selections, off-the-shelf parts purchased from vendors are selected based upon their historical performance levels. In addition, the manufacturing operations and materials used in the fabrication of customized in-house components can be evaluated and hopefully improved through the use of failure mode and effects analysis, fault tree analysis and other techniques (see Chapter 9).

Intentional redundancy in a design simply means that there are duplicate or standby subsystems that will begin to operate if the primary mechanism fails to operate. For example, if there are two emergency shutoff devices designed to operate in parallel in a system, then in an emergency, the first device is expected to shut the system down. If the first device fails to operate, the second device is then responsible for shutting the system down. If each device has an expected reliability of 99 percent, then the likelihood of the first device failing to operate is one percent, the likelihood of the second device failing to operate is one percent, and the probability that both devices will fail simultaneously is the product of these failure rates: $(0.01) \times (0.01) = 0.0001$ or 0.01 percent. Redundancy, although potentially expensive, can dramatically increase the reliability of a system (see Case History 9.3 *The United States' Worst Bridge Disaster*).

Finally, reliability can be enhanced through *preventive maintenance* operations. *Corrective maintenance* operations are least preferable, since they are performed after a failure has occurred.

11.2.4 ▨ Design for Quality

The concept of quality in products and services has been defined in many forms. Two definitions that seem to capture the essence of the concept are the following:

> Quality is the total composite product and service characteristics of marketing, engineering, manufacture, and maintenance through which the product and service in use will meet the expectations of the customer.[18]

> Quality is the totality of features and characteristics of a product or service that bear on its ability to satisfy stated or implied needs.[19]

Design for quality (DFQ) emphasizes this need to satisfy customer requirements and expectations by minimizing the potential impact of variations in a product's manufacture and operation.[20] These efforts may be reactive, in which one seeks to detect and eliminate existing flaws in a product or service, often through extensive inspection and testing operations, or proactive, in which one seeks to prevent flaws from occurring by eliminating the underlying causes. Reactive or diagnostic operations often examine samples taken from the entire population of manufactured units; if these sampled units are found to be satisfactory, the entire population is then deemed fit for distribution to the marketplace. Although largely successful if the statistical sampling and analysis are performed correctly, this approach may nevertheless allow some flawed units to enter the market unless 100 percent of the population undergo inspection and testing. In addition, there may be other types of flaws in the product that were never recognized by those who designed the quality inspection tests. For these reasons, more emphasis has been placed in recent years upon the proactive or preventative approach since it can lead to improvements in the quality of an entire production line with a corresponding reduction in the costs for reworking or discarding flawed units. Proactive strategies seek to avoid losses in production by eliminating flaws entirely, whereas reactive methods seek to limit the extent of these losses.[21]

---

18. Feigenbaum (1983).

19. ISO 9000 (1992); ISO 9000–9004 is a series of quality standards for products and services, introduced in 1987 and since adopted by many firms throughout the world.

20. Crow (1992) and Kuo and Zhang (1995).

21. Kolarik (1995).

Failure mode and effects analysis (FMEA), cause-and-effect (Ishikawa) diagrams, and fault tree analysis (FTA)—all of which were reviewed in Chapter 9—are among the many tools used in proactive quality assurance strategies. Specific DFQ strategies include:

- The *Taguchi engineering method,*[22] in which one seeks to reduce any undesirable variability in the manufacture of a system. Some differences among manufactured units of a design are inevitable because of the variations that will occur throughout the production operations and in the materials' characteristics. The challenge for the design engineer is to ensure that these variations will not substantially reduce the quality of a product.

  The Taguchi method is performed at three levels of sequential development: system design, parameter design, and tolerance design. At the system level, one focuses upon the overall functionality of the product by considering the components, materials, assembly operations, fabrication processes, and other elements of the design's manufacture and operation. This initial system level generally corresponds to the development of a conceptual design that will satisfy the needs and expectations of the customer.

  Once the overall system design has been completed, one next seeks to identify the product and process parameters that can be controlled in order to limit variations in the finished units, thereby optimizing product performance while minimizing costs. Experimentation, sampling, and statistical analysis are used to determine parametric values.

  Finally, one needs to determine the appropriate tolerance specifications for each of the designated design parameters. Narrow or small tolerances generally correspond to higher costs in production and inspection; thus, one needs to identify those critical design parameters for which tolerance ranges must be small and determine the appropriate values for each of these ranges together with the tolerance values for the other parameters in which relatively large variations among the manufactured units will be acceptable.

- *Benchmarking,* in which one examines the best practices within one's own company (internal benchmarking), at competing firms (external benchmarking), and throughout the world (generic benchmarking) in order to adapt these practices to the current set of tasks to be performed.[23] By comparing and objectively evaluating its own products against those of the competition, a company becomes more likely to recognize aspects of its operation that need improvement.

- *Quality function deployment* (QFD)[24] was developed by Technicomp, Inc., of Cleveland, Ohio, and since used by such major firms as Alcoa, Bethlehem Steel, Boeing, Caterpillar, Chrysler Corporation, and the

22. Taguchi (1986), Kolarik (1995), and Kuo and Zhang (1995).

23. Kolarik (1995), Kuo and Zhang (1995), and Zairi (1992).

24. Berger (1988), Havener (1993), Kuo and Zhang (1995), Sullivan (1986), and Walker and Boothroyd (1996).

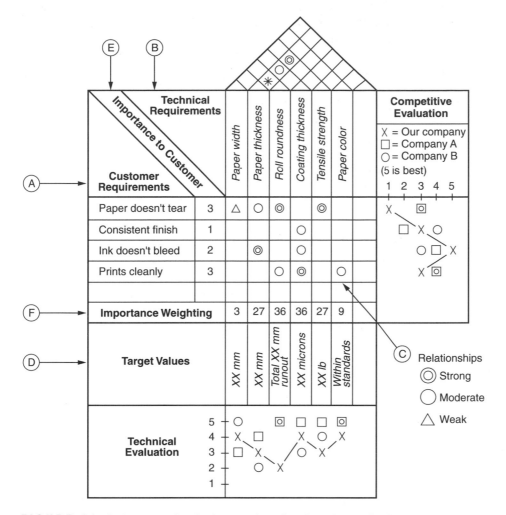

FIGURE 11.4 An example of a house of quality chart for quality function deployment. *Source:* Technicomp, Inc., Cleveland, Ohio. Reprinted with permission.

Cadillac Motor Car Company.[25] In this method, one uses customer feedback to create a "house of quality," as shown in Figure 11.4. The house of quality helps one to correlate stated customer requirements (denoted by A in Fig. 11.4) with the design's technical requirements (B). As can be seen in the figure, these correlations can be classified according to their strength (C). Target values (D) correspond to the desired specifications for each technical requirement, whereas other values indicate the importance of each requirement to the customer (E) and the importance of each technical requirement (F), obtained by multiplying each E factor

25. Kolarik (1995) and Walker and Boothroyd (1996).

by the set of correlation factors C. In Figure 11.4, a weak correlation relationship is assigned a C value of one, a moderate relationship is assigned a value of three, and a strong relationship is given a value of nine. The roof of the house of quality is used to identify relationships among the technical requirements, the relative strengths of which are indicated by different symbols. The competitive evaluation portion of the house is used to denote the relative abilities of a firm and its competitors to satisfy the customer requirements, and the technical evaluation wing of the house is used to note the relative abilities of different companies to meet the technical requirements. QFD and its house of quality provide engineers with the opportunity to identify areas in which product design and performance should be improved.

11.2.5 ▨ Design for Packaging

Many packages (e.g., containers for food products) are designed to protect the product properly during transport, storage, display, and possibly throughout its entire operating life. Moreover, these packages must be aesthetically pleasing and cost effective. *Design for packaging* (DFP) provides guidelines for achieving these objectives.[26]

A product's package is often expected to perform many different functions, some of which may appear to be in conflict, for example:

- Provide aesthetic appeal and brand name recognition for the product.
- Protect the product contents from spoilage or damage.
- Provide the customer with a broad range of product sizes and formats.
- Help to establish a set of product standards and measures of quality.
- Prevent the misuse of the product.
- Provide closures that are inexpensive to manufacture and easy to open.

Packaging requirements have acted as a catalyst to stimulate advances in technology. Moreover, technical advances in other fields have been adapted to satisfy the needs of the packaging industry. Consider, for example, the changes in beverage cans that have occurred since the 1930s.

Beverage Cans

Just prior to and during World War II, the amount of metal used in cans was reduced significantly, initially to limit U.S. dependence upon imports of tin from Malaya (a country that was then controlled by Japan) and then to reduce the amount of steel used in cans so that this metal could be devoted to more critical military products. This reduction was achieved by developing new manufacturing methods for thinner yet strong cans.

26. Hines (1995) and Stern (1981).

Before 1935 most commercial beverages were sold in bottles. In that year, cans that could withstand the high internal pressures of beer were first developed and used. This application of cans resulted in a 64 percent reduction in the amount of space required by bottles, thereby saving significant storage space in warehouses, trucks, and retail stores. The introduction of cans also produced a very significant reduction in the packaging weight of the product—a reduction of more than 50 percent when compared to bottles of equivalent size.

However, cans even stronger than those used for beer would be needed to store soft drinks because of the higher internal pressures of these carbonated beverages. Finally, in 1953 high-pressure cans were successfully developed and introduced in the market for this use.

The all-aluminum can first appeared in the market in 1958. The use of this metal simplified the can production process. Previously, tin-plated cans were manufactured in three sections (top, bottom, side), which were then sealed together to form the enclosure with a vertical seam running along the side of the can. In contrast, aluminum cans were composed of only two sections, a bottom piece directly connected to the side, and a top piece. This two-piece construction simplified production and eliminated the vertical seam along the aluminum side, providing customers with a more appealing beverage container.

Additional modifications occurred in can openings, beginning with the introduction of the pull-tab top and followed by the removable ring and finally the captive ring. Each change was necessitated by an obvious need: Pull-tabs sometimes fractured and broke off before the can could be opened, and removable rings created dangerous litter and were sometimes swallowed. Additional improvements in cans should be expected in future years.

The popularity and widespread use of plastics in product design was greatly accelerated by the packaging industry during World War II. Among the packaging applications for plastics introduced at that time were lightweight and durable water canteens constructed of ethylcellulose, firearms and other equipment wrapped in protective plastic coverings, and plastic envelopes for maps and other important documents.

Many products have relied (at least partially) on their packaging formats for success in the marketplace. The patented "In-er" seal wrapper used in Uneeda Biscuits provided the National Baking Company (Nabisco) with the opportunity to establish its crackers as the benchmark against which other crackers would be measured for freshness and cleanliness (see Section 5.3.2 for a discussion of this product). Similarly, in 1930 Kleenex tissues introduced its "Serv-a-Tissue" dispenser box , in which two interfolded rolls of tissues resulted in a continuous supply of tissues—provided one at a time—to "pop up" from the box as needed. Again, innovative packaging helped the product to become established as the leader in the marketplace.

Modern efforts have produced such innovations as tamper-evident containers, child-resistant closures, and aseptic packaging. In aseptic packaging, the container is sterilized separately from its contents and then filled in a sterile environment. Aseptic single-serving juice boxes and pudding snack containers do not require refrigeration if the package remains unopened. In Europe, fresh milk is packaged in aseptic containers and stored without refrigeration. (Supermarkets in the United States have not encouraged the use of this packaging for milk because of their desire to encourage customer traffic flow through the refrigerated areas of their stores in the hope of selling other chilled items.)

The packaging industry also must reduce waste and environmental pollution due to its products. It seeks to *reduce, reuse,* and *recycle* its products, in that order of priority. Much progress has been achieved in these efforts, from the elimination of dangerous chlorofluorocarbons (which deplete the ozone layer in the stratosphere) in aerosol containers, to the use of more efficient, biodegradable, recyclable, and/or reusable packaging. In 1990 the U.S. Environmental Protection Agency estimated that about 26 percent of all used packaging in the nation's waste stream was being recycled or reused (at that time, 63.2 percent of all aluminum cans were being recycled and 37 percent of all paper products). Many household products (such as laundry soaps and juice drinks) are now packaged in more concentrated form, thereby reducing the amount of material needed for the container. Increasingly, the containers themselves are composed of recycled materials.

Governments have also set expectations for reductions in waste (e.g., the European Union required that 50 percent of all wastes be recycled by 1998). Although some progress has been achieved, work in this area must continue if we are to become more efficient and responsible in our use of the earth's resources. Effective packaging should be incorporated into the design of manufactured products.

11.2.6 ■ Design for Maintainability, Disassembly, and Recyclability

Design for maintainability emphasizes the need to ensure that a product will continue to operate properly with only minimum maintenance and repair (perhaps through periodic preventive maintenance). *Design for disassembly* and *design for recyclability* are two strategies that focus the engineer's attention on the end of a product's operating life; they encourage engineers to select materials, fasteners, sequential assembly operations, and other design elements that will allow obsolete or deteriorated units to be safely and economically reused, recycled, or discarded.

Many manufacturers are working to ensure that their products can be safely recycled after their operating life is over, thereby protecting our environment from unnecessary wastes and reducing the use of raw materials in fabrication processes. For example, 65.4 percent of all aluminum beverage cans were recycled in 1994. If a product cannot be entirely recycled, the disposal of any of its components should be kept to a minimum and performed safely.

Single-Use Cameras

To illustrate the importance of recycling in the success of a product, consider single-use cameras: When first introduced by the Fuji Photo Film Company (in 1986) and by the Eastman Kodak Company (in 1987), these inexpensive cameras that could only be used once and then returned to the store for development of the film inside were a commercial failure. Part of the reason for this failure was the fact that the cameras could not be recycled, leading to concerns about their negative impact upon the environment.

The cameras were redesigned to be more recyclable, and they became increasingly successful in the market. By 1995, 77 percent of the single-use cameras sold by Kodak were being returned to the company by photo developers and Kodak was recycling 85 percent (by weight) of each returned camera. Recycling and design for environmental protection can lead to enormous success in the marketplace. (See Case History 1.4 *Single-Use Cameras* for further information about this product.)

In summary and to reiterate: Implementation is the final phase of the design process. Nevertheless, concurrent engineering requires that choices in the fabrication, assembly, and eventual disposal of a product should be considered simultaneously with the development of the design concept itself—and not delayed until after the final design has been selected—since many of these options will be determined by the final design.

11.3 Poor Implementation Can Lead to Failure

As we have seen throughout this text, much effort is usually expended during the initial stages of an engineering project in order to ensure that the design will perform its functions safely and successfully. It then becomes important to recognize that if the design is not fabricated according to its specifications, failure—or even tragedy—may result. A simple modification in the support structure for the hotel walkway system at the Hyatt Regency in Kansas City, Missouri, led to the 1981 collapse that claimed 114 lives; none of the engineers working on the project recognized the structural implications of this modification (see Case History 11.4 *Kansas City Walkway Disaster*).

The Citicorp Center in New York City is another example of a potential disaster that was hidden within a seemingly harmless construction modification, a disaster that was averted by the actions of one man.

Citicorp Center

The unique architecture of the Citicorp Center (Fig. 11.5) was in response to a major design constraint: St. Peter's Church was located on the northwest corner of the site for the Center, along Lexington Avenue between 53rd and 54th Streets. An accord was reached wherein the Center would rise around and above a new freestanding church structure. The 59-story tower of the Center would stand on four nine-story columns or stilts, located at the midpoints of the building's four sides rather than at the corner points as would normally be the case. The midpoint positions of the columns allowed the church to nestle under one corner of the 914-ft. tower and provided a stunning architectural design for the Center.[27]

Structural engineer William J. LeMessurier designed the innovative steel skeleton for the Center. In his design, LeMessurier used six tiers of eight diagonal braces each that would carry the loads on the building to the four main columns. The two-story structural steel members forming these braces were to be welded to interconnecting steel plates.

However, the steel erector for the building recommended that the welded joints be replaced by bolted ones in order to reduce labor costs. Subsequent calculations indicated that these bolted connections would indeed be sufficiently strong, and construction proceeded. The finished Citicorp Center opened in 1977 with the bolted connections in place.

The following year LeMessurier received a telephone call from an engineering student who was preparing a report on the Center's design. After the conversation, LeMessurier began to ponder the details of the design. He knew that the steel braces would easily resist any perpendicular winds striking the building and that the midpoint positions of the load-bearing main columns provided the strongest support against quartering winds (i.e., winds that strike at the corners of the building and increase the loads on the two adjacent sides). LeMessurier now calculated the forces acting on his diagonal braces because of the quartering winds, discovering that these winds increased the tension on half of the braces throughout the structure by 40 percent. Such an increase was insufficient to raise concern about the tensile strength of the braces; however, the bolt connections between the braces were an entirely different matter. The 40 percent increase in tension within the braces resulted in a 160 percent increase on the bolt connections between the braces. Unfortunately, those who had originally specified the number of bolts per connection had assumed that the diagonal braces were acting as *truss* members in the structural framework rather than *columns*. In fact, the braces were serving as columns, thereby requiring considerably more bolts to resist the expected wind loads than had been used in the actual construction.

LeMessurier was stunned to discover that a so-called 16-year storm (i.e., one expected to occur on average once every 16 years) would be powerful enough to shatter his building, causing it to fall on Bloomingdale's and other nearby structures.

LeMessurier contacted Citicorp and others to inform them of the problem. It was decided to weld steel plates (each weighing 200 to 300 pounds) over each of the

27. Engineering News Record (1978), Goldstein and Rubin (1996), Korman (1995a, 1995b), and Morgenstern (1995).

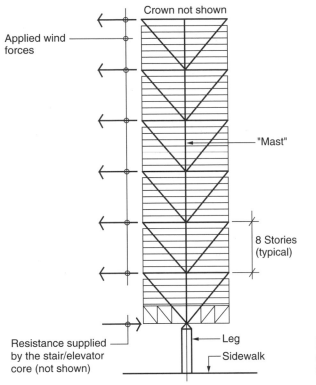

Applied wind forces

Crown not shown

"Mast"

8 Stories (typical)

Resistance supplied by the stair/elevator core (not shown)

Leg

Sidewalk

FIGURE 11.5 The design of the Citicorp Center.

bolted joints throughout the building. Work began as disaster agencies prepared for the possibility of an emergency if a powerful storm arrived in the city before the corrective work could be completed. Fortunately, a threatening hurricane named Ella moved out to sea before reaching the city, and the welded connections were completed without incident. It is estimated that with these corrections the Citicorp Center now would be able to withstand a 700-year storm (or more), if necessary, making it one of the safest structures in the world.

Today, William LeMessurier is immensely admired for his courage and candor in quickly moving forward to protect the public once he discovered the danger that was lurking within the Center. He risked both professional humiliation and financial ruin; instead, he has received many accolades for his honesty and integrity. LeMessurier explains his actions in these words: "You have a social obligation. In return for getting a license and being regarded with respect, you're supposed to be self-sacrificing and look beyond the interests of yourself and your client to society as a whole. And the most wonderful part of my story is that when I did it, nothing bad happened."[28]

28. Morgenstern (1995, p. 53).

11.4 Materials Selection in Design

The materials selected for a design often will determine the fabrication processes that can be used to manufacture the product, its performance characteristics, and its reclyclability and environmental impact. As a result, engineers should acquire a robust understanding of material characteristics and the criteria that one should use in making material selections.

11.4.1 ▦ Matching Materials to Performance and Processing Requirements

Engineers must select the most appropriate materials for their designs, that is, materials that match both the performance requirements of the product and the processing requirements for its manufacture. For example, the materials used in some baseballs have remained essentially unchanged for more than one hundred years because they satisfy some very basic performance requirements.[29]

Baseball materials	Performance requirements
Cork center	Stable spherical core
Rubber covering over cork	Necessary bounce and elasticity
Wool yarn around rubber	Proper sizing of the ball
Layer of cotton yarn	Smooth surface
Latex glue over yarn	Yarn fixed in place
Leather covers	Durable and comfortable surface

When selecting materials for a design, questions that should be asked and answered include

- What materials are available?
- How will each material, if selected, aid or hinder efforts to achieve the performance requirements and specifications (e.g., strength, shape, electrical and thermal conductivities, corrosion resistance, weight, availability of materials, minimum costs, elasticity, color, response to variations in temperature and humidity, surface finish, protection of the environment, safety of the user)?
- Which manufacturing processes are available? For each material under consideration, how would the corresponding manufacturing process affect the production of the design in terms of costs, precision of work, fabrication time, assembly, plant space and equipment, use of personnel, waste of raw materials, and so forth?

The following case describes the selection of a specific material in order to satisfy a given set of desired performance requirements.

29. *How Things Are Made* (1981).

Soy Ink for Newspaper Production

In 1979 petroleum shortages led a committee from the Newspaper Association of America to begin searching for an alternative to the petroleum-based ink that was being used throughout the industry. Any viable alternative would need to satisfy most of the following requirements:

- General availability
- General compatibility with existing press equipment
- Low toxicity and minimal hazard to the environment
- Able to produce bright, clear images and colors
- High "runability;" that is, a large number of copies per unit of ink
- Fast absorption
- Minimal bleeding through paper
- Low cost

After considering and rejecting various alternatives, the committee found that ink made from soybeans would satisfy most of these requirements.[30] In comparison to petroleum-based ink, the soy ink produced more vibrant colors, generated more printed copies per pound, emitted far fewer dangerous fumes, and was more easily available as a resource. Although its cost was slightly higher, soy ink was adopted, first by the Cedar Rapids *Gazette* and then by other newspapers. Today, most large newspapers throughout the United States use soy ink, indicating that higher cost sometimes can and should be overridden by other factors that are more important to a successful design.

In summary, the materials chosen for an engineering design should match the overall performance, environmental, processing, and financial requirements of the product. For example, metal used in a bookcase shelf should be able to support the weight applied to it without bending, yet be easy to move and assemble by hand—that is, it must be strong yet lightweight in order to satisfy these performance requirements. In addition, it must be rust-resistant or capable of being treated with a rust-preventive coating (an environmental requirement), and it must be able to acquire a smooth finish for aesthetic and safety reasons (a processing or manufacturing requirement). Moreover, both the material itself and its manufacture must be relatively low in cost if the product is to be profitable (financial requirements).

11.4.2 ▓ Material Properties

In order to achieve a successful match of materials to requirements, the design engineer should be familiar with the basic material properties to be considered.[31] By reflecting on these properties and reviewing current materials

30. Wann (1996).
31. Bolton (1994), Dieter (1991), Kalpakjian (1989), and Lindbeck (1995).

handbooks, trade journals, and other sources of data, a set of initial material candidates can be formed. Each of these candidates can then be evaluated and compared in more detail until a final selection can be made.

The basic properties that are usually considered during materials selection can be divided into such general categories as mechanical, electrical, physical, chemical, thermal, and economic.

Mechanical Properties   A material can be in compression (i.e., forces are acting to compress or squeeze it), in tension (when it is being stretched), in shear (when its faces or layers are sliding with respect to one another, possibly leading to fracture), or in torsion (when it is being twisted). The response of a material to such forces is described by such mechanical properties as strength, stiffness, ductility, toughness, hardness, and creep, as described below.

■ **Strength**   Measured in terms of the maximum (tensile, compressive, or shear) force per original cross-sectional area, a material's (tensile, compressive, or shear) strength describes its ability to resist breaking. Another quantity sometimes used to measure the resistance of a material to applied force is its yield stress (force/area), which denotes the point at which the material becomes inelastic, that is, it begins to becomes permanently deformed due to the applied force and will not return to its original unstressed state.

■ **Stiffness**   Measured in terms of the modulus of elasticity, a material's stiffness describes its ability to resist bending. This modulus is equal to the ratio of stress divided by strain in the region of elastic behavior, where stress and strain are defined by the following ratios:

Stress = force/area

Strain = $\Delta L / L$

where L is the original length of the material before the force is applied and ΔL represents the change in this length. Thus,

Modulus of elasticity = stress/strain

in the range of elastic behavior. (This elastic range corresponds to the initial linear or straight-line region of a stress-vs.-strain graph. In this range the modulus of elasticity is sometimes referred to as Young's Modulus or the tensile modulus.) Therefore, a relatively large modulus of elasticity corresponds to only small changes ΔL in a material's length under applied forces, that is, the material resists bending and can be described as stiff.

■ **Ductility**   Measured in terms of the percentage of elongation (percent elongation), a material's ductility describes its tendency to exhibit permanent deformation or plastic behavior under applied forces before breaking. Such deformation can be important as a warning signal that the material is about to break. Conversely, a brittle material such as glass will shatter without providing any such warning signal. Percentage of elongation is given by

$$\% \text{ elongation} = 100 \times (L_f - L_i)/L_i$$

where L_i is the initial length of the material, and L_f is its final (deformed) length.

- **Toughness** A material's toughness is related to its ductility and yield stress. It describes the material's ability to withstand sudden impact by absorbing the applied energy without fracturing. Toughness can be measured in several ways. For example, if one wishes to measure the energy or work required to stretch a unit volume of material until it fractures, a force-extension graph can be developed by measuring and plotting each of the applied tensile forces F and resultant extensions or elongations ΔL (where L is the original unstretched length of the material). As noted earlier, stress is equivalent to force per unit area and strain is equal to elongation per unit length, thereby allowing this force-extension graph to be easily converted into a stress-strain graph by simply dividing the forces by the value of the unit area and the extensions by the original length. Since the work W performed in stretching the material is equal to the product of the applied tensile force and the resultant elongation (i.e., Work = Force × Elongation), and given the definitions of stress and strain, we then have

$$\begin{aligned}
\text{Work/volume} &= (\text{force} \times \text{elongation})/\text{volume} \\
&= (\text{force} \times \text{elongation})/(\text{area} \times \text{length}) \\
&= (\text{force/area}) \times (\text{elongation/length}) \\
&= \text{stress} \times \text{strain}
\end{aligned}$$

Hence, the area under the curve of plotted points on the stress-versus-strain graph corresponds to the work or energy per unit volume needed to stretch a material to its breaking point. Materials with high-yield stress values remain elastic even when stretched over large distances, eventually becoming permanently deformed and finally breaking. Therefore, such materials also exhibit high values of toughness.

Toughness also can be measured via the Charpy and Izod tests in which sudden impacts are applied to a test piece until it fractures; the amount of applied energy necessary to break the material corresponds to its toughness. Since ductile (nonbrittle) materials deform plastically before finally breaking under an applied force, they also exhibit high values of toughness.

Fracture toughness denotes a material's ability to resist crack propagation, given that a crack is already present in the material.

- **Hardness** This term is used to denote a material's ability to resist abrasion or scratching. Several standardized tests (e.g., Brinell, Vickers, Rockwell, Moh) have been developed to measure this resistance.
- **Creep** If a material continues to deform under an applied constant stress (rather than reaching a final equilibrium state), this gradual deformation is the creep of the material.

Electrical Properties Among the electrical properties that can be associated with a material are conductance, resistance, and dielectric strength.

- **Conductance** is the ability of a material to conduct electric current, measured in terms of electrical conductivity. Conductors contain large numbers of electrons that are so loosely bound to individual atoms that they are relatively free to move in response to an applied voltage potential. As a conductor's temperature is increased, its conductivity decreases because of the increased kinetic vibrations of the molecules acting to block passage of electrons through the material.

  Semiconductors (such as silicon and gallium arsenide) contain very few electrons that are free to move at low temperatures. However, as the temperature of the material is increased more electrons break free of the atoms to which they are bound; these electrons and the holes or empty sites on the atoms that they leave behind are then free to move through the material. Moreover, impurities can be added to a semiconductor in order to intentionally modify its electrical conductivity by increasing the number of free electrons or holes in the material; this process is known as "doping." (Case History 11.2 *Microchip* describes the use of doping as part of a manufacturing process). Of course, the introduction of any impurity must be intentional and not accidental if the material is to behave in the desired manner.

  Insulators can conduct little or no current because they essentially lack free electrons. If the material's temperature is raised to very high levels, some electrons may break free of the atoms to which they are bound, thereby allowing a current to be transmitted through the material.

- **Resistance** describes a material's ability to oppose the passage of electric current through it; it is measured in terms of the material's electrical resistivity (the reciprocal of electrical conductivity).

- **Dielectric strength** is a measure of the highest voltage that can be applied to an insulating material without causing electrical breakdown.

Each of these quantities may be important when trying to match a material to a performance function that involves electricity.

Physical, Chemical, Thermal, and Other Properties Many other properties can be associated with a material. A single physical property such as mass might appear in several related forms: density (= mass per unit volume), specific strength (= strength/density), and so forth. Other physical properties include permeability, porosity, reflectivity, texture, transparency, and viscosity.

Among the chemical properties that may be important in materials selection are corrosion, flammability, oxidation, and toxicity.

Thermal properties include the coefficient of linear expansion (which measures the amount of expansion occurring in a material as its temperature rises), melting point (the temperature at which a change in a material's state—from solid to liquid—occurs), specific heat capacity (which denotes the amount of heat needed to raise the temperature of a unit mass of material by one degree), and thermal conductivity (a measure of a material's ability to conduct heat).

Moreover, the design engineer must consider such material characteristics as availability, color, cost, malleability, odor, and recyclability, some of which may be (at least partially) dependent upon some of the other properties already noted.

Table 11.1 summarizes some of these common properties used to match materials with functional, environmental, manufacturing, and/or economic requirements.

TABLE 11.1 Some material properties used to match materials with functional, environmental, manufacturing, and economic requirements.

Property	Interpretation	Common or Related Measure
Mechanical Properties		
Strength	Ability to resist breaking	Yield stress
Stiffness	Ability to resist bending	Modulus of elasticity
Ductility	Permanent deformation before breaking	% elongation
Toughness	Ability to withstand impact or resist breaking	Energy or work necessary to fracture material
Hardness	Ability to resist abrasion/scratching	Scores on hardness tests
Creep	Gradual, continuing deformation under an applied constant stress	Creep strength
Electrical Properties		
Resistance	Opposition to passage of electrical current	Electrical Resistivity R
Conductance	Ability to conduct electrical current	Electrical Conductivity $C = 1/R$
Dielectric strength	Highest voltage that can be applied to insulator without electrical breakdown	DS = Breakdown voltage/ insulator thickness
Thermal Properties		
Thermal conductivity	Ability to conduct heat	Amount of heat per second flowing through a temperature gradient ΔT
Specific heat capacity	Heat per unit mass needed to raise temperature T by an amount ΔT	Amount of heat/(mass x ΔT)

continued

TABLE 11.1 Some material properties used to match materials with functional, environmental, manufacturing, and/or economic requirements. (continued)

Property	Interpretation	Common or Related Measure
Thermal Properties		
Coefficient of linear expansion	Amount of expansion ΔL with increasing temperature ΔT	$\Delta L/(L \times \Delta T)$
Melting point	Temperature at which change of state from solid to liquid occurs	T_m = Melting temperature
Other Properties		

Appearance	Flammability	Odor
Availability	Mass oxidation	Texture
Corrosion	Permeability	Toxicity
Cost	Porosity	Transparency
Density	Reflectivity	Viscosity

11.5 Common Fabrication Materials

Materials are often classified into several major categories, including metals, polymers, ceramics, and composites (which include wood and reinforced concrete).[32] Many handbooks, trade magazines, manufacturers' catalogues, and computerized databases are available to guide the engineer when he or she is selecting materials that will satisfy a particular set of design specifications. Nevertheless, some general statements can be made about each of these material categories.

11.5.1 Metals

In general, most metals are

- good conductors of electricity and heat,
- ductile, and
- lustrous.

32. Bolton (1994), Dieter (1991), Lindbeck (1995), and Voland (1987).

Alloys, in which two or more metals are combined, are most often used in product manufacture to obtain a desirable set of material characteristics (e.g., high tensile strength with low ductility). It has been estimated that there are more than 40,000 metallic alloys available for use.[33] Pure metals are seldom used unless one particular attribute is critically important in the expected application (e.g., if a high degree of electrical conductivity is needed, relatively pure copper wire may be selected).

Metals and alloys are usually divided into two categories: ferrous (i.e., those containing iron) and nonferrous. Ferrous alloys include various forms of iron and steel.

- **Iron** Historically, iron is the most important of the basic metals. It is usually combined with carbon in various ratios to form ferrous alloys with desirable properties. These alloys include wrought iron (with 0–0.05 percent carbon content), steel (0.05–2 percent carbon), and cast iron (2–4.5 percent carbon). Wrought iron is a soft, ductile, and corrosion-resistant alloy often used for decorative railings and gates. Cast iron is a hard, brittle alloy to which varying amounts of carbon and other elements are added to make it more easily malleable; it is used for engine blocks, manhole covers, fire hydrants, and other cast products that must be both durable and strong.

- **Steel** Because of its range of desirable attributes, steel is the most common metal in manufacturing. Different ratios of iron, carbon, and other elements can be combined to form various types of steel alloys that exhibit different levels of strength, hardness, ductility, and resistance to corrosion.

   Carbon steels (which include mild steel, medium-carbon steel, and high-carbon steel) are composed of mostly iron and carbon, in contrast to the alloy steels that contain additional elements. Added carbon content increases both the tensile strength and hardness of carbon alloys while decreasing their ductility. Mild steel (with a relatively low percentage of carbon ranging between 0.1–0.25 percent) is used in the manufacture of wire, automobile bodies, and other products in which significant ductility is needed to increase the formability of the material. Crankshafts, gears, and other components that require moderate levels of strength and ductility are made with medium-carbon steel (0.2–0.50 percent carbon content). Wear-resistant products (e.g., hammers, razor blades, and chisels) are manufactured from high-carbon steels (a carbon content between 0.5 percent and 2 percent).

   Stainless steels contain significant (12 percent or more) amounts of chromium for greater resistance to corrosion and oxidation.

The many nonferrous metals include the following elements.

- **Aluminum** Naturally soft and relatively weak, aluminum is often alloyed with copper, manganese, nickel or other metals to increase its

33. Dieter (1991).

strength and hardness. It has become extremely popular because of its formability into various shapes, an advantage of ductility and lack of strength. Moreover, it is lightweight, corrosion-resistant, and an effective conductor of both heat and electricity. It is used in a wide range of applications, from beverage cans and cookware to chalk trays and aircraft components.

- **Copper** A metal with very high electrical and thermal conductivity that is both ductile and corrosion-resistant, it serves as a major component in such common alloys as brass (mostly a mixture of copper and zinc) and bronze (essentially copper mixed with tin).
- **Magnesium** A relatively expensive metal that will corrode without surface protection, magnesium is desirable for aircraft and spacecraft vehicles because of its toughness, formability, relatively low weight, and durability.
- **Nickel** Nickel serves as a base metal of many nonferrous, corrosion-resistant alloys, such as nickel-copper and nickel-chromium-iron.
- **Titanium** A corrosion-resistant metal with a high ratio of strength to weight, it is as strong as steel yet considerably lighter in weight. Ductile and tough but somewhat difficult to fabricate and relatively expensive, it often is used in aircraft components, pumps, and other products that must be strong yet light weight.
- **Zinc** Used to galvanize steel so that it becomes relatively corrosion-resistant, this metal also is alloyed with other metals (e.g., copper in order to form brass) to create materials with a specific set of desirable characteristics.

11.5.2 ▓ Polymers

Increasingly, plastics are being used to manufacture both new and established products that were previously formed in metal. Plastics are a mixture of polymers and other additives such as stabilizers, flame retardants, plasticizers, and fillers. When combined, these components can result in such desirable properties as:

- Plasticity, which is the property of being easily deformed, under applied pressure or heat, into a desired shape which is then retained upon the removal of the pressure or heat. Usually, only a single manufacturing operation is needed to obtain the final shape of a part.
- Lightweight.
- Low thermal conductivities when compared to metals.
- Low electrical conductivities.
- The ability to acquire smooth surface finishes.
- The ability to be produced in a wide range of colors.
- Raw materials that are relatively inexpensive, although a significant initial investment in plastics-forming equipment may be required.

This combination of material properties is the basis for the high demand for plastics.

One major disadvantage of plastics is their generally weak or low-impact strength compared to that of most metals. However, some plastics, such as polycarbonate, high-density polyethylene, and cellulose proprionate, do exhibit high-impact strength.

Plastics are formed from such natural raw materials as petroleum, water, coal, lime, and fluorspar. These materials of nature are used to manufacture chemical raw materials such as phenol, formaldehyde, acetylene, hydrogen chloride, cellulose, urea, and acetic acid. Finally, reactions involving these chemical raw materials produce monomers or simple molecules, which can then be used to produce large molecules, often composed of thousands of the monomer building blocks, known as polymers. The prefix "poly" is attached to the name of the monomer from which the polymer has been constructed. For example, polystyrene is polymerized from styrene, and polyvinyl chloride is formed from the monomer vinyl chloride, which in turn is the result of a reaction between the raw chemical materials acetylene and hydrogen chloride (formed from the natural raw materials water, coal, lime, and salt). Variations in polymeric mixtures provide the tremendous range of characteristics and properties exhibited by plastics.

There are two general categories or types of plastics: *thermoplastics* and *thermosets* (thermosetting plastics), reflecting the structural formation of the polymers. Polymeric molecules in thermoplastics are usually linked together in long linear chains that can slide past one another if the material is heated. As a result, thermoplastics can be repeatedly reformed into new and different shapes upon heating. In contrast, the chains of the polymeric molecules in thermosets are cross-linked together to form a more rigid pattern; thus, they retain their shape even when heated unless they are melted into the liquid state.

This distinction in the behavior of thermoplastics and thermosets is critical when one is trying to match a plastic material to a specific engineering function. For example, one must use a thermoset plastic in the manufacture of automobile distributor caps since the heat of the engine may sufficiently soften a thermoplastic until it can no longer function properly.

Some polymers are manufactured in both thermoplastic and thermoset forms (e.g., polyurethane, which is often used as a protective coating because of its resistance to abrasion, tearing, and chemicals, and as a filler in mattresses.)

Elastomers (such as natural and synthetic rubbers) form a third category of polymeric materials in which cross-links among the molecular chains occur only infrequently, allowing these chains to slide and untangle from one another when stretched so that the material is very elastic.

Examples of thermoplastics, thermosets, and elastomers include the following.[34]

34. Bolton (1994) and Lindbeck (1995).

Thermoplastics

Acrylics	Transparent, strong, stiff, and resistant to impacts, acrylics are used in aircraft windows, lenses, and signs. Plexiglass and Lucite are registered trademark names for two particular types of acrylic.
Cellulosics	(Cellulose acetate, cellulose nitrate, ethyl cellulose, etc.) Generally strong and resistant to scratching, cellulosics are thermoplastics used in such products as cameras, ping pong balls, toys, and packaging.
Fluoroplastics	Very resistant to chemicals, temperature variations, and electrical conduction, these thermoplastics are used in electrical and thermal insulation and to coat cooking utensils. Teflon is a brand name fluroplastic.
Nylons	(Polyamides) Exhibiting low levels of surface friction together with relatively high levels of strength and temperature resistance, these popular thermoplastics are used in bearings, gears, rollers, fishing tackle, door hinges, funnels, clothing fabrics, electric plugs, and other products. Kevlar and Nomex are brand names of two nylon polyamides.
Polyethylene	Polyethylene remains perhaps the most popular plastic, manufactured in both low-density and high-density forms. Low-density polyethylenes (LDPE) are composed of linear chains of polymers; they are used in the manufacture of such products as squeeze bottles and cable insulations. The stronger and stiffer high-density polyethylenes (HDPE) are composed of polymer chains with side branches; these are used in piping, toys, and other products.
Polypropylene	A thermoplastic used in automotive batteries, films, and machine parts because of its resistance to chemicals, strength, and flexibility.
Polystyrene	A thermoplastic that can be transparent or opaque, manufactured in bright colors, and mixed with rubber to increase its impact resistance, it is used in packaging, cups, shelving, equipment housings, toys, storage boxes, and other products that must withstand sudden impacts. Among its brand-name forms is Styrofoam.
Polyvinyl chloride	(PVC) A thermoplastic that resists moisture and chemicals, it is used in such products as windows, exterior siding for houses, bottles, hoses, piping, raincoats, and shoe soles.
Vinyl	Strong and resistant to abrasion, chemicals, and electrical conduction, vinyl can be found in fabrics, luggage, wire insulation, raincoats, packaging, hoses, and other products.

Thermosets

Phenol formaldehyde	The first synthetic plastic to be patented (under the brand name Bakelite), it is a low-cost material that is resistant to heat, scratching, chemicals, and water. It is used in the manufacture of such products as electrical plugs, switches, radios, telephones, and door handles.

| Melamine formaldehyde | Resistant to scratching, oils, and heat, this thermoset also can withstand impacts. It is used to laminate counter tops and decorative plastic sheets and to manufacture buttons, light fixtures, electrical equipment, and other products. |
| Urea formaldehyde | This is a hard, strong thermoset used in the manufacture of dishware, handles, and toys. |

Elastomers

| Natural rubber | A tree sap that undergoes vulcanization (i.e., cross-linked polymer chains are created) with the addition of sulphur. |
| Synthetic rubbers | These include butyl rubber (impermeable to gases), ethylene propylene (resistant to heat, oxygen, and ozone), neoprene (or polychloroprene, which exhibits a significant resistance to oils and is used in gasoline hoses, gaskets, and tank linings), and polyurethane rubbers (which resist abrasion and tearing and are used in tires and shoe soles). |

11.5.3 ▓ Ceramics

Ceramics are usually formed by applying high temperatures to inorganic, nonmetallic, and generally inexpensive materials, especially clay. Most ceramics are strong, nonconductive, and weather resistant. Easily formable at high temperatures, they become brittle afterwards.

Ceramic products include brickware for building and decoration, whiteware for plumbing fixtures and tiles in bathroom and other areas of high moisture, glassware, enamels for coating the metal surfaces of refrigerators, washers, kitchen utensils, and other household products, and cements mixed with water and other particles to form concrete.

11.5.4 ▓ Composites

Composites are formed from two or more constituent materials that, when combined, exhibit a combination of the individual constituents' characteristics of strength, conductivity, density, ductility, and other attributes. Thus, a substance can be obtained with a customized set of material properties. Fiberglass is an example of a composite that is customized for use in many different applications (e.g., boat hulls, aircraft, and automobile frames). Other examples include reinforced concrete (concrete in which steel rods have been embedded for added strength) and wood (a naturally occurring composite of cellulose and lignin).

Wood is a truly beautiful material with which to work. However, because such work is labor-intensive and requires a high degree of skill, wood is most often reserved for selected types of products and applications in which its warmth and beauty is so advantageous that its cost can be justified. As a result, wood is used in such applications as office paneling, household flooring, doors, windows, furniture, and trim on otherwise metal or plastic products.

TABLE 11.2 Important characteristics and example applications of some woods.

Sample hardwoods	Some significant characteristics	Example applications
Hickory	Very hard; very tough	Tool handles, gymnasium equipment
Ash	Hard; resilient	Baseball bats, cabinets, handles
Oak	Hard; tough; durable; Accepts high polish	Furniture, cabinets
Maple	Light in color; tough; Accepts high polish	Flooring, furniture, cabinets

Sample softwoods	Some significant characteristics	Example applications
Douglas fir	Hard, strong, durable	Heavy construction work, furniture
Cedar	Lightweight, very durable	Fencing, shingles, clothing chests
White pine	Very soft, easily worked; Not very strong	Interior finishing, paneling, toys

Source: Based on the discussion by Lindbeck, 1995.

Hardwoods are obtained from deciduous trees with broad leaves, such as oak, maple, ash, birch, hickory, and mahogany. Softwoods are from coniferous, needle-bearing trees, like white pine, cedar, and Douglas fir. In general, hardwoods have a higher density than softwoods, resulting in greater strength, toughness, and durability, although there are exceptions to this general rule. Table 11.2 presents certain attributes and a few common applications of some hardwoods and softwoods.

The strength and thermal expansion of wood vary with respect to the direction of the grain pattern. For example, wood can more successfully resist breaking when an external force is applied across (perpendicular to) the grain rather than along (parallel to) the grain. To increase the strength and to ensure that thermal expansion or shrinkage will occur equally in both directions, wooden sheets sometimes are stacked together with their grains oriented at right angles to one another and then laminated to form a single panel.

11.6 Importance of Materials Testing

Once the appropriate materials for converting a design concept into reality have been selected, one must carefully test or examine these materials in order to ensure that the actual substances used in the fabrication of the design do in fact satisfy the required design specifications. Materials testing can be the difference between success and failure—or between life and death, as illustrated in the following case.

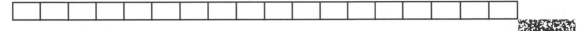

The Brooklyn Bridge

At the time of its dedication on May 24, 1883, the Brooklyn Bridge was the longest suspension bridge in the world.[35] It was an engineering marvel that combined functionality, by finally linking Manhattan with its neighboring New York borough of Brooklyn, with aesthetic beauty. For more than one hundred years, it has stood as a monument to the work and dedication of three members of the Roebling family: John, who designed the bridge but perished as its construction was about to begin; John's son Washington, who oversaw the construction following his father's death; and Washington's wife Emily, who became an expert on bridge design and served as her husband's partner during the many years of construction work after he became paralyzed during the project. The story of this bridge also illustrates the importance of materials testing in engineering practice.

John Roebling had been considering retirement from bridge building when he was asked to design and build a bridge between Brooklyn and Manhattan. Transportation between these two boroughs (the most populous in New York) had always been limited to ferries crossing the East River, and in the winter ice made this journey nearly impossible. During one tedious wintertime ferry trip of four and a half hours across the river, Roebling himself recognized the very practical need for a bridge. Some years later in 1867, public outcry finally led the city's leaders to hire Roebling to direct the bridge project.

Roebling was an expert in suspension bridge design, credited with many important innovative techniques that are still in use today. For example, he was the first to spin wire cables in the air, one strand at a time, on a wheel traveling back and forth between the anchorages of a bridge's towers. Prior to Roebling's work, cables were formed on the ground, wrapping the wires together and then lifting the completed cable into place between the towers. Roebling's spinning technique allowed designers to create larger and thicker cables, much too heavy to lift from the ground into place, that were capable of supporting massive suspension bridges. Roebling also was the first to use stays or inclined cables to connect a bridge deck directly to the towers, thereby eliminating some of the cyclic motions that can lead to collapse. Clearly, he was the experienced designer who could transform the dream of a Brooklyn bridge into a reality.

Unfortunately, Roebling would never see his design become that reality. While inspecting the location at which the Brooklyn tower was to be built, Roebling's foot was caught between the boat and a set of pilings. The injury became infected, and John Roebling—one of the premier bridge designers of all time—died of tetanus on July 21, 1869.

Washington Roebling was appointed chief engineer on the project following John's death. John had sent Washington and his wife Emily abroad to study underground

35. See Salvadori (1980) and Stirling (1966).

caissons, then in use by some European designers. Washington's first task was to build the caissons that would form the foundations for the Brooklyn Bridge's two towers.

Caissons are essentially large empty boxes opened at the bottom. A caisson first is submerged at the designated site for a bridge tower until it reaches the soft sand and soil of the river bottom. Next, compressed air is pumped into the caisson through steel pipes from above, forcing all water out of the box. Workers (known as sand hogs) then descend into the caisson chamber, passing through an air lock that allows the workers to enter the caisson without depressurizing it. The workers remove the sand and soil from the river bottom as the sharp bottom edges of the chamber's walls cut into the ground, allowing the caisson to sink deeper into the river. Eventually, the workers reach bedrock or solid soil, after which the caisson chamber is filled with concrete to serve as the tower foundation. While the underwater work is in progress, the tower itself is under construction on the roof of the caisson, thereby forcing the caisson to sink deeper as the weight of the tower increases and allowing construction to proceed in two directions (up and down).

The caissons of the Brooklyn Bridge were formed of wood (except for the steel cutting edge at the bottom of each wall) and made watertight with oakum caulking. The sides of the Brooklyn caisson were 168 feet by 102 feet with a roof thickness of twenty feet; the Manhattan caisson measured 172 feet by 102 feet with a roof that was 22 feet thick. The walls were eight feet thick at the top of the chamber, gradually decreasing to only two inches at the bottom cutting edge.

Working in one of these caissons was extremely dangerous and difficult. Among the many hazards was the risk of contracting the "bends," a deadly condition familiar to deep sea divers that results from depressurizing the body too quickly. If a worker exited from the decompression airlock too soon, nitrogen gas—dissolved in the blood under the high pressures of the caisson work chamber—could form bubbles, impeding the flow of blood. This condition results in immediate pain in the stomach (causing the person to bend over in torment), followed by agony throughout the arms and legs, possible paralysis, and sometimes death.

The bends and their underlying cause had been discovered only recently during another bridge project in St. Louis. At the time of the Brooklyn Bridge construction, it was believed that a maximum decompression rate of six pounds per minute would be sufficient to avoid the bends by allowing the body to absorb the dissolved nitrogen slowly. Unfortunately, this assumption was incorrect. (Today, a rate of one pound per minute or less is recommended.) Many of the Brooklyn Bridge workers developed the bends, among them Washington Roebling himself. He became a paraplegic, suffering in pain while gradually losing his vision and speech. He also developed an acute sensitivity to sound so that even a human voice induced great pain. Nevertheless, Washington would not give up. For the next thirteen years of the project he relied upon his wife Emily to help complete the work. Emily became the on-site overseer of the project, serving as the vital communication link between her stricken husband and the workers. The project would continue.

In 1872 the caissons settled on hard rock and sand 79 feet below the river surface. Five years later, the two towers had been completed, allowing work on the masonry anchorages and the spinning of the cables to begin.

The Roeblings insisted that each of the 5282 steel wires used in each of the four bridge cables be pretested to ensure their strength and quality. Unfortunately, the wire manufacturer tested only a single batch of high-quality wire at the factory, replacing it while in route to the bridge site with wire of mediocre quality. The high-quality wire was then returned to the factory for another round of testing and inspection, repeating the unethical pattern again and again.

On June 14, 1878, a wire strand broke, killing two workers and injuring several others. The subsequent investigation cleared the Roeblings of any wrongdoing and led to the discovery that the wire tests had been faked. The Roeblings then added enough high-quality wires to each cable to offset any inherent weakness in the wires already spun.

Once spun, the strands were wrapped together with coated steel wire to form circular cables and to protect the wire from water damage. Stays were installed and the road deck was completed. Sixteen years of design and construction under the direction of three Roeblings finally ended with the opening of the bridge in 1883.

Engineering practice sometimes requires long-term commitment to a single project. The Roebling family—John, Washington, and Emily—sacrificed much of their lives for the dream of a Brooklyn Bridge that would serve the people of New York.

11.7 Manufacturing Processes

Derived from the Latin words *manus* (hand) and *factus* (made), the word *manufacture* originally referred to objects that were literally made by hand; with the Industrial Revolution, the world entered a new era in which machines began to play more prominent roles in manufacturing processes. This revolution began around 1763 when James Watt invented a reliable steam engine.[36] Watt's engine allowed iron ore and other materials to be mined and processed in large quantities. Moreover, raw materials and processed commodities could be transported quickly by steam-driven trains and ships. Watt's engine provided the power for machines that could perform very labor-intensive manufacturing tasks with relative ease and precision.

One illustration of how important a manufacturing process can be to the successful implementation of a design concept is Lewis Howard Latimer's development of a method to fabricate durable, inexpensive carbon filaments for electric lights, as described in the following case history.

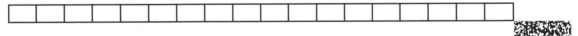

Latimer's Process for Carbon Filaments

The first incandescent electric light (based upon the Joule effect, see Chapter 5) with a carbon filament was created in France in 1838, but it proved to be impractical because it lacked a long-lasting vacuum in which the filament could glow without consuming itself. Thomas Alva Edison was the first to develop such a long-lasting vacuum, receiving patent #223,898 for his design.

36. Claypool (1984).

Although a long-lasting vacuum had been developed, long-lasting carbon filaments remained difficult to manufacture in large quantities and at low cost. Carbon filaments were created by heating pieces of paper, wood, or other fibrous materials to high temperatures; unfortunately, the resulting filaments (often broken or formed into irregular shapes by the heating process) would burn for only a few days, making electric lights too expensive and impractical for widespread use. Many people—including Edison and Hiram Maxim in the United States, England's Sir Joseph Wilson Swan and Frederick de Moleyns, and Russia's Alexandre de Lodyguine—searched for a method that would produce durable, inexpensive carbon filaments.

Finally, in 1881, Lewis Howard Latimer (an African-American and the son of former slaves who would become one of the world's foremost experts in electric power and lighting) developed the process that was being sought by so many others.[37] Latimer placed the filament blanks (the fibrous pieces of paper or wood) in cardboard sleeves or envelopes before heating them. To prevent the blanks from adhering to the sleeves, he applied a nonsticky coating to them or he separated them with tissues. When heated, the sleeves then expanded and contracted at the same rate as the blanks, protecting the filaments from suffering breakage or acquiring irregular shapes.

The filaments produced via the Latimer method were the long-lasting and inexpensive components needed to convert incandescent electric light into a commercially viable product. (A more long-lasting filament made of tungsten was later created by William David Coolidge in 1910.) Private homes, small offices, large manufacturing complexes, and city streets all could now be brightly and affordably lit. Latimer went on to codesign (with Joseph V. Nichols) a more effective connection between the carbon filament and the lead wires in a lamp. He also introduced the use of parallel circuitry in street lamp wiring, a definite improvement over the series configuration that had been used. When connected in series, all steel lights would be extinguished if one failed; with parallel circuitry, such global system failures were eliminated. Moreover, Latimer directed the initial installation of electric lighting in many major office buildings and other facilities throughout the United States and Canada. He also helped to establish many of the early electric lamp manufacturing factories and electric power plants, and he wrote the first major text on electric lighting (*Incadescent Electric Lighting: A Practical Guide of the Edison System,* D. Van Nostrand & Company, 1890), a vital resource relied upon by electrical engineers for many years. In 1918 he became a charter member of the Edison Pioneers, the highest honor that one could receive in the electrical industry.

Fabrication processes can be classified into eight categories:[38]

- **Solidification processes**   Essentially, metals, plastics, and glasses are cast into the desirable forms while in a molten state via these processes.
- **Deformation processes**   Materials are rolled, forged, extruded, or drawn into the necessary shapes.
- **Material removal processes**   Grinding, shaving, milling, turning, lapping, or polishing is used to remove material from a workpiece.

37. Haskins (1991) and Turner (1991).
38. Dieter (1994).

- **Polymer processes** Special processing techniques (e.g., injection molding, thermoforming) are used to shape polymeric materials into the desired forms.
- **Particulate processes** Pressing, sintering, hot compaction, and other techniques are used to force clusters of material particles into a specific form.
- **Joining processes** Techniques such as soldering, riveting, bolting, gluing, brazing, and welding are used to join material components together either permanently or temporarily.
- **Heat and surface treatment processes** A material's characteristics can be intentionally modified through heating, cooling, electroplating, coating, and other processes.
- **Assembly processes** Components are assembled to form the final product or system.

The selection of a manufacturing process is largely determined by the performance characteristics that must be present in the finished product. One must consider the materials to be formed into the desirable shapes, the shapes themselves, and the performance characteristics that must be present in the finished product. Only selected shapes can be achieved with a particular fabrication process, and different processes can vary widely in the precision that can be achieved in the final workpiece. Moreover, manufacturing processes can greatly affect the properties or attributes of a material. For example, if a thick steel beam is not properly heat treated, internal stresses may cause lamination to occur in which the single beam splits into many thinner layers. Australia's West Gate Bridge collapsed because of lamination.[39]

Cracks in the Liberty Ships

The impact that a fabrication process can have on an engineered system was tragically illustrated by the cracking of 400 Liberty ships during World War II.[40] Twenty of these vessels sank, including twelve that actually broke in two.

The 2700 Liberty ships were critical to the success of the Allied war effort. They were able to transport troops, vehicles, coal, or oil, and they were produced very quickly through a division of labor and the use of arc welding. Only a few specialized tasks were assigned to each worker, allowing each ship to be produced in an assembly line fashion. Welding rather than riveting the ships' steel panels together also significantly increased the speed of production. Eventually, these fabrication techniques allowed an entire Liberty ship to be built in less than five days! In March 1942, the first

39. Salvadori (1980).
40. Cartwright (1991) and Schlager (1994).

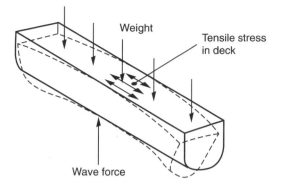

FIGURE 11.6 Stresses acting on a ship's hull.

Weight

Tensile stress in deck

Wave force

crack appeared on the deck of a Liberty ship (the *J. L. M. Currey*). By the end of that year, two other vessels had broken entirely in two. The fractures were caused by a combination of elements: applied stresses, low temperatures, and welded plates of low-grade steel. Ocean waves alternately compress and then stretch a ship's hull along its length (Fig. 11.6). These cyclic stresses of compression and tension can lead to microscopic cracks forming in the hull, particularly at points (e.g., hatch corners) and sharp discontinuities. Moreover, the relatively low-grade steel used in the Liberty ships became brittle in the cold waters of the sea, thereby reducing the toughness of the material and allowing small cracks to grow. Finally, the arc welding of the steel plates provided a continuous path for the cracks to travel around the hull, eventually leading to a partial or complete fracturing of the ship.

Once the cause of the fracturing was determined, crack arresters were installed throughout the Liberty ship fleet. These arresters (essentially connective metal strips and brackets) subdivided each hull into separate sections and prevented cracks from traveling between these sections. The sharp hatch corners on existing ships were reinforced to resist cracking, and new ships were fabricated with rounded hatch corners to reduce cracking. Since World War II, higher grade steels and better welding methods also have been developed.

Common fabrication processes involving metals include

- **Casting,** in which molten metal is poured or injected into a mold after which it is allowed to solidify into the desired form. (Some machining may be required to finish the work.) There are several different types of casting processes that vary in the materials used for the molds, the techniques used to place the molten metal in the mold, and the metals to be cast. These include sand casting, plaster-mold casting, shell-mold casting, investment (lost wax or precision) casting, die (pressure) casting, permanent-mold casting, and centrifugal-mold casting.[41]

41. See, for example, Voland (1987) for details on each of these processes.

- **Forging,** in which heated (soft but not liquid) metal is formed into the desired shape via applied pressure or impact. One can obtain greater strength and toughness in such metals as steel, brass, copper, aluminum, and bronze through forging than from the casting process; however, forging is usually more expensive than casting.
- **Powder metallurgy (sintering),** in which a fine powdered mixture of metals (and possibly ceramics) is pressed into the shape of the desired piece under pressure of 20–50 tons per square inch (psi). The product is then heated in a sintering furnace to increase its strength. This process is used for high-quantity production runs of parts for which machining costs can be saved. So-called oil-less bearings, cams, lock cylinders, gears and many other products—including jewelry—can be manufactured from different combinations of metals and other materials.

  The process produces a fine finish, precision forms, and allows one to control such attributes as weight, hardness, and porosity by varying the mixture used to form the product. Disadvantages include the need for high pressures, expensive dies, and relatively weak products.
- **Stamping and forming** processes, in which metal sheets or strips are punched, rolled, cut, drawn, perforated, bent, or otherwise formed into the desired shape.
- **Extrusion,** in which hydraulic pressure is used to push hot (soft but not liquid) metal through the opening of a die, thereby forcing the metal into the desired shape. Tubing, rods, moldings, and wires of copper, magnesium, aluminum and brass are produced via this process. (A similar process known as cold extrusion is used to form deep cans and cups from cold metals.) The process is relatively inexpensive.
- **Machining** processes, in which a part may undergo drilling, tapping, sawing, boring, grinding, lathe working, reaming, milling, planing, and shaping until it achieves the desired form. These processes are expensive and require skilled labor; because of this expense, machining often is used to first create design prototypes for testing purposes, after which one of the more inexpensive processes (such as casting or extrusion) is used to produce the design in large quantities. Machining then may be used to finish the workpieces that were produced via these other fabrication processes. (Machining is most frequently used in such finishing operations.)

Some of the fabrication processes used for polymers/plastics include:

- **Blow molding,** in which air is blown into a molten thermoplastic balloon contained within a mold, thereby forcing the plastic into the desired shape. The finished product is then cooled and removed from the mold. The process is useful for producing such items as squeezable plastic bottles.
- **Injection molding,** in which molten plastic (under very high pressures ranging from 12,000–30,000 psi) is forced through a nozzle into a cool, closed mold. The process is rapid and suitable for high-production

runs (since the initial financial investment for high-pressure equipment can be significant).

- **Laminating,** in which sheets of wood, paper, fabric, or other material is coated with plastic resin, then stacked and pressed between two highly polished steel plates in a hydraulic press under heat and pressure until cured. Low-pressure (0–400 psi) laminating is used to produce automobile bodies, luggage, boat hulls, and so forth, whereas high-pressure (1,000–2,000 psi) laminating is used to produce sheets of electrical insulation and other products. Thermosets usually are used in the formation of laminates.

- **Thermoforming** (also known as forming) in which air or mechanical forces are applied to heated thermoplastic sheets until they take the form of the mold. Variations of the process include vacuum forming, pressure forming, and slip forming, among others.

11.8 Construction Processes

Not all engineered designs are mass-produced; many are instead single large-scale systems, such as buildings, tunnels, and bridges. As is true for the fabrication of mass-produced designs, construction of large-scale systems requires knowledge of material properties, scientific principles, current engineering practices, and the system's expected environment of use. We will review three common large-scale systems (high-rise buildings, dams, and tunnels) in order to illustrate certain principles of construction.[42]

11.8.1 High-Rise Buildings

After the construction site for a high-rise building has been selected, analyzed, and prepared, a foundation must be laid that matches the constraints of the environment. For example, the type of foundation to be used will depend upon the condition of the soil at the site. Raft foundations are used in weak soil (e.g., a combination of sand and water) so that a building's weight or load can be spread over an area larger than its own base, thereby preventing the structure from sinking into the ground. Pile foundations consist of concrete and steel tubes that are driven deep into the ground until they reach bedrock or sufficiently solid soil; a concrete mat is then laid over these piles, forming the base of the building.

---

42. Some aspects of bridge construction were already described in Section 11.6.

The framework for the building is then constructed, using columns and connecting beams to provide vertical, horizontal, and rotational support for the structure. Horizontal beams are used to directly support the flooring and other loads which are then transmitted through the vertical columns to the foundation and finally into the ground. As a result, the columns are in compression, squeezed by the loads acting on them, whereas the beams are in tension and compression. Imagine a load acting directly at the midpoint of a horizontal beam that is supported at both of its ends: The beam bends downward in response to this load, causing its upper portion to become squeezed and its lower portion to become stretched, that is, the top of the beam is in compression and the bottom is in tension.[43] In contrast, a beam may be supported at only one end like a diving board; this is known as a cantilever beam, used to support balconies and similar structural features. An applied load at the unsupported end would result in that portion of a cantilever beam bending downward; the top of the beam is then stretched (in tension) and the bottom is squeezed (in compression).

Elasticity and plasticity are among the other necessary characteristics of construction materials. For example, if a beam bends slightly when a load is applied to it, it must return to its undeformed state once the load is removed (i.e., it must exhibit elastic behavior); otherwise, each subsequent loading would create further deformation of the beam, leading to stresses throughout the framework of the building. In addition, if a structural element has been loaded beyond its elastic limit, it should behave plastically (become permanently deformed) rather than break; the plastic deformation can serve as a warning that any further loading may lead to sudden fracture and collapse.

Bricks and stone can be used for columns since they are strong in compression; however, they should not be used for beams because of their inherent weakness in tension. The builders of the pyramids used limestone's lack of tensile strength to advantage by first inserting wooden wedges into small cuts in the stone and then moistening these wedges so that they expanded; the resultant tensile forces broke the stone away from the quarry walls.[44]

Wood and steel are strong in compression and in tension, making them feasible material candidates for use in beams. However, as noted in section 11.5.4, wood's strength is very dependent upon the direction of its grain. If wood is to be used as a building material, care must be taken to ensure that it will be able to resist the forces applied to it by orienting the wood properly with respect to the loads or by using laminated sheets of wood with their grains oriented at right angles to one another. For large modern structures, steel is often the material of choice for beams.

43. Also, notice that the central or neutral axis of the beam remains unstressed; it is neither stretched nor squeezed. The shape of I-beams, in which the cross-sectional area of the beam resembles the capital letter I, reflects this fact: More material is added to the top and bottom portions (the so-called flanges) of the beam in which stress will occur, and little material is used in the middle region (see Salvadori, 1980).
44. Salvadori (1990).

Concrete is strong in compression but weak in tension. In order to overcome this weakness, steel rods are inserted into the concrete before it sets, creating reinforced concrete that is strong both in compression and in tension. Reinforced concrete beams are less expensive than those made of pure steel, hence their popularity.

Prestressed concrete contains metal bands that are stretched (i.e., placed in tension) before the concrete sets; when released after the concrete dries, these bands then apply compressive forces to the concrete. When used as beams, these compressive forces counteract the tensile forces of the applied loads so that the concrete will not develop small tensile cracks during its lifetime.

Finally, brickwork, masonry, concrete, and similar materials may be used for flooring, walls, and decorative features.

11.8.2 ▓ Concrete Dams

Dams are designed and built to achieve three important objectives: (1) to control rivers and eliminate the danger of flooding; (2) to provide water for irrigation of farmlands; and (3) to provide electricity.

There are three basic types of concrete dams: gravity, arch, and buttress. (Arch and buttress designs also are known as structural dams, in contrast to the so-called massive dams of the gravity type.) A gravity dam is able to restrain water because of the dam's massive size and weight; essentially, it acts as an immovable object against the water's force. Washington's mammoth Grand Coulee Dam is an example of a gravity design.

Arch dams are designed to curve upriver into the reservoir area, thereby allowing the water collected behind the dam to press on the arched wall, forcing it against the adjoining rock walls of the canyon and creating a very strong, stable configuration. Arch dams operate in much the same way as arch bridges: loads are transmitted through the structure directly into the ground or foundation supports on each side of the bridge, thereby actually increasing the stability of the system with increasing load (although, of course, there is a limit to the load that can be supported by any structure). If an appropriate site (i.e., a relatively narrow and deep gorge through which the river runs) can be found, an arch dam can be less massive and less expensive than a gravity dam. Iran's Karun Dam is of the arch type.

Finally, water held in a reservoir by a buttress dam (also known as a hollow dam) is supported by a sloping deck, which in turn is supported by a set of buttresses. The weight of the water itself is used to increase the stability of the system by transmitting the load downward to the dam's foundation through the buttresses. (Buttress dams sometimes are described as either flat-slab or multiple-arch, referring to the type of deck surface used in a particular design.)

The first dams were of the simple gravity type, relying on their massive bulks to restrain water. The more complex buttress and arch designs have been used sporadically since the Roman Empire. For more than 500 years, the tallest dam in the world (at a height of 190 feet) was Persia's Kurit Dam, an arch design constructed during the thirteenth and fourteenth centuries.

Case History 11.5 *Hoover Dam* describes some of the challenges that can confront engineers during the construction of a dam, and the lessons that can be learned through such experiences.

11.8.3 ▨ Tunnels

Tunnels are constructed through mountain ranges for railways and highways, in cities for underground transportation systems (i.e., subways), and underwater.[45] Ingenious construction methods are used in each of these environments in order to overcome various obstacles to success.

There are a variety of cross-sectional formats that may be used for a tunnel: the common circular shape, the vertical sidewall, and the wide basket-handle in which the floors of the tunnels are flat, and the horseshoe form. Moreover, specific terminology is used to denote different locations in the tunnel; that is,

- *Crown* or *roof* for the top of the tunnel
- *Invert* for the floor
- *Sidewalls* for the tunnel sides
- *Face* for the digging end of the tunnel
- *Mouth* or *portal* for the tunnel entrance(s)

When digging a path through soft soil, workers often use air spades (air-powered shovels similar to jackhammers). To blast through rock, several drills can be mounted on a large, wheeled platform known as a jumbo. These drills are used to create holes in a specific pattern. Explosives are then inserted into the holes and ignited, breaking the rock into small pieces. A small train or a conveyor belt carries the broken rock or "muck" out of the tunnel.

Several techniques can be used to support a tunnel as it is under construction. Steel ribs or steel liner plates, bolted together to form a structural brace, can be used to hold the roofs and walls of the tunnel in place. Alternatively, small steel plates can be connected to roof bolts that have been driven deep into the rock, locking together the strata forming the rock above the tunnel. A third method is to spray concrete onto the walls and roof of the tunnel to form a primary (protective) lining as the construction proceeds. After the tunnel is completed, a secondary lining of steel-reinforced concrete is installed.

Workers enter and leave the tunnel via air locks if compressed air is used to prevent water from seeping into the tunnel work area. Exhaust and fresh-air fans are used to ventilate the tunnel during its construction and operating life.

45. See, for example, Lewis (1980) and Kelly and Park (1976).

Simplon Tunnel

Converting a design into reality can lead to the development of better implementation methods. For example, methods to provide cross-ventilation and control rock dust during tunnel construction were first developed in 1898 for the Simplon Tunnel project.[46] This tunnel, over 12 miles long and built by two teams of workers heading toward one another under the Simplon Pass, would provide a direct passage between Italy and Switzerland through the Alps. The temperature within the tunnel was expected to be as high as 130°F during construction because of the rock heat. The danger of working in such heat had become apparent some years earlier during the construction of a shorter passage (the St. Gotthard Tunnel) through the Alps: In that project 310 workers had perished and 887 had become disabled because of the deadly conditions, which included hazardous rock dust and exposure to extreme hot and cold temperatures.

As chief engineer on the Simplon project, Alfred Brandt struggled with the problems of heat and dust: How could he protect his workers from these hazards? To overcome the problem of extreme heat, he decided to create two parallel bores that would be interconnected via cross-passages. Fresh air would be blown by fans into one passage while stale air would be extracted with exhaust fans through the other passage. This double-bore, cross-ventilation scheme greatly reduced the temperatures in the tunnel.

Brandt devised another solution for the problem of hazardous rock dust: He designed a hydraulic drill that would pour water over the rock as it cut into the rock, thereby minimizing the amount of dust that would be generated during the drilling operation.

Alfred Brandt never lived to see the completion of the Simplon Tunnel in 1905; he died in the tunnel the year after construction began. However, his legacy was much more than the tunnel itself: Double-bore passages for cross-ventilation and the Brandt hydraulic drill became standard elements in tunnel construction, preserving the health and lives of thousands.

Modern tunnel construction for highways and railways often involves the use of very sophisticated and expensive machinery. For example, tunneling mole machines can be used to cut through soft rock and soil.[47] Each mole is essentially a very large cylinder, shielding the workers inside the machine and consisting of several subsystems that perform a variety of tasks. A rotating cutting head moves forward and creates a tunnel opening by grinding the soil in front of it into small pieces. The broken soil is carried by a conveyor belt inside the mole to small railcars that then travel out of the tunnel and dispose of the dirt at a nearby site. Curved concrete or steel segments are transported into the mole and bolted together into place to form the lining of the tunnel. The process continues until the tunnel is completed.

46. Stirling (1966).
47. If there is very hard rock through which to tunnel, explosives are used to break the rock into small chunks that can then be removed.

Underground highway tunnels sometimes are formed by first digging a trench. Tubular tunnel segments are then lowered into the trench, connected, and covered by soil.

Underwater tunnels also can be constructed by using preformed tubular sections. After the sections are fabricated on land, a underwater trench for the tunnel is dredged and covered with concrete. The tubular sections are then transported sequentially by barge to the trench site, submerged, and joined tightly together with other sections. A protective blanket of stone is laid over the sections. Finally, the tunnel is connected to passageways on land. Hong Kong's Harbor Tunnel was constructed using this method.

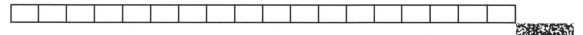

Holland Tunnel

The 1.6-mile Holland Tunnel, connecting New Jersey and New York below the Hudson River, was constructed between 1920 and 1927 via the shield method in which workers were enclosed in 400-ton protective steel cylinders as they dug underwater. As the digging proceeded, each shield was pushed forward by thirty hydraulic jacks as huge cast iron rings were bolted together to form the tunnel wall behind the shield. Compressed air was pumped through pipes in a 10-ft. thick concrete wall into the work chamber to prevent water from entering; the wall was continuously rebuilt as the shield moved forward so that the work chamber would remain constant in size. Chief engineer Clifford Holland also constructed ventilator buildings with huge fans to provide workers with fresh air and remove the stale air.[48]

SUMMARY

- Poor implementation of a design concept can lead to failure or tragedy.
- Engineers should determine the processes, materials, and equipment that will be used in fabrication, assembly, and other manufacturing operations as they develop the design concept itself (i.e., concurrent engineering should be practiced).
- Materials chosen for an engineering design must match the performance, environmental, processing, and financial requirements of the product or system.

48. Boring (1984).

- Design for X provides valuable guidelines for developing successful products; X may represent manufacturability, assembly, reliability, quality, packaging, maintainability, disassembly, and recyclability.
- DFX seeks to achieve the following objectives, all of which can be critical to a design's ultimate success or failure:
  - Superior, or at least satisfactory, performance of all design functions, that is, satisfy the customers' needs
  - Minimum cost
  - Minimum maintenance and repair
  - High levels of quality and reliability
  - Fastest time to market
  - Environmentally safe products and processes
  - Designs that can be updated and upgraded as necessary
- Numerous metals, polymers, ceramics, and composites can be used to form a particular product or system, each offering distinct advantages and disadvantages to the design engineer. These materials should be compared in terms of the relevant mechanical, electrical, physical, thermal, chemical, and other properties offered by each candidate.
- Basic principles and general guidelines for effective fabrication should be applied when converting a design concept into reality.
- Important aspects of packaging should be considered when designing a product.

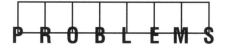

PROBLEMS

11.1 Prepare a brief report about a company in which concurrent engineering practices have been adopted. Describe the impact of these practices on the company and its product(s).

11.2 Prepare a brief report on a product that has successfully adopted one of the following strategies in its development:
- Design for manufacturability
- Design for assembly
- Design for reliability
- Design for quality
- Design for packaging
- Design for maintainability
- Design for disassembly
- Design for recyclability

11.3 Evaluate the selection of materials used in a manufactured product and prepare a report describing which material properties and fabrication processes apparently were considered in making this selection.

11.4 Select an appropriate set of materials and fabrication processes for one of the following designs and provide the rationales to support your selections:
- Computer keyboards
- Lawn mower blades
- A frame for a photograph
- Skateboards
- Rollerblades
- Computer diskettes
- Tennis rackets
- Polevaults
- Shoes

11.5 Evaluate the packaging for a consumer product. Identify all strengths and weaknesses in this packaging, and present your evaluation in tabular format. Finally, develop an improved package design for this product

11.6 Investigate examples of the following four types of bridge designs by performing appropriate library research:
- Arch
- Beam
- Cantilever
- Suspension

Compare and contrast these examples of bridge designs in terms of the materials and construction methods that were used in each case, the environment of use for which each design was developed, and the structural similarities and dissimilarities between the designs.

Describe the conditions under which each type of bridge construction might be preferable.

CASE HISTORY 11.1

Disposable Blades and Razors

The general goal of safety includes the need to protect our environment from harmful effects resulting from a product's manufacture, its operation, and/or its disposal.

Disposable products (such as certain types of razors, pens, and batteries) are designed to be used for a relatively short period of time and then simply discarded.[49] These products can be environmentally irresponsible, particularly if their manufacture and operation consume nonrenewable resources and if they are not biodegradable. Some may argue that disposable products simply provide convenience to the user while ensuring continuous sales for the manufacturer: The truth is that such products have often encouraged the user to be self-indulgent and the manufacturer to be greedy. Fortunately, many of today's engineers strive to make disposable products environmentally friendly in all ways—that is, in their manufacture, use, *and* disposal.

The steel straight-edge razor had been popular in England since the 1700s. Since the straight-edge razor could be hazardous, others—beginning with Jean-Jacques Perret around 1760—began to design various safety razors in which only the blade edge itself was exposed. However, a truly popular safety razor did not enter the market until the double-edge blade design of former salesperson King Camp Gillette appeared in 1904.

Gillette had been searching for a disposable product since the moment when his boss William Painter (the inventor of the disposable crown bottle cap) had noted that any product that is used only once and then discarded will always be in demand. (Of course, this is one of the primary attractions of disposable products to a manufacturer.)

One day in 1895 the idea for a disposable razor blade held in a protective holder occurred to Gillette while he was shaving. He then contructed a rough prototype from steel ribbon and brass and secured patent rights to the design in 1901. (Gillette's patent disclosure is considered to be one of the best ever written in that it contains a total of 98 patentable claims for the relatively simple concept of a razor blade.)

Although the concept was promising, Gillette soon discovered that manufacturing techniques had not progressed to the level where thin, tempered steel blades with a razor edge could be produced inexpensively. There are innumerable examples of such engineering designs that seem to be quite promising when first conceived, but which are then discovered to be impossible to implement because the manufacturing process needed to fabricate the design is itself unknown.

Six years passed as Gillette searched for the solution to this manufacturing dilemma. Finally, Henry Sachs, Jacob Heilborn, and William Nickerson invested $5000 to form the American Safety Razor Company with Gillette. More important, Nickerson was asked by Sachs and Heilborn to develop new fabrication techniques that would be capable of producing thin, hard, razor-sharp steel blades. Nickerson was a logical choice for this task since he had invented cereal packaging machinery, push-button systems for elevator operation, and a safety device to prevent elevator doors from opening prematurely.

Although others had claimed it would be impossible to produce steel razor blades economically, Nickerson found a way. Through careful experimentation, he developed new processes to harden and temper sheet steel, together

49. Based upon Travers (1994), d'Estaing (1985), and Jewkes et al. (1969).

with the machinery needed to produce a sharp edge. The company, now known as the Gillette Safety Razor Company, began to manufacture their double-edge blades in 1904 with market demand eventually rising to phenomenal levels after U.S. troops were issued Gillette razors during World War I.

The double-edge razor continued to dominate the marketplace until the late 1930s by which time electric razors had become popular. Jacob Schick became determined to develop a razor that required neither water nor soap following a 1910 mining expedition in British Columbia. Although various designs for electric razors had been patented since 1900, Schick created one that combined slots for holding hairs in place with moving blades that then trimmed each hair. Schick patented his design in 1928 and began to manufacture his razor in 1931.

The next major innovation in the razor industry occurred in 1975 when Baron Marcel Bich began to market the Bic disposable razor. Consisting only of a half-blade and a plastic handle, the entire Bic razor could be discarded after a few shaves. (Bic had introduced the disposable pen during the 1950s.)

Disposable products are not inherently harmful to our environment. However, we should always consider the environmental impact of such products. Just as William Nickerson extended steel manufacturing capabilities in order to produce the Gillette razor blade, so too must we engineers develop new ways to create disposable products that will be truly safe for both today's customer and all future generations.

CASE HISTORY 11.2

The Microchip

Product manufacture can be surprisingly complex, sometimes requiring many steps and real engineering ingenuity, as seen in the manufacture of microchips.[50]

Electrical circuitry once required components of substantial size, often resulting in finished products that were cumbersome and unwieldy. For example, early computers were the size of a large room and the first commercial television sets were encased in large wooden cabinets that housed the many tubes and other elements needed to produce those video images and sounds of yesteryear. Today, much more powerful computers are available in notebook size and some television sets are designed to be held in a person's hand. This

50. See Cone (1991) and Jeffrey (1996).

reduction in size of computers and televisions, together with the development of other small yet powerful electronic devices such as cordless telephones and digital watches, were made possible by the creation of the transistor in 1947 and the microchip (or monolithic integrated circuit) about ten years later.

With the development of the transistor, design engineers were able to replace large vacuum tubes—which operated at high temperatures, required significant amounts of power and space, and were easily damaged—with a low temperature, small, energy efficient, and durable alternative. The size of computers and other electronic devices could be reduced significantly. However, size reduction appeared to be limited by the need to connect these transisters and other components manually to form a finished circuit; such wiring was expensive and often unreliable.

The solution to this wiring problem, the microchip, was first recognized in 1958 by Jack St. Clair Kilby while he was working for Texas Instruments. Kilby realized that an entire electrical circuit could be embedded on a small piece of silicon, a semiconducting material through which current can travel. Switches or "gates" can be located throughout the chip; each gate acts as a sentry that will only allow an electrical pulse to exit from the gate under prespecified conditions (e.g., two input pulses must arrive at the gate simultaneously in order to generate a corresponding output pulse). These pulses denote different types of data, and the gates allow the circuit to perform various logic operations in response to the data that it is provided.

Kilby demonstrated the validity of his idea by building a working model of a microchip. Working independently of Kilby, Robert Noyce at Semiconductor Corporation designed a complete configuration for an integrated circuit. Kilby and Noyce agreed that each of them would receive partial credit for the development of the microchip.

Asked by the chairperson of Texas Instruments to develop a product that would demonstrate the commercial importance of microchip technology, Kilby created the first hand-held calculator (four inches by six inches by two inches) in 1967. The commercial version of this little electronic wonder was released to the public in 1972, thereby providing a convincing argument that integrated circuitry could be used effectively to reduce the size and costs of many electrical systems.

The manufacture of microchips or integrated circuits in large quantities is performed in several stages, reflecting the difficulty of the task. During the first stage, engineers prepare a drawing or schematic of the circuitry and components to be included in a chip. After appropriate analysis, this schematic is then reduced in size using a projection camera.

Hundreds of duplicate images of the circuit schematic are projected onto a thin sheet of glass (a so-called mask) that has been coated with a light-sensitive emulsion. The images on the mask are then transferred onto a silicon wafer, either through contact printing (in which the wafer has been prepared with a light-sensitive coating in order to capture the projection of the circuit images), projection lithography, or electron beam lithography.

The surface of the wafer is then etched in accordance with the projected circuit design. Next, the wafer is doped by adding impurities to selected areas of the wafer's surface. Such doping causes these areas to become conductive by freeing electrical charges as the impurities react with the silicon.

The wafer is coated with an aluminum alloy or other metal, and this coating is etched to form the necessary circuit connections between the doped regions. The wafer can then be tested and cut into hundreds of individual chips.

Finally, it should be noted that the manufacture of microchips is becoming more sophisticated with each passing year, using materials other than silicon and high-resolution imaging systems. The reader is encouraged to investigate current manufacturing techniques in this important industry.

CASE HISTORY **11.3**

The Panama Canal

Before the Panama Canal[51] was completed, anyone wishing to travel between the Atlantic and the Pacific oceans across the Isthmus of Panama had to cross fifty miles of rugged mountain terrain. Ships traveling between these oceans had to navigate around Cape Horn at the southern tip of South America. The completion of the canal in 1914 eliminated many thousands of miles from this route.

During the first half of the nineteenth century, the United States was invited by the government of New Granada (now Panama and Columbia) to build a canal route across the isthmus. The United States did complete a railroad across the isthmus in 1855, but France became the first nation to attempt construction of a canal beginning in 1878. Disease claimed 20,000 lives before this tragic initial effort was abandoned in 1889.

In 1903 the United States—determined to build a direct water route between the oceans—leased a five-mile wide and fifty-mile long "canal zone" from the Republic of Panama. Major William Crawford Gorgas, an Army surgeon, recognized that the workers had to be protected from the deadly diseases of yellow fever and malaria transmitted by mosquito bites. Under Gorgas's orders, stagnant pools of water throughout the region were either drained or sprayed with oil, which killed mosquito larvae before they could hatch. In addition, screens were added to the doors and windows of the workers' living quarters, and workers were required to drink quinine medicine. Gorgas's heroic efforts were successful: Yellow fever and malaria almost disappeared after the first two years of the project.

51. See Boring (1984), Jeffrey (1996), and Stirling (1966).

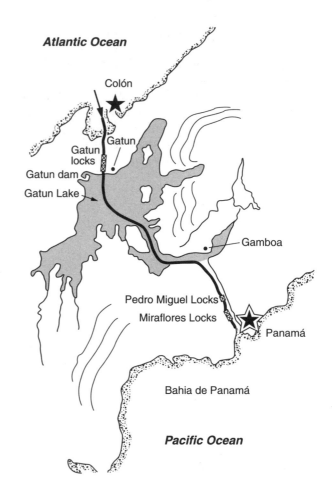

FIGURE 11.7 The Panama Canal route between the Atlantic and Pacific oceans.

In 1905 John Stevens became the chief engineer for the project. He oversaw the rebuilding of the railroad that would be needed to remove the mountains of dirt and rock from the excavations. Stevens also realized that locks would be necessary if a canal route between the oceans were to become a reality.

Travelers crossed the isthmus by navigating the Chagres River and traversing the Culebra Pass, the lowest part of the mountain range separating one ocean from the other. If the canal was to follow this same path, either the Culebra Pass would need to be excavated 235 feet down from its lowest point to sea level *or* the ships passing though the canal would need to be raised 235 feet via a system of locks. An excavation of 235 feet in the soapy clay of the region would be prone to frequent rock slides if it were too steep and would require mammoth removals of rock and dirt if such slides were to be minimized. The decision was made to build locks, which, although more complex and expensive than a simple cut through the mountains, would require far less excavation.

In 1907, after persuading the United States Congress to sponsor the construction of an expensive lock system, Stevens resigned as chief engineer (never indicating the reason for his decision). He was replaced by Colonel George Washington Goethals, an expert on locks and waterways.

Under Goethals's direction, a dam across the Chagres River was constructed at Gatun, creating the Gatun Lake at 85 feet above sea level (see Fig. 11.7). The Culebra Pass now only needed to be excavated to the depth of the lake rather than to sea level. Nevertheless, the excavation was an excruciating task, performed in temperatures that sometimes reached 130°F and hampered by numerous rock slides that brought death and destruction to the project. The Culebra Cut eventually was nine miles long with an average depth of 120 feet.

A system of three locks was constructed at Gatun to raise or lower ships sequentially through the 85-ft. difference in height of the the Atlantic Ocean and the lake. Each lock consisted of two chambers, one for each direction of travel, 1000 feet long and 110 feet wide. The gates to a chamber are closed after a ship has entered, and the ship is then lowered or raised by allowing lake water to pass out of or into the chamber through a set of sluices.

Three more locks were constructed near the Pacific end of the canal: two at Miraflores and one at Pedro Miguel. The Panama Canal opened on August 15, 1914—an immense engineering feat that now allows more than 15,000 ships to pass through the canal annually.

Implementation of an engineering design requires work and sacrifice. The Panama Canal is a stark example of this fact. Approximately 50,000 people had worked on the project, with the loss of 5,600 lives during the construction. Their sacrifices—and those of the 20,000 people who had perished during the earlier French effort—now provide travelers with direct access between two great oceans.

CASE HISTORY 11.4

The Kansas City Walkway Disaster

One of the most publicized structural disasters in modern times was the 1981 collapse of a walkway system in the Kansas City, Missouri, Hyatt Regency Hotel.[52] Three pedestrian walkways overlooked an atrium and allowed people to move between the hotel tower building (containing guest

---

52. See Levy and Salvadori (1991).

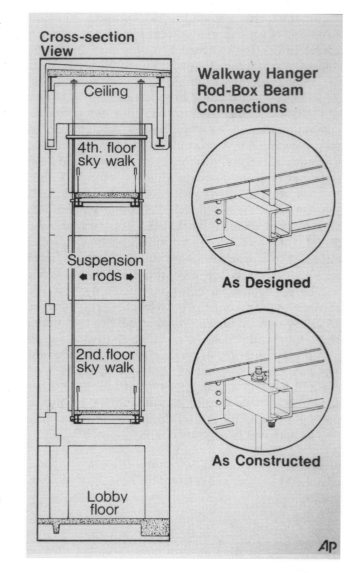

Cross-section View

Ceiling

4th. floor sky walk

Suspension ◄ rods ►

2nd. floor sky walk

Lobby floor

Walkway Hanger Rod-Box Beam Connections

As Designed

As Constructed

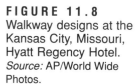

FIGURE 11.8
Walkway designs at the Kansas City, Missouri, Hyatt Regency Hotel. *Source:* AP/World Wide Photos.

rooms) and an adjoining four-story function block. Each walkway was on a different floor, with the third-floor walkway on the east side of the atrium. On the west side, the second-floor and fourth-floor walkways were both supported by the same set of steel hanger rods from the ceiling; these two walkways crashed onto the atrium floor—filled with 1600 people—on the evening of July 17, 1981. One hundred and fourteen people were killed and more than 200 were injured or maimed.

A careful investigation of the tragedy discovered that a modification had been introduced in the walkway design between its initial conceptual vision and its final implementation. Six continuous single hanger rods had been specified in the *original* working drawings for the design, rods that would support both the second- and fourth-floor walkways (see Fig. 11.8). However, in the *final* shop drawings for the walkways, each continuous single rod had been replaced by a pair of shorter rods, the first running between the ceiling and the upper walkway, the second connecting the upper and lower walkways. Nut and washer assemblies were then to connect these hanger rods directly to the walkways' box beams.

Why was such a fundamental change made in the walkway support system? The contractor apparently sought to simplify construction by using pairs of shorter rods in place of each long single connector. Tragically, the design engineers who approved this suggestion failed to realize its functional implications.

As Petroski (1992) has noted, the double-hanger design is equivalent to one person (corresponding to the fourth-floor walkway's box beam) holding onto a rope connected to the ceiling while a second individual hangs onto the first person. In such a case, the first person must support the weight of both individuals. In contrast, the original single-rod design is equivalent to each person holding directly onto the rope, supporting only his or her own weight.

Required to carry a load essentially twice that specified in the original design,[53] the box beam located at midspan of the fourth-floor walkway broke free from its connecting rod. With the loss of this support, the remaining box beams quickly broke free from their connecting rods and the two walkways crashed to the floor.

The Kansas City walkway disaster clearly demonstrates the need to carefully consider the impact of any modification in a design, particularly in terms of unexpected hazards. Every engineer—every person—has a moral responsibility to perform his or her work in the most diligent manner possible.

CASE HISTORY 11.5

The Hoover Dam

In 1901 Charles Rockwood, an engineer, and George Chaffee, an expert on irrigation, diverted some of the water from the raging Colorado River into a wide-ranging set of irrigation ditches, successfully transforming a dry desert

53. It was determined during the investigation following the disaster that the original walkway design failed to satisfy the Kansas City Building Code; nevertheless, the collapse might never have occurred if the original design had been implemented.

region into California's fertile Imperial Valley farmlands.[54] (Chaffee's earlier irrigation efforts had resulted in the creation of orange groves in Southern California.) Farmers settled throughout the valley to grow such crops as cantaloupe, alfalfa, and wheat.

Unfortunately, problems still existed: Periodically, the water level of the Colorado River dropped in response to severe droughts upriver and the irrigation system ran dry, killing the farmers' crops. Moreover, in times of flood the river would overrun its banks and flood the valley; in 1905, entire towns were lost because of a flash flood. Clearly, additional engineering work would be needed to eliminate these problems by providing farmers with a truly stable supply of water.

In response, Hoover Dam (once known as Boulder Dam) was designed and built to achieve three important objectives:

1. To control the Colorado River and thereby eliminate the danger of flooding;
2. To provide water for irrigation of the many farmlands throughout the valley, particularly in times of drought; and
3. To provide electricity to many populated areas throughout the southwestern United States.

Designed by Edmund Wattis, Hoover Dam sometimes is incorrectly perceived as an arch dam because it does indeed curve upstream into its reservoir, Lake Mead.[55] In actuality it is a curved gravity dam, so massive that even if it were straight across the canyon with no curvature it would still be able to restrain the reservoir waters. (The curvature was added to increase its strength.) A true arch dam would collapse if it were straightened, unable to resist the hydrostatic pressure of its reservoir.

Construction of the Hoover Dam began in Black Canyon in 1931. The task confronting the engineering team was indeed immense. An entire town (today's Boulder City, Nevada) first had to be built seven miles from the site to house the dam's 5250 workers and their families Next, an access road was constructed between the town and the dam site.

During the first year of activity, sixteen people died of heatstroke (summer temperatures reached 120°F). Dr. David Bruce Dill, in his efforts to save lives at the project site, discovered that salt can help the human body survive excessive heat (a very important discovery that has been used to preserve lives ever since). The deaths due to heatstroke ended once salt was added to the drinking water.

Before construction of the dam could begin, the Colorado River had to be diverted from the worksite. Four bypass tunnels (each about 4000 feet long) were formed in the canyon walls, a preliminary task that required more than two years to complete. A trestle bridge was built above the point at which the river was to be diverted. Once construction of the tunnels was

---

54. See Boring (1984), Copen et al. (1977), Jackson (1995), Lewis (1980), and Stirling (1966).
55. See Section 11.8.2 for a discussion of dam types.

completed, two dozen trucks loaded with rock and soil moved onto the trestle. Dynamite was ignited to break loose part of the rock walls of the canyon. Simultaneously, all of the dump trucks released their loads. The total accumulation of rock and soil succeeded in blocking the river, forcing it into the bypass tunnels around the dam site. Work could now begin on the dam itself.

The wall of the dam was to be made of concrete, with a thickness of 660 feet and a height of 726 feet. The wall would be 300 feet wide at its base, increasing in its width to match that of the canyon until it achieved a final value of 1225 feet at its crest (the top of the dam). The concrete blocks used in the dam wall would vary between 5 and 30 feet in height and between 25 to 60 feet in both width and depth. This mammoth edifice would require a total of 3,250,330 yds^3 of concrete, creating a new obstacle to success: Such a huge amount of concrete would need at least 125 years to dry—and perhaps as many as 700 years—because of the excessive heat emitted by the chemicals in the concrete as it hardened.

To overcome this problem, engineers Frank Crowe and Walker Young decided to pour the concrete into forms laced with piping, after which cold water could be sent gushing through the pipes. Much of the chemically generated heat in the concrete was transferred to the cold water and carried away, accelerating the drying process so that it reached completion in only 22 months.

Hoover Dam was completed on May 29, 1935, well ahead of schedule. The bypass tunnels were sealed and the river, returning to its old path but finding itself blocked by the huge dam, began to form the mammoth Lake Mead behind the wall. This manmade 115-mile-long lake then took two years to fill. The dam went into full operation near the end of 1936.

One final note on implementation decisions: As with other structures, the foundation of a dam is critical. Weak or porous foundations can result in collapse. For example, the foundation supports of an early steel buttress dam at Hauser Lake, Montana, were not properly set into the underlying bedrock, leading to the dam's collapse in 1908. The weak rock foundation of California's St. Francis Dam softened while in contact with water, leading to the dam's collapse in 1928 and the deaths of more than 400 people.

Moreover, the materials used in the dam itself must be of high quality. It is suspected that the quality of sand used in the concrete of the Gem Lake Dam, together with the cyclic freezing of the dam at high elevations, led to its gradual deterioration; a protective coating was eventually applied to the structure.

During implementation of a design, unexpected obstacles sometimes must be overcome in the field or factory. Each obstacle should be treated as a new design subproblem for which a creative solution must be developed. The solutions (e.g., water-cooled concrete, consumption of salt to avoid heat-stroke) to such implementation subproblems often become part of the standard operating procedures for similar situations. Moreover, in engineering practice, the choice of materials and the quality of workmanship can mean the difference between success and failure.

CASE PROBLEMS

1 Methodology in Design: Airfoils

Aerodynamic Lift, Thrust, and Drag Forces

When an airplane flies, four forces are always at work on it: *gravity* (or weight) pulls the airplane toward the earth; *lift* draws it away from the earth; *thrust* speeds the airplane forward; and *drag* acts as a brake (Figure CP1.1). The last three of these forces—lift, thrust, and drag—are produced by the action of the air flowing past the airplane.[1]

The lift, or upward force on the airplane, keeps the heavier-than-air aircraft aloft. Lift is the vertical component of the reaction force due to the change in momentum of the air mass flowing near the surface of the wing. Drag is the frictional resistance of the air to the passage of the airplane through it. The cross-sectional shape of the wing is designed in such a way that the drag is kept small in relation to the lift. This shape is known as an *airfoil section* (Figures CP1.2 and CP1.3).[2]

Aerodynamic Lift and Newton's Laws of Motion

The important thing about an airfoil (such as an aircraft wing) is not so much that its upper surface is *cambered* (has a slightly convex curvature) and its lower surface is nearly flat, but that it moves through the air at an angle. Even a flat plate that moves through the air at an angle will produce a downward acceleration of the air passing over it. That is, it will produce lift, although not very effectively. It should be remembered that airplanes can and do fly with wings that have the same curvature top and bottom, and airplanes with conventionally cambered wings can fly upside down. The critical condition for flight is a net downward deflection of the airstream passing over the wing. The amount of downward deflection of air produced

---

1. NEA (1942).
2. See Blackburn and Holister (1987), Weltner (1987), *The Way Things Work* (1967), NEA (1942).

F i g u r e C P 1 . 1   Four forces work on an airplane while it is in flight.

by the wing depends largely on the angle of attack and the speed of the airstream.[3]

Figure CP1.4a shows a plane rectangular plate moving through the air with velocity v in such a way that it meets the airstream at an angle α (called the *angle of attack*). For convenience, let us suppose that the plate is stationary and the air is moving past it. The direction of the airstream changes as it flows over the plate (see Fig. CP1.3) so that it experiences a change in momentum which necessitates a reaction force F_{reac} on the plate. This reaction force is predicted by Newton's third law, which states that for every action force *on* a body there is an equal and oppositely directed reaction force *by* the body. According to Newton's second law, the reaction force is equal to the time rate of change of momentum of the air:

$$F_{\text{reac}} = m\,\Delta v\,/\,\Delta t = \Delta\,(mv)\,/\,\Delta t$$

This reaction force can be resolved into two components: drag force D which is parallel to v, and lift force L which is perpendicular to v. Figure CP1.4b shows that an airfoil maximizes lift while it minimizes drag. The lift first increases with angle of attack, reaches a maximum, and then decreases (Fig. CP1.5).[4]

■ Boundary Layer Airflow

The *boundary layer* is the thin layer of air adjacent to the airfoil surface, formed as air is flowing over the airfoil, which has flow characteristics distinct from the main airflow because of friction with the surface. The airflow

3. See Fletcher (1975), Weltner (1987).
4. See Fletcher (1975), Hewitt (1974).

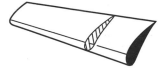

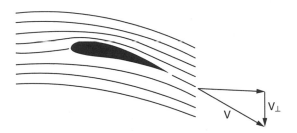

FIGURE CP1.2 The airfoil section is a shape that keeps drag small in relation to lift. *Source: The Way Things Work,* 1967.

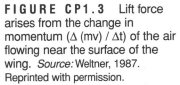

FIGURE CP1.3 Lift force arises from the change in momentum ($\Delta (mv) / \Delta t$) of the air flowing near the surface of the wing. *Source:* Weltner, 1987. Reprinted with permission.

(a)

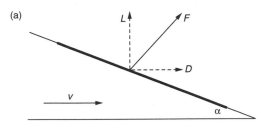

(b)

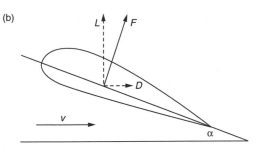

FIGURE CP1.4 (a) The change in direction of the airstream flowing over a plate, which meets the airstream at an angle α, gives rise to a reaction force, F_{rearc}, having a lift component perpendicular to the airstream and a drag component parallel to the airstream. (b) An airfoil shape maximizes the lift component of the reaction force, while it minimizes the drag component. *Source:* Fletcher, 1975. Reprinted with permission.

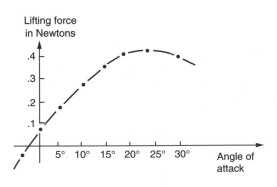

FIGURE CP1.5 The lifting force of an airfoil having an area of 0.045 m² in air with streaming velocity of 6 m/s changes with angle of attack according to the above experimental data. *Source:* Fletcher, 1975. Reprinted with permission.

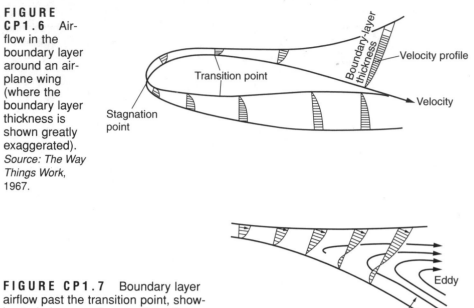

FIGURE CP1.6 Airflow in the boundary layer around an airplane wing (where the boundary layer thickness is shown greatly exaggerated). *Source: The Way Things Work*, 1967.

FIGURE CP1.7 Boundary layer airflow past the transition point, showing turbulence and separation. *Source: The Way Things Work*, 1967.

in the boundary layer around an airplane wing starts at the stagnation point, and is laminar up to the transition point, where turbulence sets in (Fig. CP1.6). In the *laminar* portion of the boundary layer, the flow is relatively smooth and moves in layers parallel to the wing. The velocity of the airflow increases steadily from zero at the surface of the wing to a maximum equal to the velocity of the main airflow. The *transition point* is located near the point of minimum pressure, approximately where the wing has its greatest thickness. At the transition point, boundary layer airflow changes from laminar to turbulent. In *turbulent* airflow, the boundary layer becomes thicker and the airflow exhibits large random motions. In this region, usually toward the trailing edge of the wing, separation may occur.

Separation means that the boundary layer airflow detaches itself from the surface of the wing. Separation occurs at the separation point. Where separation occurs, airflow immediately adjacent to the wing surface occurs in a direction opposite to the direction of the main flow, which is called *counterflow* (Fig. CP1.7).[5]

Separation usually takes place near the trailing edge of the wing, where eddies develop. If separation occurs too far forward toward the leading edge of the wing, there will be a serious loss of lift and increase in drag. The shift of the separation point toward the leading edge happens when the angle of attack exceeds the critical value called the *stalling angle*, or when the air-

5. *The Way Things Work* (1967).

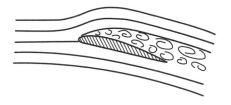

FIGURE CP1.8 The angle of attack has been increased beyond the value that produces maximum lift, so that the airflow detaches itself from the upper surface, and "stall" occurs, with loss of flying speed and lift. *Source: The Way Things Work*, 1967.

(a)

(b)

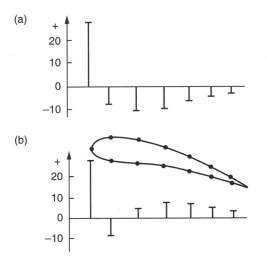

FIGURE CP1.9 (a) Pressure difference in pascal with respect to atmospheric pressure at the upper side of an airfoil. Note that the pressure is less than atmospheric pressure beyond the stagnation point. Depth of airfoil is 15 cm; velocity of air stream is 7 m/s. (b) Pressure difference in pascal with respect to atmospheric pressure at the lower side of an airfoil. Note that the pressure is greater than atmospheric pressure along most of the length of the lower side of the airfoil. *Source:* Weltner, 1987. Reprinted with permission.

speed becomes too low (Fig. CP1.8). There may also be regions of the wing in which laminar boundary layer airflow is apt to separate from the wing. To prevent premature separation, certain devices, such as small projecting plates, are installed, to break up the laminar flow and produce turbulence, which "adheres" better to the wing.[6]

Bernoulli's Equation

The airfoil shape of an aircraft wing produces a particular pressure distribution at the top surface of the wing. The pressure on the upper surface of the wing is lower than the corresponding point on the lower surface for most of the length of the wing surface (Fig. CP1.9). This pressure difference, which is often considered to produce lift on an aircraft wing, is an effect of *downwash* (the downward acceleration of air flowing over an aircraft wing), and is predicted by Bernoulli's equation.

6. Ibid.

Bernoulli's equation is an expression of the conservation of energy for an ideal fluid[7] flowing in a pipe, or streamline, or in this case the boundary layer adjacent to the airfoil surface. Consider a mass m of air moving from an initial location i, to a final location f, in the boundary layer airflow of a wing. If the mass begins at an initial altitude h_i and curves to a final altitude h_f, then the conservation of energy principle states (ignoring deviations from ideal fluid behavior):

kinetic energy$_i$ + potential energy$_i$ + work done on the fluid =

kinetic energy$_f$ + potential energy$_f$

$$(1/2)mv_i^2 + mgh_i + (p_iA_id_i - p_fA_fd_f) = (1/2)mv_f^2 + mgh_f,$$

where v_i, v_f are initial and final velocities of the fluid, d_i, d_f are distances through which the fluid moves at the initial and final locations, and p_iA_i, p_fA_f are the forces due to pressure pushing on cross-sectional areas of fluid at the initial and final locations.

Substituting density = mass/volume [that is, density $\rho = m / (A \, d)$], we obtain Bernoulli's equation for an ideal fluid:

$$(1/2)\rho v_i^2 + \rho g h_i + p_i = (1/2)\rho v_f^2 + \rho g h_f + p_f.$$

Considering the contribution made by gravitational potential energy ρgh to be negligibly small, Bernoulli's equation can be written in the simpler form:

$$(1/2)\rho v^2 + p = \text{constant.}$$

The above equation indicates that as the velocity of the flow increases, the pressure must decrease. As the airmass in the boundary layer is accelerated over the upper portion of the airfoil (due to frictional effects of the air in contact with the wing, which force the air to change direction and follow the wing's curvature), the velocity of the airmass increases with time, and consequently the pressure on this upper surface decreases.

We noted above that laminar boundary layer airstreams over smooth airfoil surfaces produce considerably less "skin friction" with the wing surface than turbulent layer airstreams. *Minimizing drag* depends on the extent of the laminar flow over the airfoil surface. *Maximizing lift* and the *behavior at high angles of attack* depend on the characteristics of the turbulent boundary layer. Generally speaking, a conventional wing cannot operate at peak efficiency with significant boundary-layer separation. The stability of the laminar layer depends on the occurence of a "favorable pressure gradient," which is a pressure decreasing in the direction of the airflow. The Liebeck wing design (Fig. CP1.10), which is an airfoil with a large fore-portion and favorable pressure gradient, achieves this design goal. The boundary layer remains laminar at the onset of pressure rise (called the adverse pressure gradient). This laminar boundary layer is thin and

7. An *ideal fluid* is a fluid that is incompressible (its density does not change), irrotational (it is not turbulent), steady (its velocity does not change over time), and nonviscous (all work applied to the fluid becomes kinetic energy). Bueche and Wallach (1994).

8. Jones (1990); Houghton and Carpenter (1993).

FIGURE CP1.10 Geometry of a low-speed high-lift airfoil, after the Liebeck design.

therefore can develop in the region of adverse pressure gradient before separation occurs. The net effect is a wing that has high lift and low drag at low speeds.[8]

This type of wing design was the basis for the wings used successfully on human-powered aircraft such as the *Daedalus*, which crossed the Aegean Sea under pedal power.[9]

Adler's Annular Airfoil Gliding Toy

First Generation: Skyro

Alan Adler, the holder of some 20 high- (and low-) tech patents, and the designer of 18-meter racing sailboats and 6-meter racing canoes, began thinking about a long-range hand-launched flying apparatus in the early 1970s. He was looking for a "technological fix for a distinctly average Frisbee arm."[10] High drag is inherent to the basic Frisbee design; it is a necessary feature for the disc's stability.

Adler's new approach to reducing drag was to use a low-profile ring design. Being knowledgeable of the principles of aerodynamics and eager to build and test prototypes, he began working on the angle of attack that the ring presented to the oncoming air. Stabilizing a flying ring depended on providing both the front and back with an identical degree of lift. Since the back of the ring is always flying in the downwash of the front, it normally feels a smaller lifting force than the front. This was a major challenge for Adler.[11]

In June of 1984 Adler obtained a U.S. patent (#4,456,265) for a gliding ring toy, which was marketed under the trademark Skyro. It was cited in the *Guinness Book of World Records* for the longest throw of an inert heavier-than-air object (857 ft. 8 in.). After years of work, Adler had succeeded . . . almost. The Skyro has extremely low drag, and is stable—but at a specific speed. This means that in the hand of an inexperienced thrower, it is somewhat unstable. However, it was an important first-generation design (recall Section 1.3 on Nth-generation design).

The Skyro is a thin, low-drag closed-form airfoil, molded from a flexible, thermoplastic elastomer. It has a circular arc top and a straight bottom (Fig. CP1.11). It has an outer diameter of 11.75 inches; an arc chord length of 1.75 inches (in Fig. CP1.12 extended chord line 5 passes through the outer and inner perimeters 3 and 4); an airfoil thickness of 0.135 inch; and a total ring weight of 3.4 ounces.[12]

---

9. Ibid.
10. Cassidy (1989, p. 28).
11. Ibid.
12. U.S. patent #4,456,265.

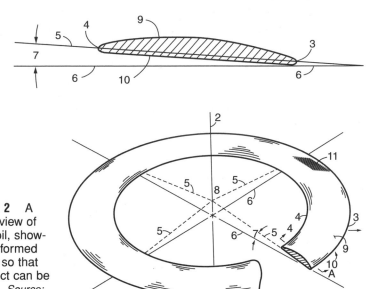

FIGURE CP1.11 An elevational sectional view of a portion of the Skyro ring airfoil shown in Figure CP1.12, taken along line A–A in the direction of the arrows. *Source:* U.S. patent #4,456,265.

FIGURE CP1.12 A cutaway isometric view of the Skyro ring airfoil, showing that the ring is formed on a conical plane so that the downwash effect can be fully compensated. *Source:* U.S. patent #4,456,265.

The ring itself has an airfoil angle, $\phi = 1.5°$ (angle 7 in Fig. CP1.11), which is the angle of the upper conical surface of an imaginary right circular cone formed by extending ring chord lines 5 radially inward to their point of intersection 8 on vertical axis 2 (Fig. CP1.11 and CP1.12). In typical level flight, the entire plane of the ring is oriented at an angle of attack α of approximately 2° to the flight path, or airflow. The ring is generally cast at a speed of 33 to 48 feet per second.

Calculations yield that the airflow over the aft airfoil area are at a 3° lower angle than the airflow over the forward airfoil area for a flat ring ($\phi = 0°$), having the above dimensions. Downwash is directly proportional to the lift coefficient of the forward airfoil. Therefore, a downwash angle of $\theta = 3°$ would result in a smaller lift on the aft portion of the airfoil, which gyroscopic precession would convert into roll, or banking. However, if airfoil angle ϕ is equal to $\theta/2 = 1.5°$, the difference in the angle of attack of the forward and aft airfoil areas is $2\phi = \theta$.

In this case, the forward and aft airfoil areas will each meet the airflow at the same angle of attack and have equal (and balanced) lift. This provides the Skyro with straight flight.[13]

The optimal airfoil angle of an annular ring can be calculated by:

$$\phi \cong (64{,}000\ W) / (V^2\ D^2),$$

where

W = weight (ounces),

13. Ibid.

V = velocity (feet per second),

D = mean diameter of annulus (inches) = $(OD + ID)/2$,

OD, ID = outer diameter, inner diameter.

Second Generation: Aerobie

In the early months of 1984, Adler designed a computer program to simulate ring flight. He sought to determine whether balanced flight was possible at *all* speeds. The result was encouraging. The computer simulation indicated that such a balanced design was possible, at least theoretically. After three unsuccessful prototypes, Adler built a fourth prototype with a "separator" lip on the outside of the rim. From the very first toss, it was clear that this design was fundamentally different from the Skyro. No matter how you threw the ring, the separator lip kept the ring level. Adler knew that this was the design he had been searching for! A systematic series of tests allowed Adler to refine the lip and the overall geometry of the outer rim. After a dozen models with different heights and angles were made and tested, one configuration was recognized as optimal. This became the "Aerobie," which went into production in December 1984.[14]

A U.S. patent (#4,560,358) was awarded to Adler in December 1985 for the Aerobie design. Its main new feature is the separator lip (8 in Fig. CP1.13) on the outer perimeter of the upper surface. This lip extends upward to a narrow peak (9 in Fig. CP1.13), which is higher than the adjacent portion of the upper airfoil surface. The lip is referred to as a *separator lip* because it is believed to cause the airflow to separate from the leading edge of the forward portion of the airfoil, and to reduce the lift slope of the forward portion of the airfoil so that it becomes balanced with the lift slope of the after portion of the ring. The *lift slope* is the rate of change of lift versus angle of attack.[15]

Life slope = $dL/d\alpha$

where L = lift, α = angle of attack.

FIGURE CP1.13 Cross-section of the Aerobie showing separator lip 8, which causes air to separate from the leading edge of the forward portion of the airfoil, and which matches the lift slopes ($dL/d\alpha$) of the forward and aft sections of the ring. This results in stability of flight for the Aerobie over a wide range of flight velocities and angles of attack. *Source:* U.S. patent #4,560,358.

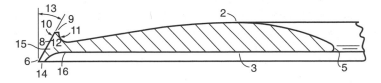

14. Cassidy (1989).

15. U.S. patent #4,560,358.

The stability of the Aerobie over a wide range of flight velocities and angles of attack suggests that the lift slope of the forward and aft sections of the ring have indeed become matched, due to the action of the separator lip.[16]

Specifications for the Aerobie airfoil perimeter and surface are as follows (refer to Fig. CP1.13):

- Separator lip peak 9 < 1 mm (in order to produce stable flight)
- Angle 12 between adjacent surfaces 10 and 11 of the airfoil < 60°
- Angle 13 formed between a line tangent to the lip's outer surface 10 and the axis of revolution of the body < 45° (in order to maintain straight flight over a wide range of velocities)
- Angle 13 ≈ 30° in the Aerobie
- Zenith of convex upper surface 2 of the airfoil section is about 1/3 of the distance from the inner to the outer perimeter (in order to produce adequate lift combined with stability and low drag)
- Flap 14 at the bottom edge of the perimeter is 1 mm below lower airfoil surface (the flap helps to maintain stable flight for the Aerobie)
- Diameter of outer perimeter = 330 mm
- Diameter of inner perimeter = 254 mm
- Maximum airfoil thickness = 3.8 mm (measured from the zenith of the convex upper surface to the lower surface immediately below it)
- Height of separator lip peak above immediately adjacent airfoil surface = 1.2 mm
- Height of separator lip peak above lower tip of flap = 3.6 mm
- Weight = 107 grams, or 3.26 g/cm² of airfoil surface.[17]

Recall that the downwash thrust of air by an airfoil depends on the adhesion and viscosity of the air passing over it. It was found that the ring flies greater distances if the upper and lower surfaces are slightly textured, to approximately the equivalent of #400 to #600 grit abrasive paper. This improves adhesion. The slightly rough surface also improves the grip for throwing and catching.

This second annular airfoil gliding toy by Adler—the Aerobie—is even more successful than the Skyro for distance. On July 8, 1986, Scott Zimmerman threw an Aerobie 1,257 feet, setting a new Guinness World Record for the longest throw in history of any heavier-than-air object. The Aerobie also has demonstrated stability over flight velocities ranging from a few meters per second to twenty meters per second. A typical flight velocity is approximately ten meters per second.[18]

Figure CP1.14 illustrates the method of manufacture of the Aerobie. A central plastic armature ring 17 is separately molded from high-impact thermoplastic. The armature ring has thin tongues (17a and 17b) with a number of through-holes 18. The armature is placed into a second mold, and a thermoplastic elastomer is injected to form the inner and outer cushions

---

16. Ibid.
17. Ibid.
18. Cassidy (1989); U.S. patent #4,560,358.

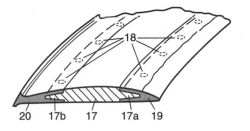

FIGURE CP1.14 An illustration of the method of manufacture of the Aerobie, showing the central armature ring 17 and attached thermoplastic elastomer cushions 19 and 20. *Source:* U.S. patent #4,560,358.

19 and 20. As the elastomer flows through holes 18, the cushion becomes linked to the armature. The ring toy is safer and more comfortable to catch because of the soft cushions.[19]

Spin and Vortices

Nothing has been said to this point on the effect of spin on a hand-launched gliding ring or disc. A body that has a high rotational spin tends to keep the axis of spin oriented in its original direction in space. This phenomenon is referred to as *gyroscopic stability*.[20] When an Aerobie or a Frisbee is launched by hand with a spinning motion, its spin helps to keep the gliding toy in stable flight by the same principle of gyroscopic stability.[21]

Also important is the effect of spin on lift. Nakamura and Fukamachi performed a study at Kyushu University in Japan on airflow past a Frisbee. The Frisbee was connected to an electric motor by means of a metal rod and the assembly was positioned in a low-speed wind tunnel. Smoke was used to detect the airflow past the Frisbee while the disc was at rest and while it was spinning at 3 rotations per second.[22]

The experiment showed that rotation greatly enhanced the formation of vortices behind the disc. A *vortex* is basically a whirlpool that can develop on a moving object. It is said to be shed when it "slides" off the object. The white streaks behind high-flying airplanes arise from condensed water vapor that becomes trapped in shed vortices. The increased *vortex shedding* by the spinning Frisbee is thought to increase downwash and consequently increase lift on the airborn disc.[23]

Design Assignment

Certain aerodynamic principles of airfoils, as well as technical information and engineering design of gliding toys, have been presented. Working in two-person teams, design, construct, and test a prototype of your own hand-

19. U.S. patent #4,560,358.
20. Scarborough (1958).
21. Walker (1977).
22. Nakamura and Fukamachi (1991).
23. May (1993); Nakamura and Fukamachi (1991).

launched gliding toy. Although your work should be original, it may be based on an existing concept.

Relate the test data that you collect to the aerodynamic properties of your invention. Submit a formal report on your work, together with your design prototype.

2 Reducing Tropospheric Ozone

Air: An Endangered Resource

Each day the average adult consumes about six pounds of food and water; in contrast, 35 pounds of air will pass through that person's respiratory system during those same 24 hours. Indeed, when we compare the human body's relative dependence upon air, water, and food, clearly air is the most vital substance for sustaining life; none of us could survive for more than several minutes without it.[1]

The *troposphere* is the lowest part of the atmosphere surrounding the earth. It starts at the earth's surface and extends upward to about 6.2 to 8.7 miles (10–14 km) at midlatitudes. It can be as little as 5 mi (8 km) high near the poles and as much as 11 mi (18 km) high near the equator. The *stratosphere* is the atmospheric layer above the troposphere, having a depth of about 20 mi (33 km) near the equator. The stratosphere is separated from the troposphere by the tropopause. The *tropopause,* a boundary layer between the troposphere and the stratosphere, prevents most (but not all) of the tropospheric gases from mixing with stratospheric gases, and vice versa.[2]

The troposphere, in its natural state, is composed of about 78 percent nitrogen (atmospheric molecular nitrogen N_2), 21 percent oxygen (O_2), one percent carbon dioxide (CO_2) and other trace gases. Any substance that contaminates and thereby impairs the use of air, water, or other natural resources is a *pollutant*. For example, if smoke and fog particles are present in sufficient quantities to visibly contaminate the air, the resulting pollutant is known as *smog. Smoke* is technically a combination of liquid and solid particles less than one micrometer in diameter, which are emitted from burning materials. Many types of smoke released into the atmosphere give rise to various types smog. *Photochemical smog*, which results from motor-vehicle exhaust, is produced by the interaction of nitrogen oxides (NO_x) and hydrocarbons (any compound composed of carbon and hydrogen, which are a subset of VOCs, volatile organic compounds) under the influence of sunlight.[3]

1. Kupchella and Hyland (1986).
2. Ibid.; National Research Council (1991).
3. Kupchella and Hyland (1986).

▨ Ozone: An Oxidizing Gas That Is Helpful Far from Us

The use of fossil fuels to generate electricity, to run automobiles, and to produce commercial petrochemicals, results in the release of nitrogen oxides and volatile organic compounds into the air. NO_x and VOCs are called *primary pollutants* because they are emitted directly into the atmosphere. Once in the atmosphere, these chemicals are transformed through hundreds of chemical reactions into ground-level or tropospheric ozone.[4] Ozone (O_3) is a reactive oxidizing gas, which, when inhaled, can be deposited along the entire respiratory tract, causing toxic effects (that is to say, it is poisonous).[5]

Kupchella and Hyland define *oxidation* generally as the loss of electrons. In other words, an atom or molecule is oxidized whenever it loses one or more electrons. In turn, *oxidants,* or oxidizing agents, are the molecules that attract these electrons. Ultraviolet and other high-energy, short-wavelength radiation can transform certain chemicals into relatively powerful oxidants; ozone is one such powerful oxidant.[6] Ozone is also called a *secondary pollutant* because it is not emitted directly into the atmosphere, but is produced photochemically in the troposphere by the breakdown of other compounds. These other compounds are called the *precursor emissions* of ozone.[7]

Ozone, which is a form of oxygen, occurs naturally in the cool air of the stratosphere. The stratosphere is home to roughly 90 percent of the earth's ozone. The remaining 10 percent of the ozone resides in the troposphere, associated with transport downward from the stratospheric ozone layer. This means that there is a naturally occurring background ozone concentration (estimated to be between 0.010 ppm and 0.020 ppm, or parts per million) in the lower atmosphere.[8]

In the stratosphere, ozone forms an effective sunscreen by absorbing much of the ultraviolet radiation from the sun. In 1985 British Antarctic Survey scientists reported data that revealed that there had been more than a 40 percent loss of ozone over Antarctica from 1977 to 1987. Other research groups confirmed the conclusions of the British report and showed that the ozone "hole" found in the lower stratosphere is about half the size of Canada. Another ozone hole, about one-third the size of the Antarctic hole, has been observed over the Arctic.[9] Furthermore, the average thickness of the stratospheric ozone layer in the northern mid-latitudes is estimated to have decreased by as much as 5 percent during

4. Mungall and McLaren (1990).
5. Horvath and McKee (1994).
6. Kupchella and Hyland (1986).
7. NRC (1991).
8. Abbatt and Molina (1993); Sillman (1993).
9. Mungall and McLaren (1990).

the 1980s.[10] As stratospheric ozone disappears, more and more ultraviolet radiation from the sun reaches the earth's surface. Such radiation is linked to a dangerous weakening of the human immune system. It is estimated that each 1 percent drop in stratospheric ozone concentration will result in a 4 to 6 percent increase in skin cancers. Ultraviolet radiation is also associated with cataract formation, which is a major cause of blindness.[11]

Chlorofluorocarbons (*CFCs*) are a highly stable and unreactive group of chemicals developed in the twentieth century. They have been used as coolants for air conditioners and refrigerators, propellants for aerosol sprays, blowing agents to produce foam and insulation, and cleaners for computer microchips and other electronic parts. Once released into the atmosphere, they are not broken down at the earth's surface, but rise slowly into the upper atmosphere. By 1987 proof was conclusive that the ozone hole was largely a result of CFC emissions.[12] At levels above 15.5 miles (25 km), which is roughly where stratospheric ozone is at its highest concentration, CFC molecules are bombarded by intense ultraviolet light and finally are broken down. When a chlorine atom (Cl) collides with an ozone molecule (O_3) in the upper atmosphere, the chlorine atom "steals" the third oxygen atom from ozone to form a chlorine monoxide radical (ClO) and an oxygen molecule (O_2). When the chlorine monoxide radical meets with a free oxygen atom (formed by photodissociation of O_3 into its component parts O_2 and O), the oxygen atom separates from the chlorine to form a new oxygen molecule O_2. This frees the chlorine atom to start its destruction of ozone anew.

In the absence of pollutants such as chlorine, the stratospheric ozone concentration would remain stable, since O_3 is formed anew by the joining of O_2 and O after the photodissociation of O_3 by ultraviolet light. As we have seen, chlorine disrupts the formation of ozone (by reacting with atomic oxygen), and what is worse, it does so *catalytically*, which means that chlorine is not changed, or "used up," in the process. The result is that a little chlorine as a free radical can be responsible for the conversion of much ozone into oxygen.[13] It is estimated that 1 free chlorine atom in the stratosphere can eliminate 100,000 molecules of ozone.[14]

With the loss of the highly reactive atomic form of oxygen in the stratosphere, and the resultant loss of stratospheric O_3, increasing amounts of ultraviolet radiation are striking the earth's surface. It happens to be the case that nitrogen dioxide, which is one of the oxides of nitrogen (NO_x) occurring in the troposphere as a result of the combustion of fuel, is a very efficient absorber of ultraviolet radiation. When NO_2 is *photolyzed* (split by light), nitric oxide (NO) and atomic oxygen (O) are produced. Subsequently, atomic

10. Sillman (1993).
11. Mungall and McLaren (1990).
12. Sillman (1993).
13. Bryner (1993); Kupchella and Hyland (1986).
14. Mungall and McLaren (1990).

oxygen, O, combines with molecular oxygen, O_2, to form ozone O_3, which can account for as much as 90 percent of the oxidizing chemicals in smog.[15]

In the absence of competing reactions, nitric oxide will react rapidly with ozone to form nitrogen dioxide (NO_2) and oxygen (O_2). Unfortunately, there are many competing reactions. As an example of annual emissions, at least six million metric tons of VOCs were emitted into the air by the transportation sector (automobiles, trucks, planes, trains, etc.) in the United States in 1990. (Even more was emitted by industrial processes.)[16] When the photochemically reactive VOCs are oxidized, they result in the formation of highly reactive *free radicals*, which convert NO into NO_2 without destroying O_3.[17] In this manner, ozone is actively generated and built up in the troposphere.

It should be noted that stratospheric ozone loss is relatively irreversible, whereas tropospheric ozone production is highly reversible. Even if CFC emission were halted immediately, the destruction of stratospheric ozone would continue to *increase* for almost a decade, and after that the destruction would *continue* for nearly a century. In contrast, ozone and its precursors have a lifetime of only a few days near the ground, and a few months in the upper troposphere. Therefore, if ozone precursor emissions were halted or drastically reduced, excessive tropospheric ozone would disappear almost immediately.[18]

■ The Effects of Ozone on Human Health

"[Ozone] O_3 is a highly toxic chemical which can adversely affect the respiratory systems of any individual depending on the conditions of exposure."[19] Due to its oxidative properties, O_3 inhaled and deposited along the human respiratory tract causes toxic effects that can lead to structural damage to the lungs as a result of lesions, a permanent impairment of lung functions. A summary of some health effects of greatest concern linked to ozone exposure are the following:

- Ozone can lead to pulmonary inflammation (i.e., inflammation of the lungs) and can cause or contribute to several pulmonary diseases, including emphysema, bronchitis, pulmonary edema (i.e., accumulation of fluid in the lungs), and asthma.
- Ozone is highly toxic when inhaled, and can break down body tissue and cells, particularly lung tissue, causing lesions in the lungs, as well as premature aging of lung tissue.[20]
- Ozone has been shown to reduce the body's resistance to disease; it can lead to alterations in red bloods cells and enzyme activity.

15. Kupchella and Hyland (1986).
16. Bryner (1993).
17. McKee (1994).
18. Sillman (1993).
19. Horvath and McKee (1994, p. 67).
20. Ibid.; Stewart (1979); Kupchella and Hyland (1986).

- Ozone can produce genetic alterations in lymphocytes (i.e., colorless corpuscles produced by the lymph glands), which circulate in the blood stream.[21]
- Acute high-level exposure to ozone can cause stress on the heart; chronic exposure can result in heart failure.[22]

Ozone exposure also causes crop loss, which will affect both human nutrition and the national economy in the long term. In 1980 the U.S. Environmental Protection Agency (EPA) initiated a field-experimental program, called the National Crop Loss Assessment Network (NCLAN), in order to assess the impact of O_3 on crops. For the years 1982 through 1987 data indicated that major U.S. crops suffered a 14 percent yield loss due to ozone exposure.[23] Cotton and soybeans have suffered the greatest loss, since they are grown in areas heavily affected by ozone. Monetary loss from crop damage in the United States is estimated to range from $1 billion to $2 billion per year.[24]

▓ National Ambient Air Quality Standards

The 1970 Clean Air Act Amendments—which were amendments to the 1963 Clean Air Act—required the EPA to establish National Ambient Air Quality Standards (NAAQS) for major pollutants. Ozone and nitrogen dioxide were counted among the major pollutants for which NAAQS were specified. The ozone concentration limit originally set in 1971 was 0.080 ppm for a one-hour average, with one expected exceedance allowed per year. In other words, the O_3 gas concentration in ambient air (i.e., the air around us) for any one-hour period, was not to exceed 0.080 ppm more than once per year. In 1979 the NAAQS for ozone was raised to 0.12 ppm for a high hourly average, due to the large number of nonattainment regions in the country at the lower limit.[25]

The number 0.12 ppm was decided upon as a result of acute respiratory problems that had been reported in one- and two-hour controlled human exposure studies.[26] However, clinical studies have shown the seriousness of lower level, but longer term, ozone exposure. Loss of lung function and inflammation were found to occur in healthy, exercising subjects at O_3 levels as low as 0.080 ppm, after approximately seven hours of exposure. This O_3 *cumulative effect* is of particular concern because heavily populated areas in the eastern United States frequently experience O_3 levels between 0.080 and 0.12 ppm for six to eight hours at a time.[27] Because of the detrimental effects of protracted O_3 exposure, a workplace Permissible Exposure Limit (PEL) of 0.10 ppm for eight hours (which is lower than the 0.12 ppm

21. Horvath and McKee (1994).
22. Coffin and Stokinger (1977); Sterling et al. (1969).
23. Tingey et al. (1994).
24. Sillman (1993).
25. See for example Chang et al. (1992), Kupchella and Hyland (1986).
26. Whitfield et al. (1994).
27. White (1994).

for one hour NAAQS) was set. Studies have demonstrated that a number of places in the United States, especially in the Northeast, violate the workplace PEL for outdoor workers.[28]

Cumulative effects have also been considered in terms of ozone seasons. The *ozone season* ranges from 4 months in two northern states to 12 months in the south. Most states have a 7-month O_3 season, and usually those months are April through October. Considering the populations of children and outdoor workers in New York and Los Angeles, experts have reported expected chronic effects of lung impairment resulting from lesions for a substantial portion of the population as a result of ozone exposure at mean levels of 0.050 ppm or higher for 10 ozone seasons.[29]

As we have seen, ozone exposure at and below the NAAQS limit still damages human health, depending on length of exposure. Even so, numerous U.S. cities (among them Houston, Chicago, Milwaukee, and the cities of the northeast corridor) have exceeded the NAAQS 5 to 10 times per year. These still do not compare to the ozone problem in southern California, where the NAAQS has been violated on one day out of three.[30] "The most critical aspect of this problem," according to the National Research Council (an organization administered jointly by the National Academy of Sciences, the National Academy of Engineering, and the Institute of Medicine), "is the formation of ozone in and downwind of large urban areas where, under certain meteorological conditions, emissions of nitric oxide and nitrogen dioxide and volatile organic compounds can result in ambient ozone concentrations up to three times the concentration considered protective of public health by the U.S. Environmental Protection Agency."[31] Depending on weather conditions, O_3 concentrations of 0.20 ppm to 0.40 ppm (200 ppb–400 ppb, or parts per billion) can occur downwind of large urban areas (recall that the NAAQS for O_3 is 0.12 ppm or 120 ppb). (An ozone concentration of 12 ppm is lethal, even for a very short time of exposure.[32]) Air quality continues to decline as millions of tons of pollutants are added annually to the atmospheric burden of waste gases.

▓ Ground-Level Ozone Control Strategies

Approximately 40 percent of *anthropogenic* (i.e., that which results from human activity) VOC emissions are produced by transportation, according to the 1985 National Emissions Data System (NEDS) and the National Acid Precipitation Assessment Program (NAPAP) emissions inventories (see Fig. CP2.1).[33] All anthropogenic VOC emissions are derived from the use of petroleum or natural gas, rather than coal.

28. NRC (1991).
29. Whitfield et al. (1994).
30. Sillman (1993).
31. NRC (1991, p. 1).
32. Kupchella and Hyland (1986).
33. NRC (1991).

It is important to distinguish anthropogenic VOC emissions from *biogenic* (i.e., that which arises, not from human activity, but from natural processes) VOC emissions. *Isoprene*, a VOC emitted primarily by deciduous trees, contributes significantly to the VOC inventories on summer days. In urban areas, isoprene may account for 15 percent of VOC emissions (anthropogenic and biogenic). Considering both urban and rural areas, isoprene can equal or exceed anthropogenic VOC emissions for part of the ozone season in the eastern United States.[34] However, by itself this biographic emission is not a pollutant. Only in the presence of anthropogenic emissions does it contribute to pollution as an ozone precursor.

Approximately 43 percent of NO_x emissions arise from mobile sources (i.e., transportation) according to the 1985 NEDS and NAPAP data. Within this transportation sector, about 50 percent of NO_x emissions come from light-duty gasoline trucks and cars and 25 percent come from heavy-duty gasoline and diesel vehicles. Of the 57 percent of NO_x emissions arising from point sources (i.e., stationary sources, primarily stack emissions), 32 percent result from utilities (primarily coal-fired power plants). (See Fig. CP2.2.) In summary, motor-vehicle emissions typically account for about 40 percent of the total ozone precursor emissions (VOCs and NO_x) in any region.[35]

Ozone forms and accumulates in a polluted urban air mass over a period of several hours as the air moves downwind. This phenomenon results in peak ozone concentrations in suburban locations 30 to 100 miles from an urban center. On occasion, plumes of high ozone have been transported much farther than 100 miles. Ozone in excess of the NAAQS has been recorded on Cape Cod, Massachusetts, associated with transport from New York; in Acadia National Park, Maine, from Boston; in northern Wisconsin, from Chicago; and in the mountainous regions of Japan, from Tokyo.[36] These areas affected by polluted urban air masses are termed transport regions (see Fig. CP2.3).

Because two chemical species, VOCs and NO_x, are needed to produce ozone, ozone production is governed by a *law of the minimum,* which means that ozone formation is limited by whatever precursor (VOCs or NO_x) is in shortest supply. In NO_x-limited regions, where the VOC/NO_x ratio is 20 or more, reduction of NO_x emissions is the more effective means of lowering ozone production. In VOC-limited regions, where the VOC/NO_x ratio is 10 or less, reducing VOC emissions is the more effective means of lowering ozone production.[37]

34. Sillman (1993).
35. NRC (1991).
36. Sillman (1993).
37. NRC (1991).

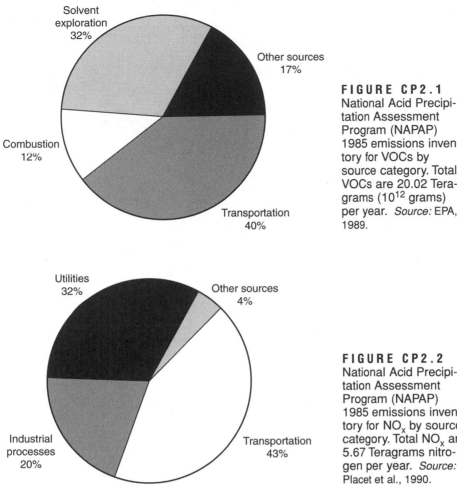

FIGURE CP2.1
National Acid Precipitation Assessment Program (NAPAP) 1985 emissions inventory for VOCs by source category. Total VOCs are 20.02 Teragrams (10^{12} grams) per year. *Source:* EPA, 1989.

Solvent exploration 32%

Other sources 17%

Combustion 12%

Transportation 40%

FIGURE CP2.2
National Acid Precipitation Assessment Program (NAPAP) 1985 emissions inventory for NO_x by source category. Total NO_x are 5.67 Teragrams nitrogen per year. *Source:* Placet et al., 1990.

Utilities 32%

Other sources 4%

Industrial processes 20%

Transportation 43%

Recent research shows that as a polluted air parcel moves downwind, its VOC/NO_x ratio *increases* steadily (unless it encounters fresh emissions with high NO_x). This typical increase in the VOC/NO_x ratio is due to the fact that NO_x has a shorter chemical lifetime than do most VOCs so that it is removed from the air parcel more rapidly. The lifetime of NO_x in a polluted air parcel is approximately six hours, whereas VOC lifetimes vary by species from less than one hour to five days. However, most VOC species are longer lived than NO_x. The phenomenon of the aging polluted air parcel becoming NO_x-limited

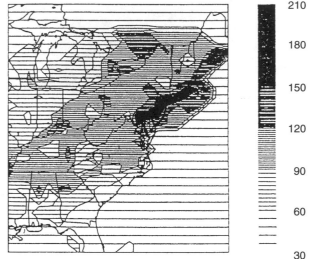

FIGURE CP2.3 Peak tropospheric ozone concentrations (in parts per billion) in the transport region that includes the Ohio River Valley and the entire Northeast corridor is shown for a severe air pollution event on June 15, 1988. (Values shown for Canada and for the Atlantic Ocean are inaccurate, since no observations were made there.) Data is based on surface observations at 350 EPA monitoring sites. *Source:* EPA/AIRS, 1991.

Ozone (ppb)

210
180
150
120
90
60
30

is summarized in *isopleth diagrams*, showing ozone production as a function of NO_x and hydrocarbon (i.e., technically a subset of VOCs, but here used synonymously with VOCs) emissions in Fig. CP2.4 and CP2.5.[38]

After an air parcel has aged for 36 hours or more, it is almost always NO_x-limited, regardless of how it started in terms of precursor concentrations. At this point, it has become insensitive to the level of VOC emissions. In locations near emissions sources (e.g., city centers), air parcels are more likely to be VOC-limited. In this case, reducing NO_x may lead to local increases in ozone, as seen in the isopleth plot (upper left quadrant of Fig. CP2.4). Nevertheless, NO_x reduction is favored for total ozone reduction, because reducing NO_x would result in reducing ozone in downwind suburban locations, rural areas, and entire transport regions. However, VOC controls are likely to be effective in a city where concentrations of both VOCs and NO_x remain high throughout the day. Figure CP2.5 shows that polluted air parcels with high initial concentrations of NO_x and VOCs (upper right quadrant) are still VOC-limited after eight hours, due to the slower rate of photochemical evolution in heavily polluted air.[39]

Los Angeles and Chicago are more likely to be VOC-limited cities, since they both have a high density of emissions and stagnant meteorology. (The atmospheric circulation over Chicago, associated with Lake Michigan, can restrict dispersal of air pollutants as the mountains do in the Los Angeles basin.) In contrast, medium-sized cities in the eastern United States, such as Boston, Cincinnati, and Atlanta, are likely to be NO_x-limited.[40]

38. Sillman (1993).
39. Ibid.
40. Ibid.

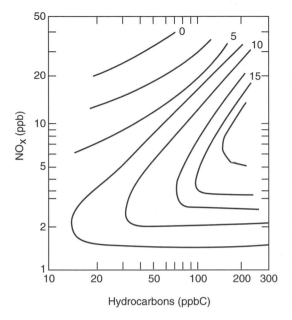

FIGURE CP2.4 Isopleth plots are given showing net rates of ozone production, ranging from 0 ppb to 17.5 ppb per hour, in 2.5 ppb-per-hour intervals, averaged over an eight-hour daytime period. Ozone production is presented as a function of NO_x and hydrocarbon (HC) concentrations. The isopleth plots may be considered three-dimensionally as the contours of a hill whose maximum height is the maximum ozone concentration. In the region of high HC:NO_x ratio (>20:1), ozone concentration increases steadily with increasing NO_x but is insensitive to increases in HC. In the region of low HC:NO_x ratio, ozone production increases with increasing HC concentration, but actually decreases with increasing NO_x. *Source:* Sillman, S., "Trophospheric Ozone: The Debate Over Control Strategies," *Annual Review of Energy and the Environment*," vol. 18, p. 41, Reprinted with permission.

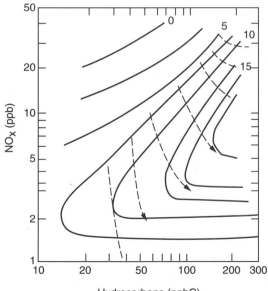

FIGURE CP2.5 The isopleth plots of ozone production are overlaid by dashed lines representing the aging of parcels of polluted urban air over an eight-hour period. The air parcels each have an initial HC:NO_x ratio of 5:1 but different total amounts of HC and NO_x. (As the urban air moves downwind [dashed lines], the calculations assume that additional HC and NO_x emissions received from suburban areas are in the same proportion as the initial concentrations in the air parcels.) The polluted air parcels reach maximum ozone concentration after some time and evolve toward higher HC:NO_x ratios. *Source:* Sillman, S., "Trophospheric Ozone: The Debate Over Control Strategies," *Annual Review of Energy and the Environment*," vol. 18, p. 41, Reprinted with permission.

Federal Test Procedure and the Problem of Super-Emitters

A common assertion is that new cars today emit only a few percent of the exhaust emissions that came from autos built before emission controls some 25 years ago. How much progress has actually been made is now in question. Two reasons for this are (a) the awareness that common episodes of inefficient catalytic converter operation and consequent high exhaust emissions are not being measured in the Federal Test Procedure (FTP) and (b) roadside inspection has found late model "super-emitters" in the in-use vehicle fleet.

In the 1970s the Federal Test Procedure was established for determining vehicle emissions as a model for new car certification. The FTP involves dynamometer[41] tests on stationary prototype cars over 50,000 miles of operation, and at the end of the production line. Exhaust is collected in bags and chemically analyzed. Evaporation of fuel from the vehicle is also measured. The emissions must meet the grams per mile standards for CO, HC, and NO_x.

The driving cycle used in the FTP incorporates only one cold start and involves no high-power operation (except, in effect, for a few cars with very low power-to-weight ratios). This means that the fuel-rich operations associated with repeated cold starts and with high-power driving are "off cycle" (i.e., they are not factored into the vehicle's emissions count). It is precisely the fuel-rich operations that generate high CO and VOC emissions. High-power operation of the vehicle is driving with the accelerator pedal virtually on the floor at wide-open throttle. This occurs for long stretches when owners of moderate-power cars find themselves keeping up with 70-mph expressway traffic or climbing hills.[42]

According to California Air Resources Board (CARB) measurements, typical CO and HC emissions in a high-power episode *of a few seconds* are 28 grams and 0.7 gram, respectively (compared with FTP limits for CO and HC of 3.4 and 0.41 grams *per mile,* driven in perhaps *one minute*). Furthermore, on the basis of data from speed sensors used in two groups of cars in Spokane, Washington, and Baltimore, Maryland, and other characteristics, it is expected that fuel-rich operation will occur about once every three miles in typical driving with modern vehicles. CARB has shown that brief high-

---

41. Fundamentally, a *dynamometer* is a device used for measuring the torque, force, or power available from a rotating shaft. The speed of the shaft (in revolutions per minute, or rpm) is measured with a tachometer. In a transmission dynamometer, the turning force or torque (in lb-ft) of the shaft is determined by measuring the elastic twist of the output shaft. In an absorption dynamometer, the power is absorbed and dissipated as heat by restraining the output shaft by means of a brake or an electromagnetic force. In this case, the torque is determined by measuring the force required to keep the restraining element from rotating. The chassis dynamometer, which is widely used in the automotive industry, has rollers that are turned by the test vehicle's drive wheels. This dynamometer is used for mileage accumulation, emissions, fuel economy, and performance testing of cars and trucks (McGraw-Hill, 1992).

42. Calvert et al. (1993); Ross (1994).

power episodes can dominate emissions from properly functioning cars.[43] This means that the dominant emissions are off FTP cycle.

In order for a three-way catalytic converter (which is the heart of the emissions control system) to work, the catalyst must be hot, and the air–fuel mixture input to the engine must be *stoichiometric* (i.e., it must have the chemically correct balance so that the catalyst can work on exhaust that has balanced residues of partially burned fuel (VOCs), nitrogen oxides (NO_x), and oxygen). Fuel injectors are routinely designed to introduce excess fuel (non-stoichiometric air–fuel ratio) into a cold engine in order to improve combustion stability. In doing this, they defeat the emissions control system.[44] Emissions from each cold start can be approximately *twice* the emissions from a high-power episode.[45] (High-power episodes also involve non-stoichiometric air–fuel ratios, as discussed above.)

The California Random Roadside Inspection Survey Program has been carried out every year since 1983 in order to evaluate the effectiveness of California's Inspection/Maintenance program (I/M). In 1989 the survey conducted by CARB was performed at 60 urban locations throughout the state and involved about 4500 vehicles. It was found that for the no-load 1000 rpm idle test, 10 percent of the vehicles were responsible for about 60 percent of the exhaust CO, and that another (although not necessarily the same) 10 percent were responsible for about 60 percent of the exhaust VOCs. (The results showed only a weak relationship between vehicles that emit large amounts of CO and those that emit large amounts of VOCs.)[46] These vehicles, the worst 10 to 20 percent of the vehicle fleet, which are responsible for 60 percent of the exhaust emissions of CO and VOCs, are termed super-emitters. Although older vehicles were expected to be among the super-emitters due to their less capable emissions control systems and mechanical deterioration, surprisingly super-emitters spanned all year models, even late-model cars.

To address this problem, the EPA has proposed an enhanced I/M program that would

- require a dynamometer test under varying engine loads;
- test evaporative emissions (i.e., emissions resulting from evaporation of fuel from various locations in the fuel tank-to-engine system), which have been found to be far higher than had been thought, especially on hot days when there is the greatest potential for high smog concentrations; and
- raise the repair cost limit, so that repair of high-emitting vehicles will be less likely to be waived.[47]

43. Ross (1994).
44. Ibid.
45. Calvert et al. (1993).
46. NRC (1991).
47. Calvert et al. (1993).

Both the 1990 Clean Air Act Amendments and CARB require that during the next few years,

- extensive on-board engine and emission control system diagnostics (OBD) capability be built into new vehicles. On-board diagnostics are combinations of sensors, computer diagnostics, and warning lights that alert the driver and maintenance personnel to malfunctions of the emissions control system.[48]

In an effort to stop the practice of cheating, either through collusion between the tester and vehicle owner or by the vehicle owner modifying the vehicle before and after scheduled testing, it has been suggested that there be established

- remote monitoring of tailpipe emissions of passing cars by monitoring sites located in positions that are invisible to the drivers and are relocated regularly, which have the capability of photographing license plates. The car owner can then be notified so that a more detailed examination can be made by an inspector, high emissions verified, problems identified, and corrective action ensured.[49] One would expect, however, that in the case of vehicle owner tampering, this procedure would have to be modified, because emission controls may be reinstated before a scheduled examination.

Modifying FTP is needed as well. In particular,

- the FTP cycle can be made more representative of actual driving cycles by including high-power episodes and more cold starts; and
- FTP tests should be performed not on pre-selected clean-running prototypes, but on vehicles picked at random from the vehicle fleet, which should produce a proportion of cars that are, for various reasons, high emitters.[50]

A Precision Air–Fuel Ratio Control System for Natural Gas–Fueled Vehicles

As indicated in the preceding section, the key to efficient use of a three-way catalytic converter (so as to produce maximum reduction of exhaust emissions) is to keep the *air–fuel* (*A/F*) *ratio* at its stoichiometric value. (A stoichiometric A/F ratio means that there will be just the right number of oxygen molecules to burn all the fuel molecules.) If the mixture is *lean* (i.e., having excess air), the catalytic converter is very good at oxidizing exhaust VOCs and CO, but almost useless at reducing NO_x. Conversely, if the mixture is *rich* (i.e., having excess fuel), the catalytic converter is good at

---

48. Ibid.
49. Ibid.
50. Ibid.

reducing NO_x, but ineffective on VOCs and CO. For low emissions, the A/F ratio of the combustible mixture supplied to the engine should be maintained to within ±1 percent of its stoichiometric value under all operating conditions. Operating conditions are either static (as in idling mode) or dynamic (as with throttle changes).[51]

Modern gasoline fuel-injection systems use an oxygen sensor in the exhaust to determine whether the combustible mixture being supplied to the engine is lean or rich. Since the air–fuel mixture takes some time to be ingested, compressed, burned, expanded, and exhausted, it takes time for the oxygen sensor to send corrections to the A/F ratio controller. The time required for the oxygen sensor to send corrections to the A/F ratio controller is called the oxygen signal lag time. After a change in torque demand (i.e., a change in power demand, as in acceleration), typical recovery time for the A/F ratio controller to arrive at stoichiometric value is three to five seconds. However, torque demand can vary from zero at idle to full torque during hard acceleration, in as little as 1/10 second. Usual driving in traffic shows demand (i.e., changes in acceleration) varying every two to five seconds, with changes being as much as half of full torque. The result is that the A/F ratio is almost never correct. In fact, excursions (i.e., departure from stoichiometric value) of the A/F ratio are as much as ±25 percent during transients. (Transient performance is the non–steady-state values of the A/F ratio, as after throttle change, before stoichiometric value is attained.) In short, emission levels are dominated by transient performance.[52]

Also noted in the preceding section, cold starts result in the highest episode of exhaust emissions for gasoline-powered vehicles. Excess fuel is injected into a cold engine in order that a certain amount may evaporate before the spark occurs. The air–fuel mixture is rich overall once the flame has ignited in order to provide adequate heat to evaporate all the fuel. This fuel-rich mixture is responsible for large amounts of exhaust VOCs and CO. However, with a gaseous fuel, no pre-spark evaporation is needed. The engine can be run, theoretically, at the stoichiometric A/F ratio from the start.[53]

The three types of carburetors in use today for natural gas (NG) vehicles are the fixed orifice carburetor, the variable restriction carburetor, and the gaseous fuel-injection system. The first type places severe demands on the accuracy of the pressure regulator which cannot be met. The second type has several drawbacks. Variations in the area of the air and gas ports have to be controlled to within tight tolerances, which makes it an expensive piece of equipment. Flow through the two orifices must track one another precisely. And the underlying assumption that pressure drop is proportional to flow squared does not hold to within 1 percent accuracy over a 30 to 1 range of flows. The third type of carburetor is modeled after gasoline fuel-injection systems, and responds to rapid changes in demand no better

51. Sibilia and Durbin (1994).
52. Ibid.
53. Ibid.

than current gasoline systems. It produces transient peak errors in the A/F ratio of ±25 percent.[54]

Researchers at Princeton University have designed and built a system for precise dynamically controlled air/fuel (DCAF) ratio in NG vehicles. Their system eliminates the flow-squared dependency of the pressure change and yields a predictable relationship between pressure and flow so that the oxygen sensor in the exhaust is not needed for dynamic control, but only for changes which occur with time constants greater than three seconds (easily within the oxygen sensor control capability). DCAF uses structures in which the pressure drop is due to the viscosity of the flow, and therefore is proportional to the rate of flow (not to the rate of flow squared).[55]

The DCAF system consists of the following five components:

- First, in order to produce laminar flow,[56] the existing automobile air filter is replaced by a new porous structure with finer pores made of very stiff material (to eliminate stretching, pore enlargement, etc.). The area is adjusted by changing the number of folds so that an appropriate pressure drop can be obtained with linearity (to rate of flow) of better than 1 percent, even under very high flow rates.

- The second element is a gas flow meter. A rectangular ceramic porous structure is used, similar to that which is used as a base element in a catalytic converter and in a diesel engine exhaust soot filter.

- Third, a gas flow controller valve was designed, using a voice coil actuator with a flexible membrane supporting the structure. It is similar in design to loudspeaker structures, except that cloth and fiber parts are replaced with metal for increased stability. The flow of air is determined by measuring the pressure drop across the air element. The flow of gas is determined by measuring the pressure drop across the gas element. If too little gas is flowing for the amount of air, the gas flow controller valve opens more; if too much gas is flowing, the valve closes accordingly. The speed of response of the servo valve is the most critical factor in diminishing A/F ratio dynamic errors. In the DCAF system, the valve is solenoid[57] operated and controls the gas flow in an annular channel. (An annular orifice minimizes the force needed to oppose the inlet supply pressures, relative to a circular orifice.)

54. Ibid.

55. Ibid.

56. A flow of viscous fluid is called laminar when the fluid moves in smooth orderly layers, called laminae, without the formation of small eddies or irregular fluctuations. Laminar flow occurs in low-density flows, as with rarefied gases; in low-velocity or "creeping" flows; with small-size bodies in a large fluid flow; or in high-viscosity fluids, as with lubricating oils. Aircraft in flight have regions of laminar flow near the leading edges of their wings (McGraw-Hill, 1992).

57. A *solenoid* is a coil of insulated wire with an electric current running through the wire. A solenoid produces a magnetic field within the coil. The magnetic field inside the coil may be used to magnetize and therefore attract a plunger or armature to a position within the coil. In such a case, the solenoid may be considered a special form of electromagnet (McGraw-Hill, 1992).

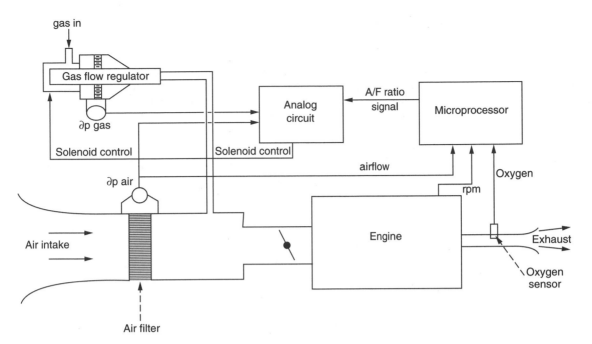

FIGURE CP2.6 A block diagram of the DCAF system for the precision control of the air-to-fuel ratio in gas-fueled vehicles. *Source:* Sibilia, M.J, and Durbin, E.J., "DECAF: A System for Dynamically Controlled Air/Fuel Ratio in Gaseous Fueled Engines (U.S. patent #5,143,111)." Princeton, NJ: Rutgers University Department of Mechanical and Aerospace Engineering, October 21, 1994. Reprinted with permission.

- Fourth, a fast inner loop of analog control circuits controls the opening of the servo valve to make the gas mass flow proportional to the air mass flow. Adjustable constants in the analog circuit allow the vehicle to run on a range of fuel compositions from pure methane to pure butane. (These adjustments are made automatically by the system.)

- Fifth, a slower outer control loop, implemented in software on a microprocessor, senses whether the A/F ratio is high or low and makes corrections to the constants in the inner loop. The microprocessor gives approximate values of the constants (beginning with a value stored in memory when the engine was last shut down) to the inner loop during engine start-up. The approximate values are refined by the microcontroller without A/F ratio sensing, until the oxygen sensor has reached several hundred degrees Celsius, and begins to give valid readings. At this point, the sensor is used to correct the approximate values.[58]

Figure CP2.6 is a block diagram of the DCAF system.

A prototype of the DCAF system was built and installed on a 1990 Plymouth Acclaim with a 2.5-liter single-point fuel-injection gasoline engine. Natural gas was supplied to the system at approximately 30 kPa (5 psi). The

58. Ibid.

same vehicle was tested for emissions, running first on gasoline and then on natural gas using the DCAF control system. For cold start, gasoline resulted in 10 percent CO emissions, which corresponds to 40 percent excess fuel. Natural gas with DCAF resulted in 3 percent CO emissions, which corresponds to 12 percent excess fuel. With gasoline, each time the throttle was opened, a pulse of CO peaking at 1.5 percent was generated, which corresponds to 6 percent excess fuel. With natural gas and DCAF, the CO peaked at 0.05 percent when the throttle was opened, which corresponds to 0.21 percent excess fuel.[59] These results demonstrate a spectacular 70 percent reduction in CO emissions and excess fuel supplied to the engine in cold starts, and a 90 percent reduction in CO emissions and excess fuel use in throttle changes, obtained by using natural gas with the DCAF control system as opposed to gasoline with fuel injection.

The 0.21 percent excursion is well within the design goal of holding the A/F ratio to ±1 percent. And the next generation of DCAF control software should reduce cold-start transient behavior to less than 30 seconds.[60]

Lacking the infrastructure of a large number of natural gas fueling stations, the first candidates for conversion to natural gas are centrally fueled vehicles. Forty-five million (or 23 percent) of the U.S. fleet of 194 million vehicles are commercial vehicles (e.g., buses, trucks, taxis, delivery vans). Vehicles used for personal transportation consume an average of 559 gallons of gasoline per year. Commercial vehicles use an average of 1200 gallons per year. Fifty-four billion gallons (or 40 percent of vehicle fuel annual consumption) are used by commercial vehicles in a year. From these figures, we see that converting a small percentage of the U.S. vehicle fleet—that is, commercial vehicles—to natural gas using DCAF could have a large effect on reducing both atmospheric pollution and oil imports.[61]

◼ Higher Fuel Economy and Lower Lifetime Emissions for the Internal Combustion Engine

The gasoline-powered internal combustion (IC) engine has given millions of Americans extraordinary mobility. It also has polluted our air and has made us, as a nation, heavily reliant upon imported oil.[62] The 194 million vehicles on the highways in the United States consume about one-third of the world's oil production, and about two-thirds of the oil we consume is imported.[63]

From 1984 to 1992, vehicle-miles in the United States grew about three and a half percent per year. Although the fuel economy (miles per gallon, or mpg) of the average car has been increasing during most of this period, a shift

59. Ibid.
60. Ibid.
61. Durbin (1993).
62. Williams (1994).
63. Durbin and Schemmann (1995).

toward the use of trucks as cars has had the effect of decreasing the fuel economy of the average in-use vehicle. The net result has been an upward creep in gasoline consumption from 1984 to 1992 of about one percent per year.[64] The expectation now is that motor-vehicle fuel use will increase as rapidly as vehicle-miles traveled, perhaps at three and a half percent per year.[65] This pattern of consumption is not sustainable in view of the limits of world natural resources, environmental preservation, and the national economy.

Under the Corporate Average Fuel Efficiency (CAFE) law, the average fuel efficiency of all vehicles sold by a particular automaker must meet a certain mileage standard. Currently, that standard is 27.5 mpg.[66] Since the gasoline-powered internal combustion engine may be the primary technology in the automobile market for the next decade or more, improving the efficiency of the IC engine in order to produce a high-mpg car may be one of the most direct ways of reducing combustion emissions, as well as the national trade deficit due to oil imports. An example of such technological improvements is the proposed High-MPG vehicle of Marc Ross.[67] High-MPG would have a fuel economy of 50 mpg (a 79 percent improvement over the standard), without reduction of vehicle size or performance relative to the average 1993 model car, which we will designate AVCAR '93. (In other words, High-MPG is not reduced in size and performance like the Geo Metro and Honda Civic VX.)

The high power of most of today's cars is a major cause of engine inefficiency because it is achieved with large engines having large frictional loads. Roughly 30 kilowatts (kW) should suffice for sustained driving in demanding situations, while the 105 kW maximum provided in the average car enables one to accelerate rapidly at speeds far above legal limits. The 89 kW maximum of High-MPG, with its 15 percent lower mass, should have plenty of power for highway driving.[68]

The overall engine efficiency is the product of two factors: thermal efficiency and mechanical efficiency. *Thermal efficiency* expresses how much of the fuel energy is converted into work moving the pistons. *Mechanical efficiency* indicates the fraction of that work that is delivered by the engine to the vehicle (the rest goes to overcome frictions in operating the engine). The *engine power output* (P_b, in kW) is the product of the mechanical efficiency (η_m), the thermal efficiency (η_t), and the fuel energy input (P_f):

$$P_b = \eta_m \, \eta_t \, P_f.$$

For AVCAR '93, $\eta_t = 38$ percent. Improving thermal efficiency from roughly 38 percent to as much as 45 percent is an important goal. This goal could be achieved, for example, through a natural gas distribution network and development of a high-efficiency engine for natural gas, since a fuel, such as hydrogen or methane, with simpler molecules and high octane would significantly increase thermal efficiency. (Natural gas is primarily methane

64. Energy Information Administration (1993).
65. Ross (1994).
66. Durbin (1993).
67. Ross (1994).
68. Ibid.

[>90 percent], with other light hydrocarbons, including ethane, ethene, propane, propene, and butane, as impurities.[69]) Or research could be directed to solving the NO_x reduction problem for lean-burn gasoline-fueled engines or automotive diesel engines, which have higher thermal efficiency due to, among other things, increased compression ratio. However, thermal efficiency is only very modestly improved in High-MPG, to 39 percent, since High-MPG uses current technology. It is likely to be impractical to try to achieve thermal efficiencies above about 50 percent for IC engines.[70]

In the U.S. Environmental Protection Agency composite cycle, $\eta_m = 52$ percent. (Mechanical efficiency is 0 when the engine is idling, providing no power output, and it is typically about 90 percent near wide-open throttle.) High-MPG attains a significantly increased mechanical efficiency of 65 percent. In contrast to thermal efficiencies, it may be practical to strive for mechanical efficiencies approaching 100 percent. Overall engine efficiency for AVCAR '93 is calculated to be $\eta_m \eta_t = 20$ percent, while High-MPG has 25 percent overall engine efficiency.[71]

In the energy flows of AVCAR '93, calculated in the EPA composite driving cycle, 48 percent of the fuel is used to overcome engine frictions; engine output is just 20 percent of the fuel energy, and *only 15 percent of the fuel energy reaches the wheels.*

- In High-MPG, engine frictions would be reduced by 15 percent, mostly through reduction of energies used for pumping and for driving engine accessories. *Pumping* refers to moving the air and vaporized fuel into the cylinders, and the combustion products out through the exhaust system.
- The engine displacement (volume *V*, in liters) is reduced by 15 percent in High-MPG by means of a reduction in inertial weight. *V* is reduced an additional 40 percent through specific-power increases from 51 to 85 horsepower per liter (hp/l), well within achievements by today's high-tech engines. Engine friction will be reduced also in proportion to reduction in engine displacement.
- Average engine speed (in revolutions per second) in High-MPG is maintained the same as AVCAR '93, even though High-MPG is a four-cylinder engine with just over half (50.5 percent) of the engine dis-

69. NRC (1991).

70. In general, a *heat engine* (which is a device that uses heat energy to do work) operates on a cycle. The typical cycle of a gasoline-fueled internal combustion engine is the four strokes per cycle of the piston: intake, compression, ignition/power, and exhaust. The *efficiency* of any heat engine is the measure of how effectively it uses the input energy: efficiency = output work / input energy. The ideal heat engine described by Sadi Carnot, called the Carnot engine, is the most efficient engine possible. Even its efficiency, however, is less than 100 percent. Carnot engine ideal efficiency = $1 - (T_c / T_h)$,

where T_c = Temperature, in °K, of exhaust gases, and
T_h = Temperature, in °K, of driving gases.

Since exhaust gases always have a temperature and pressure greater than zero, and in real engines the exhaust gases have greater internal energy than the driving gases, efficiency for internal combustion engines has an upper limit of about 50 percent (Bueche and Wallach, 1994; Ross, 1994).

71. NRC (1991).

placement of AVCAR '93. This is obtained through *aggressive transmission management* (*ATM*), which involves modifying the transmission to reduce engine speed at a given power output. Honda, for example, has implemented ATM by building more gears and lower gear ratios, so that one shifts up at lower engine speeds than is typical. Manufacturers may be reluctant to reduce engine speeds since the engine may not run as smoothly at low speed, and relatively high power is not immediately available at low engine speeds (shifting up is needed). However, these concerns may not be considered overly important to a large segment of the public, who are aware of the environmental, societal, and personal benefits of fuel economy, and who are supported by a system of fees and rebates based on fuel economy.

Implementing these design changes in High-MPG results in a reduction of fuel used to overcome engine friction by a dramatic 58 percent. A well-designed 1.4-liter engine in High-MPG could provide 22 kW (30 hp) at 1800 rpm, and 30kW (40 hp) at 2400 rpm. With its 15 percent lower mass, High-MPG would have plenty of power in this engine speed range for highway driving.[72]

High-MPG is still not at the limit of fuel economy for a petroleum-fueled IC engine vehicle. For example, inertial weight could be reduced by 30 percent or more (instead of the assumed 15 percent) by the use of lighter materials, such as the fiber-reinforced composites under development, and aluminum. A practical technology promising fuel economy of 80 mpg or more is a hybrid car, based on an IC engine, which uses an energy storage device, such as a flywheel[73] or an electric capacitor,[74] that emphasizes power capability over energy capacity. In such a hybrid vehicle, the IC engine would operate only in the interval of its high mechanical efficiency and would be turned off much of the time. This fuel-based hybrid would retain the sustained-power and peak-power capabilities, as well as the range and quick-refueling capabilities, of today's car.[75]

72. Ibid.

73. The energy stored in a flywheel is based on the physical principle described by Sir Isaac Newton's first law of motion: A body at rest will remain at rest, and a body in motion will remain in motion, provided that in neither case is the body acted upon by an unbalanced force. A *flywheel* is a wheel designed to have high inertia (i.e., it is massive) so that once put in motion by an applied force, it will remain in motion for a long time after the force is removed. Swiss trolleys were designed to operate on the principle of flywheel energy storage. They were equipped with an electric motor, but it was not connected to an overhead power-line all of the time, as in the United States. The trolley tapped into the electric outlet only at stations, at which time the electric motor spun a heavy flywheel on board the trolley. Although the motion of the flywheel was opposed by friction, enough energy was stored in the flywheel to take the trolley to the next station (Koff, 1961).

74. A *capacitor* is an electrical device that is capable of storing electrical energy. It generally consists of two metal plates insulated from each other by a dielectric. Capacitors come in a variety of sizes and shapes and employ various dielectric materials. For example, with air, gas, or a vacuum as the dielectric, capacitors are constructed with flat parallel metalic plates or concentric metallic cylinders. Capacitors using plastic film as the dielectric generally have wound-roll construction. The *capacitance* of a capacitor is that property which determines how much charge can be stored in it for a given potential difference across its terminals (*McGraw-Hill*, 1992; McKechnie, 1983).

75. Ross (1994).

In the midst of technological research in alternative vehicle-energy systems for the longer term, we should not neglect research in fundamentals. Technological opportunities to create, for the nearer term, IC engine cars with high fuel economy, with hybrid propulsion, and with much lower lifetime emissions are good. Support for this work could be generated through

- stronger fuel-economy standards,
- stronger emissions standards based on reasonable predictions of lifetime emissions, and
- incentives geared to shift the new car market toward these social goals.[76]

Engelhard Corporation's PREMAIR Catalyst System

A different approach to tropospheric ozone reduction was taken by engineers at Engelhard Corporation of Iselin, New Jersey. Engelhard Corporation is the inventor and a major manufacturer of the three-way catalytic converter used to reduce automotive exhaust emissions. The three-way catalyst used on cars today reduces exhaust VOCs by 97 percent, NO_x by 90 percent, and CO by 96 percent (when the catalyst is hot and the air–fuel ratio is stoichiometric).[77]

Engelhard Corporation's new system, called PREMAIR, is another catalyst system. Specifically, it is a platinum-based coating that is to be applied to the radiator and air conditioner condenser of a car. (The PREMAIR catalyst will use as much, or more, platinum as is used in automotive three-way catalytic converters, because of the superior catalytic activity of this metal.) The PREMAIR catalyst is designed to reduce, not ozone precursor emissions—VOCs and NO_x—but ambient ozone itself. As the radiator fan draws in ambient air, PREMAIR will convert ozone to oxygen, and carbon monoxide into carbon dioxide:

$$O_3 \rightarrow O_2 + O$$

$$CO + O \rightarrow CO_2.$$

According to laboratory and on-road testing in Los Angeles, the PREMAIR catalyst system converts about 90 percent of the O_3 and CO passing through it.[78]

Some benefits of the PREMAIR catalyst system are

- its relatively low cost (Engelhard estimates that the PREMAIR system may cost less than $1000 per car); and
- the fact that it would not require major vehicle redesign (radiator fans would be modified).[79]

Engelhard Corporation retained Systems Applications International (SAI), an independent company specializing in atmospheric modeling, to assess the

76. Ibid.
77. Engelhard (1995).
78. Ibid.
79. Ibid.

potential benefits to air quality that might be achieved using PREMAIR. If the nine million vehicles traveling 266 million miles per day in Los Angeles were equipped with PREMAIR catalyst systems, there would be, according to SAI, a potential to reduce peak levels of ozone concentration in Los Angeles by 0.0045 ppm.[80] (Recall from the "National Ambient Air Quality Standards" section that high ozone concentrations are in the range of 0.20–0.40 ppm). An advantage to the Engelhard PREMAIR catalyst system is that it could be used to supplement, but not replace, other ozone reduction strategies.

The Fuel Cell Electric Vehicle of the Future

Going beyond the "age of fire," in which the internal combustion engine has been employed, the next generation of personal transportation vehicles is likely to introduce the "age of electrochemistry," in which electric vehicles (EVs) will be used extensively. By removing combustion, with its necessary moving parts, in energy conversion, the EV could virtually eliminate automotive air pollution, and simultaneously provide a leap in energy efficiency of 250 to 300 percent relative to a gasoline-fueled IC engine vehicle. At present, the fuel cell EV (FCEV), powered by hydrogen or methanol, promises cleaner transportation than a battery-powered EV (BPEV), supported by a mix of coal and nuclear power plants for electricity generation.[81]

The fuel cell was developed initially for the space program. It provided electrical power on board the Gemini space mission in the 1960s. It was used also in the Apollo orbiters, space probes, and the space shuttle.[82]

Fuel cell operation involves a *process* rather than a *cycle*, so it is not limited by the Carnot cycle efficiency (see footnote 70). Consequently, fuel cells can operate at relatively low temperatures, with energy efficiencies of 45 percent and higher. The main products of the process are water, electricity, and heat. In actual practice, fuel cell vehicles are not *ZEVs (zero-emission vehicles)*, but rather *ULEVs (ultra-low emission vehicles)*. They produce nearly zero nitrogen oxides (NO_x), and about half the carbon dioxide (CO_2) of gasoline engines.[83]

Of the four kinds of fuel cells that are candidates for motor vehicles (namely, phosphoric-acid, alkaline, solid-oxide, and proton-exchange membrane) many researchers believe that the proton-exchange membrane (PEM) fuel cell is best suited for light-duty motor vehicles in the short term. The process of energy conversion in a PEM fuel cell involves the ionization of hydrogen and oxygen, and the formation of water, in the following way:

- Hydrogen, stored either as a compressed gas or liquid on board the vehicle, or produced by reforming methanol into hydrogen and carbon

80. Ibid.
81. Williams (1994).
82. Lee et al. (1993).
83. Ibid.; Communications Division, North American Operations Research and Development Center, General Motors Corporation, Warren, Michigan (1998).

monoxide, is delivered to an electrode (the anode) where, with the help of a platinum catalyst, it is ionized. Molecular hydrogen separates into hydrogen ions and electrons: $H_2 \rightarrow 2H^+ + 2e^-$.

- The voltage drives the electrons released through an external load, where they perform the useful work of turning an electric motor, to the opposite electrode (the cathode).
- The hydrogen ions, which are protons, are transported by an ion-conducting membrane (i.e., the proton-exchange membrane) to the cathode. Here the protons combine with oxygen from the atmosphere, as well as the electrons returning from the motor, to form water: $2H^+ + (1/2)O_2 + 2e^- \rightarrow H_2O$. Last, the water is removed from the fuel cell (see Fig. CP2.7).[84]

Not only are FCEVs nearly emission-free and two and a half to three times more energy efficient than comparable gasoline-fueled cars (depending on whether the FCEV operates on methanol or directly on hydrogen), but the fuel (hydrogen) has virtually unlimited supply compared to petroleum-based fuels and natural gas. Hydrogen can be derived from natural gas, urban wood wastes and municipal solid wastes, biomass (grains, grasses, wood stock), coal, or by the splitting of water electrolytically (potentially using electricity generated by wind, photovoltaic sources, or some other renewable supply).[85]

Fuel cell technology for transportation is advancing. Among a number of projects being undertaken worldwide are the following:

- Ballard Power Systems of Vancouver, Canada, has built and field-tested a 30-foot bus, powered by hydrogen and PEM fuel cells.
- Ballard also has entered a joint venture with the German automaker Daimler-Benz to develop compact PEM fuel cells for automobiles.
- General Motors, with Opel, has established a Global Alternative Propulsion Center (GAPC) in Russelsheim, Germany, to research and develop automotive fuel cell technology.

General Motors' first electric car, EV1, powered by 26 nickel metal hydride (NiMH) battery modules, was put into production in 1996. Now the EV1 has been elongated to accommodate a methanol fuel cell. EV1 carries tanks of methanol and water and fuel cell stacks. A multistage fuel processor converts the methanol into a hydrogen-rich mixture for the PEM fuel cell. Having built and demonstrated concept cars, GM is working on reducing the size of the fuel cell elements, and expects to have fuel cell vehicles in production by 2004.[86]

84. International Energy Agency (1993); Williams (1994).

85. Ibid.

86. Ibid.; Communications Division, North American Operations Research and Development Center, General Motors Corporation, Warren, Michigan (1998).

How a Fuel Cell Works

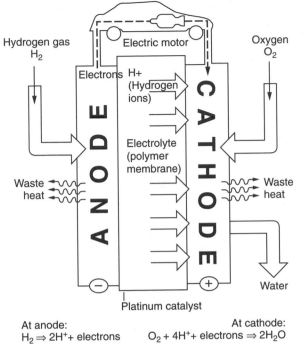

Hydrogen gas H₂

Electric motor

Oxygen O₂

Electrons

H+ (Hydrogen ions)

Electrolyte (polymer membrane)

ANODE

CATHODE

Waste heat

Waste heat

Water

Platinum catalyst

At anode:
$H_2 \Rightarrow 2H^+ + electrons$

At cathode:
$O_2 + 4H^+ + electrons \Rightarrow 2H_2O$

FIGURE CP2.7 In a proton-exchange membrane (PEM) fuel cell, hydrogen fuel is ionized at the anode. The electrons are collected into an external circuit to turn an electric motor and then released to the cathode. Hydrogen ions are transported by the PEM to the cathode, where they combine with oxygen and the electrons to form water. *Source:* Williams, R.H., "Fuel-Cell Vehicles: The Clean Machine," *Technology Review*, vol. 97, no. 3, pp. 20–30, April 1994. Designed by Perugi.com. Reprinted with permission.

In the closing years of the twentieth century, there is a growing consensus in the technical community that it is possible, practical, and urgent that the United States administration and the nation's automobile producers remove the automobile from the list of the nation's environmental and energy-security problems.

▨ Design Assignment

■ When asked what single event was most helpful in developing the theory of relativity, Albert Einstein is reported to have answered, "Figuring out how to think about the problem."[87] Make a list of the different ideas concerning tropospheric ozone implicit in the different approaches used to resolve the problem, as presented in this case problem.

■ Trevor Kletz once asked, "Why are we so much better at answering questions than at answering the right questions? Is it because we are trained at school and university to answer questions that others have asked? If so, should we be trained to ask questions?"[88] (That is, should

87. Sylvan (1994).
88. Kletz (1994, p. 195).

we be trained to ask the right questions?) In view of Trevor Kletz's questions, formulate a question that would help to clarify the issue of reducing tropospheric ozone. Prepare a formal design proposal in which you recommend and defend a particular approach (i.e., methodology) for answering this question.

3   Addressing the Global Landmine Crisis

■ The Changing Aspect of Warfare

It is estimated that more than one million people have suffered the terrible pain of a landmine explosion since 1975, according to the United Nations Childrens Fund [UNICEF].[1] According to Handicap International, the French humanitarian organization, one person is killed or injured by landmines every 10 minutes in daylight hours, every day of the year. That is, at least 2,160 people killed or injured every month, or more than 26,000 people every year.[2] The vast majority of these victims, approximately 80 percent, are civilians or noncombatants.

Moreover, at least 30 times more landmines are being laid everyday than are being cleared, according to the U.N. demining expert General Patrick Blagden. Newly laid landmines are added daily to the 110 million unexploded landmines already strewn in more than 60 countries. Meanwhile, another 100 million landmines are believed to lie in stockpiles ready to be deployed, while 5 to 10 million more mines are being produced each year.[3]

The overwhelming majority of antipersonnel landmines are being used as offensive weapons against civilian populations to empty territory, create refugees, block food supplies, and generally spread terror. The whole concept of maneuver warfare has changed. Antitank mines, which were developed during World War I (1914–1918), and antipersonnel mines, which were heavily used during World War II (1939–1945), at one time were deployed to halt or detour an advancing army or to protect important installations. Today, as internal armed conflict builds, the theater of operations often expands to encompass an entire national territory. Once a "defensive tactical battlefield weapon," the role of landmines has shifted to a "theaterwide weapon of mass civilian destruction," according to former U.N. Secretary General Boutros Boutros-Ghali.[4]

---

1. *The Global Landmine Crisis,* Hearing before a Subcommitte of the Committee on Appropriations, United States Senate, One Hundred Third Congress, Second Session, Superintendent of Documents #Y4.AP 6/2: S.HRG. 103-666, May 13, 1994.

2. Handicap International, *Landmines Special Report,* the Internet, January 16, 1996.

3. *The Global Landmine Crisis* (1994).

4. Ibid, p. 97.

Landmines are used by both regular military forces and by guerilla forces. The word *guerilla,* derived from the Spanish word *guerra*, means *little war.* This term dates from the period of the French occupation of Spain (1807–1813). The Spanish word for a fighter in such warfare is *guerrillero.* A guerilla is commonly understood to be a member of an irregular, usually politically motivated, armed force that combats a stronger regular force, such as the army or police. Some guerilla movements have been devoted to expelling an invader, such as the overthrow of the German occupation forces in Yugoslavia by Tito's partisan movement, or the attempts of the Polisario to expel the Moroccans from the Western Sahara. Other guerilla groups have been formed in countries where basic human rights are routinely violated, or major social injustices, such as the lack of fair ownership of land, are tolerated.[5]

Guerillas rely heavily on mobility. Their security depends on their ability to disappear among the local population. Their weapons, therefore, must be portable, and generally handheld, so conventional artillery as a rule is excluded. Rural guerilla fighters today use primarily firearms and landmines. Landmines are weapons that are cheap, easy to manufacture and use, small, lighweight, and easy to conceal. Both regular armies and guerilla forces are able to purchase landmines, which are mass-produced; the smaller mines, approximately the size of a can of shoe polish, cost as little as $3 per unit. On the other hand, landmines are difficult to detect and are expensive and dangerous to remove.[6]

Nations Most Plagued by Landmines

Africa is the most mined region in the world, with approximately 18 to 30 million uncleared mines in 18 countries. Angola may be suffering the greatest hardship, with hospitals reporting up to 24 landmine casualties per day in 1994. According to 1993 statistics of the International Committee of the Red Cross (ICRC), one Angolan in 470 is a landmine amputee.

Between 17 and 24 million mines litter eight countries in the Middle East. In East Asia, eight countries collectively contain between 15 and 23 million uncleared mines; between 13 and 25 million mines have been sown in South Asia. In Cambodia there may be 4 to 5 million unexploded mines dispersed among a total population of eight million people. Approximately one out of every 236 people is a landmine amputee, which makes Cambodians the most disabled population on a percentage basis. Cambodians, who have suffered the full horror of landmines, call these weapons "eternal sentinels," for they continue to kill and maim long after hostilities have ceased.[7]

5. Gander (1989).
6. Ibid.; *The Global Landmine Crisis* (1994).
7. *The Global Landmine Crisis* (1994).

Domestic activities such as collecting firewood, drawing water, and grazing cattle, which are primarily carried out by women and children, have become hazardous, if not deadly, in many lands because of the sowing of landmines in areas of regular civilian use. A 1993 study of the ICRC found that child casualties from landmines in Kabul, Afghanistan, were particularly high because the fields around the city where children grazed the flocks were heavily mined.[8]

It is estimated that the number of those who die of landmine injuries equals the number of those who survive. In the case of a fatality, the victim either is killed immediately or die in agony before medical treatment can be given. (In some areas, Afghanistan for example, the ICRC reports that the majority of mine victims dies before reaching a hospital.) Children, in particular, suffer catastrophic consequences from a landmine explosion, since their small bodies are closer to the center of blast and cannot withstand either the terrible injuries the mines inflict or the great loss of blood.

For landmine victims who do survive, only a small percentage receive adequate rehabilitation. The victims frequently live in regions where there is very little rehabilitative medical care or prostheses and where there are few jobs for the disabled. One estimate is that only 20 percent of mine-injured amputees in Cambodia, for instance, will find employment. In many societies, young amputees, especially young women, are not regarded as marriageable. Disabled people are often treated as outcasts and reduced to begging or stealing in order to survive.

Farmers and herders in agrarian and pastoral economies, often depending on the bicycle as the primary mode of transportation, rely heavily on the use of eyes, hands, legs, and feet for their livelihoods and the support of their families. "For a peasant," says Boutros Boutros-Ghali, "amputation or blindness is often the end of working life."[9]

International Impact of Landmines

Of international concern, mined borders block the repatriation of refugees and so worsen the world refugee problem. Most of the nearly 20 million refugees in the world want to return home, but repatriation is severely hampered because of uncultivated landmines, according to the United Nations Secretary General. On the Zambia–Zimbabwe border, one million acres of heavily mined land is almost deserted.

The widespread practice of mining fertile agricultural land has turned much scarce arable land into unsown minefield wastelands. The Cambodian government's demining agency, CMAC, estimates that landmines place 20 percent of agricultural land out of production. In Angola starvation and famine have resulted from the mining of agricultural land. In Afghanistan,

8. Ibid.
9. Ibid, p. 94.

during the war with the former Soviet Union, livestock owners shot and killed entire herds because they could not graze them safely. As much as 35 percent of the land in Afghanistan is now unusable due to mines.[10] Moreover, when an agricultural and pastoral population cannot work its land because of landmines, the people are compelled to gather in cities where work opportunities are scarce and housing conditions are poor. Thus able people, prevented from cultivating their farms, take up lives of poverty in the city.

Mined roads and bridges, water plants, power lines, railroad networks, and waterways impede the circulation of goods and labor, the flow of commerce, and the delivery of aid. The sowing of landmines has led to increased disease, housing and sanitation problems, malnutrition, and famine. They have slowed down national recovery, intensified economic crises, and threatened political stability. In addition, the current proliferation of landmines in the environment is destroying vegetation, killing animal life, introducing poisonous substances into the soil, and increasing the vulnerability of soil to water and wind erosion by destroying soil structure. The central crisis, however, occurs in the lives of men, women, and children who suffer the most extreme injuries from these devices, and the plight of families, locales, and communities which are compelled to bear an increasingly heavy medical and social burden.

In a joint statement before the U.S. Senate, former Secretary of State Cyrus Vance and former U.S. Ambassador to the United Nations Herbert Okun observed: "Landmines have and continue to cause a human and economic catastrophe of monumental proportions."[11] Robert Muller, Executive Director of the Vietnam Veterans of America Foundation, observed: "Sown in such vast numbers, the lowly landmine has become a weapon of mass destruction in slow motion."[12] And a news release (December 21, 1993) from the Office of the Assistant Secretary of Defense announced, "Demining programs have become an important mission of the DOD in promoting stability and in preventing the recurrence of conflict in nations just recovering from years of war."[13]

The Long Lifetime of Landmines

Once laid, antipersonnel landmines can remain active for as long as 50 years. In 1984 the Associated Press reported that in the Netherlands a team of 90 demolition experts continued to handle about 2000 World War II ordnance disposal cases annually, and the team could not keep up with the demand for its services. Still, an average of 12 people per year in the Netherlands are victims of World War II landmines. Poland especially is still

10. Ibid.
11. Ibid, p. 9.
12. Ibid, p. 73.
13. Ibid, p. 112.

burdened with a vast number of explosive remnants from World War II. Since 1945 Poland has disposed of more than 88 million items of explosive ordnance, of which about 15 million have been mines. The task continues with more than 200,000 ordnance items being located and destroyed each year. Victims of World War II explosive ordnance included 13,000 Polish civilians between 1945 and 1977.[14]

In the various North African campaigns during World War II, Germany, Italy, the United Kingdom, and France laid between 5 and 19 million landmines, mostly antitank mines. In Libya, for example, during the decades since World War II, it continues to be a fairly common occurrence for shepherds to step on old buried landmines and be killed in the explosion that follows.[15] The ICRC describes landmines as "fighters that never miss, strike blindly, do not carry weapons openly, and go on killing long after hostilities are ended."[16]

Landmines and Modern Munitions Systems

Landmines and other unexploded ordnance are an ever greater problem because of the increased delivery capability of modern conventional munitions systems. Scatterable mines are dispersed from ground vehicles, rockets, artillery shells, helicopters, and planes by the thousands, or even tens of thousands, in a matter of minutes. Landmines may be made of plastic, so that they are undetectable by metal detectors. They blend in with the soil or ground cover and are quickly obscured by a layer of dust or vegetation. Mines have become sophisticated weapons systems with electronic fuses and sensor systems that make them even more deadly. They can now detect the footstep patterns, body heat, sound, and signal of deminers—any of which can make them explode. Still, the inexpensive, conventional, hand-deployed mine, like the Soviet PMN or the Chinese Type 72, is most commonly used and will continue to pose the greatest threat for the foreseeable future. Technical advances in the manufacture of landmines and the inability of states to control their use has moved the issue beyond military considerations and has made it a "threat to humanity," similar to those of bacteriological and chemical weapons, warns John Lloyd, Refugee Affairs Consultant to the International Federation of Red Cross and Red Crescent Societies (IFRC).[17]

An effort was made to restrict the use of landmines to wartime by designing and manufacturing self-neutralizing mines. Self-neutralizing mines become inert when their batteries expire or through some other mechanical, chemical, or electronic means. However, it is usually not known whether such mines, in fact, have become neutralized or which mines in a minefield are self-neutralizing. Therefore, both conventional and self-neutralizing mines are assumed to be active by deminers and are given

---

14. Westing (1985); *The Global Landmine Crisis* (1994).
15. Westing (1985).
16. *The Global Landmine Crisis* (1994, p. 13).
17. Ibid., p. 130.

the same painstaking care in removal before an area is declared demined. Furthermore, the explosive charge in self-neutralizing mines remains in the ground and can become dangerous over time as it degrades chemically. Self-neutralizing mines also can be dug up by combatants, refitted, and resold. According to a May 1994 UNICEF report, civil and military personnel indicate that the failure rate of the self-neutralizing mechanism of hand-laid landmines is 25 to 50 percent and even higher for scatterable mines. So the threat they pose is not eliminated.

Another attempt at limiting the long-term destructive effect of landmines is fitting landmines with self-destruct mechanisms. During their active lifetimes, however, they, like conventional mines, cannot distinguish between a civilian and a soldier. They also are much more expensive than conventional mines, so that they are not used by combatants in civil wars in poorer nations. In fact, they have the contrary effect of worsening the global landmine crisis. While self-destructing and self-neutralizing mines have allowed wealthier nations to argue for the proliferation and use of "safe" mines, by equity, poorer nations have tenaciously supported the distribution and use of conventional mines. Furthermore, UNICEF reports that the self-destruct mechanisms are believed to have a 10 to 15 percent failure rate. When scattered by the thousands, the many defective mines, acting like conventional mines, pose a danger to civilian populations for more than a generation.[18]

In addition, the United States has advocated that the Landmines Protocol of the 1980 United Nations' Convention on Conventional Weapons require all mines to have substantial metallic content, so that they will be detectable by metal detectors. However, such a strengthened U.N. Landmines Protocol, even if internationally adopted, would not be enforceable. Adequate inspection of landmine stockpiles to verify the absence of minimal metal or all-plastic mines would be impossible. "The Landmines Protocol is a complete failure and has been routinely ignored by nearly all mine users. It will not be possible to strengthen the Protocol in any significant way. It must be replaced with a ban...," advised Rae McGrath.[19] A veteran in demining, McGrath is the director of Mines Advisory Group (MAG), the British nongovernmental organization that draws on the expertise of British military veterans to provide training in ordnance disposal in Afghanistan, Kurdistan, Mozambique, Angola, Cambodia, and Laos.

The crucial question, it is said, is not whether landmines are a useful military weapon, but whether they are an *essential* military weapon. To date, the U.S. Department of Defense has not undertaken a serious, thorough cost–benefit analysis of antipersonnel landmines, according to the Arms Project of Human Rights Watch.[20]

Landmines are said to violate two generally accepted rules of armed conflict, namely, discrimination and proportion. Most nations have agreed that weapons of war should discriminate between military objectives and

18. Ibid.
19. Ibid, p. 136.
20. Ibid.

locations of purely civilian and commercial activity. Landmines do not distinguish between soldiers and civilians, nor between friendly and enemy forces. General Gray, former Commandant of the U.S. Marine Corps, has said that American landmines have killed more Americans than anyone else.[21] Most nations have also agreed that there should be a certain proportion between the expected military utility of a weapon and its anticipated human toll. Therefore, according to international agreement, weapons that are indiscriminate or exact a disproportionately high human toll for their nominal military utility are called inhumane. By definition, inhumane weapons of war should be banned from military use.

Designating antipersonnel mines as inhumane weapons of war, 20 countries are now on record calling for an immediate ban on these mines.[22] They have announced a commitment to stop production and export of all antipersonnel mines, renounce the use of all antipersonnel mines, and begin the destruction of stockpiles. As of 1996, 43 nations had placed a ban, moratorium, or other restriction on the export of antipersonnel mines.[23]

◼ Types of Landmines

Pressure-activated blast mines are the most common form of landmine. Such mines are normally detonated by the pressure of a foot, and the resulting explosion will kill or severely injure their victims. The PMN blast mine, manufactured by the former Soviet Union and Iraq, is considered the most extensively used antipersonnel land mine in the world. It has a cylindrical, bakelight body with rubber cover, with a height of 56 mm, diameter of 112 mm, and weight of 550 g. It is actuated by a pressure force of 5 to 8 kg. Its large explosive content of 200 trotyl is very often fatal.[24]

The infamous small green "butterfly" mines, designed with wings to allow them to float to the ground without exploding, were scattered from helicopters or launched from artillery throughout the duration of the war in Afghanistan. Hundreds of thousands, perhaps millions, of these mines were strewn, often claiming children as victims. One of the "wings" of the mine contains liquid explosive. When pressure is applied, the explosive is forced into contact with the fuse and the mine detonates.[25]

21. During the Vietnam War, many U.S. Marine Corps casualties were caused by landmines or booby traps (60–70 percent during 1965), a large number of which resulted when U.S. troops were forced to advance or retreat through areas mined by their own side. (A booby trap is a device fitted to a mine that will set it off when it is tampered with. A booby trap may be an apparently harmless object, which is not hidden, and is fitted with an explosive device.) (*The Global Landmine Crisis,* 1994; Anderberg, 1985).

22. They are Austria, Belgium, Cambodia, Colombia, Denmark, Estonia, Iceland, Ireland, Laos, Malaysia, Mexico, Mozambique, New Zealand, Nicaragua, Norway, Peru, Philippines, Slovenia, Sweden, and Uruguay. In addition, France has banned production and trade of antipersonnel mines.

23. *Handicap International* (1996).

24. U.S. Department of State (1994).

25. *The Global Landmine Crisis,* (1994).

Directed fragmentation mines, often known as "Claymore mines," were introduced by the United States in 1960, and have been widely copied. The M18A1, manufactured by the United States, Chile, and South Korea, consists of a rectangular body of fiberglass-reinforced plastic, having length of 216 mm, width of 35 mm, height of 83 mm, and weight of 1.58 kg. The mine is supported by two pairs of folding scissorslike legs, and can be fired by a tripwire or command detonation. Its explosive content of 682 g C-4 can propel 700 steel ball bearings forward in a 60 degree arc. The impact can kill at up to 50 meters, and maim at up to 100 meters. About 70 percent of these mines remain effective for over 20 years.[26]

Bounding mines are another type of fragmentation mine. An example is the Italian Valmara 69, widely deployed in northern Iraq. When triggered, it leaps 45 centimeters into the air before shattering into more than 1000 metal splinters. The lethal casualty radius is at least 25 meters.

Minimum metal landmines, developed in the 1970s, contain as little as 1/30 ounce of metal, often only a metal firing pin, spring, or seal. This makes them almost impossible to detect with a metal detector. In addition, highly sensitive metal detectors will give positive readings for every buried bottlecap, discarded food can, or scrap of ordnance in a former battlefield, thus burdening deminers with an overwhelming number of "false positives" or type II errors. The Falkland Islands are proving exceedingly difficult to demine since the 1982 conflict, because of a large number of minimal metal landmines.

The field of statistics gives a method for testing the validity of one hypothesis (or claim) against another hypothesis (usually an opposing claim). The two hypotheses are called the null hypothesis (H_0) and the alternative hypothesis (H_1), respectively. A type I error is committed if, by using the method, one rejects H_0 as false when in fact it is true. Conversely, a type II error is committed if one accepts H_0 as true when in fact it is false.[27] For mine detection, we may take the null hypothesis H_0 to be the presence of a landmine. We commit a type I error (false negative) if we declare a region to be clear when it contains, for example, one or more plastic mines. A type II error (false positive) is committed when we mark an area as the location of a landmine when it is, for instance, the location of a buried piece of shrapnel.

Even though type II errors do not explode, since what is unearthed is a nonexplosive scrap of metal, still the danger and cost of type II errors should not be underestimated. Colonel G. C. McDowall reports on his demining experience in Afghanistan that an average of 1000 metal fragments are extracted by deminers for every mine removed from a battlefield. Every metal fragment is laboriously located and identified as though it were a mine, wasting time and manpower. The large number of type II errors also is responsible for laxity in applying safety precautions among deminers. McDowall recounts that one deminer located metal fragment after metal fragment until he became exhausted, and developed a headache wearing his helmet and

---

26. U.S. Department of State (1994).
27. Walpole and Myers (1978).

visor. He removed his helmet, but the next metal object he encountered was a PMN mine, whose explosion left him blind. McDowall estimates that reduction of type II errors, that is, discrimination between mines and metal fragments, would eliminate about two-thirds of deminer casualties and reduce the severity of injury of many of the rest.[28]

Truly *plastic mines,* with chemical fuses, have been developed more recently by some 20 countries. All-plastic mines, manufactured in former Yugoslavia, posed the greatest challenge to demining efforts in Kuwait in 1993 after the Gulf War.[29]

▓ Demining Efforts

Artillery mine-scattering systems can deploy more than 1700 landmines per minute. Remote delivery systems (including aircraft, helicopter, artillery, rocket, and mortar) have expanded the role of antipersonnel landmines so that mapping is no longer reliable and, indeed, is sometimes not even attempted. At the same time research and development continue to be directed toward ever more efficient methods of dispersal and ever more sophisticated landmines. In contrast, landmine removal is a slow, laborious, and dangerous task.

In 1994 Thomas McNamara, Deputy Assistant Secretary of State for Political and Military Affairs, reported that the U.S. Department of State and the Department of Defense were making plans to develop new technologies to assist affected nations in humanitarian detection and clearance of land-mines.[30] Paralleling military technology, Colonel Richard Johnson, U.S. Army, Retired, suggested to a subcommittee of the U.S. Senate Committee on Appropriations, that a less rugged version of the Airborne Stand-off Mine-field Detection Sytem (ASTAMIDS) would be appropriate for humanitarian demining.[31] ASTAMIDS consists of an imaging sensor (based on laser diode technology) mounted on an unmanned aerial vehicle (UAV), with a data link to a ground control station (GCS). Minefield imagery transmitted to the GCS will be processed in near real time by a very high-speed parallel processor. ASTAMIDS will be a highly reliable, but high-tech and expensive, mine detection system.[32]

The appropriations figure for humanitarian mine-clearing programs in the U.S. Defense Department was $25 million for fiscal year 1994–1995. U.S. personnel are not expected to lift or clear the mines themselves, but provide technical assistance and support for landmine clearing. If appropriate technology is not developed and deployed in this area, mineclearing may continue claiming arms, legs, and lives.[33]

28. U.S. Department of Commerce (1994).

29. *The Global Landmine Crisis* (1994).

30. Ibid.

31. Ibid.

32. Foss and Gander (1995).

33. *The Global Landmine Crisis* (1994).

Initially in mine clearance operations, civilian contractors with extensive experience and specialized equipment at their disposal may have to be employed to clear the most urgent routes. Although a high clearance rate, subject to quality control, can be achieved relatively quickly by this means, commercial demining is very expensive. For example, Kuwait was heavily sown with landmines. However, this country had the financial resources to respond to the problem. Within months after the end of the Gulf War, the Kuwaiti government had awarded contracts for mine clearing that totaled $700 million. In one year, that country spent more on mine clearance than the rest of the world combined during the same period. Experience in Afghanistan and Cambodia, according to the U.N. Development Program, has shown that responsibility for long-term mine clearance is transferred eventually to the country concerned. Mine clearance is a very long-term process. Mine-clearing technology should facilitate a national capacity to manage and maintain the mine-clearance process over the long term.

A clear distinction must be made between military minefield breaching and humanitarian area mine clearance. The military is interested primarily in clearing a path through a minefield. Since speed is often critical for military operations, their methods do not ensure total clearance and casualties are expected. Military explosive hoses, flails, and tank ploughs, commonly used for clearing mines, do not achieve adequate humanitarian clearance. Humanitarian demining requires that at least 99 percent, and preferably 99.9 percent, of the mines be cleared. However, technology for minefield breaching, with quality control to assure a high clearance rate, may be an important part of the work of humanitarian demining. A typical example is a road punched through a minefield in Cambodia by military minefield-breaching techniques, which allowed nearby villagers to haul water from the local river for the first time in over 20 years.[34]

Local, nonprofessional demining efforts have been undertaken by necessity. In northern Iraq, for example, young children who are shepherds burn parts of minefields in order to clear the land for grazing. This method of demining is common in the region, although it is not foolproof. Driving livestock over a minefield is another local method, although the loss of livestock may be a major sacrifice. Like burning, this is not reliable.

Prodding is the most widely used method of locating landmines, and it is employed by trained humanitarian deminers. The deminer is required to lie on his or her stomach and gently but firmly prod the ground in front with a metal rod or a bayonet inserted at a 30-degree angle. The ground is prodded every 2 to 5 centimeters, or at least 400 times per square meter. Metal detectors are often used prior to prodding, but they get readings on old tin cans and spent shells and fail to detect the latest plastic mines. Moreover, due to limitations in funding and personnel, metal detectors may not be practical for demining vast stretches of land.

34. Ibid.

Clearing unexpected ordnance (UXO) has many features in common with clearing landmines. During the 14 years from 1961 to 1975, the United States dropped on countries of Indochina approximately 20 million bombs of various sizes and fired some 230 million artillery (including naval) shells. In addition, countless millions of rockets, mortar shells, and grenades were expended. U.S. artillery shells equipped with the standard point-detonating fuse failed to explode 2.5 percent of the time when set in the super-quick mode and from 5 to 50 percent of the time when set in the delay mode. U.S. mortar shells did not detonate 10 to 20 percent of the time during the dry season and 30 percent or more of the time during the wet season. U.S. hand grenades were duds 15 to 25 percent of the time during the dry season, and 40 to 50 percent of the time during the wet season. These and similar statistics for bombs and other munitions indicate that millions of unexploded ordnance, the explosive remnants of war, litter the landscape or lie buried in the soil within reach of the blade of a plow or a shovel in southeast Asia.[35]

For example, for over 20 years, cluster bombs have been exploding upon impact as farmers plow the land, or villagers plant gardens, or even with the movement of being picked up by a child. Even lighting a fire to cook a meal has proved hazardous because the heat could cause buried bombs to explode. Cluster bombs belong to a category of weapons known as submunitions. Technically, a *submunition* is any piece of ordnance carried to its target or point of dissemination within a larger munition—usually a rocket, bomb, or shell. In the case of Laos, the cluster bombs were generally dropped from airplanes in large bomb containers. The containers opened in mid-air, and the small bomblets spread out during their descent, covering a wide area. The bomblets are about the size of a tennis ball and were designed to detonate upon impact or shortly thereafter via internal random time-delay fuses. Many of the bomblets failed to explode as designed, due to either production defects or improper deployment.[36] "Indeed, there are large portions of Indochina where there seems to be no peasant family that cannot recount a personal tragedy—whether of death or maiming—caused by previously unexploded munitions," reported Arthur Westing of the Stockholm International Peace Research Institute (SIPRI).[37]

The Xieng Khouang Unexploded Ordnance Project in Laos (a project of the Mennonite Central Committee U.S.) reports that its clearance with metal detectors is limited to specific areas where building projects are taking place, or in schoolyards, garden plots, and other areas that warrant intensive searches.[38]

In this kind of high-reliability clearance for UXO, a plot of land is divided into lanes 1 meter wide with brightly colored marking strips. Two bomb technicians are assigned to each lane. The technician with the metal detector enters the lane and searches until the detector indicates metal. He

35. Westing (1985).
36. *The Global Landmine Crisis* (1994).
37. Westing (1985, p. 7).
38. *Mennonite Central Committee* (1995).

marks the spot and retreats from the lane. The second technician then enters the lane, lies on the ground, and gently probes the ground with a long metal rod, searching for the metal. When the metal is found, it is exposed. If it is a cluster bomb, it is clearly marked. The technician with the detector then enters the lane again to continue the search. This continues back and forth in this manner until all the lanes in a plot of ground are searched.

Once all the explosive ordnance is found and marked, countercharges are placed beside each piece of ordnance, blasting caps and detonating cord are set in place, and the entire lot is blown up at once.[39]

Even using much care, prodding is very dangerous for deminers. In Afghanistan, where the soil is difficult, U.N. deminers suffered one accident per 648 mines recovered. Where the soil is easier, as in Cambodia, the ratio drops to one accident per 2300 mines recovered. In an operation in the 1980s, highly trained munitions disposal units in Egypt experienced one fatality for every 7000 landmines cleared. It has been suggested that in general one deminer will be killed and two injured for each 5000 landmines removed.[40] Deminers are paying a great human toll for their work. This will not change until technology is used to clear landmines remotely, with the deminer's hands and feet well away from the mines.

■ Devices for Humanitarian Demining

At a 1994 international workshop in Sweden, sixty-eight demining experts from the United Nations, the European Union, and ten nations[41] arrived at certain conclusions and offered recommendations for humanitarian demining. Demining experts are aware that global mine-clearing capability must increase dramatically, perhaps by as much as a factor of 50, if we are to have more landmines being removed in a year than are being emplaced. However, it was estimated that funding for humanitarian demining would allow only a five-fold increase in cost. This means that cost-effectiveness would have to increase by a factor of 10. To achieve this, there would have to be a greatly increased use of technology in place of manual demining. (Although it is likely that there will always be areas in which manual demining techniques will have to be used.) Furthermore, the technology would have to be inexpensive, yet effective. The cost criterion was calculated by considering that a team of deminers can manually clear about 2300 square meters of land per day at a cost of about 500 US dollars. This averages to a cost of about 22 US cents per square meter per day for demining. A 10-fold increase in cost-effectiveness would require a technology that operates at a cost of 2 to 3 US cents per square meter.[42]

---

39. Peachey (1996).

40. *The Global Landmine Crisis* (1994); Westing (1985).

41. Namely, Afghanistan, Australia, Austria, France, Germany, Japan, Norway, Sweden, the United Kingdom, and the United States.

42. U.S. Department of Commerce (1994).

The following design criteria for demining technology were offered at the workshop:

- Operating cost for the device should be 2 to 3 US cents per square meter of land cleared.
- Local labor should be able to both operate and maintain the device.
- The device should clear mines to a depth of 30 cm.
- The device should provide crew protection against all explosive charges up to a 12-kg explosive force antitank mine.
- The device should continue to operate after the detonation of any antipersonnel landmine.
- The device should clear all antipersonnel landmines (including minimum metal and plastic mines) to at least 95 percent probability. (The remaining 5 percent are usually deeply buried or tree mounted, and will be handled by more specialized or sophisticated technology.)
- The device should be operable in 90 percent of soil conditions.
- The device should be operable in undergrowth of up to 50 mm in diameter.
- The device should be operable in temperature ranges from 5° C to 45° C.
- The device should be easily transportable (moving at a rate of at least 20 km per hour).
- The device should be able to operate 8 hours per day, 6 days per week.
- The mean time between failures of the device should be no less than 300 hours.
- The design and construction of the device should allow field servicing.
- The device should be operable by a maximum crew of one person.[43]

It is true that mechanical systems for mine clearance would not result in a 99.9 percent probability of clearance for landmines. This is their main drawback. However, 95 percent probability of clearance is a tolerable first pass as stated above, until detection systems could be brought to a region. Also, the environmental damage caused by flails, but especially by ploughs, is a concern. One of the deminer task groups thought a roller, or perhaps a vibrating roller, to be the best design for a mechanical mine-clearing device.[44]

In 1994 humanitarian deminer Daniel Wolf of San Diego, California, patented a design for an electromechanical remotely operated mine-exploding device (called Transport Platform and Mine Exploder), which is intended to satisfy the needs of humanitarian demining. Figure CP3.1 is a top view of the device (patent #5,313,868). A remotely controlled tractor (20) pulls behind it a disk system (21) which acts as a mine exploder. The tractor is equipped with ground support mass (15) designed to traverse varied terrain (hard surface, vegetation overgrowth, sand, marsh, etc.). The "footprint" of ground support (15) is broad enough so that it does not provide enough pressure to trigger antipersonnel landmines when passing over them. Remote control receiver

43. Ibid.
44. Ibid.

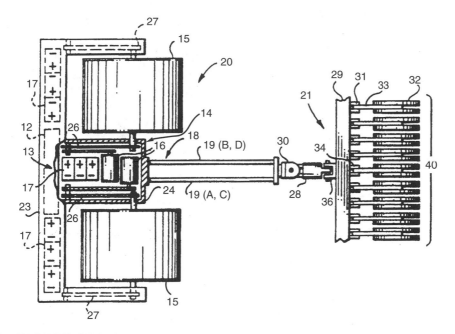

FIGURE CP3.1 Top view of Transport Platform and Mine Exploder, which is an electromechanical remotely controlled tractor (20) pulling a disk system (21) that acts as an antipersonnel landmine exploder. Pairs of massive disks (32) are rotatingly coupled to support bars (33), and are free to pivot up and down as their support bars pivot about their pivotal couplings (31). *Source:* U.S. patent #5,313,868 by Daniel Wolf, May 24, 1994.

(12), power source (13), battery power packs (17), and drive motors (16) direct and move tractor (20). Motors (16) are drive coupled to the wheels of ground support mass (15) by drive chains (26). Swing bars (27) are spring loaded and are designed to absorb and expend the energy of explosive shock transmitted to ground support mass (15), thereby relieving stress and preventing or minimizing damage.

Central arm (18), which connects tractor (20) with disk system (21), is comprised of four torsion bars (19 A, B, C, D). The torsion bars allow the central arm to twist under the impact of an explosion and afterwards resume its normal shape and function.

Disk system (21) is ruggedly constructed in order to survive impact from explosions. Its rounded surfaces deflect debris and shrapnel impelled by mine explosions. A row of pivotal couplings (31) are attached to tow bar cross member (29), and one end of each disk support bar (33) is pivotally coupled to couplings (31). Each disk support bar (33) supports a pair of disk (32). Only at coupling (34) is disk support bar (33) fixedly coupled so that it supports without pivoting its pair of disks (32). Therefore, disk system (21), towed by tractor (20), is supported primarily by a pair of disks (32) rotatingly coupled to the disk support bar (33), which is fixedly coupled to the cross

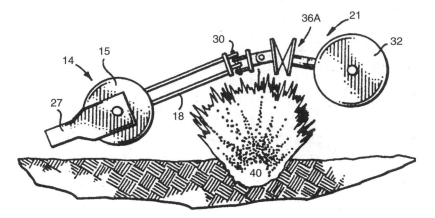

FIGURE CP3.2 The hinge arragangement (36A) absorbs shock after the Transport Platform and Mine Exploder has encountered and actuated a landmine buried in the ground at (40). *Source:* U.S. patent #5,313,868 by Daniel Wolf, May 24, 1994.

member (29) at coupling (34). All of the other disks are free to pivot up and down as their support bars pivot about their pivotal couplings.

Disks (32) are massive, like barbell disks in terms of shape and weight, in order to assure that all the ground over which disk array (40) passes will be subjected to sufficiently high bearing pressure to actuate any antipersonnel landmines buried therein. However, this mass will not be sufficient to actuate any antitank mines it may pass over.

A shock-absorbing hinge (36), which is a simple pivotal pin coupling, located in tow bar (28), also constitutes a shock-absorbing reaction component to decrease the likelihood of damage to disk system (21) upon mine explosion. Figure CP3.2 shows a mine, originally buried in the ground at (40), exploding and propelling both the tractor and the disk system into the air. Hinge coupling (36A) expands, absorbing shock and reducing the probability of damage elsewhere in the device. In a test at a military facility, such an explosion thrust disk system (21) into the air and rotated it at least 90 degrees. When uprighted again, torsion bars (19A–D) had absorbed the forces without damage and could resume their working positions.[45]

This remotely controlled tractor and disk system is designed for clearing landmines along pedestrian trails through forests and mountains. It will also be useful for demining farm lands to be cultivated.[46] This device satisfies the expressed needs of the demining experts in terms of the priority of areas to be cleared. One short prioritized list for humanitarian demining is

45. U.S. patent #5,313,868.
46. Ibid.

- roads and paths
- water sources
- villages
- agricultural areas[47]

Summary

Frederick Downs, Director of Prosthetic and Sensory Aid Service for the U.S. Veterans Administration, who was injured himself in the Vietnam War by an American-made bounding landmine, summarized his goal for this country in the following way:

> You know, I have met communists, I have met Moslems, I have met Christians, I have met people who worship trees and the Earth, I have met people who have no religion at all. Their politics are that, their religions are these, it makes no difference, because once they step on a landmine, once they become disabled, then their dignity as a human being is in jeopardy and they have lost it. They have lost a certain part of their independence as human beings and being able to take care of themselves. They have lost their mobility, their ability to...do things, work, play, get married, support families....One of the things we do best, I think, as a country, is when we try to make things right—we try to do something that will start a program that over a long, long term...will give some hope to them....Great changes...have always started with the actions of one person who believes in the right thing, and it grows over a period of time. And it may not happen in a year or a decade, and it may take many decades, but it is certainly a worthwhile goal.[48]

Design Assignment

Consider the issue of humanitarian demining and develop a problem statement. The specific goals are particularly important in demining efforts.

- Rank-order your goals.
- Use the design process to develop one or more design solutions.
- Develop a decision matrix by which designs are evaluated and compared, or perform Kepner–Tregoe decision analysis of alternative designs.
  Note: You may take Wolf's patented design as one of your alternative designs.
- Perform Kepner–Tregoe potential problem analysis of alternative designs. Consider all risks associated with each design, together with their probabilities of occurrence and relative severities. (Estimate these probabilities and severity levels, and include an explanation of your estimates.)
- Identify and discuss your choice of the best alternative.

---

47. U.S. Department of Commerce (1994).
48. *The Global Landmine Crisis* (1994), pp. 64–65.

4 Pulp for Paper Production: Manufacturing Process and Material Options

The News of Paper Goes West

The process of making paper dates from at least A.D. 100, when the Chinese developed a paper manufacturing process using rags, bark fiber, and bamboo. (Some historians believe that the Chinese were making paper from as early as 120 B.C.) The process of making paper consisted in liberating vegetable fibers, usually by soaking or boiling them and working them with a forked pole, dispersing them in liquid, and forming them into a web by the use of a cloth-bottomed box-shaped mold, which acted as a sieve. After drying in the sun, the sheet of paper could be removed easily from the mold.

Paper was being made in Persia by the mid-eighth century. Moors from the Mediterranean area introduced this technology into Europe in the early twelfth century.[1]

Wood Fiber for Paper: A Diminishing Supply

With the chronic shortage of rags and linen for papermaking in Europe and the isolation of cellulose by the French chemist, Anselme Payen, in 1839, by treating wood with nitric acid, the door was opened to manufacturing paper from wood by commercial delignification methods. Lignin is a polymer in wood that binds the plant fibers together. Most commercial pulping plants use strong alkaline or acidic reagents[2] to delignify wood effectively (that is, remove the lignin from the wood chips or bolts), with a minimum dissolution of hemicellulosic and cellulosic fibers.[3]

The United States, Canada, Sweden, and Finland have been the world's major suppliers of conifer forest products. The primary hardwood suppliers have been the United States, Brazil, Malaysia, China, Indonesia and India. Although physical supply is not diminishing drastically—in some areas it is even increasing—the availability of timber for industrial purposes is declining, while demand for paper is increasing rapidly. The conifer supply region in the southern United States suffered from decreasing growth rates through the 1970s and 1980s. Hardwood growth in eastern North America is dispersed and difficult to procure in

1. McGinnis and Shafizadeh (1980); Hunter (1930).

2. In our case, a *reagent* is a substance that chemically converts one substance (e.g., lignin) into another substance by means of the reaction that it causes (McKechnie, 1983).

3. McGinnis and Shafizadeh (1980); Glasser (1980).

commercial volumes. In the northwestern United States, extensive forest management has changed the operation of production forestry. The five-state region comprised of Washington, Oregon, California, Idaho, and Montana have experienced a decrease in availability of trees for harvesting from harvest levels of the early 1990s.[4]

In traditional production forestry, the forest was viewed as being available to serve multiple purposes. Timber production was to be carried out in such a way as to preserve wildlife, as best as possible, and to permit recreational use. This is no longer the case. Forests now are either extensively or intensively managed. In *extensively managed forests,* wildlife or ecosystem preservation and recreation outweigh the interests of timber production. In *intensively managed forests,* genetic engineering and technology is used to produce uniform, low-cost (that is, high per-hectare yield) raw material for industrial wood uses.

Industrial plantations are supplying an increasing proportion of the world's softwood and hardwood. Growth in industrial plantations has been primarily in the Southern Hemisphere (Brazil, Indonesia, Chile, Argentina, and Venezuela are leaders in this development). By means of intensively managed forests, the annual conifer cut for Australia, Chile, and New Zealand is expected to exceed that of the entire Pacific Northwest by the year 2010. The anticipated availability of industrial plantation hardwood is even larger than that of conifer softwood. Eucalyptus plantations alone could supply up to one-quarter of the world's hardwood consumption by the year 2010, to which can be added widespread acacia, gmelina, and albizia (hardwood) plantations.[5]

Still the worldwide virgin fiber shortage is not resolved. Anticipating a considerable use of secondary fiber so that paper pulp is estimated to contain only 55 percent virgin fiber, demand for virgin fiber by 2010 has been calculated to run about 230 percent of the estimated demand in the year 2000. Industrial plantations require a major investment of capital, and a considerable lead time—10 to 15 years is the average age of forests worldwide that can be used for industrial production. Land itself, moreover, is at a premium in industrialized nations. Domestically, agronomists and members of the pulp and paper industry are looking to nonwood fiber crops to supplement forest production for the following reasons:

- They are harvested annually, thereby satisfying immediate demand for fiber, but not the same long-term capital investment of forestry.
- In favorable climates, they have potentially higher fiber production per square meter than forests, which would reduce fiber costs and land area needed for fiber production.
- Large areas of southern United States are suitable and available for such crops.[6]

4. Hagler (1996).
5. Ibid.
6. Kaldor (1992).

Policies concerning renewable resources often encompass the idea of "sustainable development." In terms of fiber procurement, natural resources are to be managed in such a way so as to satisfy the need for fiber of present and future generations of people while maintaining or restoring the sustainability of ecosystems.[7] In forestry production, sustainable development is being advanced through *green certification,* or what some people call "stewardship forestry." Green certification represents an effort to verify that timber companies are producing lumber in a "sustainable" way. According to The Forest Partnership, a Vermont-based nonprofit group which distributes green wood and tracks certifiers, "(Green certification is) the most revolutionary change in the forest industry in its history. This is the bridge between the environmental community and the forest products industry."[8]

Stewardship forestry is nothing new, however, to the 8,500-member Menominee tribe, who have been practicing it on their 234,000-acre forest in northeastern Wisconsin since 1850. Menominee Tribal Enterprises (a certified "green" lumber producer) has "always let the forest dictate how much wood will be harvested in a given year."[9] Tribal forester, Marshall Pecore, states, "If we maintain a healthy, vibrant forest, over the long term it will sustain the people and the land itself."[10]

Pulps for Papermaking

Except for chemically produced synthetic paper, paper production relies on photosynthetic natural resources. Various words are used to describe these resources. *Phytomass* is a general term that refers to all plant fiber. *Agromass* is somewhat more limited. It commonly refers to the fiber of plants grown in a field, which excludes forests. Agromass, therefore, typically refers to crop fiber.

Four considerations are important in determining the suitability of a particular phytomass for paper production: availability and dependability of supply, certain fiber and plant characteristics, new industrial techniques, and economic and environmental considerations.

Availability and dependability of supply   To ensure availability and dependability of fiber supply, one must explore what is now grown and what can be grown in each geographic area of the nation and the world.

While there is a question about the increased availability and dependability of forest production in the United States, the U.S. Department of Agriculture, mandated by Congress, is promoting an effort to expand U.S. farm goods to include agricultural products of use to industry, which are not necessarily food crops.[11] With sufficient demand, an avenue may be created for American farmers to diversify into industrial fiber crops.

7. Werber and Hamilton (1997).
8. Bryce (1994, p. 9).
9. Ibid.
10. Ibid.
11. Werber and Hamilton, (1997).

Fiber and plant characteristics As pulp and paper expert Raymond Young notes, "Practically any plant material can be digested to yield a pulp for papermaking."[12] However, such fiber and plant characteristics as yield, strength, suitability to climate, growth rate, productivity, requirements of harvesting, collection, and extent of contamination, must be considered in determining the best plants for a particular farm location, pulping plant, and paper grade.

Given the vast variety of agromass, a large spectrum of fiber properties are available for selection in making paper-manufacturing decisions. Such fiber properties as length, diameter, chemical composition, density, and morphology bear on what processes can be used, and paper qualities of strength, brightness, and opacity.[13]

New industrial techniques The use of nontraditional fibers in existing paper industries will require appropriate technologies for separating, sorting, shipping, and handling fibers, as well as adapted process manufacturing techniques. In the United States, machines have been developed to separate the long outer bast fibers from the short inner core fibers of kenaf stems. Such engineering advances are critical for the production of wood fiber pulp.[14]

Economic and environmental considerations Economic and environmental considerations of scale, pollution control, energy requirement, water use, processing, disposal, life-cycle assessment, and recyclability are important factors to be considered in selecting a particular phytomass for pulp and paper production.[15]

Estimated statistics for 1998, on a worldwide basis, indicate that a sheet of paper was about 12 percent agromass, on average. That percentage drops to 2 percent nonwood fibers on average for paper made in the United States. The trend, however, on a global scale is toward agromass as the preferred fiber source for paper production. Agromass is a largely untapped reserve of fibrous raw materials, annually renewable, and available worldwide. Among the most plentiful are bamboo, the straw of cereal crops, bagasse,[16] and the stalks of corn and sorghum. At least 25 countries depend on agromass as the fiber source for more than 50 percent of their paper pulp production; among them are China (87 percent), Venezuela (75 percent), Greece (86 percent), and Egypt (100 percent).[17]

Biermann lists the following as agro-based fiber that can be used for pulp production:

- Straws such as wheat, rye, rice, and barley
- Grasses such as esparto and papyrus
- Canes and reeds such as bamboo, bagasse, corn stalks, and kenaf

---

12. Young (1997, p. 149).
13. Ibid.; Werber and Hamilton (1997).
14. Werber and Hamilton (1997); White and Cook (1997).
15. Werber and Hamilton (1997).
16. *Bagasse* is the fibrous residue of sugar cane left after the cane has been crushed and the sugar extracted (Misra, 1980).
17. Young (1997); White and Cook (1997); Biermann (1996).

- Bast (rope material) such as flax (linen), hemp, jute, and ramie
- Seed hairs such as cotton[18]

There are several advantages in using agromass for paper production. Its availability is a major attraction. Agromass can be grown in areas that have insufficient rainfall or soil quality to support trees. Nonwood fiber, depending on the type, may be a by-product of agriculture, and in a sense free. Agromass harvesting results in annual crop tonnages generally in the range of 1 to 2 tons per acre on the low side, to 8 to 10 tons per acre on the high side. Its productivity potentially can surpass most species of trees. In addition, nonwoody plants can be harvested within a year or two, unlike trees that normally take at least 10 years and as much as 20 years to grow large enough for use in the paper industry.[19]

Another important factor is that fiber crops often have only half the lignin content of wood. This means that pulping time for agromass may be expected to run about one-quarter to one-third of the time required for wood pulping. In addition, chemical and semichemical agromass pulps are relatively easy to bleach, which results in a reduction of pollution relative to hard-to-bleach pulps.[20]

There are also some disadvantages associated with the use of agromass for paper production. Its harvesting is typically seasonal. In some areas it may be limited to a short span of time in summer or fall. Care would be required that expensive harvesting equipment would not lie idle for most of the year. The bulkiness of agromass complicates transportation and storage. Nonwood fiber may be damaged by microbiological degradation, especially when wet, within a few days, and, thus, made unfit for pulping. Hence, it may be best to store this fiber under shelter. In addition, costs of handling agromass could be higher than costs of handling wood.[21]

In terms of processing agromass, its high content of ash and silica poses a problem. In the chemical recovery stages of alkaline pulping processes, silica deposits can coat the inner walls of the evaporator tubes, inhibiting heat transfer. Elsewhere silica may accumulate on tube surfaces, inhibiting the passage of flue gases. Silica also may impede the settling rate in clarifiers. The silica is not only internal, but also external to the fiber; that is, it is present in contaminating dirt. Cleaning the fiber before it goes into the digester would reduce the presence of silica. The internal silica content is extremely variable, even for a single type of crop, and depends on the species and soil conditions in which the crop is grown. Certain processing methods have been developed for dissolution of silica.[22]

18. Biermann (1996, p. 47).
19. Bublitz (1980).
20. Young (1997).
21. Bublitz (1980).
22. Young (1997).

Processes Used in Making Paper Pulp

The separation of wood and plant fibers results in *pulp,* which is eventually processed into paper and paperboard. Two broad classifications for pulps are mechanical and chemical pulps.

In *mechanical pulping,* the chemical constituents of the fibrous material remain unchanged, except for removal of water solubles. For example, groundwood, which is produced by purely mechanical means, retains all its lignin content. Therefore, its fibers do not exist as individual entities, but as fiber bundles and fragments of fibers. *Chemical pulping* involves the dissolution of lignin, which is the substance that binds the fibers together. However, dissolving lignin reduces yield, which is a measure of how much cellulosic fiber is retained by the end of the process.[23]

In terms of yield, there is an advantage in mechanical over chemical pulping processes. In general, *yield* indicates the amount of material recovered after a certain process relative to the starting amount of material before the process. Yield in pulping operations is a percentage given by the ratio of the oven-dry pulp mass to the oven-dry fiber mass.[24] Mechanical pulping processes tend to have higher yields than chemical pulping processes. Bleaching generally reduces yield for both mechanical and chemical pulps. There is also a correlation between yield and pollution—higher-yield pulping processes are less polluting than lower-yield processes.[25]

Burst and tear strengths are also important characteristics of paper pulps. Due to its high lignin content, mechanical pulps tend to have high stiffness and low strength, since the lignin interferes with hydrogen bonding between fibers. It is the lignin content of the paper that is responsible for the well-known phenomenon of yellowing with exposure to air and sunlight. The higher yield of mechanical pulps makes them lower-cost pulps.[26]

In *thermomechanical pulp (TMP)* processing, raw wood chips are first steamed at a low pressure, for a short period of time, to soften the lignin, and then mechanically refined. TMP allows greater hydrogen bonding between fibers, and therefore improves the strength of mechanical pulps. Thermomechanical pulp was developed for the newsprint industry. It has significantly higher bulk than chemical pulp, which is an advantage when show-through becomes important, as with low-weight paper grades. TMP can replace significant amounts of the more expensive semi-bleached kraft (pulp resulting from a full chemical pulping process described below) without adversely affecting the runability of the paper machine. TMP optimizes pulp strength for softwoods among mechanical pulping processes.[27]

Chemithermomechanical pulp (CTMP), which is now the most common chemimechanical pulping process, is more effective for hardwoods than

23. McGovern (1980); Daniell (1980).
24. Biermann (1996).
25. Kurdin (1980).
26. Biermann (1996).
27. Kurdin (1980); McGovern (1980).

TMP. CTMP involves sodium sulfite treatment of fiber prior to refining. This moderate chemical treatment results in a decrease in yield of only 2 to 3 percent relative to TMP processing. The waste from CTMP, however, has high color and high *biological oxygen demand, (BOD)*[28] which needs to be treated in order to reduce environmental pollution.[29]

Mechanical pulps, on the whole, account for a relatively small percentage of pulps produced in the United States. They amounted to approximately one-ninth of U.S. paper pulp production in 1990.[30]

In *full-chemical pulping* wood chips are cooked in appropriate pulping liquors in order to achieve nearly complete delignification and fiber liberation. Chemical pulps are about 95 to 97 percent lignin free.[31] See Figure CP4.1 for a flow diagram of wood utilization and preparation for paper pulp.

The term *kraft* (from the German and Swedish words for strong) is applied to the *sulfate process* (an alkaline process for pulp production using sodium sulfide; a compound of sulfur) because it produces very strong pulps. Kraft pulping accounted for as much as 80 percent of the U.S. pulp production from virgin wood in 1990.[32]

In chemical pulping processes, an important parameter is the *chemical charge,* which is the ratio of the dry mass of chemical used relative to the dry mass of material treated. For example, kraft pulping is done with about 24 percent alkali on wood (for softwoods). In alkaline pulping, the cost of the chemicals is too high for economic operation unless they are recovered and reused. For kraft pulping, with recovery, sodium losses in a kraft mill may run 20 to 50 kg per ton of pulp produced (which is a sodium loss of about 2.25 percent to 5.5 percent of the weight of pulp produced). For a full-sized paper pulp mill that produces 1,000 tons per day, this represents a loss of 22 to 55 tons per kraft mill per day in sodium alone. See Figure CP4.2 for an illustration of the kraft chemical recovery process.[33]

Some advantages of the kraft process are the following:

- It is useful for both hardwoods and softwoods.
- It produces a high strength pulp.
- It has an efficient energy cycle.

28. "The United States Environmental Protection Agency (USEPA): broadly classifies pollutants into eight types: oxygen-demanding substances, disease-causing agents, synthetic organic compounds, plant nutrients, inorganic chemicals and mineral substances, sediments, radioactive substances, and thermal discharges" (Springer and Peterson, 1980, p. 1177). Organic matter falls into the category of oxygen-demanding substances. In order to determine the amount of organic matter in waste, one usually measures the amount of oxygen required to oxidize the organic matter in a waste sample. This is accomplished by a BOD test. In the BOD test, seed bacteria are allowed to work on biodegradable organic matter (Springer and Peterson, 1980).

29. Kurdin (1980); Biermann (1996).

30. Biermann (1996).

31. Glasser (1980); Biermann (1996).

32. Biermann (1996); Bryce (1980).

33. Ibid.

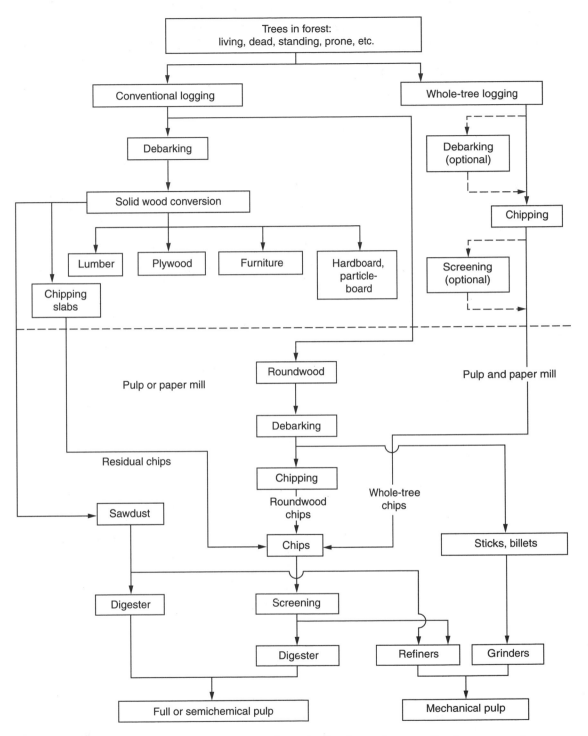

FIGURE CP4.1 A flow diagram overview of wood utilization and preparation for paper pulp.
Source: Bublitz, J. W., in Casey J. P. (Ed.), *Pulp and Paper: Chemistry and Chemical Technology,* vol. 1, 3d ed., New York: John Wiley and Sons, 1988. Reprinted with permission.

Some disadvantages of the kraft process are the following:

- It has low yield (due to fiber dissolution), which makes it a high-cost pulp.
- It produces emissions bearing sulfur in its reduced forms (which have a bad odor).
- The pulp is difficult to bleach (for example, it requires about twice as much bleaching chemicals as sulfite pulp—an important consideration since conventional bleaching results in toxic and environmentally damaging chlorinated organic effluents).
- Significant amounts of chemicals are consumed.[34]

The *soda process* is the second major alkaline process for pulp production. Soda pulping, an older process than sulfate pulping, is sulfur-free and therefore less polluting. However, it produces a weaker pulp than kraft pulping. Advances also have been made in soda-oxygen pulping and other types of nonsulfur soda chemical and semichemical pulping.[35]

The *sulfite pulping process,* which in certain ways is intermediate between CTMP and kraft processes, has higher yield than kraft pulping for softwoods. It produces a medium strength pulp with low lignin content, and it is easily bleached (requiring only about half the amount of bleaching chemicals as the kraft process). The sulfite process is a full chemical process, which uses mixtures of sulfurous acid and/or its alkali salts.[36]

Environmental Considerations

In the kraft process paper mill, condensates from the relief and blow system, and the multiple-effect evaporators contain substantial amounts of hydrogen sulfide (H_2S) and methyl mercaptan (CH_3SH), and smaller amounts of dimethyl sulfide (CH_3SCH_3). These units discharge a portion of these condensates into the atmosphere, causing a bad odor in the vicinity of the mill. The portion of these sulfur-containing by-products remaining in the condensate also contributes to toxicity in the mill effluent.[37]

It has been noted that effluents from the bleaching process are the most serious problem in the pulp and paper industry. As a measure of pollution, the chemical oxygen demand (COD)[38] of the liquid effluent from reductive bleaching of mechanical pulp is about 13 kg O_2 per ton of pulp. This is less than the COD of the liquid effluent from chlorine bleaching of kraft pulp, which ranges from 23 to 49 kg O_2 per ton of pulp, measured at various stages of the bleaching process.[39]

34. Ibid.

35. Ibid.

36. Biermann (1996).

37. Bryce (1980).

38. The chemical oxygen demand test is similar to the BOD test, but whereas the BOD test is based on natural processes involving active seed bacteria, the COD test relies on chemical oxidation of the organic matter in the waste sample (Springer and Peterson, 1980).

39. Loras (1980).

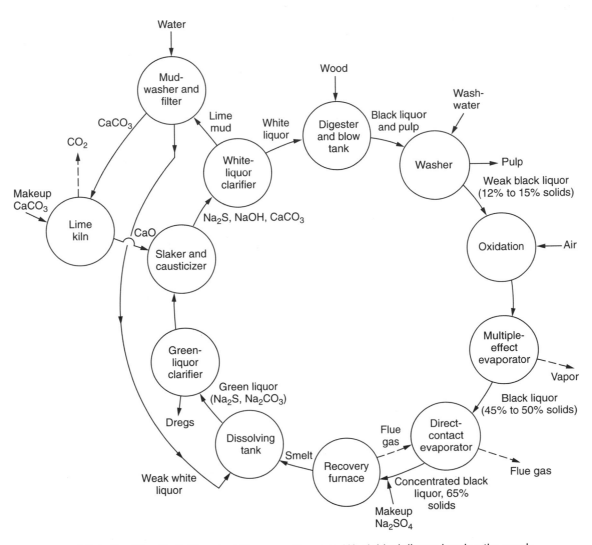

FIGURE CP4.2 The Kraft Chemical Recovery Process. Weak black liquor leaving the washers contains about 12 percent to 15 percent solids. It is concentrated by means of two evaporation processes to a strong black liquor with solids concentration of 62 percent to 65 percent. It then is burned in a furnace in order to produce the desired end-products of Na_2S and Na_2CO_3 in the smelt, and CO_2, H_2O and SO_2 in the flue gas. The smelt from the furnace is dissolved to form green liquor, then recausticized with slaked lime to produce $NaOH$, before it can be reused in the digestion process. *Source:* Bryce, J. R. G., in Casey, J. P. (Ed.), *Pulp and Paper: Chemistry and Chemical Technology,* 3d ed., New York: John Wiley and Sons, 1980. Reprinted with permission.

In terms of the liquid waste from the bleach plant, *chlorinated aromatic compounds,* so-called chlorotoxins, are dissolved in liquid in the chlorination and alkaline extraction stages of the bleaching process. These chlorinated compounds are carcinogenic and mutagenic.[40] *Dioxin,* one of the chlorotoxins resulting from the kraft bleaching process, is perhaps best known for its use in Agent Orange, the defoliant used in the Vietnam War. It has been said to be "one of the most potent poisons known to man."[41] In total, there are as many as 1,000 toxic chemicals of concern in kraft paper mill effluents.[42]

Pulp mills commonly have a wastewater treatment plant to filter solids from its liquid effluent before the liquid is discharged into a river. The solids, or sludge, that is filtered out is often put into landfills or lagoons, which poses environmental threats. This sludge may contain dioxin. Chlorine compounds do not biodegrade easily, so they accumulate in increasing concentrations in the food chain (they "bioaccumulate"). For example, dioxin concentrations of up to 24 parts per trillion were measured in the sludge produced at one paper mill in New Hampshire. Brook trout and horned pout sampled downstream of the mill had dioxin concentrations high enough to compel the state to advise limits on their consumption, and urge pregnant and nursing women to avoid them altogether.[43]

In addition, lagoons used as dump sites for sludge from paper plants should be dredged regularly. If they are not, they can emit hydrogen sulfide gas, wich has been known to cause illness to nearby residential populations, and a burning sensation in the eyes.[44]

Sulfur pollution control has also been achieved by the *alcell process* (a name derived from the words *alcohol* and *cellulose*), which uses alcohol and water, instead of sulfur, for delignification. Another attraction of alcell technology is that it can be used in a mini-mill, which can have a production capability of about 350 tons of pulp per day. Building kraft pulping mills that produce 1,000 tons per day has become exceedingly expensive in terms of steel, and in terms of their requirement of massive amounts of wood stock feed.

In response to congressional pressure to eliminate chlorine discharge,[45] Louisiana-Pacific Company, headquartered in Portland, Oregon, developed a process that uses oxygen, caustics, and hydrogen peroxide—but no chlorine—to produce paper. This pulp produces paper that is only slightly less bright, and the cost of the pulp so far is only about 10 percent higher. A smaller, but still significant step toward reducing chlorine-related pollution is *elemental chlorine-free (ECF)* technology. In this technology, chlorine

40. Ibid.
41. Seabrook (1989, p. B6).
42. Braile (1990).
43. Ibid.; Knickerbocker (1994).
44. Braile (1990).
45. The Chlorine Zero Discharge Act, a 1994 House initiative which garnered more than 40 cosponsors, would require complete phaseout of chlorine or other chlorinated oxidizing agents in the paper industry over a five-year period (Knickerbocker, 1994).

dioxide compound still is used, but elemental chlorine gas (from which toxic dioxins and furans form) is not used. EFC technology is used by International Paper, headquartered in Westchester County, New York.[46]

Quite a different approach to chlorine-free paper production has been taken by researchers in the area of biological applications. Fungal pretreatment of fiber has been used to break down lignin and reduce energy consumption in refining the fiber.[47] At the University of Georgia, a biochemical technique, which they dubbed the *Enzone process,* uses enzymes to remove lignin. Such biotechnology has negligible negative environmental impact. It can be used to produce fully bleached pulps with high strength for quality paper manufacturing.[48]

A means of wood fiber and paper conservation, as well as sulfur-free pulp production, has been given by Adam E. Kaliski, a chemist from East Windsor, New Jersey, by his patented formula for *microgels*,[49] Microgels are mineral salts made up of aluminum and silica. When mixed with calcium chloride it produces a gel, which can replace the polymers usually added to the pulp to congeal it.[50]

By replacing acid or polymers with microgels as the thickener, Kaliski observes that paper production costs could be reduced in several ways. Paper mills could operate at approximately twice the speed, because microgels ensure even dispersal of all the ingredients, so that the pulp is stronger and more uniform. The conveyor belt can run faster with the new formula pulp, and still produce high-quality, uniform paper. Kaliski's method has made paper that has tripled the opacity of the best photocopier paper, without increasing its weight of thickness. With this improved opacity, both sides of the page can be printed on, without show-through. Also the increased absorbency of pulp thickened with microgels means that more inexpensive clay can be used as a filler, saving expensive wood pulp, and less water and dye are needed.[51]

Saving water in the papermaking process is very important. Consider the case of Union Camp's plant in Franklin, Virginia. Union Camp Corporation, based in Wayne, New Jersey, ranks among the top dozen U.S. paper companies. It operates a large and profitable fine-paper mill along the bank of the sleepy Blackwater River. At the Franklin plant, six paper machines, each a city block long, run almost continuously, making paper out of a mixture of Southern pine and hardwoods. This mill requires more than 30 million gallons of water per day to be used in making paper. The chlorine bleaching process makes the used water too corrosive to metals to be used in other parts of the papermaking process, so it is discharged into the river. However, the level of the Blackwater River is too low most of the year to

46. Ibid.
47. Young (1997).
48. Knickerbocker (1994).
49. U.S. patent #5,240,561 (August 31, 1993).
50. *New York Times* (October 25, 1993).
51. Ibid.

absorb the effluent. Consequently, plant managers store billions of gallons of water in ponds covering approximately 1,500 acres of land. The pond water is discharged into the river during the period of the year when the water level is low in order to aid the river in absorbing the dark-colored water that results from the bleaching process.[52]

Union Camp has three bleaching lines: a conventional chlorine bleaching line, an oxygen bleaching line (in which oxygen is used during the first bleaching stage), and their own improved *ozone bleaching line.* (Ozone, O_3, is a form of oxygen.) Union Camp has found a way to bleach moist wood pulp by exposure to ozone, without weakening the wood fibers while brightening them, by carefully limiting the exposure time. Although chlorine dioxide is still used to brighten the pulp, a number of chlorotoxins are removed, along with elemental chlorine, in this bleaching process. A further advantage to ozone bleaching is that the residues from this process are not corrosive, so that the used water can be recycled and chemicals can be recovered.[53]

Another water problem arises if a paper plant owner builds a hydro-electric project on the river abutting the mill in order to ensure a cheaper and more reliable source of electricity. A hydroelectric plant can reduce by two-thirds the flow of river water (measured in cubic feet per second), further compromising the ability of the river to absorb the mill effluent.[54]

▨ Nonwood Fibers for Pulp

Three general categories of nonwood fibers for pulp are agricultural and agro-industrial waste, plants growing in the wild, and cultivated fiber crops. Bleached pulps produced from bagasse, bamboo, cereal straw, grass, and reeds have been used in the production of all types of paper grades from high-quality writing and printing paper to poster paper, glassine, tissue, and towelling.[55]

In the United States, the following crops can supply fiber for paper pulp on a reasonably large scale. Sugar cane bagasse, cereal straws, rice straw, grain sorghum stalks, and seed grass straw are agricultural crops whose residues can be used for pulp; kenaf and crotalaria (or sunn hemp) are crops grown specifically for pulp and paper use.[56]

Essentially the same processes are used to produce pulp from nonwood plants as are used on wood. These include various chemical processes, as well as soda, sulfate, and sulfite processes.[57]

52. Holusha (1993); Union Camp Corporation (1998).
53. Ibid.
54. Braile (1990).
55. Misra (1980).
56. Atchison (1995).
57. Misra (1980).

Furthermore, certain crops are multipurpose. This opens up the possibility of creating *integrated mills.* In India, for example, an integrated mill produces pulp, paper, and sugar. Also, kenaf core fibers, because of their excellent absorption qualities, have been used to make oil absorbents, mats for seeding and bank erosion control, poultry litter, potting media, and so forth.[58]

USDA and Kenaf

In southeastern United States (for example, Mississippi), kenaf is grown as a source of fiber competitive with hardwood pulp for papermaking. Kenaf is an annual fiber crop that is native to East-Central Africa. It looks like corn, but it is, instead, a fast-growing member of the Malvaceae family (same family as cotton). The kenaf crop is very sensitive to frost, and therefore is best suited to southern states from Florida to Texas, and California. It thrives in fertile, well-drained soil, and increases in height until daylight becomes less than 12 1/2 hours per day, at which point it does not grow significantly taller and flowering begins. It can reach heights of 12 to 18 ft.[59]

In early 1991 a farmer's cooperative was formed in Mississippi and 2,000 acres of kenaf were planted. One of the intended commercial uses of kenaf would be to substitute for woodchip and woodyard operations in a pulp mill.

As a first step in proving the feasibility of a kenaf-based pulp and paper industry, the cooperative has successfully carried out agricultural production and fiber processing of kenaf. This development was possible due to the work of the Mississippi Agricultural and Forestry Experiment Station (MAFES).[60]

With the help of funding from the U.S. Department of Agriculture's (USDA's) Agricultural Research Service (ARS), the MAFES began research on kenaf in 1989. MAFES conducts research at Mississippi State University and several of its branch stations.[61] Research has ranged from evaluating varieties of kenaf seed, weed science, nematodes, row spacing, and fertility to the processing and marketing of the separated fiber. The result of the MAFES work has been to establish a system for efficiently producing, harvesting, and transporting kenaf in the Mississippi area.

More recent research by the USDA-ARS has shown that kenaf pulped with hardwood kraft, soda, or neutral sulfite pulps results in a pulp that is superior to standard commercial hardwood pulps. Also, kenaf pulped with softwood sulfite pulps is superior to commercial softwood sulfite pulps.[62]

58. White and Cook (1997); Young (1997).
59. Bublitz (1980); White and Cook (1997); Horn et al. (1992).
60. Kaldor et al. (1992).
61. MAFES (1994).
62. Myers and Bagby (1995).
63. Horn et al. (1992).

Newsprint can be made successfully from a blend of 95 percent or 82 percent kenaf CTM pulp, and 5 percent or 18 percent kraft, according to research conducted by the USDA Cooperative State Research Service (CSRS).[63] Specifically, a kenaf-based peroxide chemithermomechanical pulp has been shown to produce a high-quality newsprint on a commercial scale.[64]

Newspaper print is a large contributor to the growing problem of solid waste. In order to address this major concern, it is desirable to use recycled newsprint in the production of newspaper. The result of research by the USDA Forest Service and the CSRS showed that an acceptable newsprint can be made from 75 percent recycled newsprint and 25 percent kenaf CTM pulp for reinforcement.[65]

Design Assignment

- Analyze and report the merits and demerits of a specific paper pulp manufacturing process of your choice, for a specific geographic location of your choice.
- Discuss the market advantages of your chosen fiber sources in terms of availability, price, and performance.
- What improvements can you recommend for your manufacturing process? List and describe general goals and specific goals to be satisfied by your manufacturing process.
- Discuss your product in terms of its quality and expected economic viability.
- What environmental impact is your plant likely to have?

64. Kugler (1988).
65. Horn et al. (1992).

ADDITIONAL CASE PROBLEM TOPICS AND PROJECT THEMES

Each of the following case problem topics can serve as the basis for a design project.[1] Since we are only describing situations in which engineering design solutions might be helpful, readers will need to develop an understanding of the background and current status of each situation by performing appropriate research; moreover, they will need to develop specific problem statements. (The four full case problems in this text can serve as examples of the type of background research that should be performed for each topic.)

Many of the following situations are so broad in scope and difficulty that students will need to focus upon only one aspect or portion of the entire situation. For instance, one's efforts could be directed towards the needs of the population group that is most affected by a particular problem. For example, falls are the most common type of unintentional injury leading to death among people 75 years of age and older, whereas motor-vehicle accidents are the leading cause of death due to unintentional injury among those aged 1 year to 74 years. Hence, a focus upon the needs of older people might be natural if one is trying to develop a solution that will protect people from unintentional falls or slips in the home.

Sports-Related Injuries

1. **Basketball injuries**   Approximately 761,000 people were treated in hospital emergency rooms during 1994 for injuries suffered while playing basketball. Design a solution to this problem or to part of this problem.
2. **Golf injuries**   Nearly 38,000 people were treated in hospital emergency rooms during 1994 for injuries suffered while playing golf. In addition, another 6,372 injuries involved golf carts. Design a solution to one or both of these problems (or parts thereof).
3. **Tennis injuries**   More than 30,000 people were treated in hospital emergency rooms during 1994 for injuries suffered while playing tennis. Design a solution to this problem or to part of this problem.

1. Unless otherwise noted, all statistical data in this section are taken from the National Safety Council's Accident Facts, 1995 Edition.

4. **Injuries while fishing**   Nearly 76,000 people were treated in hospital emergency rooms during 1994 for injuries suffered while fishing. Design a solution to this problem or to part of this problem.

5. **Boating deaths and injuries**   Recreational boating accidents (as reported to the U.S. Coast Guard) during 1994 involved more than 4,000 injuries and 784 deaths. Design a solution to this problem or to part of this problem.

6. **Deaths and injuries while hunting**   Nearly 1,100 injuries and 101 deaths were reported by the Hunter Education Association during 1993; about one-third of these injuries were self-inflicted. Design a solution to this problem or to part of this problem.

7. **Swimming fatalities**   It is estimated that 1,500 people perished while swimming during 1994. Design a solution to this problem or to part of this problem.

Vehicle-Related Accidents

8. **Tractors**   It is estimated that more than 350 people lost their lives during 1994 in accidents involving farm tractors. Most often these accidents involved overturning of a tractor, accounting for 52 percent of fatalities. The second most common type of incident involved people being run over by a tractor (29 percent). Design a solution to this problem or to part of this problem.

9. **Child restraints in automobiles**   The lives of 286 children were saved by child restraints in motor-vehicle accidents during 1993. Unfortunately, more than 200 lives were lost that year because such restraints were not used. Design a solution to this problem or to part of this problem.

10. **Motorcycle deaths**   Only about two percent of the nearly 200 million motor vehicles registered in the United States in 1994 were motorcycles; however, during that same year nearly seven percent of the deaths due to motor-vehicle accidents were those of motorcycle riders. Another measure of the relative danger associated with motorcycles is captured in the mileage death rate from 1994: The number of deaths of motorcycle riders per travel miles was *17 times* that of riders in other types of motor vehicles. Design a solution to this problem or to part of this problem.

11. **School buses**   During the single school year 1993–1994, 14,000 people were injured and 115 were killed in school bus related accidents. Design a solution to this problem or to part of this problem.

12. **Speeding while driving**   The most common factor (19.5 percent during 1994) in fatal automobile accidents is excessive speed. Design a solution to this problem or to part of this problem.

13. **Pedestrian deaths and injuries**   Injuries to most pedestrians (53.5 percent in 1994) occur while they are crossing or entering a street,

particularly between intersections (37.5 percent). Design a solution to this problem or to part of this problem.

14. **Sleepy drivers**   Many motor-vehicle accidents occur each year due to sleepy drivers whose cars swerve across lanes or off the road. Design a solution to this problem or to part of this problem.

15. **Accidents involving emergency vehicles**   During 1993, 8,300 police cars, 3,300 ambulances, and 2,000 fire trucks were involved in traffic accidents while their emergency lights or sirens were operating. Moreover, 53 police cars, 16 ambulances, and 12 fire trucks were involved in fatal accidents while their emergency lights or sirens were operating. Design a solution to this problem or to part of this problem.

16. **Cyclist injuries**   More than 604,000 people were treated in hospital emergency rooms during 1994 for injuries suffered while riding bicycles. Design a solution to this problem or to part of this problem.

17. **Forklift injuries**   More than 18,000 injuries occurred during 1992 in which forklifts were involved. This type of accident results in the deaths of about 100 people annually. Common subcategories of these fatal accidents include overturning (accounting for 24 percent of the fatalities), workers being struck by materials (17 percent), workers either falling from the forklift (14 percent) or being struck by the forklift (14 percent), and workers being pinned between objects (11 percent). Design a solution to this problem or to part of this problem.

▩ Product-Related Injuries

18. **Scissors**   It is estimated that 34,602 people were injured in accidents involving scissors during 1992. Design a solution to this problem or to part of this problem.

19. **Television sets and stands**   It is estimated that 42,000 people were injured in accidents involving television sets and stands during 1992. Design a solution to this problem or to part of this problem.

20. **Toys**   It is estimated that 177,061 people were injured in accidents involving toys during 1992. Design a solution to this problem or to part of this problem.

21. **Fireworks**   Each year (particularly around the Fourth of July) many people—especially children—are injured or killed by fireworks. Although the sale or use of fireworks is illegal in many states, it has been estimated that up to 12,000 people are treated each year in hospital emergency rooms for such injuries.[2] Design a solution to this problem or to part of this problem.

2. Dowdy, Z. R., "Police Seize Motorists' Fireworks," *The Boston Globe,* July 2, 1996, pp. 17 and 20. According to the U.S. Consumer Product Safety Commission's National Electronic Injury Surveillance System (NEISS), fireworks were involved in 13,263 injuries during 1992.

22. **Clothing**   It is estimated that 142,457 people were injured in accidents involving clothing during 1992. Design a solution to this problem or to part of this problem.

23. **Fans**   It is estimated that more than 17,000 people were injured in accidents involving electric fans during 1992. Design a solution to this problem or to part of this problem.

24. **Glass bottles and jars**   It is estimated that 63,170 people were injured in accidents involving glass bottles and jars during 1992. Design a solution to this problem or to part of this problem.

25. **Glass doors and windows**   It is estimated that more than 216,000 people were injured in accidents involving glass doors, windows, and panels during 1992. Design a solution to this problem or to part of this problem.

26. **Heaters**   It is estimated that nearly 38,000 people were injured in accidents involving heating stoves and space heaters during 1992. Design a solution to this problem or to part of this problem.

27. **Hoists, lifts, and jacks**   It is estimated that 16,562 people were injured in accidents involving hoists, lifts, and jacks during 1992. Design a solution to this problem or to part of this problem.

28. **Lawn mowers**   It is estimated that more than 85,000 people were injured in accidents involving lawn mowers during 1992. Design a solution to this problem or to part of this problem.

29. **Playground equipment**   It is estimated that more than 290,000 people were injured in accidents involving playground equipment during 1992. Design a solution to this problem or to part of this problem.

30. **Chain saws**   It is estimated that 38,692 people were injured in accidents involving chain saws during 1992. Design a solution to this problem or to part of this problem.

▨ Other Injuries

31. **Lightning**   During the 1940s, the average number of people who died each year after being struck by lightning was 329, whereas during the 1980s this number had dropped to 73. Although these numbers clearly reflect progress in reducing the hazards associated with lightning, people continue to be injured or killed each year; for example, 43 people died because of lightning during 1993 and another 295 were injured. Design a solution to this problem or to part of this problem.

32. **Firearms**   Between June 1992 and May 1993, 19,727 unintentional nonfatal injuries due to firearms were treated in hospital emergency rooms. In addition, 1,409 deaths were recorded during that time. (These numbers exclude injuries and deaths due to assaults and suicides.) Design a solution to this problem or to part of this problem.

33. **Railroad deaths and injuries**   During 1994, 1,226 people died in railroad accidents (572 of these deaths occurred at rail–highway public-grade crossings). Moreover, 16,812 nonfatal injuries occurred during

that year (of which 1,829 occurred at rail–highway crossings). Design a solution to this problem or to part of this problem.

34. **Stairs, ramps, and floors** It is estimated that more than 1,879,000 people were injured in accidents involving stairs, ramps, landings, and floors during 1992. Design a solution to this problem or to part of this problem.

35. **Injuries in the home** More disabling injuries (about 7,000,000 in 1994) occur in the home than in motor-vehicle accidents and work-place accidents combined. During 1994, 26,700 people died because of injuries suffered in the home. Such hazards as falls, poisons, fires, and suffocation due to unintentionally ingested or inhaled objects contributed to these deaths. Design a solution to this problem or to part of this problem.

General Themes

Develop an original and improved design for one of the following types of products or situations. Be certain to review any current designs that are available in the marketplace. If possible, also review the patent records for any designs that may have been developed in the past and either failed to survive in the market or were never produced. Consider the specific goals that should be achieved by the design as you develop your solution.

36. **Sports equipment** Select a particular type of sports product (tennis racket, baseball glove, golf ball, etc.) and develop an improved design. Be aware of official requirements or specifications that must be satisfied by your design.

37. **Home damage** Severe weather conditions (in the form of excessive water, ice, wind, heat, and dust) can damage the exterior of a house. Develop a design that will minimize or eliminate such damage and loss.

38. **Egg containers** Egg cartons sometimes fail to prevent breakage and loss. Design a cost-effective and environmentally safe container that will prevent such breakage.

39. **Gardening equipment** Select a particular type of gardening tool (e.g., shovels, tillers, wheelbarrows, rakes, trimmers) and develop an improved design.

40. **Ladders** Develop an improved design for a ladder.

41. **Food preservation** Develop an improved method for preserving particular types of food. Consider the specific conditions under which the food must be stored, transported, and preserved.

42. **Pet feeders** Design an automatic feeder that will provide a pet with an appropriate amount of food without the danger of overfeeding the animal.

43. **Measuring devices** Develop an improved device or method for measuring distances, speeds, weights, volumes, or any other quantity of your choice. (The device does not need to measure more than one quantity, although it may do so.)

44. **Cleaning systems** Develop an improved system for cleaning specific types of material or equipment. Your design must be environmentally safe and cost effective.

45. **Waste streams** Select a particular type of manufacturing process waste stream. Develop an improved method for managing this waste stream in an environmentally safe and economically viable manner. Methods might include increased recovery of valuable chemical elements that may be in the stream, greater recycling of other wastes, and more effective disposal or elimination of particularly hazardous materials. Consider the manufacturing process to determine if it can be modified to reduce the costs and hazards associated with its waste stream.

46. **Garage door openers** Develop an improved garage door opener.

47. **Raw materials handling systems** Develop an improved system for transporting and handling a specific type of raw material (e.g., liquids, sand, soil, gases). Consider the conditions under which such handling may occur.

48. **Tree stump removers** Develop an improved and environmentally safe method for removing tree stumps of various sizes.

49. **Bicycles** Develop an improved bicycle design.

50. **Filling machines** Develop an improved design for accurately filling containers with a specific type of material (e.g., liquids, powders, etc.).

THE NSPE CODE OF ETHICS FOR ENGINEERS

Preamble

Engineering is an important and learned profession. As members of this profession, engineers are expected to exhibit the highest standards of honesty and integrity. Engineering has a direct and vital impact on the quality of life for all people. Accordingly, the services provided by engineers require honesty, impartiality, fairness and equity, and must be dedicated to the protection of the public health, safety, and welfare. Engineers must perform under a standard of professional behavior that requires adherence to the highest principles of ethical conduct.

I. Fundamental Canons

Engineers, in the fulfillment of their professional duties, shall:

1. Hold paramount the safety, health and welfare of the public.
2. Perform services only in areas of their competence.
3. Issue public statements only in an objective and truthful manner.
4. Act for each employer or client as faithful agents or trustees.
5. Avoid deceptive acts.
6. Conduct themselves honorably, responsibly, ethically, and lawfully so as to enhance the honor, reputation, and usefulness of the profession.

II. Rules of Practice

1. Engineers shall hold paramount the safety, health, and welfare of the public.
   a. If engineers' judgment is overruled under circumstances that endanger life or property, they shall notify their employer or client and such other authority as may be appropriate.
   b. Engineers shall approve only those engineering documents that are in conformity with applicable standards.

   c. Engineers shall not reveal facts, data or information without the prior consent of the client or employer except as authorized or required by law or this Code.
   d. Engineers shall not permit the use of their name or associate in business ventures with any person or firm that they believe are engaged in fraudulent or dishonest enterprise.
   e. Engineers having knowledge of any alleged violation of this Code shall report thereon to appropriate professional bodies and, when relevant, also to public authorities, and cooperate with the proper authorities in furnishing such information or assistance as may be required.
2. Engineers shall perform services only in the areas of their competence.
   a. Engineers shall undertake assignments only when qualified by education or experience in the specific technical fields involved.
   b. Engineers shall not affix their signatures to any plans or documents dealing with subject matter in which they lack competence, nor to any plan or document not prepared under their direction and control.
   c. Engineers may accept assignments and assume responsibility for coordination of an entire project and sign and seal the engineering documents for the entire project, provided that each technical segment is signed and sealed only by the qualified engineers who prepared the segment.
3. Engineers shall issue public statements only in an objective and truthful manner.
   a. Engineers shall be objective and truthful in professional reports, statements, or testimony. They shall include all relevant

and pertinent information in such reports, statements, or testimony, which should bear the date indicating when it was current.

   b. Engineers may express publicly technical opinions that are founded upon knowledge of the facts and competence in the subject matter.

   c. Engineers shall issue no statements, criticisms, or arguments on technical matters that are inspired or paid for by interested parties, unless they have prefaced their comments by explicitly identifying the interested parties on whose behalf they are speaking, and by revealing the existence of any interest the engineers may have in the matters.

4. Engineers shall act for each employer or client as faithful agents or trustees.

   a. Engineers shall disclose all known or potential conflicts of interest that could influence or appear to influence their judgment or the quality of their services.

   b. Engineers shall not accept compensation, financial or otherwise, from more than one party for services on the same project, or for services pertaining to the same project, unless the circumstances are fully disclosed and agreed to by all interested parties.

   c. Engineers shall not solicit or accept financial or other valuable consideration, directly or indirectly, from outside agents in connection with the work for which they are responsible.

   d. Engineers in public service as members, advisors, or employees of a governmental or quasi-governmental body or department shall not participate in decisions with respect to services solicited or provided by them or their organizations in private or public engineering practice.

   e. Engineers shall not solicit or accept a contract from a governmental body on which a principal or officer of their organization serves as a member.

5. Engineers shall avoid deceptive acts.

   a. Engineers shall not falsify their qualifications or permit misrepresentation of their or their associates' qualifications. They shall not misrepresent or exaggerate their responsibility in or for the subject matter of prior assignments. Brochures or other presentations incident to the solicitation of employment shall not misrepresent pertinent facts concerning employers,

employees, associates, joint venturers, or past accomplishments.

   b. Engineers shall not offer, give, solicit or receive, either directly or indirectly, any contribution to influence the award of a contract by public authority, or which may be reasonably construed by the public as having the effect of intent to influencing the awarding of a contract. They shall not offer any gift or other valuable consideration in order to secure work. They shall not pay a commission, percentage, or brokerage fee in order to secure work, except to a bona fide employee or bona fide established commercial or marketing agencies retained by them.

III. Professional Obligations

1. Engineers shall be guided in all their relations by the highest standards of honesty and integrity.

   a. Engineers shall acknowledge their errors and shall not distort or alter the facts.

   b. Engineers shall advise their clients or employers when they believe a project will not be successful.

   c. Engineers shall not accept outside employment to the detriment of their regular work or interest. Before accepting any outside engineering employment they will notify their employers.

   d. Engineers shall not attempt to attract an engineer from another employer by false or misleading pretenses.

   e. Engineers shall not actively participate in strikes, picket lines, or other collective coercive action.

   f. Engineers shall not promote their own interest at the expense of the dignity and integrity of the profession.

2. Engineers shall at all times strive to serve the public interest.

   a. Engineers shall seek opportunities to participate in civic affairs; career guidance for youths; and work for the advancement of the safety, health and well-being of their community.

   b. Engineers shall not complete, sign, or seal plans and/or specifications that are not in conformity with applicable engineering standards. If the client or employer insists on such unprofessional conduct, they shall notify the proper authorities and withdraw from further service on the project.

c. Engineers shall endeavor to extend public knowledge and appreciation of engineering and its achievements.

3. Engineers shall avoid all conduct or practice that deceives the public.
   a. Engineers shall avoid the use of statements containing a material misrepresentation of fact or omitting a material fact.
   b. Consistent with the foregoing, Engineers may advertise for recruitment of personnel.
   c. Consistent with the foregoing, Engineers may prepare articles for the lay or technical press, but such articles shall not imply credit to the author for work performed by others.

4. Engineers shall not disclose, without consent, confidential information concerning the business affairs or technical processes of any present or former client or employer, or public body on which they serve.
   a. Engineers shall not, without the consent of all interested parties, promote or arrange for new employment or practice in connection with a specific project for which the Engineer has gained particular and specialized knowledge.
   b. Engineers shall not, without the consent of all interested parties, participate in or represent an adversary interest in connection with a specific project or proceeding in which the Engineer has gained particular specialized knowledge on behalf of a former client or employer.

5. Engineers shall not be influenced in their professional duties by conflicting interests.
   a. Engineers shall not accept financial or other considerations, including free engineering designs, from material or equipment suppliers for specifying their product.
   b. Engineers shall not accept commissions or allowances, directly or indirectly, from contractors or other parties dealing with clients or employers of the Engineer in connection with work for which the Engineer is responsible.

6. Engineers shall not attempt to obtain employment or advancement or professional engagements by untruthfully criticizing other engineers, or by other improper or questionable methods.
   a. Engineers shall not request, propose, or accept a commission on a contingent basis under circumstances in which their judgment may be compromised.

b. Engineers in salaried positions shall accept part-time engineering work only to the extent consistent with policies of the employer and in accordance with ethical considerations.
   c. Engineers shall not, without consent, use equipment, supplies, laboratory, or office facilities of an employer to carry on outside private practice.

7. Engineers shall not attempt to injure, maliciously or falsely, directly or indirectly, the professional reputation, prospects, practice, or employment of other engineers. Engineers who believe others are guilty of unethical or illegal practice shall present such information to the proper authority for action.
   a. Engineers in private practice shall not review the work of another engineer for the same client, except with the knowledge of such engineer, or unless the connection of such engineer with the work has been terminated.
   b. Engineers in governmental, industrial, or educational employ are entitled to review and evaluate the work of other engineers when so required by their employment duties.
   c. Engineers in sales or industrial employ are entitled to make engineering comparisons of represented products with products of other suppliers.

8. Engineers shall accept personal responsibility for their professional activities, provided, however, that Engineers may seek indemnification for services arising out of their practice for other than gross negligence, where the Engineer's interests cannot otherwise be protected.
   a. Engineers shall conform with state registration laws in the practice of engineering.
   b. Engineers shall not use association with a nonengineer, a corporation, or partnership as a "cloak" for unethical acts.

9. Engineers shall give credit for engineering work to those to whom credit is due, and will recognize the proprietary interests of others.
   a. Engineers shall, whenever possible, name the person or persons who may be individually responsible for designs, inventions, writings, or other accomplishments.
   b. Engineers using designs supplied by a client recognize that the designs remain the property of the client and may not be duplicated by the Engineer for others without express permission.

c. Engineers, before undertaking work for others in connection with which the Engineer may make improvements, plans, designs, inventions, or other records that may justify copyrights or patents, should enter into a positive agreement regarding ownership.

d. Engineers' designs, data, records, and notes referring exclusively to an employer's work are the employer's property. Employer should indemnify the Engineer for use of the information for any purpose other than the original purpose.

As Revised July 1996

"By order of the United States District Court for the District of Columbia, former Section 11(c) of the NSPE Code of Ethics prohibiting competitive bidding, and all policy statements, opinions, rulings or other guidelines interpreting its scope, have been rescinded as unlawfully interfering with the legal right of engineers, protected under the antitrust laws, to provide price information to prospective clients; accordingly, nothing contained in the NSPE Code of Ethics, policy statements, opinions, rulings or other guidelines prohibits the submission of price quotations or competitive bids for engineering services at any time or in any amount."

Statement by NSPE Executive Committee

In order to correct misunderstandings which have been indicated in some instances since the issuance of the Supreme Court decision and the entry of the Final Judgment, it is noted that in its decision of April 25, 1978, the Supreme Court of the United States declared: "The Sherman Act does not require competitive bidding."

It is further noted that as made clear in the Supreme Court decision:

1. Engineers and firms may individually refuse to bid for engineering services.
2. Clients are not required to seek bids for engineering services.
3. Federal, state, and local laws governing procedures to procure engineering services are not affected, and remain in full force and effect.
4. State societies and local chapters are free to actively and aggressively seek legislation for professional selection and negotiation procedures by public agencies.
5. State registration board rules of professional conduct, including rules prohibiting competitive bidding for engineering services, are not affected and remain in full force and effect. State registration boards with authority to adopt rules of professional conduct may adopt rules governing procedures to obtain engineering services.
6. As noted by the Supreme Court, "nothing in the judgment prevents NSPE and its members from attempting to influence governmental action . . ."

NOTE: In regard to the question of application of the Code to corporations vis-à-vis real persons, business form or type should not negate nor influence conformance of individuals to the Code. The Code deals with professional services, which services must be performed by real persons. Real persons in turn establish and implement policies within business structures. The Code is clearly written to apply to the Engineer and items incumbent on members of NSPE to endeavor to live up to its provisions. This applies to all pertinent sections of the Code.

Reprinted by courtesy of the

National Society of Professional Engineers
1420 King Street
Alexandria, Virginia 22314-2794
703/684-2800
Fax: 703/836-4875
NSPE World Wide Web site: http://www.nspe.org
Publication date as revised: July 1996
Publication #1102

REFERENCES

Abbatt, J. P. D., and Molina, M. J., "Status of Stratospheric Ozone Depletion," *Annual Review of Energy and the Environment,* vol. 18, (1993).

Acohido, B., "El-Al Boeing 747-200 Crash, Amsterdam, The Netherlands," in N. Schlager (ed.), *When Technology Fails,* Detroit, MI: Gale Research, (1994a).

_____, "United Airlines Boeing 747 Explosion, Hawaii," in N. Schlager (Ed.), *When Technology Fails.* Detroit, MI: Gale Research, (1994b).

Adams, J. L., *Conceptual Blockbusting: A Guide to Better Ideas.* San Francisco: W. H. Freeman, 1974.

Alger, J. R. M., and Hays, C. V., *Creative Synthesis in Design.* Englewood Cliffs, NJ: Prentice-Hall, 1964.

Anderberg, B., "Explosive Remnants of War on Land: Technical Aspects of Disposal," in Westing, A. H. (Ed.), *Explosive Remnants of War,* 1985.

Anderson, D. M., *Design for Manufacturability.* Lafayette, CA: CIM Press, 1990.

Anderson, R. M., et al., *Divided Loyalties.* West Lafayette, IN: Purdue Univ. Press (1980).

Applegate, J., "An Itch to Know More Leads to Success," *Los Angeles Times,* D3:1, Feb. 12, 1993.

Ardis, S. B., *An Introduction to U.S. Patent Searching: The Process.* Englewood, CO: Libraries Unlimited, 1991.

Arnett, E. C., "However This Cookie Crumbles, Recipe's Now All Duncan Hines, *Washington Post,* C 1:2, Sept. 13, 1989.

Atchison, J. E., "Nonwood Fiber Could Play Major Role in Future U.S. Papermaking Furnishes," *Pulp and Paper,* (July 1995).

_____, "Making the Right Choices for Successful Bagasse Newsprint Production: Part 1," *TAPPI Journal*, vol. 75, no. 12, (Dec. 1992).

Baile, R., "Paper Mill Offers Sludge Recycling Plan," *Boston Globe* (July 15, 1990).

Barach, A. B., *Famous American Trademarks.* Washington, DC: Public Affairs Press, 1971.

Baram, M. S., "Trade Secrets: What Price Loyalty?" *Harvard Business Review,* Nov. Dec. 1968.

Basic Facts About Patents. Washington, DC: U.S. Patent & Trademark Office.

Baumeister, T., and Marks, L. S., *Standard Handbook for Mechanical Engineers*, 7th ed. New York: McGraw-Hill, 1967.

Bayes, K., and Franklin, S., *Designing for the Handicapped*, London: G. Godwin, 1971.

Behr, P., "An Idea That Sparked a Workplace Revolution," *Washington Post,* WBIZ, 9:1, Feb. 13, 1995.

Bennett, E., Degan, J., and Spiegel, J., *Human Factors in Technology.* New York: McGraw-Hill, 1963.

Berger, R. W., "Achieving Quality," in Veilleux, R. F., and Petro, L. W. (Eds.), *Tools and Manufacturing Engineers Handbook: Manufacturing Management,* vol. 5, Dearborn, MI: Society of Manufacturing Engineers, 1988.

Berget, A., *The Conquest of the Air.* London: Heinemann, 1909.

Bickford, W. R., *A First Course in the Finite Element Method,* 2nd ed. Burr Ridge, IL: Irwin, 1991.

Biermann, C. J., *Handbook of Pulping and Papermaking,* 2nd ed., Boston: Academic Press, 1996.

Bignell, V., Peters, G., and Pym, C., *Catastrophic Failures.* Milton Keynes, England: The Open Univ. Press, 1997.

Billah, K. Y., and Scanlon, R. H., "Resonance, Tacoma Narrows Bridge Failure, and Undergraduate Physics Textbooks," *American Journal of Physics,* vol. 59, no. 2, Feb.

Bisplinghoff, R. L., "Why Did Flight 191 Lose an Engine?" *Mechanical Engineering,* May 1986.

Blackburn, D., and Holister, G., *G. K. Hall Encyclopedia of Modern Technology.* Boston, MA: G. K. Hall, 1987.

Block, V., "Accubraille Offering Credit Cards for Blind," *American Banker,* 18:1, July 14, 1994.

Boffey, P. M., "Teton Dam Verdict: Foul-up by the Engineers," *Science,* vol. 195, 21 Jan. 1977.

Boring, M., *Incredible Constructions and the People Who Built Them.* New York: Walker, 1984.

Breedon, R. L. (Ed.), *Those Inventive Americans.,* Washington, DC: The National Geographic Society, 1971.

Broad, W. J., "Space Errors Share Pattern: Skipped Tests," *New York Times,* C 1:4, June 11, 1991.

Brodeur, P., *Outrageous Misconduct: The Asbestos Industry on Trial,* New York: Pantheon, 1985.

Brown, A. E., and Jeffcott, H. A. Jr., *Beware of Imitations!* New York: Viking, 1932; reprinted in 1970 as *Absolutely Mad Inventions,* New York: Dover.

Brown, B., "Details of the Failure of a 90-Foot Molasses Tank," *Engineering News-Record,* May 1919.

Brown, T., *Historical First Patents: The First U.S. Patents for Many Everyday Things.* Metuchen, NY: Scarecrow, 1994.

Bryce, J. R. G., "Alkaline Pulping" in Casey, J. P., (Ed.), *Pulp and Paper: Chemistry and Chemical Technology,* 3rd ed., vol. 1. New York: Wiley, 1980.

Bryce, R., " 'Green' Lumber Ties Forest Products, Environmentalists," *Christian Science Monitor,* 9:1, Apr. 12, 1994.

Bryner, G. C., *Blue Skies, Green Politics: The Clean Air Act of 1990* Washington, DC: CQ Press, 1993.

Bublitz, W. J., "Pulpwood" in Casey, J. P., (Ed.), *Pulp and Paper: Chemistry and Chemical Technology,* 3rd ed., vol. 1, New York: Wiley, 1980.

Bueche, F., and Wallach, D. L., *Technical Physics,* 4th ed., New York: Wiley, 1994.

Burgess, J., "The Newton: PC Prophesy, or a Pratfall?" *Washington Post,* H1:1, Oct. 24, 1994.

Calvert, J. G., et al., "Achieving Acceptable Air Quality: Some Reflections on Controlling Vehicle Emissions." *Science,* vol. 261, 2 July 1993.

Camps, F., "Warning an Auto About an Unsafe Design," in A. F. Westin (Ed.), *Whistle-Blowing! Loyalty and Dissent in the Corporation.* New York: McGraw-Hill, 1981.

Carlton, J., "Apple's Sales Data Suggest to Analysts That New Message Pad Is Floundering," *Wall Street Journal,* B 3:1, Jan. 25, 1994.

Carlyle, T., *On Heroes, Hero-Worship and the Heroic in History,* London: Oxford Univ. Press, 1956.

Carr, F. K., *Patents Handbook,* Jefferson, NC: McFarland, 1995.

Carter, C. M., "Trade Secrets and the Technical Man," *IEEE Spectrum,* vol. 6, no. 2, Feb., 1969.

Cartwright, D. J., "Designing Against Disaster," in J. G. Truxal (Ed.), *Quantitative Examples* (Monograph Series of the New Liberal Arts Program). Stony Brook, NY: Research Foundation of State Univ. of New York, Stony Brook, 1991.

Cassady, S., *Spanning the Gate.* Mill Valley, CA: Squarebooks, 1979.

Cassidy, J., *The Aerobie Book.* Palo Alto, CA: Klutz Press, 1989.

Champanis, A., and Lindenbaum, L. E., "A Reaction Time Study of Four Control-Display Linkages," *Human Factors,* vol. 1, no. 4, 1959.

Chang, T. Y., et al., "Urban and Regional Ozone Air Quality: Issues Relevant to the Automobile Industry," *Critical Reviews in Environmental Control,* vol. 22, nos. 1 & 2, 1992.

Chapman, W. L., Bahill, A. T., and Wymore, A. W., *Engineering Modeling and Design.* Boca Raton, FL: CRC Press, 1992.

"A Cheaper Method for Paper Making" *New York Times,* Oct. 25, 1993.

Chemtech, vol. 13, no. 11, 1983, p. 655.

Clark, D., "Intel Balks at Replacing Pentium Chip Without Asking Owners Any Questions," *Wall Street Journal,* A3: 2, Dec. 14, 1994.

Claypool, J., *Manufacturing,* New York: Franklin Watts, 1984.

Clothier, W., "Good Models Never Die," *Design Graphics World,* (Dec. 1986).

Coffin, D. L., and Stokinger, H. E., "Biological Effects of Air Pollutants," in Stern, A. C., (Ed.), *Air Pollution,* 3rd ed., New York: Academic Press, 1977.

Cohen, I. B., "Florence Nightingale," *Scientific American,* Mar. 1984.

Cone, R. J., *How the New Technology Works: A Guide to High-Tech Concepts.* Phoenix, AZ: Oryx Press, 1991.

Copen, M. D., et al., "Design of Concrete Dams," in A. R. Golze (Ed.), *Handbook of Dam Engineering.* New York: Van Nostrand Reinhold, 1977.

Copp, N., and Zanella, A., *Discovery, Innovation, and Risk: Case Studies in Science and Technology.* Cambridge, MA: MIT Press, 1993.

Cramer, G. L., and Jensen, C. W., *Agricultural Economics and Agribusiness,* 5th ed., New York: Wiley, 1991.

Crane, H. R., "The Quartz Analog Watch: A Wonder Machine," *Physics Teacher,* vol. 31, Nov. 1993.

Cross, M., and Moscardini, A. P., *Learning the Art of Mathematical Modeling,* New York: Wiley, 1985.

Cross, N., *Engineering Design Methods.* New York: Wiley, 1989.

Crow, K. A., "Concurrent Engineering," in Bakerjian, R. and Mitchell, P. (Eds.), *Tools and Manufacturing Engineers Handbook: Design for Manufacturability,* vol. 6, Dearborn, MI: Society of Manufacturing Engineers, 1992.

Cullen, F. T., Maakestad, W. J., and Cavender, G., *Corporate Crime Under Attack: The Ford Pinto Case and Beyond.* Cincinnati, OH: Criminal Justice Studies, 1987.

Damon, A., Stoudt, H. W., and McFarland, R. A., *The Human Body in Equipment Design.* Cambridge: Harvard Univ. Press, 1966.

Daniell, W. F., "Stone Groundwood" in Casey, J. P., (Ed.), *Pulp and Paper: Chemistry and Chemical Technology,* 3rd ed., vol. 1, New York: Wiley, 1980.

Darr, K., and Rakich, J. S., *Hospital Organization and Management: Text and Readings.* Owings Mills, MD: National Health, 1989.

David, F. W., *Experimental Modeling in Engineering.* Boston, MA: Butterworth, 1982.

Davidson, J., "Design: Days of Irksome Childproof Caps Are Numbered," *Wall Street Journal,* B 1: 6, June 14, 1995.

Davie, M., *The Titanic,* London: The Bodley Head, 1986.

Davy, M. J. B., *Aeronautics: Heavier-Than-Air Aircraft, Part II.* London: Her Majesty's Stationery Office, 1949.

Day, K., "Pentium Chip Had Flaws, Intel Says," *Washington Post,* D 2:1, Nov. 24, 1994.

d'Estaing, V.-A. G., *The World Almanac Book of Inventions.* New York: World Almanac, 1985.

d'Estaing, V.-A. G., and Young, M., *Inventions and Discoveries 1993.* New York: Facts On File, 1993.

Dieter, G. E., *Engineering Design: A Materials and Processing Approach,* 2nd ed., New York: McGraw-Hill, 1991.

Diffrient, N., Tilley, A. R., and Bardagjy, J. C., *Humanscale 1/2/3 Manual.* Cambridge, MA: MIT Press, 1974.

Doebelin, E. O., *Systems Modeling and Response,* New York: Wiley, 1980.

Dreyfuss, H., *The Measure of Man.* New York: Whitney Library of Design, 1967.

Duncker, K., "On Problem Solving," *Psychological Monographs,* vol. 58, no. 5, Comprising the Whole (entire issue) no. 270, 1945.

Durbin, E. J. "How to Spur Our Economy, Reduce Atmospheric Polution and Enhance National Security," *Natural Gas Fuels,* (June 1993).

Durbin, E. J., and Schemmann, G. S., "NGV Conversions: Increasing the Market," *Natural Gas Fuels* (Oct. 1995).

Earle, J. H., *Engineering Design Graphics,* 5th ed., Reading, MA: Addison-Wesley, 1987.

Eddy, P., Potter, E., and Page, B., *Destination Disaster: From the Tri-Motor to the DC-10, The Risk of Flying,* New York: Quadrangle, 1976.

Edel, D. H., Jr. (Ed.), *Introduction to Creative Design.* Englewood Cliffs, NJ: Prentice-Hall, 1967.

Emshwiller, J. R., and McCarthy, M. J., "Sticky Situation: Coke's Soda Fountain for Offices Fizzles, Dashing High Hopes," *Wall Street Journal,* A 1: 6, June 14, 1993.

Energy Information Administration, Monthly Energy Review (any issue), Washington, DC, 1993.

Engelhard Corp., *PremAir Catalyst Systems Fact Sheet.* Iselin, NJ: Engelhard Corp., 1995.

"Engineer's Afterthought Sets Welders to Work Bracing Tower," *Engineering News Record,* (Aug. 17, 1978).

EPA (U.S. Environmental Protection Agency) *The 1985 NAPAP Emissions Inventory. (Version 2): Development of the Annual Data and Modeler's Tapes.* EPA-600/7-89-012a. Research Triangle Park, NC: U.S. EPA, Office of Research and Development (Nov. 1989).

EPA *Aerometric Information Retrieval System (AIRS)* Research Triangle Park, NC: U.S. Environmental Protection Agency, 1991.

Ertas, A., and Jones, J. C., *The Engineering Design Process.* New York: Wiley, 1993.

Farquharson, F. B., "Collapse of the Tacoma Narrows Bridge," *Scientific Monthly,* Dec. 1940.

Farquharson, F. B. (Ed.), *Aerodynamic Stability of Suspension Bridges,* Univ. of Washington Engineering Experimental Stations Bull., no. 116, parts 1–5 (1949–1954).

Feder, B. J., "A Can Top for a New Day?" *New York Times,* 3, 8: 5, Oct. 30, 1994.

Feigenbaum, A. V., *Total Quality Control,* 3rd ed., New York: McGraw-Hill, 1983.

Feinberg, R., *Peculiar Patents.* New York: Carol Publishing Group, 1994.

Feldman, A., and Ford, P., *Scientists and Inventors.* London: Aldus Books, 1979.

Ferguson, E. S., *Engineering and the Mind's Eye,* Cambridge, MA: MIT Press, 1992.

Ferris, B., *How to Fly,* London: Nelson, 1953.

Fisher, L. M., "Intel Earnings Decline 37% on Charge for Pentium Flaw," *New York Times,* D 4: 3, Jan. 18, 1995.

Fitzgerald, N., "Teaching with Cases," *ASEE Prism.* Washington, DC: American Society for Engineering Education, (Mar. 1995).

Flatow, I., *Rainbows, Curve Balls and Other Wonders of the Natural World Explained.* New York: Harper & Row, 1988.

_____, *They All Laughed . From Light Bulbs to Lasers: The Fascinating Stories Behind the Great Inventions That Have Changed Our Lives.* New York: HarperCollins, 1992.

Fletcher, N. H., "Mechanics of Flight," *Physics Education,* (July 1975).

Fogler, H. S., and Le Blanc, S. E., *Strategies for Creative Problem Solving,* Englewood Cliffs, NJ: Prentice-Hall, 1995.

Ford, D. N., "MGM Grand Hotel Fire," in N. Schlager (Ed.), *When Technology Fails,* Detroit, MI: Gale Research, 1994.

Foss, C. F., and Gander, T. J. (Eds.), *Jane's Military Vehicles and Logistics,* 16th ed. Alexandria, VA: Jane's, 1995.

Fowler, F. D., "Failure to Warn: A Product Design Problem," in *Proceedings of the Symposium: Human Factors and Industrial Design in Computer Products,* Medford, MA: Tufts Univ., 1980.

Frank, N., "Murder in the Workplace," in S. L. Hills (Ed.), *Corporate Violence,* Totowa, NJ: Rowman & Littlefield, 1987.

Franz, W. L., and Child, J. S., Jr., "Good Habits Before Filing a Patent Application," *IEEE Transactions on Professional Communication,* PC-22, June 1979.

French, T. E., Vierck, C. J., and Foster, R. E., *Engineering Drawing.* New York: McGraw-Hill, 1992.

Frye, R., The Great Molasses Flood," *Reader's Digest,* Aug. 1955.

Fuller, M. L., *Inventors Guidebook.* Tarzana, CA: ILMA, 1984.

Gajda, W. J., Jr., and Biles, W. E., *Engineering: Modeling and Computation.* Boston, MA: Houghton Mifflin, 1978.

Gander, T., *Guerilla Warfare Weapons.* New York: Sterling, 1989.

General Information Concerning Patents, Washington, DC: U.S. Patent & Trademark Office.

Gibbs-Smith, C. H., *A History of Flying.* London: Batsford, 1953.

Giesecke, F. E., et al., *Engineering Graphics,* 5th ed. New York: MacMillan, 1993.

Glasser, W. G., "Lignin" in Casey, J. P., (Ed.), *Pulp and Paper: Chemistry and Chemical Technology,* 3rd ed., vol. 1, New York: Wiley, 1980.

"The Global Landmine Crisis," Hearing before a Subcommittee of the Committee on Appropriations—United States Senate, One Hundred Third Congress, Second Session, Superintendent of Documents #Y4.AP 6/2: S.HRG. 103-666 (May 13, 1994).

Godson, J., *The Rise and Fall of the DC-10.* New York: David McKay, 1975.

Gold, M., "Who Pulled the Plug on Lake Peigneur?" *Science,* vol. 81, Nov. 1981.

Gold, R. J., *Eureka! The Entrepreneurial Inventor's Guide to Developing, Protecting, and Profiting from Your Ideas.* Englewood Cliffs, NJ: Prentice-Hall, 1992.

Goldsmith, S., *Designing for the Disabled.* 2nd ed., New York: McGraw-Hill Book, 1967.

Goldstein, S. H., and Rubin, R. A., "Engineering Ethics," *Civil Engineering,* Oct. 1996.

"Goodyear Proposes Expandable Structures as Space Stations," *Missiles and Rockets* **8** (May 29, 1961).

Gordon, W. J. J., *Synectics.* New York: Harper & Row, 1961.

Grandjean, E., *Ergonomics of the Home.* London: Taylor & Francis, 1973.

_____, *Fitting the Task to the Man.* London: Taylor & Francis, 1971.

Greer, T. J., *Writing and Understanding U.S. Patent Claims.* New York: Michie Bobbs-Merrill, 1979.

Guide on Professionalism and Ethics in Engineering Curricula, Accreditation Board for Engineering and Technology, July 1989.

Haber, L., *Black Pioneers of Science and Invention,* New York: Harcourt, Brace & World, 1970.

Hagler, R. W., "The Global Wood Fiber Equation- A New World Order?" *TAPPI Journal,* vol. 79, no. 1, (Jan. 1996).

Halamka, J. D., *Espionage in the Silicon Valley.* Berkeley, CA: Sybex, 1984.

Hamilton, D. H., and Meehan, R. L., "Ground Rupture in the Baldwin Hills," *Science,* vol. 172, 23 Apr. 1971.

Hamilton, D. P., "Japanese Focus on Simpler Camcorders," *Wall Street Journal,* B1: 4, May 3, 1993.

Hamilton, J. A. B., *Britain's Greatest Rail Disaster,* London: Allen & Unwin, 1969.

Handicap International, "Landmines Special Report," Internet (Jan. 16, 1996).

Harper, R. F. (Trans.),*The Code of Hammurabi.* Chicago: Univ. of Chicago Press, 1904.

Haskins, J., *Outward Dreams: Black Inventors and Their Inventions.* New York: Walker, 1991.

Havener, L., "Improving the Quality of Quality," *Quality Progress,* 26: 11, 1993.

Hewitt, P. G., *Conceptual Physics.* A New Introduction to Your Environment, 2nd ed., Boston, MA: Little, Brown, 1974.

Higgins, J. S., et al., "Identifying and Solving Problems in Engineering Design," *Studies in Higher Education,* vol. 14, no. 2, 1989.

Hill, G. C., "First Hand-Held Data Communicators Are Losers, But Makers Won't Give Up," *Wall Street Journal,* B1:3, Feb. 3, (1994a).

_____, "Getting Personal," *Wall Street Journal,* R 6:1, Feb. 11 (1994b).

_____, "Computers: Despite Furor, Most Keep Their Pentium Chips," *Wall Street Journal,* B 1: 3, Apr. 13, 1995

Hill, P. H., et al., *Making Decisions: A Multidisciplinary Introduction.* Reading, MA: Addison-Wesley, 1979.

Hiltzig, M. A., "Case Gives Rare Glimpse of Silicon Valley Intrigue," *Los Angeles Times*, part I, p. 1, 22 Mar. 1982.

Hine, T., *The Total Package: The Evolution and Secret Meanings of Boxes, Bottles, Cans, and Tubes.* Boston, MA: Little, Brown, 1995.

Holmes, G. S., "Baldwin Hills Dam Failure, Los Angeles, California," in N. Schlager (Ed.), *When Technology Fails.* Detroit, MI: Gale Research, 1994.

Horn, R. A., Wegner, T. H., and Kugler, D. E., "Newsprint from Blends of Kenaf CTMP and Deinked Recycled Newsprint," *TAPPI Journal,* vol. 75, no. 12, (Dec. 1992).

Holusha, J., "Paper Maker Turns a Cleaner Page," *New York Times* (Oct. 20, 1993).

Horvath, S. M., and McKee, D. J., "Acute and Chronic Health Effects of Ozone" in McKee, D., (Ed.), *Tropospheric Ozone: Human Health and Agricultural Impacts.* Ann Arbor: Lewis; CRC Press, 1994.

The How It Works Encyclopedia of Great Inventors and Discoveries. London: Marshall Cavendish Books, 1978.

How Things Are Made. Washington, DC: National Geographic Society, 1981.

Hunter, D., *Papermaking through Eighteen Centuries.* New York: William Edwin Rudge, 1930.

Hunter, T. A., *Engineering Design for Safety.* New York: McGraw-Hill, 1992.

Husted, B., "Little Chip Makes Big Problems for Intel," *Atlanta Constitution,* E 1: 2, Dec. 13, 1994.

Huxley, E., *Florence Nightingale.* New York: G. P. Putnam's Sons, 1975.

Ingalls, L., "Toxic Vapor Leak, Bhopal, India," in N. Schlager (Ed.), *When Technology Fails.* Detroit, MI: Gale Research, 1994.

International Energy Agency, *Electric Vehicles: Technology, Performance, and Potential.* Paris: Organisation for Economic Co-operation and Development, 1993.

ISO 9000, Geneva, Switzerland: International Organization for Standardization, 1992.

Jackson, D. C., *Building the Ultimate Dam: John S. Eastwood and the Control of Water in the West.* Kansas: Univ. Press of Kansas, 1995.

James, P. P., *The Real McCoy: African-American Invention and Innovation, 1619–1930.* Washington, DC: Smithsonian, 1989.

"Japan Does It Again with Fax Machines," *Boston Globe,* 31: 1, July 3, 1989.

Jefferson, D. J., "Building a Better Mousetrap Doesn't Ensure Success," *Wall Street Journal,* B 2: 3, Nov. 18, 1991.

Jessup, W. E., "Baldwin Hills Dam Failure," *Civil Engineering,* Feb. 1964.

Jewkes, J., Sawers, D., and Stillerman, R., *The Sources of Invention,* 2nd ed. New York: Norton, 1969.

Joenk, R. J., "Patents: Incentive to Innovate and Communicate—An Introduction," *IEEE Transactions on Professional Communication,* PC-22, June 1979.

Jones, S. V., *Inventions Necessity Is Not the Mother of: Patents Ridiculous and Sublime.* New York: Quadrangle/New York Times Book Co., 1973.

Jones, T., and Sullivan, B., "Pentium Flap Shakes Consumers—and Intel," *Chicago Tribune,* 1, 1: 2, Dec. 14, 1994.

Jordanoff, A., *Your Wings and How to Win Them.* New York: Funk & Wagnalls, 1936.

Judson, H. F., *The Search for Solutions.* Baltimore, MD: Johns Hopkins Univ. Press, 1987.

Kaldor, A. F., Brasher, B. S., and Fuller, M. J., "A Strategy for the Development of a Kenaf-Based Pulp and Paper Industry," *TAPPI Journal,* vol. 75, no. 1, Jan. 1992.

Kalpakjian, S., *Manufacturing Engineering and Technology,* Reading, MA: Addison-Wesley, 1989.

Kaminetzky, D., *Design and Construction Failures: Lessons from Forensic Investigations.* New York: McGraw-Hill, 1991.

Kamm, L. J., *Real-World Engineering: A Guide to Achieving Career Success.* New York: IEEE, 1991.

Kantowitz, B. H., and Sorkin, R. D., *Human Factors: Understanding People–System Relationships,* New York: Wiley, 1983.

Keeton, W. P., et al., *Products Liability and Safety: Case and Materials.* 2nd ed., Westbury, NY: Foundation Press, 1989.

Kelly, J. E., and Park, W. R., *The Tunnel Builders.* Reading, MA: Addison-Wesley, 1976.

Kepner, C. H., and Tregoe, B. B., *The New Rational Manager.* Princeton, NJ: Princeton Univ. Press, 1981.

Kivenson, G., *The Art and Science of Inventing.* New York: Van Nostrand Reinhold, 1977.

Kletz, T. A., *Cheaper, Safer Plants,* 2nd ed. Rugby, UK: Institution of Chemical Engineers, 1985.

———, *Learning from Accidents in Industry.* London: Butterworth, 1988.

———, *An Engineer's View of Human Error,* 2nd ed. Rugby, Warwickshire, UK: Institution of Chemical Engineers, 1991.

———, *HAZOP and HAZAN: Identifying and Assessing Process Industry Hazards,* 3rd ed. Rugby, Warwickshire, UK: Institution of Chemical Engineers, 1992.

———, *Analog Science Fiction* (Jan. 1994).

Knickerbocker, B., "Chlorine-Free Milling: Making Paper Safer for the Environment," *Christian Science Monitor* (Apr. 12, 1994).

Koff, R. M., *How Does It Work?* New York: Signet; New American Library, 1961.

Kolarik, W. J., *Creating Quality: Concepts, Systems, Strategies, and Tools,* New York: McGraw-Hill, 1995.

Konz, S., *Work Design: Industrial Ergonomics,* 2nd ed., Columbus, OH: Grid, 1983.

Korman, R., "LeMessurier's Confession," *Engineering News Record,* Oct. 30, 1995.

Kugler, D. E., *Kenaf Newsprint: Realizing Commercialization of a New Crop After Four Decades of Research and Development.* Special Projects and Programs Systems, Cooperative State Research Service, U.S. Dept. of Agriculture (June 1988).

Kuo, T. C., and Zhang, H. C., "Design for Manufacturability and Design for 'X': Concepts, Applications, and Perspectives," *Proceedings of the 17th International IEEE/CPMT Electronics Manufacturing Technology Symposium,* Austin, TX, 1995.

Kupchella, C. E., and Hyland, M. C., *Environmental Science: Living Within the System of Nature.* Boston: Allyn & Bacon, 1986.

Kurdin, J. A., "Refiner Mechanical and Thermomechanical Pulping," in Casey, J. P., (Ed.), *Pulp and Paper: Chemistry and Chemical Technology.* 3rd ed., vol. 1, New York: Wiley, 1980.

Landis, R. B., *Studying Engineering: A Road Map to a Rewarding Career.* Burbank, CA: Discovery, 1995.

Langan, F., "Paper Made Cheaper and Cleaner: Canadian Firm Uses Alcohol Rather Than Sulphur in Pulp Process." *Christian Science Monitor* (Oct. 23, 1995).

Lasson, K., *Mousetraps and Muffling Cups: One Hundred Brilliant and Bizarre U.S. Patents.* Arbor House, 1986.

Lave, C. A., and J. G. Mar., *An Introduction to Models in the Social Sciences,* New York: Harper & Row, 1975.

Learmont, D., *Flight International,* vol. 42 (Jan. 1990).

LeCornu, J., *La Navigation Aerienne,* 6th ed. Paris: Librairie Vuibert, 1913.

Lee, D. D., *Sergeant York: An American Hero.* Lexington, KY: Univ. Press of Kentucky, 1985.

Lee, J. H., Swan, D. H., and Lalk, T. R., "A Spreadsheet Model for Air Fuel Cell Stacks," in *Electric Vehicle Power Systems: Hybrids, Batteries, Fuel Cells,* Warrendale, PA: Society of Automotive Engineers, 1993.

Lee, K., *Principles of CAD/CAM/CAE Systems.* Reading, MA: Addison Wesley Longman, in press.

Levy, M. and Salvadori, M., *Why Buildings Fall Down: How Structures Fail.* New York: Norton, 1992.

Levy, R. C., *Inventing and Patenting Sourcebook: How to Sell and Protect Your Ideas,* Detroit, MI: Gale Research, 1990.

Lewis, A., *Super Structures.* New York: Viking, 1980.

Lewis, P. H., "So Far, The Newton Experience Is Less Than Fulfilling," *New York Times,* 3, 10:1, Sept. 26, 1993.

Lewis, P. H., "IBM Deals Blow to a Rival as It Suspends Pentium Sales," *New York Times,* A 1: 4, Dec. 13, 1994.

Lindbeck, J. R., *Product Design and Manufacture.* Englewood Cliffs, NJ: Prentice-Hall, 1995.

Loras, V., "Bleaching," in Casey, J. P., (Ed.), *Pulp and Paper: Chemistry and Chemical Technology*, 3rd ed., vol. 1. New York: Wiley, 1980.

Lord, W., *A Night to Remember.* New York: Holt, 1976.

Lumsdaine, E., and Lumsdaine, M., *Creative Problem Solving: An Introductory Course for Engineering Students.* New York: McGraw-Hill, 1990.

Luzadder, W. J., and Duff, J. M., *Fundamentals of Engineering Drawing, 10th ed.,* Englewood Cliffs, NJ: Prentice-Hall, 1989.

Macaulay, D., *The Way Things Work,* Boston, MA: Houghton Mifflin, 1988.

MacDonald, A. L., *Feminine Ingenuity: How Women Inventors Changed America.* New York: Ballantine, 1992.

Machlis, S., et al., "Engineering Megatrends," *Design News,* vol. 50, no. 16, Aug. 28, 1995.

MAFES (Mississippi Agricultural and Forestry Experiment Station), Vernor Hurt, *A Summary of Kenaf Production and Product Development Research 1989–1993,* Bull. 1011, Mississippi State, MS (May 1994).

Mark, R., "American Airlines DC-10 Crash, Chicago, Illinois," in N. Schlager (Ed.), *When Technology Fails.* Detroit, MI: Gale Research, (1994a).

_____, "United Airlines DC-10 Crash, Sioux City, Iowa," in N. Schlager (Ed.), *When Technology Fails.* Detroit, MI: Gale Research, (1994b).

Markoff, J., "Marketer's Dream, Engineer's Nightmare," *New York Times,* 3, 1:4, Dec. 12, 1993.

_____, "Apple's Newton Reborn: Will It Still the Critics?" *New York Times,* D 1:3, Mar. 4 (1994a).

_____, "In About Face, Intel Will Swap Its Flawed Chip," *New York Times,* A 1: 1, Dec. 21 (1994b).

_____, "Intel's Crash Course on Consumers," *New York Times,* D 1: 3, Dec. 21 (1994c).

Martin, F. F., *Computer Modeling and Simulation.* New York: Wiley, 1968.

Martin, J. C., *The Successful Engineer: Personal and Professional Skills–A Sourcebook.* New York: McGraw-Hill, 1993.

Martin, M. W., and Schinzinger, R., *Ethics in Engineering,* 2nd ed. New York: McGraw-Hill, 1989.

Marx, W., *Acts of God, Acts of Man.* New York: Coward, McCann & Geoghegan, 1977.

May, M., "Unsteady As She Goes," *American Scientist,* vol. 81, (Mar.-Apr. 1993).

McCormick, E. J., *Human Factors Engineering,* 3rd ed., New York: McGraw-Hill, 1970.

McFarland, M. W. (Ed.), *The Papers of Wilbur and Orville Wright.* New York: McGraw-Hill, 1953.

McGill, DC, "7-Up Gold: The Failure of a Can't-Lose Plan," *New York Times,* A 35: 4, Feb. 11, 1989.

McGinnis, G. D., and Shafizadeh, F., "Cellulose and Hemicellulose" in Casey, J. P., (Ed.), *Pulp and Paper: Chemistry and Chemical Technology*, 3rd ed., vol. 1, New York: Wiley, 1980.

McGovern, J. N., "Introduction to Pulping" in Casey, J. P., (Ed.), *Pulp and Paper: Chemistry and Chemical Technology*, 3rd ed., vol. 1. New York: Wiley, 1980.

McGraw-Hill Encyclopedia of Science and Technology, 7th ed., New York: McGraw-Hill, 1992.

McKechnie, J. L., (Supervising Ed.), *Webster's New Universal Unabridged Dictionary,* 2nd ed., New York: Simon & Schuster, 1983.

McKee, D. J., (Ed.), *Tropospheric Ozone: Human Health and Agricultural Impacts.* Ann Arbor, MI: Lewis; CRC Press, 1994.

Mennonite Central Committee, *Laos; War Legacy* and *Laos: Safe Villages.* Akron, PA: Mennonite Central Committee, 1995.

The Merriam-Webster Dictionary. Springfield, MA: Merriam-Webster, 1994.

Middendorf, W. H., *Design of Devices and Systems,* 2nd ed. (revised and expanded). New York: Marcel Dekker, 1990.

Miller, A. R., and Davis, M. H., *Intellectual Property: Patents, Trademarks, and Copyrights in a Nutshell.* St Paul, MN: West, 1983.

Millman, N., "Chipmaker Reels as IBM Torpedoes High-Risk 'Intel Inside' Campaign," *Chicago Tribune,* 5, 2: 3, Dec. 20, 1994.

Mischke, C. R., *Mathematical Model Building,* 2nd ed. Ames, IA: Iowa State Univ. Press, 1980.

Misra, D. K., "Pulping and Bleaching of Nonwood Fibers," in Casey, J. P., (Ed.), *Pulp and Paper: Chemistry and Chemical Technology,* 3rd ed., vol. 1. New York: Wiley, 1980.

Morgan, C. T., et al., *Human Engineering Guide to Equipment Design.* New York: McGraw-Hill, 1963.

Morgan, H., *Symbols of America.* New York: Steam Press, 1986.

Morgenstern, J., "The Fifty-Nine Story Crisis," *The New Yorker,* May 29, 1995.

Mungall, C., and McLaren, D. J., (Eds.), *Planet Under Stress: The Challenge of Global Change.* New York: Oxford Univ. Press, 1990.

Myers, G. C., and Bagby, M. O., "Feasibility of Using Kenaf Chemithermomechanical Pulp in Printing and Writing Paper," *TAPPI Journal,* vol. 78, no. 5, May 1995.

Nakamura, Y., and Fukamachi, N., "Visualization of the Flow Past a Frisbee," *Fluid Dynamics Research* (North Holland), vol. 7, 1991.

National Inventors Hall of Fame. Washington, DC: U.S. Patent & Trademark Office, 1984.

NEA (National Education Association), Dept. of Supervision and Curriculum Development, *Building America,* vol. 4. New York: American Corp., 1942.

Newhouse, J., *The Sporty Game.* New York: Knopf, 1982.

Nightingale, F., *Notes on Hospitals,* 3rd ed. London: Longman, Green, Longman, Roberts, and Green, 1863; Reprinted in *Florence Nightingale on Hospital Reform,* C. E. Rosenberg (Ed.), New York: Garland, 1989.

Norman, D. A., *The Psychology of Everyday Things.* New York: Basic, 1988.

NRC (National Research Council), *Rethinking the Ozone Problem in Urban and Regional Air Pollution.* Washington, DC: National Academy Press, 1991.

Nussbaumer, H. J., "Patents and the Engineer," *IEEE Transactions on Professional Communication,* PC-22, June 1979.

Official Gazette of the U.S. Patent & Trademark Office. Washington, DC: U.S. Government Printing Office.

Osborn, A. F., *Applied Imagination.* New York: Scribner's, 1957.

Parnes, S. J., *Creative Behavior Guidebook.* New York: Scribner's, 1967.

Parnes, S. J., Noller, R. B., and Biondi, A. M., *Guide to Creative Action* (rev. ed. *Creative Behavior Guidebook).* New York: Scribner's, 1977.

Penenberg, A. L., "Slick Solutions to an Environmental Scourge." *New York Times,* 3, 11: 1, Aug. 15, 1993.

Perrow, C., *Normal Accidents: Living with High-Risk Technologies.* New York: Basic, 1984.

Peters, M. S., and Timmerhaus, K. D., *Plant Design and Economics for Chemical Engineers,* 4th ed., New York: McGraw-Hill, 1991.

Peterson, S. R., *Patents, Getting One . A Cost-Cutting Primer for Inventors.* Rutland, VT: Academy Books, 1990.

Petroski, H., *The Evolution of Useful Things,* New York: Knopf, 1992.

_____, *To Engineer Is Human: The Role of Failure in Successful Design.* New York: Vintage, 1992.

_____, *Design Paradigms: Case Histories of Error and Judgment in Engineering.* Cambridge: Cambridge Univ. Press, 1994.

Placet, M., et al., "Emissions Involved in Acidic Deposition Processes," *National Acid Precipitation Assessment Program–State of Science/Technology Report 1.* Washington, DC: U.S. Government Printing Office, (Dec. 1990).

Plagenhoef, S., *Patterns of Human Motion*, Englewood Cliffs, NJ: Prentice-Hall, 1971.

Pollack, A., "Device to Save Infants Lifts Medical Hope, and Investors," *New York Times,* D 1: 1, Dec. 12, 1991.

Proctor, R. W., and Van Zandt, T., *Human Factors in Simple* and *Complex Systems.* Boston, MA: Allyn & Bacon, 1994.

Pugh, S., *Total Design.* Wokingham, England: Addison-Wesley, 1991.

Redmon, T., and Redmon, G., *The Inventor's Handbook on Patent Applications.* New York: Vantage, 1993.

Regenstein, L. G., *Cleaning Up America the Poisoned,* Lakewood, CO: Acropolis, 1993.

_____, "Love Canal Toxic Waste Contamination, Niagara Falls, New York," in N. Schlager (Ed.), *When Technology Fails,* Detroit, MI: Gale Research, (1994a).

_____, "Dioxin Contamination, Times Beach, Missouri," in N. Schlager (Ed.), *When Technology Fails.* Detroit, MI: Gale Research, (1994b).

Reidy, C., "Waiting on Pines and Needles," *Boston Globe,* 41: 2, Dec. 22, 1994.

Report by the U.S. Presidential Commission on the Space Shuttle Challenger Accident (Rogers Report). Washington, DC: U.S., Government Printing Office, June 6, 1986.

Richardson, L. F., *Arms and Insecurity.* Pittsburgh, PA: Borwood, 1960.

Richardson, R. O., *The Weird and Wonderous World of Patents.* New York: Sterling, 1990.

Roberts, R. M., *Serendipity: Accidental Discoveries in Science,* New York: Wiley, 1989.

Robison, R., "Tacoma Narrows Bridge Collapse," in N. Schlager (Ed.), *When Technology Fails.* Detroit, MI: Gale Research, (1994a).

_____, "Point Pleasant Bridge Collapse," in N. Schlager (Ed.), *When Technology Fails.* Detroit, MI: Gale Research, (1994b).

_____, "Mianus River Bridge Collapse," in N. Schlager (Ed.), *When Technology Fails.* Detroit, MI: Gale Research, (1994c).

Rogers Commission: *Report of the Presidential Commission on the Space Shuttle Challenger Accident.* Washington, DC: U.S. Government Printing Office, June 6, 1986.

Rosenberg, P. D., *Patent Law Fundamentals*. New York: Clark Boardman, 1980.

Ross, M., "Automobile Fuel Consumption and Emissions: Effects of Vehicle and Driving Characteristics," *Annual Review of Energy and the Environment* vol. 19, 1994.

Roy, E. P., *Cooperatives: Today and Tomorrow*. Danville, IL: Interstate Printers, 1964.

Rubinstein, M., *Engineering Problem Solving*. Englewood Cliffs, NJ: Prentice-Hall, 1985.

Saaty, T. L., and Alexander, J. M., *Thinking with Models,* Oxford: Pergamon, 1981.

Sacharow, S., *Symbols of Trade: Your Favorite Trademarks and the Companies They Represent*. New York: Art Direction, 1982.

Salvadori, M., *The Art of Construction: Projects and Principles for Beginning Engineers and Architects*. Chicago, IL: Chicago Review Press, 1990.

_____, *Why Buildings Stand Up: The Strength of Architecture*. New York: Norton, 1980.

Samuelson, R. J., "Industrial Espionage–Not to Worry," *Los Angeles Times*, part II, p. 13, (15 July 1982).

Sanders, M. S., and McCormick, E. J., *Human Factors in Engineering and Design,* 6th ed. New York: McGraw-Hill, 1987.

Scanlon, R. H., and Tomko, J. J., "Airfoil and Bridge Deck Flutter Derivatives," *Journal of Engineering Mechanics,* vol. 97, 1971.

Scarborough, J. B., *The Gyroscope: Theory and Applications,* New York: Interscience, 1958.

Schlager, N. (Ed.), *When Technology Fails,* Detroit, MI: Gale Research, 1994.

Seabrook, C., "State May Ease Limits on Dioxins: Pulp Industry Lobbying Sets Stage for Changes," *Atlanta Constitution,* p. B6 (6 Dec. 1989).

Sears, F. W., Zemansky, M. W., and Young, H. D., *College Physics,* 4th ed. Reading, MA: Addison-Wesley, 1974.

Serling, R. J., "Turkish Airlines DC-10 Crash, Ermenonville, France," in N. Schlager (Ed.), *When Technology Fails*. Detroit, MI: Gale Research, 1994.

Shaw, G., "Bureau of Reclamation Harshly Criticized in New Report on Teton Dam Collapse," *Los Angeles Times,* 4 June 1977, part I, p. 3.

Shrivastava, P., *Bhopal: Anatomy of a Crisis,* 2nd ed., London: Paul Chapman, 1992.

Sibilia, M. J., and Durbin, E. J., "DCAF: A System for Dynamically Controlled Air/Fuel Ratio in Gaseous Fueled Engines (U.S. Patent #5,143,111)" (unpublished report). Princeton, NJ: Dept. of Mechanical and Aerospace Engineering, Oct. 21, 1994.

Sillman, S. "Tropospheric Ozone: The Debate Over Control Strategies," *Annual Review of Energy and the Environment,* vol. 18, 1993.

Simon, J., "Chemical Plant Explosion, Flixborough, England," in N. Schlager (Ed.), *When Technology Fails,* Detroit, MI: Gale Research, 1994.

Singer, C. J., et al., *A History of Technology*. London: Oxford Univ. Press, 1958.

Smalley, H. E., and Freeman, J. R., *Hospital Industrial Engineering*. New York: Reinhold, 1966.

Smith, A. M., *Reliability-Centered Maintenance*. New York: McGraw-Hill, 1993.

Sowers, G. B., and Sowers, G. F., *Introductory Soil Mechanics and Foundations,* 3rd ed., New York: Macmillan, 1970.

Springer, A. M., and Peterson, R. C., "Environmental Control," in Casey, J. P., (Ed.), *Pulp and Paper: Chemistry and Chemical Technology*, 3rd ed., vol. 1. New York: Wiley, 1980.

Stanley, A., *Mothers and Daughters of Invention*. Metuchen, NJ: Scarecrow, 1993.

Starfield, A. M., Smith, K. A., and Bleloch, A. L., *How to Model It*. New York: McGraw-Hill, 1990.

Sterling, T. D., Pollack, S. V., and Weinkam, J., "Measuring the Effect of Air Pollution on Urban Morbidity," *Archives of Environmental Health,* vol. 18, 1969.

Stern, W. (Ed.), *Handbook of Package Design Research*. New York: Wiley, 1981.

Stewart, C. T., *Air Pollution, Human Health, and Public Policy*. Lexington, MA: D. C. Heath, 1979.

Stirling, N., *Wonders of Engineering*. Garden City, NY: Doubleday, 1966.

Strobel, L. P., *Reckless Homicide? Ford's Pinto Trial,* South Bend, IN: And Books, 1980.

Sullivan, L. P., "Quality Function Deployment," *Quality Progress*. 19: 6, 1986.

Sutton, C., with Anderson, D. M., *How Do They Do That? Wonders of the Modern World Explained.* New York: William Morrow, 1981.

Swasy, A., "Diaper's Failure Shows How Poor Plans, Unexpected Woes Can Kill New Products," *Wall Street Journal,* B1: 3, Oct. 9, 1990.

Sylvan, P., "Notes on Creative Problem Solving" (unpublished memo. to Prof. Clayton Dillon, Dept. of Electrical and Computer Engineering, Northeastern Univ.), May 1994.

Taguchi, G., *Introduction to Quality Engineering.* Asian Productivity Organization, Dearborn, MI: American Supplier Institute, 1986.

"T-E Packaging: Watchful Consumers Make It Work," *Packaging,* vol. 38, Oct. 1993.

Thompson, J. D., and Goldin, G., *The Hospital: A Social and Architectural History.* New Haven, CT: Yale Univ. Press, 1975.

Tingey, D. T., et al. "Effects of Ozone on Crops," in McKee, D., (Ed.), *Tropospheric Ozone: Human Health and Agricultural Impacts* Ann Arbor, MI: Lewis; CRC Press, 1994.

Travers, B. (Ed.), *World of Invention.* Detroit, MI: Gale Research, 1994.

Truxal, J. G. (Ed.), *Quantitative Examples* (Monograph Series of the New Liberal Arts Program). Stony Brook, NY: Research Foundation of State Univ. of New York, Stony Brook, 1991.

Turner, G. T., *Lewis Howard Latimer.* Englewood Cliffs, N.J.: Silver Burdett, 1991.

"Tylenol Case Raises Questions About Capsules," *Packaging,* vol. 31, Apr. 1986.

Urdang, L., (Ed.), *The Random House Dictionary of the English Language: College Edition.* New York: Random House, 1968.

U.S. Dept. of Commerce, *International Workshop of Technical Experts in the Framework of International Demining Operations,* National Technical Information Service, PB95-222907, Sept. 1994.

U.S. Dept. of State, *Hidden Killers: The Global Landmine Crisis* 1994 Report to the U.S. Congress on the Problem with Uncleared Landmines and the U.S. Strategy for Demining and Landmine Control, Dept. of State Publication #10225, Bureau of Political-Military Affairs (Dec. 1994).

U.S. Patent (Gliding Ring) #4,456,265, Alan J. Adler (Inventor) (Dec. 24, 1985).

U.S. Patent (Gliding Ring) #4,560,358, Alan J. Adler (Inventor) (Dec. 24, 1985).

Van der Zee, J. *The Gate: The True Story of the Design and Construction of the Golden Gate Bridge.* New York: Simon & Schuster, 1986.

Voland, G., *Control System Modeling and Analysis.* Englewood Cliffs, NJ: Prentice-Hall, 1986.

Voland, G. G. S., *Modern Engineering Graphics and Design.* St. Paul, MN: West, 1987.

Von Oech, R., *A Whack on the Side of the Head.* New York: Warner, 1983.

Wade, W. C., *The Titanic: End of a Dream.* New York: Penguin, 1980.

Wadsworth, H. M., Jr., Stephens, K. S., and Godfrey, A. B., *Modern Methods for Quality Control and Improvement.* New York: Wiley, 1986.

Wald, M. L., "Battery Makers Are Betting on Lithium-Polymer Cells," *New York Times,* 3, 8: 1, Oct. 30, 1994.

Walker, J., *The Flying Circus of Physics with Answers,* New York: Wiley, 1977.

Walker, J. M., and Boothroyd, G., "Product Development," in J. M. Walker (Ed.), *Handbook of Manufacturing Engineering.* New York: Marcel Dekker, 1996.

Walpole, R. E., and Myers, R. H., *Probability and Statistics for Engineers and Scientists,* 2nd ed. New York: MacMillan, 1978.

Walton, J. W., *Engineering Design: From Art to Practice,* St. Paul, MN: West, 1991.

Wann, D., *Deep Design: Pathways to a Livable Future,* Washington, DC: Island Press, 1996.

Wasserman, A. S., "Redesigning Xerox: A Design Strategy Based on Operability," in E.T. Klemmen (Ed.), *Ergonomics: Harness the Power of Human Factors in Your Business.* Norwood, NJ: Ablex, 1989.

The Way Things Work: An Illustrated Encyclopedia of Technology, New York: Simon & Schuster, 1967.

Weber, R. J., *Forks, Phonographs, and Hot Air Balloons—A Field Guide to Inventive Thinking.* New York: Oxford Univ. Press, 1992.

Weingardt, R., "Molasses Spill, Boston, Massachusetts." in N. Schlager (Ed.), *When Technology Fails.* Detroit, MI: Gale Research, 1994.

Weinstein, A. S., et al., *Products Liability and the Reasonably Safe Product: A Guide for Management, Design, and Marketing.* New York: Wiley, 1978.

Weldon, J., "Jet Pilots Don't Use Rearview Mirrors," *Chemtech,* vol. 13, no. 9, Sept. 1983.

Wellstead, P. E., *Introduction to Physical System Modeling.* New York: Academic, 1979.

Weltner, K., "A Comparison of Explanations of the Aerodynamic Lifting Force," *American Journal of Physics,* vol. 55, no. 1, (Jan. 1987).

Werber, F., and Hamilton, T. E., "Sustainable Fiber Supply" in R. M. Rowell, R. A. Young, J. K. Rowell, (Eds.), *Paper and Composites from Agro-Based Resources,* Boca Raton, FL: CRC Press, 1997.

Westing, A. H., (Ed.), *Explosive Remnants of War,* Philadelphia: Taylor & Francis, 1985.

"When the Bubble Burst," *Chemtech,* Feb. (1992).

Wherry, T. L., *Patent Searching for Librarians and Inventors.* Chicago, IL: American Library Association, 1995.

White, G. A., and Cook, C. G., "Inventory of Agro-Mass" in R. M. Rowell, R. A. Young, J. K. Rowell, (Eds.), *Paper and Composites from Agro-Based Resources,* Boca Raton, FL: CRC Press, 1997.

White, R. H., "Ozone Health Effects-A Public Health Perspective" in McKee, D., (Ed.), *Tropospheric Ozone: Human Health and Agricultural Impacts.* Ann Arbor, MI: Lewis; CRC Press, 1994.

Whitfield, R. G., et al. "Health Risk Assessment of Ozone" in McKee, D., (Ed.), *Tropospheric Ozone: Human Health and Agricultural Impacts* Ann Arbor: Lewis; CRC Press, 1994.

Williams, D. J., *Manufacturing Cells: Control, Programming and Integration.* Oxford: Butterworth-Heinemann, 1991.

Williams, R. H. "Fuel-Cell Vehicles: The Clean Machine" *Technology Review,* vol. 97, no. 3, Apr. 1994.

Witherell, C. E., *Mechanical Engineering Avoidance: Strategies and Techniques,* New York: McGraw-Hill, 1994.

Wolgalter, M., Desaulniers, D., and Godfrey, S., "Perceived Effectiveness of Environmental Warnings," *Proceedings of the Human Factors Society 29th Annual Meeting,* Santa Monica, CA: Human Factors Society, 1985.

Woodham-Smith, C., *Florence Nightingale 1820–1910.* New York: McGraw-Hill, 1951.

Woodson, T. T., *Introduction to Engineering Design.* New York: McGraw-Hill, 1966.

Woodson, W. E., and Conover, D. W., *Human Engineering for Equipment Designers,* 2nd ed., Berkeley: Univ. of California Press, 1964.

Woog, Adam, *Sexless Oysters and Self-Tipping Hats.* Seattle, WA: Sasquatch Books, 1991.

Yenne, B., *100 Inventions That Shaped World History.* San Francisco, CA: Bluewood Books, 1993.

Young, R. A., "Processing of Agro-Based Resources into Pulp and Paper," in Rowell, R. M., Young, R. A., and Rowell, J. K. (Eds.), *Paper and Composites from Agro-Based Resources.* New York: CRC Press, 1997.

Zairi, M., "The Art of Benchmarking: Using Customer Feedback to Establish a Performance Gap," *Total Quality Management,* 3: 2, 1992.

Zorpette, G., "Breaking the Enemy's Code," *IEEE Spectrum,* Sept. 1987.

INDEX

Page references in italics indicate illustrations and tables.

Flight 811, 350–352, *351*
Urea formaldehyde, 411
U.S. Patent and Trademark Office
    attorneys, 191
    fees, 188–189
    history of, 183–184
    and patent searches, 192–194
    trademark definition, 179
    Web page, 77
Usefulness, 190
User characteristics, 127–128, 234
User friendliness, 114–116, 145
Uses, unanticipated, 294–295
Utility patents, 183–191

V

Vacuum forming, 420
Values, 315, 316, 342–343
Vaporization, 80
Variability. *See* Tolerance
Variables, 82, 220–221
Variation, dimensions of, 214
Velcro, *193,* 250–251
Vending machines, 156
Ventilation, 233
Versatility, 355
Vibrations, 202–203, 272–273, 318, 338
Vickers test, 403
Villari effect, *176*
Vinyl, 410
Virtual corporations, 11
Viscosity, 404
Visual acuity, 126, *131, 132, 166.*
    See also Eyeglasses
Voice synthesizer, 122–123
Voisin brothers, 27
Von Oech, Richard, 274
Vortex noise, 249
Vortices, 203
Vulcanization, 247, 268–270

W

Walkway disaster, 433–435
Wall Street Journal, 11–12
Warnings
    inadequate, 287, 344–345, 349
    label requirements, 292, 326, 330–332
    lights, 344–345
Warranties, 291–292
Waste. *See also* Disposable products

improper disposal disasters, 137–139, 307–308
    jokes for trash, 274
    and packaging, 396
    radioactive, 318
    recovery, 193
Watches, 168–169, 184
Water, Perrier, 143
Water supply, 243
Water tanks, 90–91, 315
Water-injection systems, 139–142, *142*
Watson-Watt, Sir Robert, 266
Wattis, Edmund, 436
Wave phenomena, 162–168, 266
Wedge devices, 157
Weight factors, 124, 126
Weighting factors, *357,* 357–360, *359, 361,*
    364–366
Welch, Charles K., 273
Wertheim effect, *176*
Weyerhauser Corp., 118
What-if scenarios, 274
Wheeling Suspension Bridge, 204
Whistle blowing, 302–303, 343–344
Whitney, Eli, 385
Whittle, Frank, 38
Wiedemann effect, *176*
Willey, Ronald, 366–367, 366 n.6, *367, 368*
Wind
    and bridges, 200–205
    and buildings, 98–100, 398–399
    and water towers, 315
Windshields, 111–112
Wind-tunnel testing, 98, 205, 219
Wings, 41
Winslow, Samuel, 183
Winterscheid, Loren, 60
Wolf, Charlie, 251–252
Wood, 134–135, 411–412, *412,* 414–415, 421
Wood, John C., 111
Work, in physics, 157
Work documents, 282
Work hardening, *176,* 295–296
Workers, protecting, 300–301
Workstations, manufacturing, 386
Wrapping materials, 118
Wright brothers, 28–42, *30*

X

Xerography, 197–198
Xerox Corp., 234–235